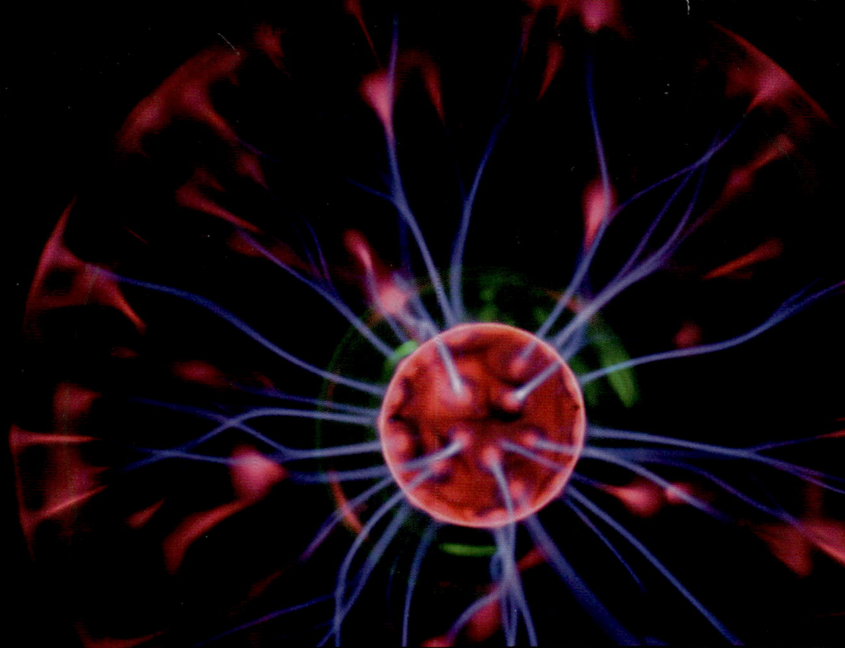

Discovery EDUCATION | SCIENCE TECHBOOK

PHYSICAL SCIENCE
FLORIDA EDITION

Log in to Science Techbook at www.DiscoveryEducation.com

ISBN 13: 978-156331-113-0

Library of Congress Cataloging-in-Publication Data available upon request from the Library of Congress.

Printed in the United States of America.

1 2 3 4 5 6 7 8 9 10

Discovery EDUCATION™

800-323-9084

One Discovery Place, Silver Spring, Maryland 20910

Table of Contents

Letter to the Student

Dear Student,

You are about to experience science like you never have before! In this class, you'll be using Science Techbook™—a comprehensive, digital science program developed by the educators and designers at Discovery Education. Science Techbook is full of Explorations, videos, Hands-On Activities, digital tools, reading passages, animations, and more. These resources will help you learn scientific concepts and procedures, and apply them to the world around you. Science Techbook allows you to work at your own pace and investigate questions you may have related to science. You'll even be able to monitor your progress in real time using the Student Learning Dashboard.

The *Core Text Companion* is a print resource that accompanies the digital Science Techbook. With this companion, you have access to Science Techbook's core text—the key ideas and details about each scientific concept—even when you do not have access to a device or the Internet. You can use this resource to explore important ideas, make connections to the digital content, and develop your own scientific understanding.

This print resource is organized by concept and includes the following:

- OVERVIEW: What's it all about? Lesson Questions, Lesson Objectives, and key vocabulary will help you prepare for each science concept.
- ENGAGE: What do you already know about the topic? Follow a link to uncover your prior knowledge about each concept.
- EXPLORE: What are the main ideas in the concept? The Explore pages include core text and images to help you answer each of the concept's Lesson Questions. Use evidence to write a scientific explanation and answer questions to check for understanding.
- STEM IN ACTION: How is science used in the real world and in STEM careers? Read more in this section to find out how the knowledge you're building applies to real-world situations—both today and in the future.

Throughout this resource, you'll find QR codes that take you to the corresponding online section of Science Techbook for that concept. For instance, the QR code in the Explore section provides a direct link to the Explore tab and its content. Once you are inside Techbook, you'll have access to simulations, interactive Explorations, and other digital resources to help you investigate each concept's most important ideas.

Letter to the Student

To use the QR codes, you will need a QR reader. Readers are available for almost any device. When your QR reader is open, hold it so that the QR code on a page is visible in the screen. The reader will scan the code and direct you to the correct page or resource in Science Techbook. Remember, you will need to log in with your username and password first.

Enjoy this deep dive into the exciting world of science!

Sincerely,

The Discovery Education Science Team

Use this resource as you explore the digital Techbook:

- ENGAGE for reviewing what you already know
- EXPLORE for reviewing each concept's essential information and evidence
- STEM IN ACTION for investigating real-world applications
- QR CODES for connecting to Techbook

Letter to the Parent/Guardian

Dear Parent/Guardian,

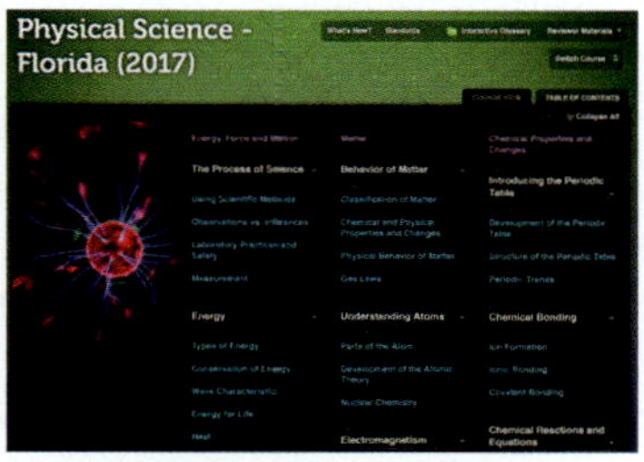

This year, your student will be using Science Techbook™, a comprehensive, digital science program developed by the educators and designers at Discovery Education. Science Techbook is an innovative program that offers engaging, real-world problems to help your student master key scientific concepts and procedures. In class, students experience dynamic content, Explorations, videos, digital tools, and game-like activities that inspire and motivate scientific learning and curiosity.

As a print resource accompanying the digital Science Techbook, the *Core Text Companion* allows students to explore the core Techbook content when the Internet is not available. Students are encouraged to use this resource to read about key ideas, seek connections, think about scientific questions, and develop their own scientific understanding.

Science Techbook is divided into units, and each unit is divided into concepts. Each concept online has five tabs: Engage, Explore, Explain, Elaborate with STEM, and Evaluate. The *Core Text Companion* is organized by concept and includes the following:

- LESSON OVERVIEW: Students preview a concept's Lesson Questions, Lesson Objectives, and key vocabulary to help them make connections to the science content.
- ENGAGE: Students answer questions to activate their prior knowledge of a concept's essential ideas and begin making connections to the Explain Question.
- EXPLORE: Students deepen their understanding of the concept by exploring the core text related to each Lesson Question. Online, students have access to additional resources, Hands-On Activities, and interactives. They will also complete scientific explanations and answer questions to check for understanding.
- STEM IN ACTION: Students connect the skills and knowledge they are building in each concept with real-world applications. Online, they can explore related videos and resources, and complete additional activities.

Within this resource, you'll find QR codes that take you and your student to a corresponding section of Science Techbook. Once in Techbook, students will have access to the Core Interactive Text of each concept, as well as thousands of resources and activities that build deep conceptual scientific understanding. Additionally, tools and features such as the Interactive Glossary and text-to-speech functionality allow Science Techbook to target learning for students of a variety of abilities.

Letter to the Parent/Guardian

To use the QR codes, you'll need a QR reader. Readers are available for phones, tablets, laptops, desktops, and virtually any device in between. Most use the device's camera, but there are some that scan documents that are on your screen. Download a free QR reader in the App Store or Google Play. To access Science Techbook resources, follow these steps:

1. Open the QR code reader on your device.
2. Hold your device so the QR code is visible within your device's screen. One of two things will happen:
   - The device may automatically scan the code; or,
   - The device will scan the code when you press a button, similar to taking a picture.
3. Once scanned, the QR code will direct you to a page or resource on the Internet.
4. For resources in Science Techbook, you'll need to log in with your student's username and password the first time you access a QR code. After that, you won't need to log in again, unless you sign out or remain inactive for too long.

Scan this QR code to access a video that provides a deeper introduction to Science Techbook:

We encourage you to support your student in using the online interactive materials in Science Techbook, as well as the core text and questions in the *Core Text Companion*. Together, may you and your student enjoy a fantastic year of science!

Sincerely,

The Discovery Education Science Team

Science Techbook and *Core Text Companion*: Introduction and Guide

How to Use Discovery Education Science Techbook

Discovery Education Science Techbook is a complete digital science basal resource designed to engage students in real-world, investigative-based science. Discovery Education Science Techbook provides teachers with powerful tools for engagement, inquiry, exploration, and evaluation. Unlike a textbook copied into digital format, Discovery Education Science Techbook uses a variety of digital resources, including video, audio, text, and interactive and hands-on experiences, to provide engaging science content while meeting the needs of students with different learning styles.

From any browser, navigate to www.DiscoveryEducation.com. Input the username and password that were provided to you.

Course Page

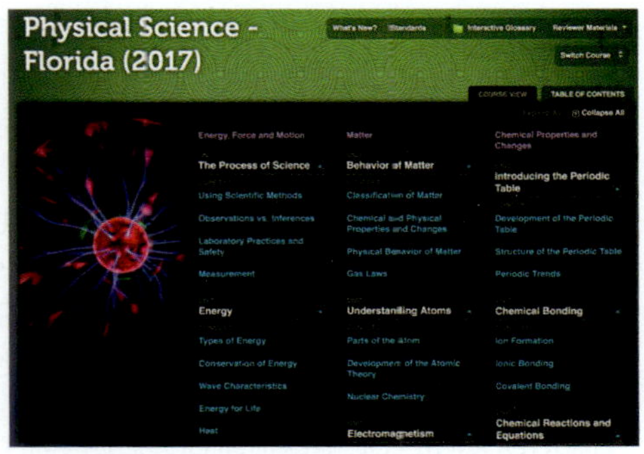

Upon sign in, you will land on the course page. On this page, you'll find links to Curriculum Standards, the Interactive Glossary, and Reviewer Materials.

The course page allows you to navigate the content contained within each course and provides quick links to helpful information. From this page, you can access all of a course's units and concepts.

Concept Pages: The 5Es

The concept structure is set up to accommodate the 5E Instructional Model, with recommended resources for each of the 5E tabs: Engage, Explore, Explain, Elaborate with STEM, and Evaluate. Both students and teachers can access the resources for instruction and assessment from the concept page by clicking on each E tab. On any tab, Teacher mode can be turned on or off.

ENGAGE

Science Techbook introduces concepts to students as real-world phenomena. Scientists observe phenomena occurring in the real world and dig deeper to uncover the science behind the observation. This is the opportunity to engage students and inspire them to ask questions they want to investigate.

Students can interact with a variety of text and multimedia resources to match their diverse learning styles. Interactive formative assessments are embedded and provide feedback to teachers on prior knowledge and common misconceptions students may be bringing to the concept. These assessment items feed directly into the Student and Teacher Dashboard.

The Explain Question at the bottom of the page provides a framing question that students should be able to answer, as a claim with evidence, at the end of their learning progression. This question will display for students again during the Explain portion of the lesson. Students can also construct their own questions for investigation in Explore.

EXPLORE

Students will obtain and analyze information to collect evidence to support their claim related to the Explain Question, or to an investigation question of their own choosing. You'll notice in this tab, and in others, how the text is constructed to support Universal Design for Learning. We call this the Core Interactive Text, or CIT. Terms on each page are linked to the Interactive Glossary, which uses animation, text, video, and images to help students comprehend the scientific language. Students can also toggle between English and Spanish versions of the text. Finally, students can highlight, take notes, and use our text-to-speech feature to have a section read aloud.

Throughout the text, students can access multimedia resources, such as videos and images. These resources help them make sense of complex topics and can be assigned to individual students or to a class. Science Techbook includes Virtual Labs, and nearly 700 Hands-On Activities allow students to connect to science and engineering practices of designing and carrying out investigations and using models.

- Virtual Labs provide a simulation where students can collect and analyze data, use computational thinking, construct explanations, and evaluate information.
- In the Hands-On Labs and Activities, students can design and conduct scientific investigations.

EXPLAIN

Students use evidence collected during Engage and Explore to construct scientific explanations. They can analyze both empirical and textual evidence to make a claim and provide reasoning on how the evidence supports the claim. After completing their explanation, students present it to their peers, critique others' explanations, and refine their own through various modes of media. This can all be done through online entry or printed for student use.

ELABORATE with STEM

Students can extend their understanding through additional real-world connections. There's an emphasis on problem solving and real-world connections to math and literacy, including careers related to the science concepts.

Students can work through the resources provided, or teachers can assign specific projects located in the STEM Project Starters section. These project starters clearly connect science to technology, engineering, and/or mathematics applications.

EVALUATE

Students can review concepts with self-assessments graded in real time. When students answer a question incorrectly, they will have the chance to explore remedial resources on the topic. In the teacher view, teachers can also assign a constructed response item or create their own assessment. Teachers can also utilize the summative concept assessment to determine student's scientific understanding. In addition to accessing content through the Model Lesson and 5Es, both teachers and students can search for content through keyword and filtered searches.

What Is the Discovery Education *Core Text Companion*?

Online, the Core Interactive Text (CIT) serves as a student pathway through each concept. The *Core Text Companion* is designed to work with Techbook and includes the core text, images, and captions from each concept's Engage, Explore, and STEM in Action pages. When a device or an Internet connection is not available, students can still access key information, consider Lesson Questions, answer each concept's Explain Question, and explore real-world science careers and applications.

LESSON OVERVIEW

For each concept, students will first encounter a Lesson Overview, which lists key vocabulary terms and previews the concept's Lesson Questions and Lesson Objectives. Students can scan the QR code on this page to connect directly to the digital version.

ENGAGE

In the Engage section, questions prompt students to activate their prior knowledge of the topic or related topics. A QR code links to the Engage page online, where they can explore the Core Interactive Text and other resources. As in the digital Techbook, the Explain Question appears at the end of this section, presenting a question for students to consider, and later answer with a scientific explanation.

EXPLORE

In the Explore section, the core text guides students through each of a concept's Lesson Questions, explaining key details of the science and providing relevant examples and models. A QR code at the beginning of this section connects students to the Explore tab online, where they can supplement their reading with multimedia resources, Virtual Labs, and Hands-On Activities. At the end of each section, students can scan QR codes to complete their scientific explanations online, or to check for understanding by answering a brief constructed-response question.

STEM IN ACTION

In the STEM in Action section, students can read about the application of science to a real-world problem or experience. STEM in Action sections conclude with questions to assess students' understanding of the connections between a concept's focus and its use in the real world or in an experimental setting. Students can explore more information and resources in the digital Techbook by scanning this section's QR code.

Throughout the *Core Text Companion*, images and captions, graphs, and diagrams support students' reading and encourage them to pose and answer additional questions. Students should combine this resource with their explorations of the digital Techbook to deepen and expand their understanding of critical science topics.

Physical Science Florida Standards

When you use this product in combination with the digital Techbook, you will find coverage for each standard in the concepts listed.

Florida Standard	Unit	Concept
Earth and Space Science		
SC.912.E.7.1 Analyze the movement of matter and energy through the different biogeochemical cycles, including water and carbon. (Cognitive Complexity/Depth of Knowledge Rating: High)	Energy	Energy for Life
	Water and Solutions	Water
Life Science		
SC.912.L.18.D The unique chemical properties of carbon and water make life on Earth possible.	Water and Solutions	Water
SC.912.L.18.7 Identify the reactants, products, and basic functions of photosynthesis. (Cognitive Complexity/Depth of Knowledge Rating: Moderate)	Energy	Energy for Life
SC.912.L.18.8 Identify the reactants, products, and basic functions of aerobic and anaerobic cellular respiration. (Cognitive Complexity/Depth of Knowledge Rating: Moderate)	Energy	Energy for Life
SC.912.L.18.12 Discuss the special properties of water that contribute to Earth's suitability as an environment for life: cohesive behavior, ability to moderate temperature, expansion upon freezing, and versatility as a solvent. (Cognitive Complexity/Depth of Knowledge Rating: Moderate)	Behavior of Matter	Chemical and Physical Properties and Changes
	Water and Solutions	Water

Florida Standard	Unit	Concept
Nature of Science		
SC.912.N.1.A Scientific inquiry is a multifaceted activity; The processes of science include the formulation of scientifically investigable questions, construction of investigations into those questions, the collection of appropriate data, the evaluation of the meaning of those data, and the communication of this evaluation.	The Process of Science	Using Scientific Methods
	The Process of Science	Measurement
	Energy	Types of Energy
	Energy	Wave Characteristics
	Energy	Energy for Life
	Energy	Heat
	Force	Fundamental Forces
	Force	Electric Forces
	Motion	Using Vectors and Scalars to Describe Motion
	Motion	Understanding and Describing Motion
	Motion	Solving Motion Problems
	Motion	Newton's First Law of Motion
	Motion	Newton's Second Law of Motion
	Motion	Newton's Third Law of Motion
	Motion	Applying Newton's Laws of Motion
	Electromagnetism	Electric Circuits
	Electromagnetism	Conductors and Insulators
	Water and Solutions	Water
SC.912.N.1.B The processes of science frequently do not correspond to the traditional portrayal of "the scientific method."	The Process of Science	Using Scientific Methods
SC.912.N.1.C Scientific argumentation is a necessary part of scientific inquiry and plays an important role in the generation and validation of scientific knowledge.	The Process of Science	Using Scientific Methods
SC.912.N.1.D Scientific knowledge is based on observation and inference; it is important to recognize that these are very different things. Not only does science require creativity in its methods and processes, but also in its questions and explanations.	The Process of Science	Using Scientific Methods
	The Process of Science	Observations vs. Inferences
SC.912.N.1.2 Describe and explain what characterizes science and its methods. (Cognitive Complexity/Depth of Knowledge Rating: Moderate)	The Process of Science	Using Scientific Methods
	Behavior of Matter	Gas Laws
	Understanding Atoms	Development of the Atomic Theory
SC.912.N.1.3 Recognize that the strength or usefulness of a scientific claim is evaluated through scientific argumentation, which depends on critical and logical thinking, and the active consideration of alternative scientific explanations to explain the data presented. (Cognitive Complexity/Depth of Knowledge Rating: Low)	The Process of Science	Using Scientific Methods
	Electromagnetism	Conductors and Insulators
	Chemical Reactions and Equations	Reaction Rate

Florida Standards

Florida Standard	Unit	Concept
SC.912.N.1.4 Identify sources of information and assess their reliability according to the strict standards of scientific investigation. (Cognitive Complexity/Depth of Knowledge Rating: High)	The Process of Science	Using Scientific Methods
	Chemical Reactions and Equations	Chemical Reactions and Equations
	Chemical Reactions and Equations	Reaction Rate
SC.912.N.1.5 Describe and provide examples of how similar investigations conducted in many parts of the world result in the same outcome. (Cognitive Complexity/Depth of Knowledge Rating: Moderate)	The Process of Science	Using Scientific Methods
	Chemical Reactions and Equations	Chemical Reactions and Equations
SC.912.N.1.6 Describe how scientific inferences are drawn from scientific observations and provide examples from the content being studied. (Cognitive Complexity/Depth of Knowledge Rating: Moderate)	The Process of Science	Using Scientific Methods
	The Process of Science	Observations vs. Inferences
	Force	Electric Forces
	Behavior of Matter	Gas Laws
	Understanding Atoms	Development of the Atomic Theory
SC.912.N.1.7 Recognize the role of creativity in constructing scientific questions, methods and explanations. (Cognitive Complexity/Depth of Knowledge Rating: Low)	The Process of Science	Using Scientific Methods
	The Process of Science	Observations vs. Inferences
	The Process of Science	Laboratory Practices and Safety
	The Process of Science	Measurement
	Energy	Types of Energy
	Energy	Conservation of Energy
	Energy	Wave Characteristics
	Energy	Heat
	Force	Fundamental Forces
	Force	Gravity
	Force	Electric Forces
	Motion	Using Vectors and Scalars to Describe Motion
	Motion	Understanding and Describing Motion
	Motion	Solving Motion Problems
	Motion	Newton's First Law of Motion
	Motion	Newton's Second Law of Motion
	Motion	Newton's Third Law of Motion
	Motion	Applying Newton's Laws of Motion
	Motion	Work and Power

Florida Standard	Unit	Concept
SC.912.N.1.7 (*continued*)	Behavior of Matter	Classification of Matter
	Behavior of Matter	Chemical and Physical Properties and Changes
	Behavior of Matter	Physical Behavior of Matter
	Behavior of Matter	Gas Laws
	Understanding Atoms	Parts of the Atom
	Understanding Atoms	Development of the Atomic Theory
	Understanding Atoms	Nuclear Chemistry
	Electromagnetism	Electric and Magnetic Fields
	Electromagnetism	Electricity and Magnetism
	Electromagnetism	Electric Circuits
	Electromagnetism	Conductors and Insulators
	Introducing the Periodic Table	Development of the Periodic Table
	Introducing the Periodic Table	Structure of the Periodic Table
	Introducing the Periodic Table	Periodic Trends
	Chemical Bonding	Ion Formation
	Chemical Bonding	Ionic Bonding
	Chemical Bonding	Covalent Bonding
	Chemical Reactions and Equations	Chemical Formulas
	Chemical Reactions and Equations	Chemical Reactions and Equations
	Chemical Reactions and Equations	Mathematics of Formulas and Equations
	Chemical Reactions and Equations	Reaction Rate
	Chemical Reactions and Equations	Thermochemistry
	Water and Solutions	Water
	Water and Solutions	Acids, Bases, and Salts
SC.912.N.2.3 Identify examples of pseudoscience (such as astrology, phrenology) in society. (Cognitive Complexity/Depth of Knowledge Rating: Low)	Understanding Atoms	Nuclear Chemistry
SC.912.N.2.4 Explain that scientific knowledge is both durable and robust and open to change. Scientific knowledge can change because it is often examined and re-examined by new investigations and scientific argumentation. Because of these frequent examinations, scientific knowledge becomes stronger, leading to its durability. (Cognitive Complexity/Depth of Knowledge Rating: High)	Understanding Atoms	Development of the Atomic Theory
	Chemical Reactions and Equations	Mathematics of Formulas and Equations

Florida Standard	Unit	Concept
SC.912.N.2.5 Describe instances in which scientists' varied backgrounds, talents, interests, and goals influence the inferences and thus the explanations that they make about observations of natural phenomena and describe that competing interpretations (explanations) of scientists are a strength of science as they are a source of new, testable ideas that have the potential to add new evidence to support one or another of the explanations. (Cognitive Complexity/Depth of Knowledge Rating: High)	Understanding Atoms	Development of the Atomic Theory
	Introducing the Periodic Table	Development of the Periodic Table
	Chemical Bonding	Covalent Bonding
SC.912.N.3.2 Describe the role consensus plays in the historical development of a theory in any one of the disciplines of science. (Cognitive Complexity/Depth of Knowledge Rating: Moderate)	Understanding Atoms	Development of the Atomic Theory
	Chemical Reactions and Equations	Mathematics of Formulas and Equations
SC.912.N.3.3 Explain that scientific laws are descriptions of specific relationships under given conditions in nature, but do not offer explanations for those relationships. (Cognitive Complexity/Depth of Knowledge Rating: Moderate)	Behavior of Matter	Gas Laws
	Understanding Atoms	Development of the Atomic Theory
	Chemical Reactions and Equations	Mathematics of Formulas and Equations
SC.912.N.3.5 Describe the function of models in science, and identify the wide range of models used in science. (Cognitive Complexity/Depth of Knowledge Rating: Moderate)	Understanding Atoms	Development of the Atomic Theory
	Chemical Bonding	Covalent Bonding
SC.912.N.4.1 Explain how scientific knowledge and reasoning provide an empirically-based perspective to inform society's decision making. (Cognitive Complexity/Depth of Knowledge Rating: Moderate)	Force	Gravity
	Force	Electric Forces
	Motion	Work and Power
	Understanding Atoms	Nuclear Chemistry
	Electromagnetism	Electric and Magnetic Fields
	Electromagnetism	Electric Circuits
	Electromagnetism	Conductors and Insulators
	Chemical Reactions and Equations	Chemical Reactions and Equations
SC.912.N.4.2 Weigh the merits of alternative strategies for solving a specific societal problem by comparing a number of different costs and benefits, such as human, economic, and environmental. (Cognitive Complexity/Depth of Knowledge Rating: High)	Understanding Atoms	Nuclear Chemistry
Physical Science		
SC.912.P.8.A A working definition of matter is that it takes up space, has mass, and has measurable properties. Matter is comprised of atomic, subatomic, and elementary particles.	The Process of Science	Measurement
	Force	Fundamental Forces

Florida Standard	Unit	Concept
SC.912.P.8.B Electrons are key to defining chemical and some physical properties, reactivity, and molecular structures. Repeating (periodic) patterns of physical and chemical properties occur among elements that define groups of elements with similar properties. The periodic table displays the repeating patterns, which are related to the atom's outermost electrons. Atoms bond with each other to form compounds.	Introducing the Periodic Table	Development of the Periodic Table
	Chemical Bonding	Covalent Bonding
SC.912.P.8.C In a chemical reaction, one or more reactants are transformed into one or more new products. Many factors shape the nature of products and the rates of reaction.	Behavior of Matter	Classification of Matter
	Chemical Reactions and Equations	Chemical Reactions and Equations
SC.912.P.8.1 Differentiate among the four states of matter. (Cognitive Complexity/Depth of Knowledge Rating: Moderate)	Behavior of Matter	Classification of Matter
	Behavior of Matter	Chemical and Physical Properties and Changes
	Behavior of Matter	Gas Laws
SC.912.P.8.3 Explore the scientific theory of atoms (also known as atomic theory) by describing changes in the atomic model over time and why those changes were necessitated by experimental evidence. (Cognitive Complexity/Depth of Knowledge Rating: High)	Understanding Atoms	Development of the Atomic Theory
SC.912.P.8.4 Explore the scientific theory of atoms (also known as atomic theory) by describing the structure of atoms in terms of protons, neutrons and electrons, and differentiate among these particles in terms of their mass, electrical charges and locations within the atom. (Cognitive Complexity/Depth of Knowledge Rating: High)	Force	Fundamental Forces
	Force	Electric Forces
	Understanding Atoms	Parts of the Atom
	Understanding Atoms	Development of the Atomic Theory
	Introducing the Periodic Table	Periodic Trends
SC.912.P.8.5 Relate properties of atoms and their position in the periodic table to the arrangement of their electrons. (Cognitive Complexity/Depth of Knowledge Rating: Moderate)	Understanding Atoms	Development of the Atomic Theory
	Introducing the Periodic Table	Structure of the Periodic Table
	Introducing the Periodic Table	Periodic Trends
SC.912.P.8.6 Distinguish between bonding forces holding compounds together and other attractive forces, including hydrogen bonding and van der Waals forces. (Cognitive Complexity/Depth of Knowledge Rating: Moderate)	Chemical Bonding	Covalent Bonding
SC.912.P.8.7 Interpret formula representations of molecules and compounds in terms of composition and structure. (Cognitive Complexity/Depth of Knowledge Rating: Moderate)	Chemical Reactions and Equations	Chemical Formulas
	Chemical Reactions and Equations	Chemical Reactions and Equations
	Chemical Reactions and Equations	Mathematics of Formulas and Equations
	Water and Solutions	Water

Florida Standard	Unit	Concept
SC.912.P.8.8 Characterize types of chemical reactions, for example: redox, acid-base, synthesis, and single and double replacement reactions. (Cognitive Complexity/Depth of Knowledge Rating: Moderate)	Chemical Reactions and Equations	Chemical Reactions and Equations
	Water and Solutions	Acids, Bases, and Salts
SC.912.P.8.9 Apply the mole concept and the law of conservation of mass to calculate quantities of chemicals participating in reactions. (Cognitive Complexity/Depth of Knowledge Rating: High)	Chemical Reactions and Equations	Mathematics of Formulas and Equations
SC.912.P.8.11 Relate acidity and basicity to hydronium and hydroxyl ion concentration and pH. (Cognitive Complexity/Depth of Knowledge Rating: Moderate)	Water and Solutions	Acids, Bases, and Salts
SC.912.P.8.12 Describe the properties of the carbon atom that make the diversity of carbon compounds possible. (Cognitive Complexity/Depth of Knowledge Rating: Moderate)	Chemical Bonding	Covalent Bonding
SC.912.P.10.A Energy is involved in all physical and chemical processes. It is conserved, and can be transformed from one form to another and into work. At the atomic and nuclear levels energy is not continuous but exists in discrete amounts. Energy and mass are related through Einstein's equation $E=mc^2$.	Energy	Conservation of Energy
	Motion	Work and Power
	Understanding Atoms	Nuclear Chemistry
	Chemical Reactions and Equations	Thermochemistry
SC.912.P.10.B The properties of atomic nuclei are responsible for energy-related phenomena such as radioactivity, fission and fusion.	Force	Fundamental Forces
	Understanding Atoms	Nuclear Chemistry
SC.912.P.10.C Changes in entropy and energy that accompany chemical reactions influence reaction paths. Chemical reactions result in the release or absorption of energy.	Chemical Reactions and Equations	Thermochemistry
SC.912.P.10.D The theory of electromagnetism explains that electricity and magnetism are closely related. Electric charges are the source of electric fields. Moving charges generate magnetic fields.	Energy	Wave Characteristics
	Force	Fundamental Forces
	Electromagnetism	Electric and Magnetic Fields
	Electromagnetism	Electricity and Magnetism
	Electromagnetism	Conductors and Insulators
SC.912.P.10.E Waves are the propagation of a disturbance. They transport energy and momentum but do not transport matter.	Energy	Wave Characteristics
	Electromagnetism	Electricity and Magnetism
SC.912.P.10.1 Differentiate among the various forms of energy and recognize that they can be transformed from one form to others. (Cognitive Complexity/Depth of Knowledge Rating: Moderate)	Energy	Types of Energy
	Energy	Conservation of Energy
	Electromagnetism	Electric Circuits
	Chemical Reactions and Equations	Thermochemistry
SC.912.P.10.2 Explore the Law of Conservation of Energy by differentiating among open, closed, and isolated systems and explain that the total energy in an isolated system is a conserved quantity. (Cognitive Complexity/Depth of Knowledge Rating: High)	Energy	Conservation of Energy
	Energy	Heat

Florida Standard	Unit	Concept
SC.912.P.10.3 Compare and contrast work and power qualitatively and quantitatively. (Cognitive Complexity/Depth of Knowledge Rating: Moderate)	Motion	Work and Power
SC.912.P.10.4 Describe heat as the energy transferred by convection, conduction, and radiation, and explain the connection of heat to change in temperature or states of matter. (Cognitive Complexity/Depth of Knowledge Rating: High)	Energy	Heat
SC.912.P.10.5 Relate temperature to the average molecular kinetic energy. (Cognitive Complexity/Depth of Knowledge Rating: Moderate)	Energy	Wave Characteristics
	Energy	Heat
	Behavior of Matter	Gas Laws
	Chemical Reactions and Equations	Thermochemistry
SC.912.P.10.6 Create and interpret potential energy diagrams, for example: chemical reactions, orbits around a central body, motion of a pendulum. (Cognitive Complexity/Depth of Knowledge Rating: High)	Behavior of Matter	Physical Behavior of Matter
	Chemical Reactions and Equations	Chemical Reactions and Equations
	Chemical Reactions and Equations	Thermochemistry
SC.912.P.10.7 Distinguish between endothermic and exothermic chemical processes. (Cognitive Complexity/Depth of Knowledge Rating: Moderate)	Chemical Reactions and Equations	Thermochemistry
SC.912.P.10.9 Describe the quantization of energy at the atomic level. (Cognitive Complexity/Depth of Knowledge Rating: Moderate)	Understanding Atoms	Development of the Atomic Theory
SC.912.P.10.10 Compare the magnitude and range of the four fundamental forces (gravitational, electromagnetic, weak nuclear, strong nuclear). (Cognitive Complexity/Depth of Knowledge Rating: Moderate)	Force	Fundamental Forces
SC.912.P.10.11 Explain and compare nuclear reactions (radioactive decay, fission and fusion), the energy changes associated with them and their associated safety issues. (Cognitive Complexity/Depth of Knowledge Rating: High)	Force	Fundamental Forces
	Understanding Atoms	Nuclear Chemistry
SC.912.P.10.12 Differentiate between chemical and nuclear reactions. (Cognitive Complexity/Depth of Knowledge Rating: Moderate)	Understanding Atoms	Nuclear Chemistry
	Chemical Reactions and Equations	Chemical Formulas
SC.912.P.10.13 Relate the configuration of static charges to the electric field, electric force, electric potential, and electric potential energy. (Cognitive Complexity/Depth of Knowledge Rating: High)	Energy	Conservation of Energy
	Electromagnetism	Electric and Magnetic Fields
SC.912.P.10.14 Differentiate among conductors, semiconductors, and insulators. (Cognitive Complexity/Depth of Knowledge Rating: Moderate)	Electromagnetism	Conductors and Insulators
SC.912.P.10.15 Investigate and explain the relationships among current, voltage, resistance, and power. (Cognitive Complexity/Depth of Knowledge Rating: High)	Electromagnetism	Electric and Magnetic Fields
	Electromagnetism	Electric Circuits
	Electromagnetism	Conductors and Insulators

Florida Standards

Florida Standard	Unit	Concept
SC.912.P.10.16 Explain the relationship between moving charges and magnetic fields, as well as changing magnetic fields and electric fields, and their application to modern technologies. (Cognitive Complexity/Depth of Knowledge Rating: High)	Energy	Conservation of Energy
	Force	Fundamental Forces
	Electromagnetism	Electric and Magnetic Fields
	Electromagnetism	Electricity and Magnetism
	Electromagnetism	Conductors and Insulators
SC.912.P.10.17 Explore the theory of electromagnetism by explaining electromagnetic waves in terms of oscillating electric and magnetic fields. (Cognitive Complexity/Depth of Knowledge Rating: High)	Energy	Wave Characteristics
	Electromagnetism	Electric and Magnetic Fields
SC.912.P.10.18 Explore the theory of electromagnetism by comparing and contrasting the different parts of the electromagnetic spectrum in terms of wavelength, frequency, and energy, and relate them to phenomena and applications. (Cognitive Complexity/Depth of Knowledge Rating: High)	Energy	Wave Characteristics
	Understanding Atoms	Nuclear Chemistry
	Chemical Reactions and Equations	Thermochemistry
SC.912.P.10.20 Describe the measurable properties of waves and explain the relationships among them and how these properties change when the wave moves from one medium to another. (Cognitive Complexity/Depth of Knowledge Rating: High)	Energy	Wave Characteristics
SC.912.P.10.21 Qualitatively describe the shift in frequency in sound or electromagnetic waves due to the relative motion of a source or a receiver. (Cognitive Complexity/Depth of Knowledge Rating: Moderate)	Energy	Wave Characteristics
SC.912.P.12.A Motion can be measured and described qualitatively and quantitatively. Net forces create a change in motion. When objects travel at speeds comparable to the speed of light, Einstein's special theory of relativity applies.	Motion	Understanding and Describing Motion
	Motion	Solving Motion Problems
	Motion	Newton's Second Law of Motion
	Motion	Applying Newton's Laws of Motion
SC.912.P.12.D Gases consist of great numbers of molecules moving in all directions. The behavior of gases can be modeled by the kinetic molecular theory.	Behavior of Matter	Physical Behavior of Matter
SC.912.P.12.E Chemical reaction rates change with conditions under which they occur. Chemical equilibrium is a dynamic state in which forward and reverse processes occur at the same rates.	Chemical Reactions and Equations	Reaction Rate
SC.912.P.12.1 Distinguish between scalar and vector quantities and assess which should be used to describe an event. (Cognitive Complexity/Depth of Knowledge Rating: High)	Motion	Using Vectors and Scalars to Describe Motion
	Motion	Solving Motion Problems

Florida Standard	Unit	Concept
SC.912.P.12.2 Analyze the motion of an object in terms of its position, velocity, and acceleration (with respect to a frame of reference) as functions of time. (Cognitive Complexity/Depth of Knowledge Rating: High)	Motion	Understanding and Describing Motion
	Motion	Solving Motion Problems
SC.912.P.12.3 Interpret and apply Newton's three laws of motion. (Cognitive Complexity/Depth of Knowledge Rating: High)	Motion	Understanding and Describing Motion
	Motion	Newton's First Law of Motion
	Motion	Newton's Second Law of Motion
	Motion	Newton's Third Law of Motion
	Motion	Applying Newton's Laws of Motion
SC.912.P.12.4 Describe how the gravitational force between two objects depends on their masses and the distance between them. (Cognitive Complexity/Depth of Knowledge Rating: Moderate)	Energy	Conservation of Energy
	Force	Fundamental Forces
	Force	Gravity
	Motion	Understanding and Describing Motion
SC.912.P.12.7 Recognize that nothing travels faster than the speed of light in vacuum which is the same for all observers no matter how they or the light source are moving. (Cognitive Complexity/Depth of Knowledge Rating: Low)	Energy	Wave Characteristics
	Motion	Applying Newton's Laws of Motion
	Electromagnetism	Electricity and Magnetism
SC.912.P.12.9 Recognize that time, length, and energy depend on the frame of reference. (Cognitive Complexity/Depth of Knowledge Rating: Low)	Energy	Conservation of Energy
	Motion	Understanding and Describing Motion
SC.912.P.12.10 Interpret the behavior of ideal gases in terms of kinetic molecular theory. (Cognitive Complexity/Depth of Knowledge Rating: High)	Behavior of Matter	Gas Laws
SC.912.P.12.11 Describe phase transitions in terms of kinetic molecular theory. (Cognitive Complexity/Depth of Knowledge Rating: Moderate)	Energy	Heat
	Behavior of Matter	Physical Behavior of Matter
	Chemical Reactions and Equations	Thermochemistry
SC.912.P.12.12 Explain how various factors, such as concentration, temperature, and presence of a catalyst affect the rate of a chemical reaction. (Cognitive Complexity/Depth of Knowledge Rating: High)	Chemical Reactions and Equations	Reaction Rate

Using Scientific Methods

photo: Getty Images

LESSON OVERVIEW

Lesson Questions

- What skills do scientists use in their investigations?
- What types of investigations do scientists do?
- What are the variables and controls in an investigation?
- How are observations from an investigation recorded?
- How is data from an investigation measured and graphed?
- How are results from an investigation analyzed?
- How is data from an investigation interpreted in order to formulate conclusions?
- What relationship exists between hypotheses, theories, and laws?

Key Vocabulary

Which terms do you already know?

- ☐ **analysis**
- ☐ **conclusion**
- ☐ **control**
- ☐ **data**
- ☐ **experiment**
- ☐ **hypothesis**
- ☐ **law**
- ☐ **scientific methods**
- ☐ **theory**
- ☐ **variable**

Lesson Objectives

By the end of the lesson, you should be able to:

- Understand the scientific method, and use the steps of the scientific method to analyze data.
- Identify controls and variables in an experiment.
- Articulate and test hypotheses in an experiment.
- Record observations in an organized, clear manner.
- Measure and graph data points collected during an experiment.
- Critically analyze experimental results.

ENGAGE

What happens when you try to determine the relationship between two different things? Visit the Engage page online to learn more.

Explain Question

What are scientific methods, and how are experimental controls and variables observed, measured, recorded, and analyzed to test hypotheses and to formulate conclusions?

EXPLORE

What Skills Do Scientists Use in Their Investigations?

Methods used by scientists apply critical thinking and systematic inquiry to determine how nature works. The approach can be applied only to phenomena that are testable or falsifiable. Anything that cannot be observed or cannot be shown to be false cannot be investigated using science.

Using Scientific Methods

Scientists use a wide range of skills and practices to conduct their work. These include:

- observing and describing
- recording, in notebooks and a variety of electronic formats
- asking and redefining questions
- researching information in books, journals, and online
- developing hypotheses
- planning and conducting experiments and other investigations
- designing and using physical and mathematical (often computer) models
- analyzing and interpreting data
- reasoning and arguing about evidence and explanations
- defining and solving problems
- communicating through discussion, presentation, illustration, and writing
- revising ideas and theories

Scientists may spend many years researching one question. They often work in research teams. The traditional idea that all scientists work alone in a laboratory is not the modern reality.

What Types of Investigations Do Scientists Do?

Scientists conduct their investigations using a wide range of approaches. There are three major approaches to a scientific investigation: experimental investigation, comparative investigation, and descriptive investigation:

1. Experimental investigations most closely mirror the traditional steps in the scientific method. They involve a fair test. This is an experiment in which variables are controlled, manipulated, and measured. This type of investigation starts with a research question. This may result from an observation about the physical world. Alternatively, a researcher may read about a phenomenon that he or she wishes to investigate, or simply have an idea he or she wants to test. A hypothesis is developed from the question, then an experiment is designed to test the hypothesis. Data is collected and analyzed, then evaluated.

2. Comparative investigations involve collecting data on samples that may differ in some way to better understand their similarities and differences, without conducting a controlled experiment. For example, a biochemist hypothesized that the bacteria he studied can coexist with some fungi but not others because the latter produce inhibitory chemicals. He compared the chemicals that each fungus secreted into the environment, using a form of chemical analysis called mass spectrometry. This allowed him to identify an antibacterial compound secreted by one fungus type but not the other, supporting his original hypothesis.

3. Descriptive investigations, like experimental investigations, involve collecting data (both qualitative and/or quantitative) to draw conclusions about a natural or human-made system. However, descriptive investigations do not have a hypothesis, and there are no experimental comparisons between variables. For example, Dmitri Mendeleev, an early 20th-century Russian chemist, developed a descriptive investigation that led to the development of the periodic table of the elements. He collected information about all of Earth's chemical elements that were known at that time and organized this information based on how the chemicals are related to each other. The relationships between chemicals was based on their atomic weights, their valences, and their known or predicted chemical properties. Mendeleev's research was purely descriptive, yet the patterns revealed by his periodic table allowed him to predict the existence of eight elements that at the time had not yet been discovered. It is perhaps the most useful discovery made in the field of basic chemistry.

Comparative and descriptive investigations are particularly important when scientists are working outside the laboratory setting. These scientists are said to work in the field. Field investigations allow scientists to observe the real world as it is, often without manipulating any variables. For example, an environmental chemist might study the impact of industrial activities on nearby rivers and streams by taking field measurements of various chemical pollutants and other indicators of water quality. These days, there are numerous portable analytical instruments suitable for field research.

Research often involves multiple investigations, and these can be of more than one type. A descriptive study that serves to catalog the information known about the structure, function, and appearance of all artificial chemical sweeteners could lead to a comparative study about their relative sweetness and reactivity with the sweet receptor, and also to controlled laboratory experiments on toxicity, mutagenicity, and other safety concerns.

What Are the Variables and Controls in an Investigation?

Designing a Controlled Experiment

Experiments can be difficult to design. They often involve careful control of variables. Variables are factors that could change the outcome of an experiment. They can be controlled or measured in some way. To better understand what this means, read about the following example.

A student, Michael, has read that when baking soda is mixed with vinegar, the two react to produce carbon dioxide bubbles. He learns that baking soda is sodium bicarbonate, and that vinegar is a weak solution of acetic acid. He decides to investigate whether the baking soda in Irish soda bread reacts with another ingredient to allow the bread to rise without yeast.

His research question is: *Does the recipe for Irish soda bread include an acidic ingredient plus the baking soda that together make bubbles that allow the bread to rise?* His hypothesis is: *An acidic ingredient in Irish soda bread, when exposed to the baking soda in the batter, will allow the bread to rise*. He learns that the buttermilk in the recipe is acidic and decides to test if a loaf of bread made without buttermilk will rise. At this point, he realizes that he has a problem. How can he be sure that he is measuring how leaving out buttermilk impacts the reaction of baking soda with buttermilk and ultimately bread rising? Perhaps there are unknown variables that are impacting his experiment—the temperature, duration of cooking, or other ingredients. How can he be sure that he has controlled all of these unknown variables? His solution is to set up a second experiment. This second experiment will be the same as the first with one exception. It will contain all of the recipe's ingredients, including the buttermilk. The second loaf of Irish soda bread will be baked in the same oven as the first loaf of Irish soda bread. That way, he can ensure that it is the presence of the buttermilk that is reacting with the baking soda to make the bread rise, and the absence of the buttermilk that yields a hard loaf. This second experiment is called a control. By using a control, all of the variables, other than the presence or absence of buttermilk, are held constant.

Controls are used in many ways. For example, some chemists help create new medicines. When testing the effectiveness of new medicines, tests are set up with carefully matched groups (in terms of health, age, sex, and other traits). One group gets the medicine in a pill. The other group gets a pill, but it does not contain the medicine under test; it is a placebo. To prevent experimental bias, neither the patients, nor the scientists, nor the doctors know who is in each group or what they are getting. The only difference between the two groups is whether they receive the medicine or not. One is the treatment group. The other is the control group.

Identifying Variables

There are different types of variables. The following example illustrates the different types of variables and how they are treated.

Sara is looking for an experiment to conduct for the science fair. She notices that on a hot day, water in her cat's saucer evaporates much faster than water in a cup left in the same kitchen. She decides that this observation is worth investigating further and develops a research question: *Why does water in a dish evaporate faster than water in a cup?* Sara spends some time in the library investigating the factors that could alter the evaporation rate of water—factors such as temperature, wind, humidity, and the material from which the cup is made. After doing this, she comes up with an educated guess: *Water in a dish evaporates faster than water in a cup because the water in the dish has a larger surface area. She refines this further: The evaporation rate of water is related to the surface area of water exposed to the air. This is her hypothesis.*

With a hypothesis to test, Sara designs and conducts an experiment to prove or disprove the hypothesis. She notices that her experiment has different types of *variables*. There are variables that she must hold constant called *controlled variables*. She identifies these as the volume of water, the time for evaporation (1 day), the location (to standardize climatic conditions), and the material from which the dish is made. In her experiment, she decides to alter one variable—the surface area of the water. This is the *independent variable*. To accomplish this, she uses five open containers with different diameters. The remaining variable, the amount of water evaporated, is the *dependent variable*. This is what Sara measures.

The careful control of variables is important in all experiments. If Sara had not controlled the climatic conditions for the experiment shown, the outcomes may have been totally different, and she would not have been able to determine whether changes in evaporation rate were due to surface area or another variable—temperature, perhaps. In any experiment, it is important to identify all the variables and control those that are not engaged by the research question.

Researching, Evaluating, and Referencing Sources

As previously stated, one of the general steps of the scientific method is conducting background research. While scientific research is often conducted by individuals, science is ultimately a communal effort. Scientists conduct background research by reviewing the existing body of scientific literature. Science often involves a lot of troubleshooting, so performing this background research allows researchers to learn from what has already been tried by other scientists. Background research also provides a solid foundation for adding to and building on the existing body of research.

When a scientist performs background research, he or she reviews the results of previous studies and the methods used to obtain them. This information usually comes from articles (or papers) published in professional scientific journals. As with all forms of research, scientists performing background research must evaluate sources for objectivity and validity. Scientists might consider the reputation of the scientist who wrote the paper, as well as the publication in which the paper appears.

An article in a scientific journal commonly includes an introduction, which includes important background information from existing scientific literature or previous studies. The introduction may also explain why the research described in the paper is important, the research questions the study was designed to answer, hypotheses the researchers developed in the earlier stages of the study, and other information to set the stage for the rest of the paper. Whenever ideas, facts, or conclusions are presented that are the result of someone else's research, the authors must state where the information came from to avoid plagiarism. This can be done in a variety of formats depending on the journal, but it typically consists of either a parenthetical or numerical citation. For example, a paper may state that something has been shown to be true based on previous findings. If using parenthetical citations, the sentence would then end with the author(s) and publication year of the publication(s) in which those findings were reported. A more detailed reference to each publication would then be provided in the reference section, which would include the author(s), publication date, title, journal name, journal volume and issue number(s), and page numbers.

The reference section of a paper is a list of all of the other publications used as background information for the research. If using numerical citations, the sentence would end with a number in superscript. This number corresponds to the order in which references are listed in the reference section. For example, the sentence would end with a 6 in superscript if it made reference to the publication that appears sixth in the list of references at the end of the paper. References are usually listed alphabetically by the first author's last name.

How Are Observations from an Investigation Recorded?

Recording Data

To be successful, an experiment requires accurate observation and data collection. An observation is the process of gaining information by using one or more of the senses. Observations can also be made by detectors that detect phenomena outside the range of our senses.

Early scientists had to record their results by hand. Today, sophisticated and sometimes remote equipment allows accurate recording of observations and measurements made during an experiment. Digital photographs enable detailed image analysis. Videos capture dynamics of a changing system. Probes and sensors make and record precise measurements at intervals. However, even with sophisticated equipment, an observant experimenter can provide insight and perspective that a machine cannot. The human experimenter might notice the small details or differences that could be the keys to forming a better hypothesis. For this reason, a conscientious researcher always carefully records the details of an investigation in a notebook.

How Is Data from an Investigation Measured and Graphed?

Organizing Data Points

All scientific questions involve observation and the collection of data. Normally, data includes measurements of some type. Scientists usually use SI units (International System Units). SI units are internationally recognized and are closely related to metric units. They are easy to convert because they are in base 10 and have an agreed international standard. Scientists display data in tables, charts, and graphs.

Prefixes Used with SI Units		
Prefix	**Symbol**	**Meaning**
tera-	T	1,000,000,000,000 or 10^{12}
giga-	G	1,000,000,000 or 10^{9}
mega-	M	1,000,000 or 10^{6}
kilo-	k	1,000 or 10^{3}
deci-	d	0.1 or 10^{-1}
centi-	c	0.01 or 10^{-2}
milli-	m	0.001 or 10^{-3}
micro-	μ	0.000001 or 10^{-6}
nano-	n	0.000000001 or 10^{-9}
pico-	p	0.000000000001 or 10^{-12}

Prefixes are used to indicate multiples or fractions of base SI units. What does the prefix "micro" indicate?

Scientists organize data in a number of ways to make it easy to analyze and interpret. For example, using the results from the previous section, Sara organized the data from her experiment in the following chart:

Container	1	2	3	4	5
Surface area (cm^2)	78.5	314.2	707.0	1,256.6	1,963.5
Volume at start (ml)	200.0	200.0	200.0	200.0	200.0
Volume after one day (ml)	197.5	190.0	176.9	157.4	135.1
Change in volume (ml)	2.5	10.0	23.1	42.6	64.9

Notice how she was careful to specify the units she used to measure the variables.

How Are Results from an Investigation Analyzed?

Data Analysis

Once the data are collected, they need to be analyzed and interpreted. Methods of data analysis vary greatly. The main objective is to find a relationship between the independent and dependent variables. The simplest form of data analysis is to check for a linear relationship between the independent variable and the dependent variable. Sara plotted a graph to help her analyze the data. She followed the convention of plotting the independent variable on the *x*-axis and the dependent variable on the *y*-axis. What relationship would her graph show?

How Is Data from an Investigation Interpreted in Order to Formulate Conclusions?

Interpreting and Sharing the Data

Scientists must interpret the results of an investigation and make a conclusion. A conclusion is an inference or deduction of whether the results support or refute (that is, fail to support) the hypothesis. However, in many situations, results are not easily interpreted. Scientists often use statistics as a tool to determine whether the results support the hypothesis. They may find some results that support the hypothesis and other results refute it. These discrepancies are discussed in the conclusion to establish whether the experiment or the hypothesis was erroneous. Accompanying this discussion, solutions are proposed to resolve the issue. Does the data from Sara's experiment support her original hypothesis: Water in a dish evaporates faster than water in a cup because the water in the dish has a larger surface area?

Scientists also must publish their results so that other scientists can study and attempt to replicate the investigation and confirm the results. This process is called peer review. Peer review is helpful not only in providing others the ability to replicate the work of a scientist, but also in presenting an opportunity to evaluate a scientist's methodology and disseminate knowledge gained from an investigation.

Analyzing Chemical Data

Modern chemistry, particularly the field of analytical chemistry, focuses on the detection, identification, and description of the composition and properties of substances, as well as chemical interactions with the physical universe. These objectives are accomplished by the collection and analysis of data. After collecting raw data, chemists analyze the measured values by summarizing and organizing them in tables and graphs. The ultimate purpose is to promote clearer visualization, interpretation, and understanding.

Mean

The most common method for summarizing data is to calculate the mean. This is also called the arithmetic average of a data set. The formula for the mean is simple and probably familiar to you. The mean of a data set is represented by the variable $\bar{x}$ and is calculated using the formula:

$$\bar{x} = \frac{\sum x_i}{n}$$

where x_i represents the value of an individual data point, n = the number of data points, and Σx_i represents the total of all data points added together. The mean provides a value that represents the center of the data set. Usually, there is not a data point that sits right on the mean.

Calculating Mean: Sample Problem

Foam insulation in homes has been known to release formaldehyde (CH_2O). Apex Supplies Inc. has introduced a new product called WinterWarm, which it claims releases formaldehyde at a level beneath the federal standard of 1,000 part per million volume (ppmv). After testing WinterWarm for five years, Apex Supplies Office of Research and Testing collected data on 13 model homes. The following table lists the levels of formaldehyde released from these model homes. Calculate the mean formaldehyde level in this data set.

Home Number	Formaldehyde (CH_2O) Release (ppmv)
1	924
2	897
3	996
4	1,047
5	1,150
6	1,060
7	1,126
8	915
9	982
10	844
11	1,105
12	1,281
13	1,005

$$\bar{x} = \frac{\sum x_i}{n} = \frac{924+897+996+1,047+1,150+1,060+1,126+915+982+844+1,105+1,281+1,005}{13}$$

$$= 1,025.54$$

Standard Deviation

Along with a measure of center such as the mean, scientists need a measurement that indicates how the data is spread around the center. The standard deviation is the value most often used. The standard deviation represents the average distance the data points lie away from the mean. The calculation of the standard deviation is much more complex than that for the mean, so most people use computers or calculators. For the bold of heart, the formula is:

$$s = \sqrt{\frac{1}{n}\sum_{i-1}^{n}(x_i - \bar{x})^2}$$

where *s* = standard deviation.

Let's try this on the chlorine concentrations from the formaldehyde calculations in the following table.

$$s = \sqrt{\frac{1}{13}[(924 - 1{,}025.54)^2 + (897 - 1{,}025.54)^2 + (996 - 1{,}025.54)^2 + \ldots + (1{,}281 - 1{,}025.54)^2 + (1{,}005 - 1{,}025.54)^2]}$$
$$= 115.69 \text{ ppmv}$$

The Standard Normal Distribution

When data are graphed, a common trend often occurs. Numerical data values often cluster symmetrically around the mean. This is called a normal distribution. When these data values are graphed in a dot plot, a shape called the bell curve appears. Because there are variations in data, the curves may look slightly different, but they are the same general shape, with the largest number of data points in the center.

Characteristics of a Normal Distribution

In a normal distribution, the mean is the peak of the curve, and the curve is symmetrical. Half of the data lie above (to the right of) the mean, and half below (to the left of) the mean. Looking at the standard deviations above and below the mean, the data are separated into intervals. About 68 percent of the data lie within one standard deviation of the mean, 95 percent within two standard deviations, and 99.7 percent within three standard deviations of the mean. Because data can vary, these percentages are not always exact, especially with smaller sample sizes. This is called the empirical rule, 68-95-99.7.

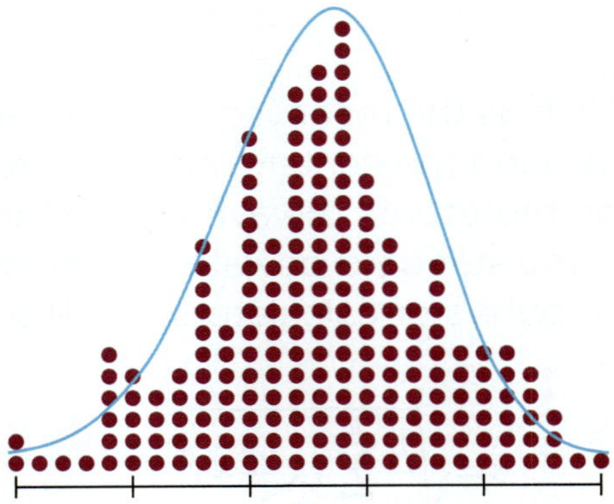

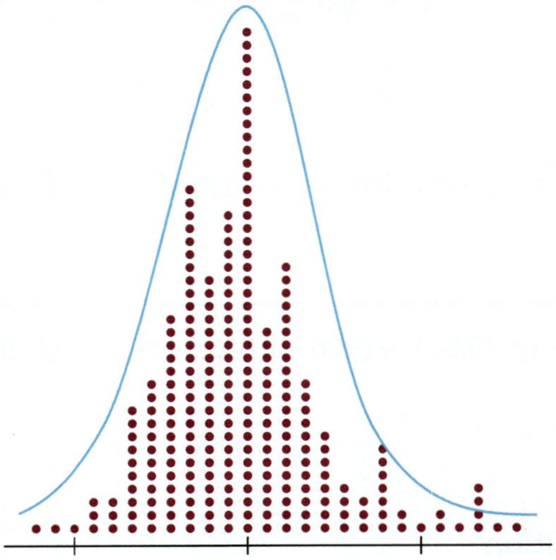

These normal distributions both form bell curves, but they look different. One is broader at the base and the other is higher at the center.

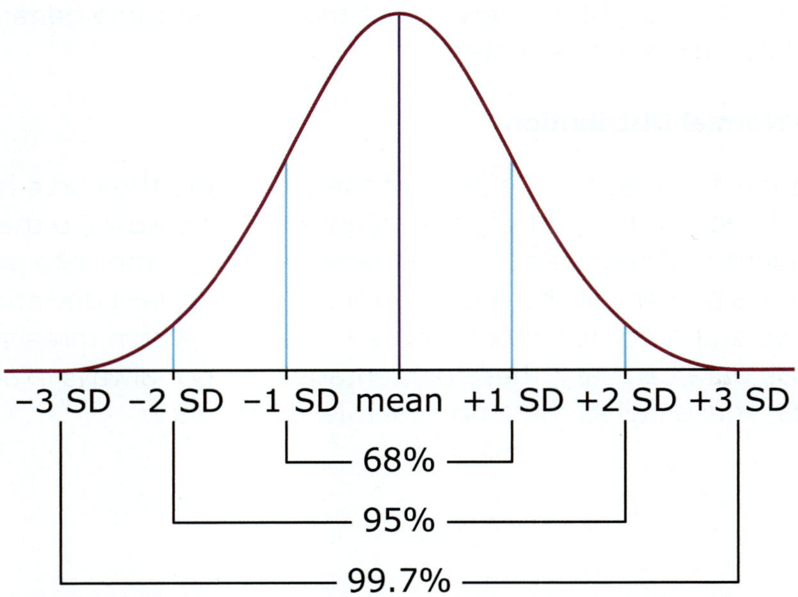

In a perfectly normal distribution, the middle 68% of the data points lie between –1 and +1 standard deviations from the mean. How much of the data lie between –2s and +2s?

The *z*-score is a measure of the position of a data value on the standard normal curve. It tells you how many standard deviations the value is above or below the mean.

The formula for the *z*-score for any data value (x) in a normal distribution is:

$$z = \frac{x - \overline{x}}{s}$$

Let's suppose that we measure the volume of formaldehyde released in another home not in our sample shown, and the value is 1,212 parts per million volume (ppmv). The *z* value for this home would be:

$$z = \frac{1{,}212 - 1{,}025.54}{115.69} = 1.61$$

This tells us that the formaldehyde release from this home is 1.61 standard deviations away from the center of the distribution. Furthermore, because the *z* value is positive, we know the data point lies to the right of the mean. In other words, the data point is 1.61 standard deviations greater than the mean.

Now, locate this *z* value on the standard normal curve shown, which illustrates the empirical rule. You can see that a *z* value of +1.61 puts this home in the middle 95 percent of our distribution.

Let's look at another example. Suppose we measure the volume of formaldehyde released in yet another home and find the reading to be 950 ppmv.

$$z = \frac{950 - 102.54}{115.69} = -0.65$$

The negative *z* value tells us that this sample has a formaldehyde release rate to the left of the mean. In other words, the formaldehyde release in this sample is less than the mean of the original data set. Using the empirical rule (shown), this reading is in the middle 68 percent of our distribution.

The Five-Number Summary

Another convenient way to summarize data and obtain a quick indication of its distribution and spread is by use of the five-number summary. In this case, the numbers are obtained by first sorting the data points in ascending numerical order, then calculating the quartiles of the data set. The first quartile is the value below which 25 percent of the data points lie. The second quartile is the value below which 50 percent of the data points lie. The third quartile is the value below which 75 percent of the data points lie. The fourth quartile is the value below which 100 percent of the points lie, so there is no actual value corresponding to this quartile.

The second quartile is also known as the median. In calculating quartiles, it is important to understand that we aren't locating the mean as the middle value. Instead, we are simply counting the number of data points and locating the value that divides the range into groups by the number of data points. For example, if your data set contains 11 data points, the median, or second quartile, is the sixth data point. Don't forget that you must first sort the data in numerical order.

The five-number summary results in the following values:

Five – Number Summary
Minimum
Q1
Median (Q2)
Q3
Maximum

Let's suppose you collect the following data set by recording the heights of mountains in the Cascade Mountain Range in the western United States.

Mountain Name	Height (Thousands of Feet)
Mount Rainier	14.4
Mount St. Helens	8.3
Mount Hood	11.2
Mount Shasta	14.1
Mount Adams	12.3
Lassen Peak	10.5
Mount Baker	10.8
Glacier Peak	10.5
Mount Jefferson	10.5
Mount Bachelor	9.1
Newberry Volcano	8.0

To determine the five-number summary, begin by sorting the data in ascending order.

Mountain Name	Height (Thousands of Feet)
Newberry Volcano	8
Mount St. Helens	8.3
Mount Bachelor	**9.1**
Lassen Peak	10.5
Glacier Peak	10.5
Mount Jefferson	**10.5**
Mount Baker	10.8
Mount Hood	11.2
Mount Adams	**12.3**
Mount Shasta	14.1
Mount Rainier	14.4

The value that lies in the very middle of the table is the median. This corresponds to the height of Mount Jefferson, 10,500 feet. The first quartile is the point that lies in the very middle of the lowest five values, which in this case is 9,100 feet for Mount Bachelor. Finally, the third quartile is 12,300 feet for Mount Adams.

For most applications, the quartiles of interest are the middle two quartiles, referred to as the interquartile range. It stretches from the first quartile to the third quartile. It is useful because it gives a rough indication of how the data is spread. By plotting the five-number summary on a graph referred to as a box plot, we can quickly see if our data deviates from a normal distribution. From a box plot of the Cascade Mountains data, you can observe that the points are distributed approximately equidistant on both sides of the median. This data, then, is approximately normally distributed. Notice that the blue box contains the interquartile range. Also observe that though three points in our data set have the same value (10.5 thousand feet), all of them are represented by a single point.

What Relationship Exists between Hypotheses, Theories, and Laws?

Hypotheses

A hypothesis is an educated guess about the answer to a question. It must have the property of being able to be proved true or false. Science involves proving a hypothesis wrong just as much as proving it right. Developing a useful hypothesis often involves researching literature on the subject. If, when tested by experiment, a hypothesis fails, then the researcher reviews his or her results and then develops a new hypothesis. Most of the time, a hypothesis describes a causal relationship. Scientifically, the cause is the independent variable. The effect is the dependent variable because it depends on the cause. In an experiment, the independent variable is manipulated. The dependent variable is what the experimenter is measuring.

Suppose a chemist asks the question: Can alcohol be used to remove ink stains due to a permanent marker? During her background research, the chemist learns that some inks are water soluble, some inks are water insoluble, and permanent marker stains are considered permanent and water insoluble. However, she also discovers that some chemicals that are insoluble in water are soluble when alcohol is used as the solvent. Based on her research, she develops the following hypothesis: Pure ethanol, when applied to a permanent marker ink stain, will remove the stain from cotton material. Her next move is to design and conduct her experiment. Her experiment involves comparing a treated group of 100 ink-stained cotton handkerchiefs with a control group of the same size. For this experiment, there are two possible outcomes, one that agrees with the hypothesis and one that does not.

Scientific Theories

Scientific theories are different from everyday theories that we may have. A scientific theory is an explanation that is strongly supported by scientific evidence. Scientific theories are usually validated by repeated observation and experimentation. In short, scientific theories describe why things happen. Scientific theories are based upon one or more hypotheses that have been verified and accepted to be true by the scientific community. These scientific theories are valid, as long as there is no scientific evidence to dispute them. This is why scientific theories can sometimes be disproven. As new observations or experimental results become available, scientific theories can be reshaped, adapted, reformed, or reaffirmed in order to reflect the strongest scientific evidence.

Often, theories are the result of many scientists working independently over long periods of time. For example, the kinetic molecular theory explains everything we know about the properties of different states of matter in terms of the energy and forces of matter's component particles. It was described in the 19th century by a theoretical scientist, James Clerk Maxwell (1831–1879), but is based on the cumulative work of numerous earlier scientists, including Robert Boyle (1627–1691) and his experiments on the pressure and volume of gases; Antoine-Laurent de Lavoisier (1743–1794) and his experiments on gas temperature versus pressure, water composition, matter transformations, and quantitative chemistry; Jacques Charles (1746–1823) and his work on gas volume; Pierre-Simon Laplace (1749–1827) and his work on gas pressure versus temperature; the experimental chemist John Dalton (1766–1844) and his work on atomic theory; Joseph Louis Gay-Lussac (1778–1850) and his work on gas volume versus temperature; and other scientists across many disciplines.

Laws of Nature

A law of nature is a statement that describes a natural phenomenon that always happens the same way under the same set of circumstances. Laws are descriptive. They do not explain the science behind the phenomenon. For example, all gases can be measured to behave similarly at the same temperature and pressure. This is a statement of fact, and it is described by the ideal gas law. In this example, the ideal gas law does not describe why this phenomenon occurs. Instead, it only describes what occurs. The kinetic molecular theory offers an explanation.

The Relationship between Hypotheses, Theories, and Laws of Nature

A scientist might develop a hypothesis that can account for descriptions of a phenomenon and then use this hypothesis to test the scientific basis for what happens. If enough experimental evidence and observational data support the hypothesis, the hypothetical explanation can become a theory. At this point, it is accepted as the most likely explanation for why something happens. A theory is generally accepted as true until the time comes that some other experiment or body of hard evidence refutes the explanation.

Therefore, a law of nature describes the way things are. A hypothesis is an educated guess, yet to be proven, to explain the reason for the law of nature. A hypothesis can become a theory to explain the law if the explanation is repeatedly confirmed by sufficient evidence and the scientific method.

Consider the Explain Question:

What are scientific methods, and how are experimental controls and variables observed, measured, recorded, and analyzed to test hypotheses and to formulate conclusions?

Go online to complete the scientific explanation.

Check for Understanding:

Researchers are studying the chemical composition of the water in a hot spring. What are the steps the researchers should use to ensure that their methods are scientific?

STEM in Action

STEM and the Use of Scientific Methods

Material science engineers are often challenged to come up with solutions to new problems, such as finding innovative ways to combat terrorism. Triacetoneperoxide (TATP) is a powerful explosive used by terrorists worldwide, such as the infamous "shoe bomber."

TATP, which was discovered in the late 1800s, is an unstable chemical that can be made from common household ingredients such as hydrogen peroxide, acetone, and an acid such as vinegar. Classified as a type of organic peroxide, TATP is highly explosive. Even small amounts pack enough explosive power to bring down an aircraft.

Due to its chemical composition, TATP is very hard to detect, which partly explains its frequent use in explosive devices. What kinds of technology could be used to detect TATP? What is the smallest amount of TATP that could be detected? Scientists used the scientific method to answer these questions and find a solution.

photo: Getty Images

The shoe bomber smuggled an explosive chemical aboard an aircraft, hidden in his shoe. How can technology improve the way explosives are detected?

In late 2010, scientists reported preliminary success in the design of a device that could detect minute amounts of TATP. The goal was to develop detection technology sensitive enough to spot TATP in concentrations of only two parts per billion in a matter of minutes. That's like spotting two friends in a crowd of two billion people.

The scientists' research used a nickel-sized sensor that contains a catalyst and dyes. When TATP fumes and the catalyst come into contact, a reaction produces compounds that, in turn, react with the dyes. The resulting colors enable screeners to quickly establish the presence of TATP. The scientific method began with a prediction about the reaction of a catalyst and TATP and then the products of that reaction with specific dyes. The scientist tested their hypothesis and got results that were consistent with their prediction. The results, in turn, enabled the TATP detector to be developed into a suitable design. Scientists will again rely on the scientific method to build upon this discovery. For example, future experiments will investigate the possibility of designing a thermodynamic sensor to detect the ingredients that make up TATP.

photo: Getty Images

ETDs work by detecting vapors and residues of explosives in collected samples. What problem did engineers solve with the invention of these devices?

Domestic explosives-related incidents in the United States have increased over time. The results shown are the number of incidents reported between 2004 and 2012, based on Bomb Arson Tracking System data. Use the data in the table to answer the questions.

Year	No. of Explosives Incidents	No. of Injuries	No. of Fatalities
2012	4,033	37	1
2011	5,219	36	5
2010	4,897	99	22
2009	3,886	57	4
2008	4,198	97	15
2007	3,143	60	15
2006	3,797	135	14
2005	4,031	148	19
2004	3,919	263	36

1. Based on the table, what can you conclude about the rates of explosives-related incidents, injuries, and fatalities between 2004 and 2012?
2. Observe the number of injuries from 2004 to 2006, and then from 2007 to 2012. Based on the data, construct a hypothesis about the cause of this change.
3. What type of background research topics would support what you observe from the data table?

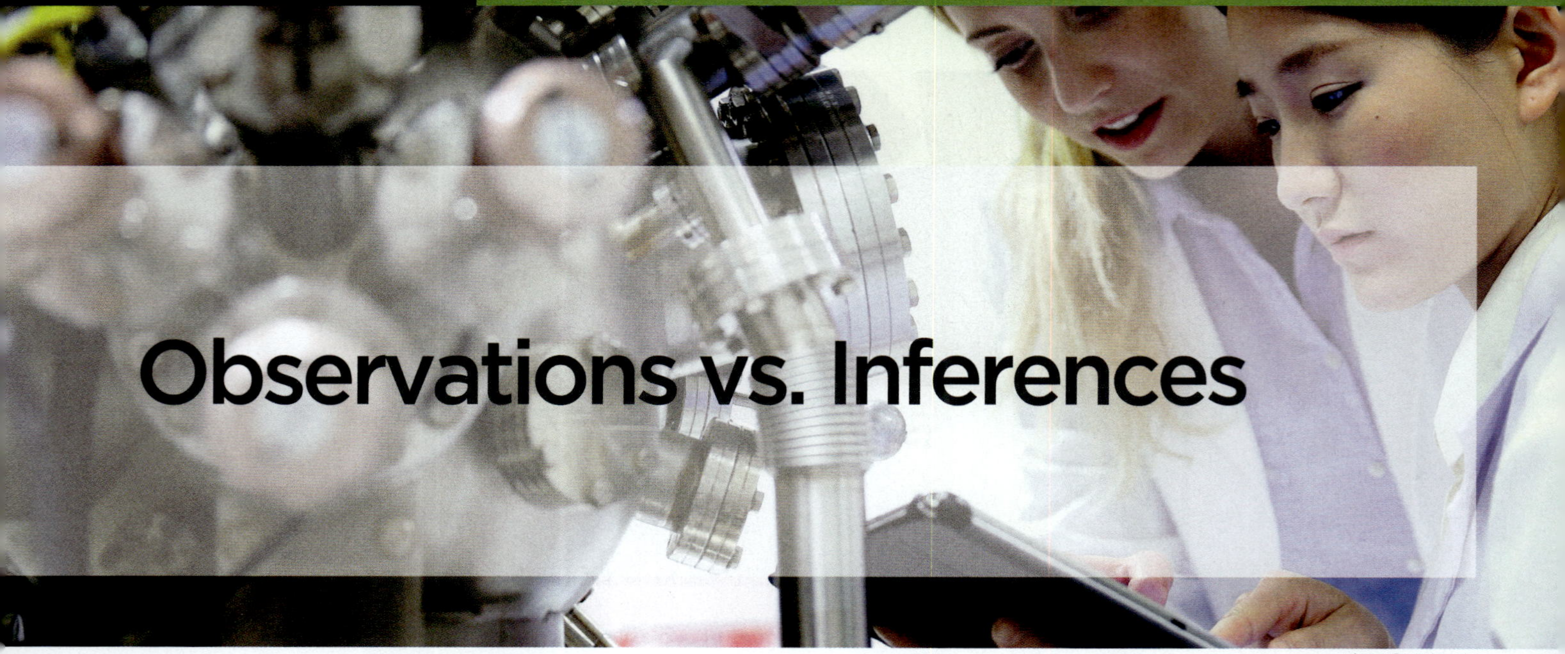

Observations vs. Inferences

photo: Getty Images

LESSON OVERVIEW

Lesson Question

- What is the difference between an observation and an inference?

Lesson Objective

By the end of the lesson, you should be able to:

- Distinguish between observation and inference.

Key Vocabulary

Which terms do you already know?

- ☐ **data**
- ☐ **inference**
- ☐ **observation**

ENGAGE

How do your senses enable you to experience different situations? Have you ever used reasoning when formulating a conclusion to a problem? Visit the Engage page online to learn more.

Explain Question

What are two examples of observations you might make, and two examples of inferences you might make, when looking at logs burning in a campfire?

EXPLORE

What Is the Difference between an Observation and an Inference?

Observations

A critical part of a scientific method is the planning and carrying out of experiments to test a hypothesis. Before scientists conduct these experiments, they must decide what types of data they need to collect. Observations and the collection of the data are critical to the success of the experiment.

All observations are made using the five senses—sight, smell, touch, taste, and hearing. The senses are used to make both qualitative and quantitative observations. Qualitative observations may involve shapes, colors, odors, or changes in these qualities. The presence of a precipitate and the release of a gas are qualitative observations, as is the generation or removal of heat. Seeing the color of a sample change during a chemical reaction and hearing the crackle of logs burning are examples of qualitative observations.

Quantitative observations involve amounts or numbers, and they often require taking measurements. Scientists may record time; measure the temperature, weight, or length of a material; count objects; or count recurrences of an event. These are all quantitative measurements. Instruments such as thermometers, rulers, stopwatches, and scales are useful when making quantitative observations.

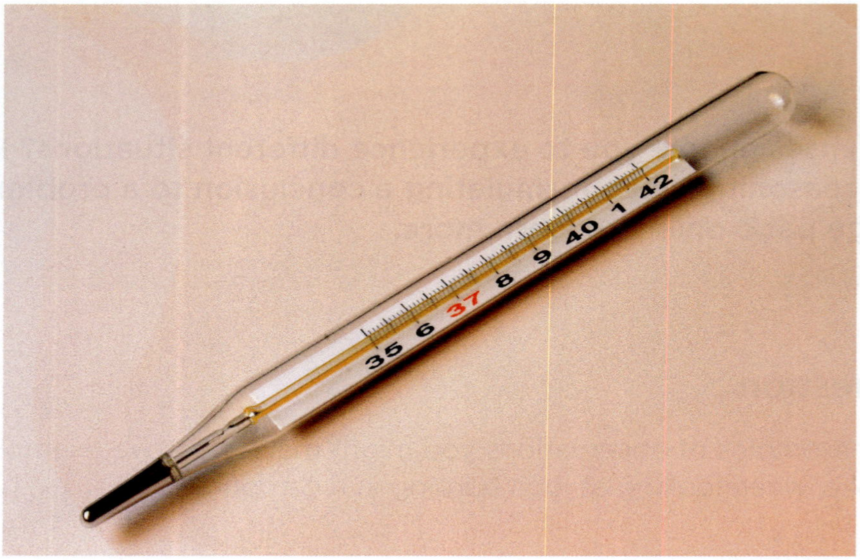

photo: Pixabay

A thermometer can be used to make quantitative observations during a reaction. How do Celsius, Fahrenheit, and Kelvin temperature scales compare?

A scientist performs an investigation that includes mixing two clear liquids that have an initial temperature of 20.5°C. After the reaction occurs, the scientist makes the qualitative observation that the product is a light blue liquid. In addition, the temperature of the product liquid is 28.2°C. The new temperature is a quantitative observation.

Inferences

Twenty years ago, hikers in the Alps found a mummified body dating over 5,000 years old. The scientists studying the mummy's remains could not identify why the man was found in a remote region in the high altitudes of the Alps; yet, they could determine how the mummy had died—an arrow to the back. This determination was made through an inference or a reasonable assessment of the evidence. The evidence that led to the inference was finding an arrowhead in the mummy's back.

More recently, scientists were able to sequence the DNA of the Iceman. The observation of his genetic code let researchers make the inferences that he was probably brown-eyed and lactose intolerant. He also seemed to be infected with the bacterium that causes Lyme disease. Comparing the Iceman's DNA with other populations enabled researchers to infer that he may have been from the Mediterranean island of Sardinia, a long way from where he was found. This added still more questions to the mystery surrounding the Iceman.

Scientists use the ability to infer to draw conclusions from the evidence being investigated during an experiment. After an experiment is conducted and observations are made, the data can be organized and analyzed. Inferences can then be made based on the data. Remember that an inference is a prediction that is supported by observations. Inferences are not always correct, but they must flow logically from the data.

Observations vs. Inferences

Let's examine how inferences are made using data collected during an experiment. The following salt substances dissolve in water: KBr, NaCl, and KNO_3. The solubility of each substance, or the amount of each substance that dissolves in water, is recorded at three different temperatures: 0°C, 60°C, and 100°C. These data points are then plotted onto a graph. This data is presented in the following table and graph.

Formula	g/100 mL at 0°C	g/100 mL at 60°C	g/100 mL at 100°C
KBr	54	86	104
NaCl	36	37	39
KNO_3	14	109	245

This table shows the solubility of different salts in water. How does solubility change as temperature increases?

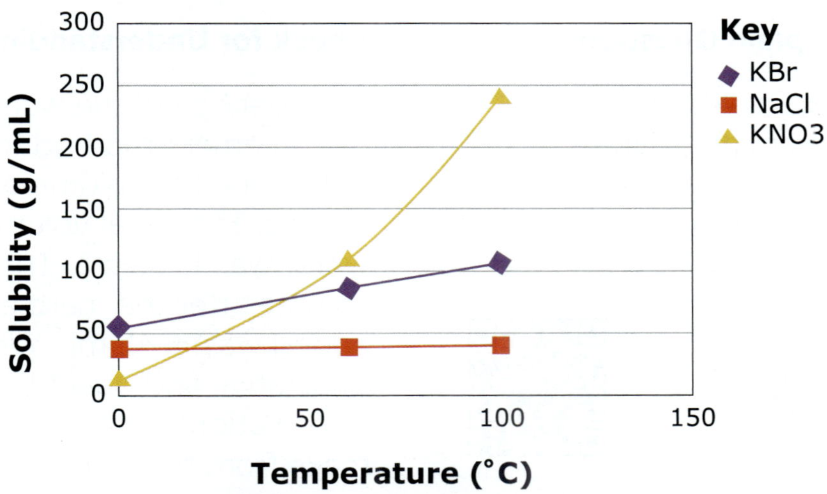

A graph shows the solubility of potassium bromide (KBr), sodium chloride (NaCl), and potassium nitrate in water at different temperatures. How does graphing the data make connections more apparent?

Many observations can be made from this collection of solubility data. One of these observations is that at 0°C, KBr is more soluble in water than KNO_3. At 100°C, however, KNO_3 is more soluble than KBr. Another observation is that the solubility of each substance increases as the temperature increases. These observations are made by looking at the data in either the table or the graph.

Many predictions, or inferences, can be made regarding the solubility of each substance at a temperature that is not recorded. For example, the graph shows that the solubility of KBr at 80°C is approximately 95 grams per 100 milliliters of water. Another inference might be that at 150°C, the solubility of KNO_3 will be greater than both the solubility of KBr and NaCl. These are both logical inferences based on the observations recorded during this experiment.

Making Observations and Drawing Inferences

Because inferences are made from careful observations, it is important to be able to distinguish between the two. For example, someone looking at a photograph might make the following observations:

1. There is a collection of dark clouds forming in the sky.
2. Part of the sky is lighter.
3. The ground is covered with green and brown plants.

These observations support several inferences about what was occurring when the picture was taken. Because part of the sky is light, this picture was most likely taken in the daytime. The dark clouds indicate that a storm is approaching or passing. The bright green grass or growing crops indicate that the season is spring or early summer. These inferences are all made based on careful observation of the photograph.

Consider the Explain Question:

What are two examples of observations you might make, and two examples of inferences you might make, when looking at logs burning in a campfire?

Go online to complete the scientific explanation.

Check for Understanding:

A student used the water displacement method to determine the volume of a marble. The initial volume of water was 12.33 mL. The final volume after the marble was added was 15.42 mL. What qualitative and quantitative observations could you make from these results?

STEM in Action

STEM and Observations vs. Inferences

Spectrometrists are chemists who work in areas such as the pharmaceutical industry to analyze the structures and other properties of pharmaceutical compounds. They may analyze small molecules that have the potential to treat illnesses. Because some drugs work by specifically targeting proteins, pharmaceutical spectrometrists are often interested in the structure of proteins or protein-drug interactions.

Pharmaceutical spectrometrists often analyze complex data obtained from intricate experiments. Mass spectrometers, nuclear magnetic resonance (NMR) spectrometers, surface plasmon resonance (SPR), or ultraviolet/infrared (UV/IR) spectrophotometers are common instruments used in this field. These instruments allow the spectrometrist to make and analyze both qualitative and quantitative observations. The pharmaceutical spectrometrist uses these observations to make inferences that can help answer complex questions about the systems that they study, such as how a drug works or how it might be improved.

photo: Getty Images

Mass spectrometers provide information on the structure of complex substances. What are some observations and inferences that scientists can make using a mass spectrometer?

Now, complete the activity that follows.

A pharmaceutical spectrometrist places a sample of a pain reliever into a mass spectrometer and obtains data about different compounds within the sample. The image shows a mass spectrograph of the data. The relative intensity, shown on the vertical axis, indicates the relative amount of each compound in the sample. The *m/z* (mass to charge) ratio, shown on the horizontal axis, is related to the molecular mass of each compound.

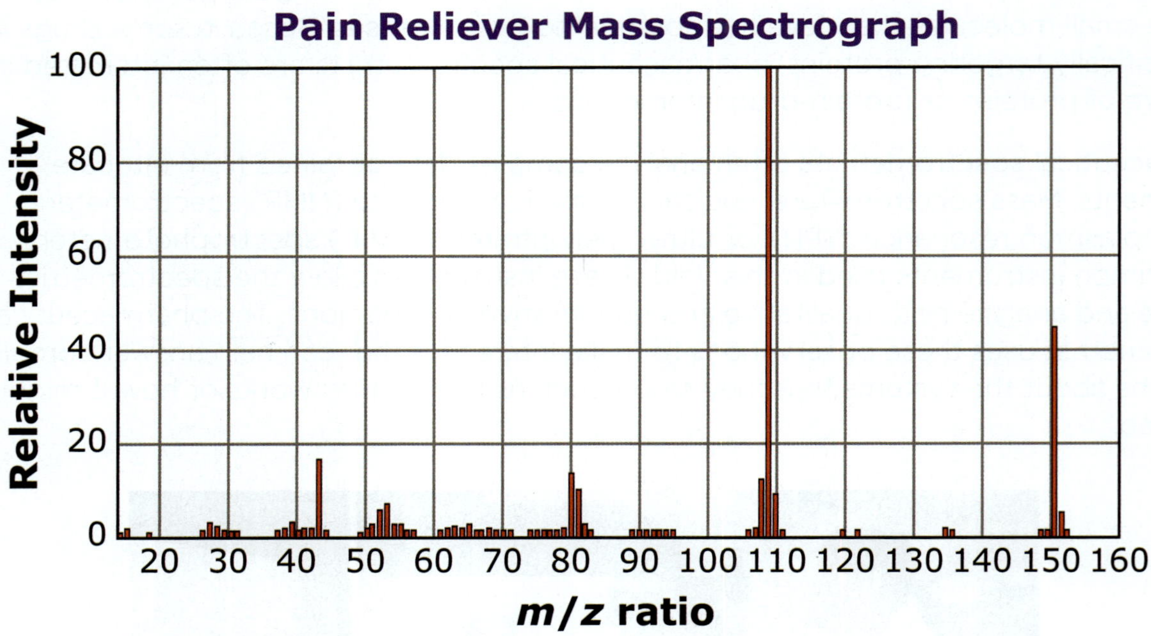

1. Describe two observations you can make from the graph.
2. What inference can you make about the number of atoms in molecules of the most common compounds in the pain reliever sample compared with the number of atoms in molecules of the least common compounds in the sample? Identify the observation that allows you to make this inference.

Laboratory Practices and Safety

photo: Getty Images

LESSON OVERVIEW

Lesson Questions

- How should basic laboratory equipment be used?
- How is safety ensured during laboratory and field investigations?
- What basic safety procedures are important in a chemistry laboratory?
- Where is safety information about a chemical found, and what useful information is provided there?
- How is safety equipment in the laboratory used?
- What procedures should be followed in emergency situations?

Key Vocabulary

Which terms do you already know?

- ☐ **MSDS (material safety data sheet)**
- ☐ **safety**

Laboratory Practices and Safety

Lesson Objectives

By the end of the lesson, you should be able to:

- Demonstrate proper use of basic laboratory equipment.
- Demonstrate safe practices during laboratory and field investigations.
- Identify all safety equipment in the laboratory and explain or demonstrate how it is properly used.
- Demonstrate knowledge of basic laboratory safety procedures.
- Locate SDS (previously called MSDS) and identify important information pertinent to health and safety for the use of a chemical.
- Explain or describe procedures to follow in emergency situations such as fire, bodily injury, and chemical spills.

ENGAGE

What personal safety rules have you followed when cooking your favorite meal or treat? Why is it important to follow safety rules in the laboratory? Visit the Engage page online to learn more.

Explain Question

How do you implement safe practices in a science laboratory?

EXPLORE

How Should Basic Laboratory Equipment Be Used?

Laboratory Equipment

photo: National Institutes of Health

This old laboratory is neat and organized, but it has little else in common with modern laboratories in terms of safety equipment. What are some modern tools and equipment that are present in the laboratory to enhance safety measures?

Most equipment in the laboratory is harmless when used properly. Thermometers, scales, and rulers are common, and they are easy to use. However, there are many pieces of equipment that are fragile. Glass beakers, flasks, test tubes, and other glassware must be treated with care. Some items, such as sensors or microscopes, can be permanently damaged if they are not used properly. More importantly, student safety is at risk when equipment is not used as intended.

Some lab equipment can be particularly dangerous. Bunsen burners are attached to a fuel source and produce a flame. If not used properly, Bunsen burners can cause severe burns. They can also be a source of gas leaks, and unattended flames can cause fires.

Laboratory equipment should be used only in the way intended. Always ask a teacher for guidance when learning to use a piece of equipment. In addition, make sure to read all equipment instructions before moving or using it. Inspect all lab equipment before using it. If a piece of equipment is damaged or not working properly, report it to the teacher immediately.

No matter what equipment is being used, be sure to read the procedures carefully before beginning any investigation. Follow all of the teacher's directions, and carefully observe all teacher demonstrations of equipment. Use common sense, and treat all equipment as potentially dangerous.

Handling Glassware

Glass beakers, flasks, graduated cylinders, and other fragile tools are used frequently in chemistry investigations. Glass is hard, withstands changes in temperature fairly well, and does not react with most chemicals. These tools must be used with caution, however, because glass is so fragile, and broken pieces are dangerous.

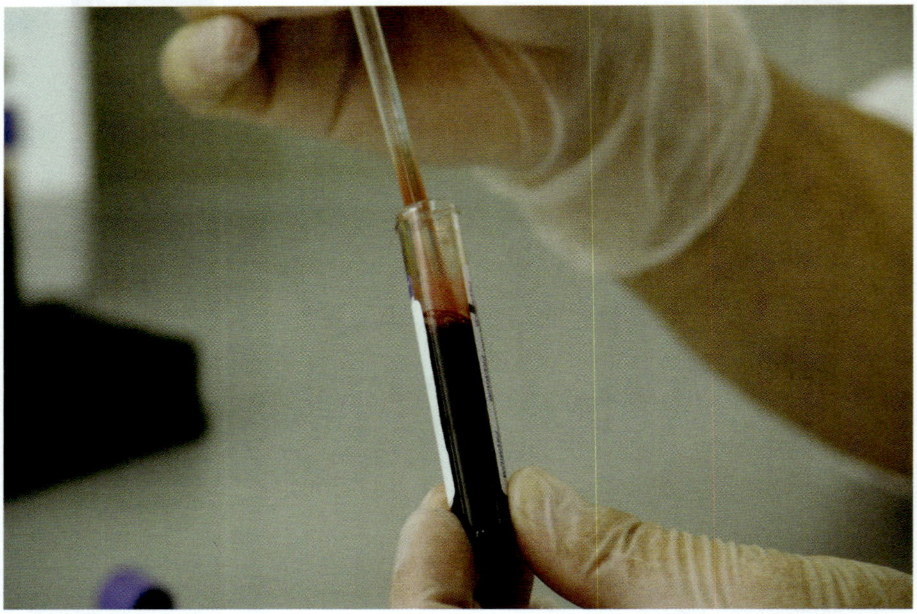

photo: Pixabay

Chemical glassware is often used in the laboratory. What precautions can you take to prevent injuries when using glassware?

To ensure safety when working with glass, follow these procedures:

1. Inspect glass items before using them. Never use chipped or broken glassware, as it is more likely to break completely if the temperature changes.
2. Do not immerse hot glassware into cold water. The instant change in temperature can cause breakage.
3. Use "hot hands" (rubber mitt that protects your hand when handling warm glassware) or tongs to handle hot beakers, and use test tube holders when heating test tubes. Always place test tubes in a test tube rack when making observations.
4. Do not touch broken glass. In many schools, students are not allowed to clean up broken glass, so accidents must be immediately reported to teachers or custodians. If that is not possible, be sure to use a broom and dustpan to pick up sharp objects.
5. Do not put broken glass into the trash, as it may pierce plastic bags and injure someone. Dispose of broken glass in the sturdy "broken glass" or "sharps" container found in the lab.

6. Don't assume that heated glassware has cooled. Approach the container slowly with the back of your hand. Do not touch glass that feels warm as you get near it.

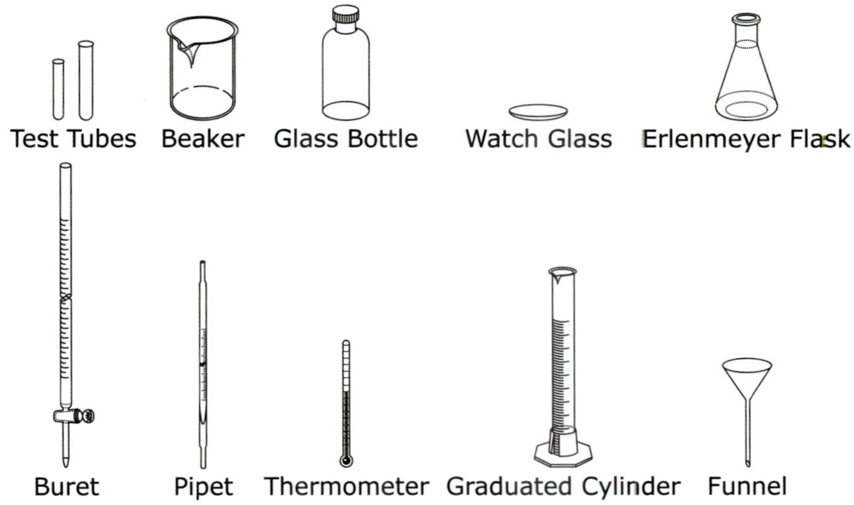

There are many types of basic glassware used in the introductory chemistry laboratory. What are the functions of the pictured glassware?

How Is Safety Ensured during Laboratory and Field Investigations?

Basic Safety Practices

Depending on the location in which an investigation is being conducted, different safety rules might be required. Some types of equipment or chemicals used in an investigation might require special handling. No matter where a scientist is working, however, certain safety procedures must always be used.

First, make sure that your body is protected from burns or other injuries:

- Always wear protective equipment, such as closed-toe shoes, eye protection, an apron, and gloves.
- Do not wear contact lenses.
- Tie back long hair and loose clothing so it is not caught in equipment and does not fall into chemicals.
- Do not eat or drink anything in the lab, whether you brought it with you or not.
- Never taste or touch any chemicals.
- Do not run, play, or swing your arms in the laboratory.
- Do not enter the laboratory unless a teacher is present.
- Do not take anything out of the laboratory without a teacher's permission.
- Always wash both hands before leaving the lab.

Field investigations are sometimes performed outside of the laboratory. These investigations may take place in the natural environment, such as a wildlife preserve or park, or they may take place in areas where people are present, such as a shopping mall or other public area.

During field investigations, many of the previously listed rules are likely to apply. In addition, proper clothing, specialized eye protection, gloves, and sun protection might be required. Use common sense regarding hairstyles, contact lens use, and food or drinks. If in a public area, do not stray from the class. When in a natural area, do not approach wildlife. Do not leave any litter, and do not disturb any plants. Be familiar with dangerous plants, such as poison ivy or poison oak, so that you can avoid them. Report any unsafe conditions or inappropriate student behavior to the teacher.

What Basic Safety Procedures Are Important in a Chemistry Laboratory?

Safe Handling of Chemicals

Sometimes, the chemicals used in a chemistry laboratory are dangerous. There are some general rules to keep in mind when dealing with any chemicals that will help ensure safety.

First, make sure to avoid contact between chemicals and skin:

- Always use a spatula or scoopula to transfer chemicals between two containers, and do not touch the chemical with your fingers.
- Never smell chemicals directly. If appropriate, use a wafting motion to carefully direct fumes toward the nose, but only if directed to do so in the lab.
- Do not look directly down into a test tube. If the chemicals in the test tube suddenly react, your eyes and skin could be damaged by fumes or debris.
- Do not return unused chemicals back to the original container. If a chemical is contaminated, it might skew later experimental results or create dangerous reactions.
- Never put solid or liquid material in the sink unless instructed to do so in the lab procedures. Some substances create toxic chemicals when mixed with water.
- When disposing of excess chemicals after the completion of a laboratory investigation, never pour them down the drain unless instructed to do so. Always follow the directions of your instructor as to the proper disposal of these chemicals.

As a general rule, handle concentrated acids and bases with care. Always add acid into a container of water; never add water to the container of acid. When water is added to large amounts of acid, the reaction creates intense heat and can be dangerous. It may bubble up and overflow or even explode. If there is a spill, be sure to dilute acids or bases with a neutralizing solution before cleaning them up.

Always read the labels on bottles carefully and handle the chemicals inside as instructed. Handle flammable liquids cautiously, and keep them away from sparks and flames. Remember to place bottles in their proper storage location immediately after use.

Where Is Safety Information about a Chemical Found, and What Useful Information Is Provided There?

Safety Data Sheets

Because it is so important to have information on hand about the chemicals used in a laboratory (or even in everyday activities), chemical manufacturers and suppliers are required to provide safety information about their products. This information usually includes safe handling and cleanup instructions. The labels of some bottles also have graphics indicating if they are flammable, toxic, or if they pose some other hazard.

A safety data sheet (SDS) or material safety data sheet (MSDS) contains information about a chemical's properties. SDS documents are provided by the chemical manufacturer and are often available online as well. In most labs, there is a file or binder readily accessible that contains an SDS for every chemical present.

photo: Getty Images

What types of physical properties do you think are listed in a material safety data sheet?

The SDS will clearly indicate any hazards associated with a particular chemical. Fire and explosion hazards, health risks, and reactivity information is included. The personal protection equipment needed when handling the product is also specified.

How Is Safety Equipment in the Laboratory Used?

Safety Equipment

Laboratories are required to have the proper equipment needed to deal with an emergency. These items include fire extinguishers, an eyewash station, an emergency spill kill, a fire blanket, and a first aid kit. To clean up broken glass, a broom, a dustpan, and a sturdy "broken glass" container must always be used. Make sure you know where each piece of emergency equipment is stored and how to use it.

Some equipment is designed to prevent emergencies. Fume hoods safely remove dangerous vapors from the air, and they help limit the spread of dangerous gases in case of an accident.

What Procedures Should Be Followed in Emergency Situations?

Emergency Situations

In case an accident does occur in the lab, it is important to keep safety in mind at all times. First, make sure there is clear access to emergency exits, fire extinguishers, and eyewash stations. Never leave book bags or other personal items lying on the floor where other students need to walk.

If there is a fire, it is important not to panic. Turn off all flames, Bunsen burner gas lines, and hot plates, if possible. If the fire is small, cover it with an empty beaker or damp cloth. For larger fires, use a fire extinguisher. Be sure to first pull out the pin that prevents accidental activation of the extinguisher. Before pressing the handle to activate the extinguisher, aim the nozzle of the extinguisher at the base of the fire, not at the top. Try to put out the fire without blowing glassware and other articles off the lab table. If clothing or hair catches on fire, remember to stop, drop, and roll. If a classmate catches on fire, tell them to stop, drop, and roll, and quickly get the fire blanket if one is available; otherwise, use the fire extinguisher or the safety shower.

If an alarm sounds noting a fire or other safety hazard in another part of the building, immediately turn off all gas sources and extinguish all flames. Set all equipment down on the lab table, closing any open chemical containers. Then, follow all normal procedures for evacuation or shelter-in-place, depending on directions from the teacher. It is especially important to remain calm and listen for specific directions during a crisis situation.

Make sure you know where first aid supplies are located. All injuries should be reported to the teacher immediately. If the injury is serious, more advanced treatment, such as stitches, may be required, so professional medical help is important.

If any substances get in the eyes, use the eyewash station or an eyewash bottle. Open the valve or squeeze the bottle so that a plentiful supply of water is directed at the eye. Have a student help hold the eyelids open to rinse out any chemicals that may be under the eyelids. It is important to wash the eyes immediately and continue washing for 15 minutes to flush out all harmful chemicals. Notify the teacher immediately but do not wait to begin flushing the eyes with water. Clean up chemical spills immediately, but use caution. In general, do not clean up broken glass; leave that for the teacher or custodial staff to take care of. Acids and bases, if spilled, must be neutralized before they are cleaned up. (Use acids to neutralize bases and bases to neutralize acids.)

If any chemicals get on skin or clothing, wash the area with a large amount of water and report the accident to the teacher. If you should splash a chemical solution on a wide area of skin or clothing, you must use the safety shower. Alert the teacher if possible, but immediately stand beneath the shower head and pull the chain or ring that activates the shower. Be sure the water thoroughly soaks and washes every place that the solution touched. Continue the shower for at least 1 minute. Safety is much more important than any embarrassment.

Laboratory accidents are rare. Following safety procedures and instructions for investigations will help ensure that most accidents and injuries are avoided. It will also make sure that every student is able to learn and enjoy the investigative activity.

Consider the Explain Question:

How do you implement safe practices in a science laboratory?

Go online to complete the scientific explanation.

Check for Understanding:

You splashed about 250 ml of dilute sulfuric acid solution on your clothing. What should you do? Explain why.

STEM in Action

STEM and Laboratory Practices and Safety

Environmental safety engineers find ways to reduce the impact that industrial activities have on the environment. They may be industrial hygienists who work to anticipate and prevent safety hazards in the workplace. Or, they may be environmental science consultants to whom others turn when designing a new building that must meet or exceed environmental regulations. These engineers use technology and knowledge of science to solve problems, such as the removal of asbestos from buildings.

During the 1970s, the Occupational Safety and Health Administration (OSHA), a government organization, set limits on how much asbestos a person could be exposed to in the workplace. Asbestos is a type of fiber that is a great insulator and fireproofing agent, and it was widely used in wall insulation and ceiling tiles. However, exposure to asbestos can lead to lung cancer. These days, you will not find newly constructed buildings with asbestos in the walls or ceilings. But what about office buildings that were constructed prior to the 1970s?

Most often, environmental engineers are called out when accidents occur where workers are exposed to asbestos while working in an old building. Sometimes, a worker may accidentally remove a ceiling tile or begin tearing down a wall without realizing asbestos is present. During such accidents, an environmental engineer may be called in to help remove the asbestos, following safety protocols that minimize everyone's exposure to the substance. To determine the health risk, engineers will give workers personal air sample pumps that have a filter used to collect any asbestos fiber. With the help of a computer, they use these pumps to analyze how much asbestos a worker is exposed to. Using this information, they can create a plan to make sure workers wear the right safety equipment, such as masks and suits, or possibly tell them to stop working if it is very unsafe.

Fiber in Disguise

As part of a study throughout the 1980s, 19,000 air samples were collected from several different workplace buildings and then analyzed for their asbestos content. The graph shows the average concentration of asbestos, in fibers per cubic centimeter, present during each year. Refer to the graph to answer the questions that follow.

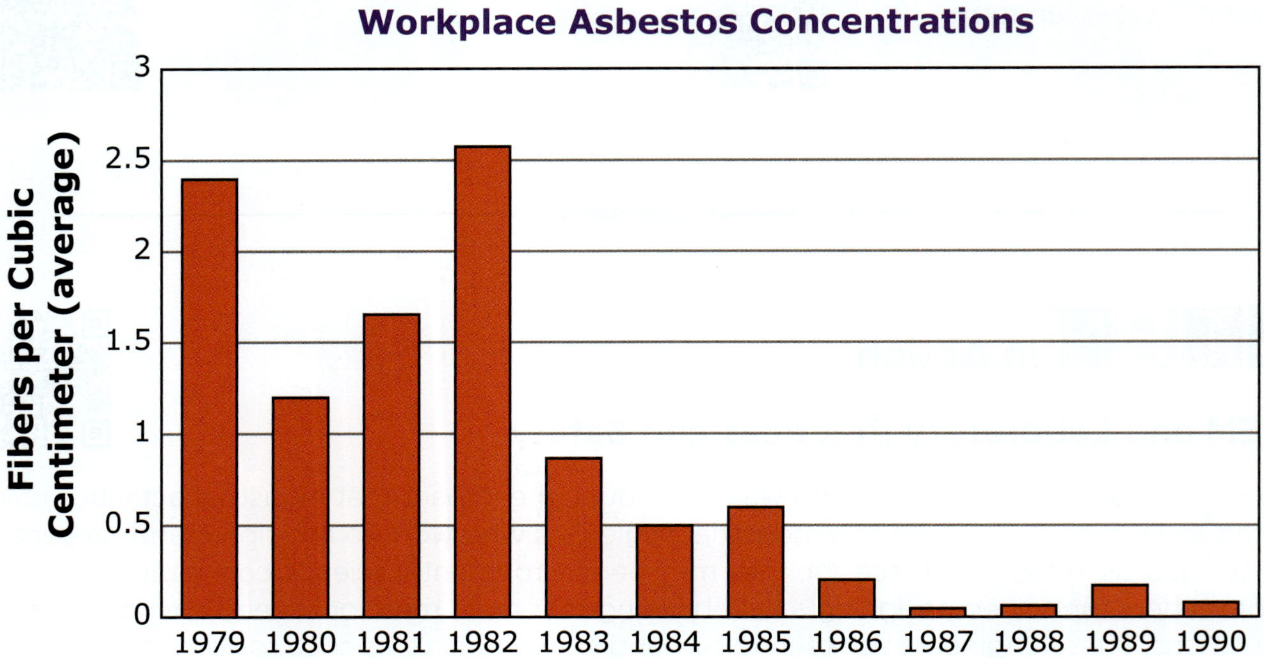

Source: EWG Action Fund, compiled from Occupational Safety and Health Administration health inspection data (1979 – 1998). Data includes 19,000 samples from 670 industries.

1. Based on the graph, describe the trend in asbestos exposure from 1979 to 1990.
2. What do you suppose contributed to the decrease in asbestos exposure during the late 1980s?
3. How much did asbestos exposure decrease during the years 1982 and 1983?

Measurement

photo: Getty Images

LESSON OVERVIEW

Lesson Questions

- What is the difference between accuracy and precision?
- How can you make and express measurements both accurately and precisely?
- What is the difference between mass and weight?

Lesson Objectives

By the end of the lesson, you should be able to:

- Accurately express measurements with appropriate certainty based on a given measurement scale.
- Distinguish between precision and accuracy in measurement.
- Explain how mass differs from weight.

Key Vocabulary

Which terms do you already know?

- ☐ **accuracy**
- ☐ **English System**
- ☐ **experiment**
- ☐ **gravity**
- ☐ **International System**
- ☐ **mass**
- ☐ **measurement**
- ☐ **precision**
- ☐ **scientific notation**
- ☐ **significant digit**
- ☐ **temperature**
- ☐ **volume**
- ☐ **weight**

ENGAGE

Why is measurement important in science? What tools can you use to make accurate and useful measurements? Visit the Engage page online to learn more.

Explain Question

How do physicists use measurements to understand the natural world?

EXPLORE

What Is the Difference between Accuracy and Precision?

Measurements

In physics, a measurement is an observation of a physical quantity of an object. A physical quantity might be the height of a street pole, the volume of a liquid, the mass of a car, the speed of an airplane, or the temperature of a pool. In order for measurements to be meaningful in the scientific community, scientists use measuring instruments with standardized units. For example, they typically express the length of objects in units of meters. Similarly, mass is typically expressed in kilograms, speed is expressed in meters per second or kilometers per hour, and so on. A measurement includes both a numerical quantity and a unit. For example, the measured speed of a car might be 30 kilometers per hour.

Measurement	Unit
Length	Meter (m)
Mass	Kilogram (kg)
Time	Second (s)
Electric current	Ampere (A)
Temperature	Kelvin (K)
Luminous intensity	Candela (cd)
Amount of substance	Mole (mol)

The International System of Units (SI) allows scientists to communicate effectively worldwide. What other measurements and units in SI do you know?

To make a measurement, different measuring tools can be used. Measuring tools usually display the different increments of a particular unit. For example, a ruler is a tool for measuring length that may display increments of centimeters and millimeters. A triple beam balance is a tool for measuring mass that may display increments of kilograms and grams.

While the units of tools may seem rather obvious, they are an extremely important aspect of measurements in science, engineering, and technology. If an engineer uses a ruler to measure a machine part, he or she would not simply write the number 15 on a piece of paper. The engineer would need to write the accompanying units to avoid confusion about the units used. If this does not happen, another engineer may think that the part is 15 inches long, or 15 centimeters long, or 15 millimeters long! This can cause serious problems. In fact, NASA lost the $125 million Mars Climate Orbiter when one engineering team expressed the orbit altitude of the device in English system units while another team thought that the data was being expressed in SI units. As a result, the orbiter moved off course and failed upon entering the atmosphere of Mars.

Accuracy and Precision of Measurements

It is important to remember that all measured values are interpretations of readings on measuring instruments. Because of this, measurements can contain quite a bit of uncertainty. In order to analyze the validity of data in an experiment, a scientist must be aware of how different measurement problems or limitations of measuring devices might affect the results of the experiment.

The classic idea behind the measurement process is that objects have some particular properties, and scientists use measuring instruments to determine these properties. In other words, objects have some "actual," "true," or "accepted" properties, and measurements are taken to try to obtain the value of these properties. A key question that scientists want to know when they take measurements is: *How close are my measured values to the true values?* The answer to this question is defined as the accuracy of a measurement. If a measured value is very close to the actual or expected value, then it is a very accurate measurement. On the other hand, a measured value that is far off from the expected value is considered inaccurate. Different factors might cause a measurement to be inaccurate. A common factor is when there is a problem with the measuring device or experimental setup.

Another question that may arise in the measurement process is: *If I take a measurement of the same property over and over, how much will my measurements vary from one to the next?* The answer to this question is defined as the precision of a measurement. If repeated measurements of a property do not vary much from one to the next, they are very precise. However, if repeated measurements vary a lot, the measurements are not precise. The precision of measurements typically relates to the limitations of the measuring tool. If a measuring tool has very small increments (such as a ruler with millimeters), it can take more precise measurements. If a measuring tool has larger increments (such as a ruler with only centimeters), it will take less precise measurements. In fact, measuring tools with small increments are said to be "precise tools" because they can take more precise measurements.

It must be noted that the philosophy of measurement in modern physics is a bit different from the discussion following. Einstein's theory of relativity suggests that the measured values of objects will differ depending on who is making the measurement. Furthermore, quantum mechanics suggests that objects may not have definite properties until they are measured. The question arises: *Do objects have true properties before they are measured, or do these properties only exist once objects have been measured?* Do not be worried if this concept seems confusing. In fact, the philosophy of measurement is still being debated among scientists and philosophers today.

The Difference between Accuracy and Precision

Accuracy and precision are easy to mix up, especially because the two words are often use interchangeably in common language. A good analogy for distinguishing the two is the analogy of a bull's-eye target.

Think of the true or expected value of a property as the center of a bull's-eye target. If a person throws darts at the target, the accuracy of the darts is how close the darts get to the bull's-eye. The precision of the darts is how close repeated throws are to one another.

An excellent player might throw the darts directly in the center of the bull's-eye. These would be accurate throws. Also, each dart might land directly next to the last, making the throws very precise. In this case, the darts are both accurate and precise. However, there are also cases where the darts are precise but not accurate. This would happen if the player throws many darts into one specific area of the target that is far from the bull's-eye. Because the darts are far from the target, they are not accurate. However, the repeated dart throws are all close to one another, so they are precise.

These darts are close to one another, so they have high precision; however, they are not near the bull's-eye, so they are not accurate. What might the target look like with low accuracy and low precision dart holes?

Another situation can arise where the darts are accurate but not precise. This would happen if the darts were thrown at different locations around the circumference of the bull's-eye. In this case, each dart is close to the bull's-eye, so the throws are accurate. However, the darts are rather far apart from one another, so they are not precise.

These darts are far apart from each other, so they have low precision; however, they are all near the bull's-eye, so they are accurate. What might the target look like with dart holes of high accuracy and high precision?

How Can You Make and Express Measurements Both Accurately and Precisely?

To better understand the concept of accuracy, it is good to consider real-world examples of measurements. Consider a pile of rough-sawn two-by-four lumber pieces used to construct a shed. The name of the wood pieces is commonly understood to mean the pieces are each 2 inches wide and 4 inches in depth. Supposedly, these are the true, or expected, dimensions of the wood. A person might use a ruler to measure the dimensions of a two-by-four. The person might measure the width of the lumber five times and obtain the following data in inches:

$$\{2.12, 1.88, 1.94, 2.06, 2.12\}$$

To analyze these measurements for accuracy, you can compare each measured value to the expected value: 2 inches. The first measurement is 2.12 inches. This is 0.12 inches greater than the expected value. The second measurement is 1.88 inches. This is 0.12 inches less than the expected value. The third measurement is 1.94 inches, which is 0.06 inches less than the expected value. A similar process can be used to analyze the remaining measurements. You can see that the greatest difference between any measured value and the expected value is 0.12 inches. It might be said that the measurements are reasonably accurate; although, this depends on the particular standard being used.

To analyze the measurements for precision, you can determine how much the measurements vary from each other. The greatest measured value is 2.12 inches, and the smallest measured value is 1.88 inches. The difference between these values is 0.24 inches, which is nearly a quarter of an inch. It can be said that the measurements are not very precise; although, this also depends on the particular standard being used.

photo: Pixabay

A caliper allows more precise measurement than a ruler. When might this increased precision be needed?

A subsequent measurement of the lumber could be made with a more precise measuring tool, such as a caliper. Calipers can measure lengths to the nearest one thousandth of an inch (0.001 inches). A person might repeat five measurements of the lumber width using a caliper to obtain:

$$\{2.001, 2.002, 1.999, 1.998, 2.000\}$$

In this case, the measurements differ from the expected value by 0.002 inches at most. This is much more accurate than the measurement with the ruler, which was accurate within 0.12 inches. Furthermore, the measurements vary by a maximum of 0.004 inches. This is much more precise than the precision of the ruler measurements, which was 0.24 inches. So, the caliper yields much more accurate and precise measurements of the lumber.

Acceptable ranges for the accuracy and precision of measurements depend on the situation. For example, a person constructing a shed may not be too concerned if a cut piece of wood is 0.12 inches longer than it should be. However, an engineer cutting tiny screws for a machine might be very concerned about a piece that is 0.12 inches too long. When one asks "how accurate" or "how precise" measurements should be, the answer is that it simply depends on the particular need.

In general, using more precise measuring tools can yield more accurate and precise measurements. Scientists usually say that the precision of a measuring tool is the smallest increment that the tool can measure. For example, a ruler with millimeter increments is said to be "precise to the nearest millimeter." Furthermore, the accuracy of measurements can be increased by eliminating any sources of error in the measuring device or observer. If the first centimeter of a ruler is worn down, a measurer can start measurements at the second centimeter. Then, he or she can subtract a centimeter from the final measured value. This would eliminate any inaccuracies in the measurement as a result of a faulty measuring tool.

What Is the Difference between Mass and Weight?

A common measurement of objects is a measurement of mass or of weight. In everyday language, mass and weight are considered interchangeable, but in science, they are not. For scientists, these are distinctly different properties of an object.

The mass of an object is a measure of the amount of matter an object contains. In other words, it is the amount of "stuff" in an object. Mass is measured in kilograms or grams in the metric system or slugs in the English system. A balance is typically used to measure the mass of objects.

The weight of an object is a measure of the gravitational force affecting an object. It is given by the equation: $w = mg$. In this equation, m is mass and g is the acceleration due to gravity. A spring scale is typically used to measure the weight of objects. Weight is expressed in units of Newtons in the metric system and pounds in the English system. This is a measure of how much force a spring in the scale must exert to hold up an object against the force of gravity. On Earth, g is 9.8 m/s^2. On the moon, g is 1.6 m/s^2. As a consequence, an object will weigh different amounts on Earth and the moon. However, the object's mass remains unchanged.

Consider the Explain Question:

How do physicists use measurements to understand the natural world?

Go online to complete the scientific explanation.

Check for Understanding:

Describe a situation in which a set of measurements is precise but not accurate. Explain what makes them precise but not accurate.

STEM in Action

STEM and Measurement

The accuracy and precision of measurements relies in a large part on the precision of the measuring tool. If the tool can measure in extremely small increments, the measurements are likely to be highly accurate and precise. One of the most precise measuring instruments is an atomic clock. Atomic clocks provide standardized time for military units and electricity grids. They are also highly important in the functioning of GPS navigational systems. Atomic clocks are used by engineers at the National Aeronautics and Space Administration (NASA) to precisely locate satellites in orbit. Anyone who uses GPS (Global Positioning System) relies on atomic time. GPS satellites use atomic clocks to triangulate time signals from orbit. Pilots rely on the atomic clocks located in these satellites to ensure their planes are flying on the right path.

To understand how atomic clocks work, remember that all clocks count the oscillations of a resonating object. For example, a classic pendulum clock counts the rate at which a pendulum oscillates back and forth. Gears move the second hand forward each time the pendulum completes one oscillation. The most important part of the clock mechanism measures how precisely the object oscillates. In atomic clocks, the clock measures the oscillation of electromagnetic radiation from atoms. This radiation can oscillate at extremely precise frequencies. For example, a cesium atom will always resonate radiation at consistent frequencies. Cesium resonates 9 billion times every second. The resonant frequency of cesium atoms controls a quartz crystal, which then oscillates 9 billion times per second. The variation in this resonance is extremely small. Atomic clocks, which are based on atomic resonance, count oscillations very precisely. This means that the measured time on atomic clocks is extremely accurate and precise.

Let us say you are working in a crash test facility. The car you are testing hits the wall, the dummy is slumped over the wheel, and broken glass is everywhere. There is just one problem: your radar that measures the car's speed was turned off!

You know that the length of your track is 15 meters. You also know that the car crossed the entire track in 0.5 seconds because your high-speed cameras recorded the entire event. Use that information to answer the following questions:

1. What was the average velocity of the car during its test on the track?
2. Given the formula for impact force,

$$F = \frac{2 \cdot m \cdot v}{t}$$

   where *F* is force, *m* is mass, *v* is velocity, and *t* is duration of impact, what other data do you need in order to determine the force of the car at impact?
3. How might you measure this missing data?

Types of Energy

photo: Getty Images

LESSON OVERVIEW

Lesson Questions

- What are two main types of energy, and what are some examples of these types of energy?
- How does the kinetic energy of an object relate to its mass and velocity?
- What are some examples of conversions between kinetic energy and potential energy in the world around you?

Key Vocabulary

Which terms do you already know?

- ☐ **chemical potential energy**
- ☐ **elastic potential energy**
- ☐ **elastic rebound**
- ☐ **electric potential energy**
- ☐ **energy**
- ☐ **gravitational potential energy**
- ☐ **kinetic energy**
- ☐ **mass**
- ☐ **potential energy**

Lesson Objectives

By the end of the lesson, you should be able to:

- Analyze devices that convert one form of energy into another form of energy, such as conversions between kinetic and potential energy.
- Solve problems involving the calculation of kinetic energy.
- Compare examples of conversions between kinetic energy and potential energy.
- Construct models of energy depicted as either motions of particles or energy stored in fields.
- Design solutions to real-world problems by using the engineering method.
- Evaluate design solutions according to constraints such as cost, safety, reliability, and aesthetics, and possible social, cultural, and environmental impacts.

ENGAGE

What do you feel like when you have a lot of energy? What are some objects or forces that have a lot of energy? Visit the Engage page online to learn more.

Explain Question

What steps are involved in converting potential energy to kinetic energy, or kinetic energy to potential energy?

EXPLORE

What Are Two Main Types of Energy, and What Are Some Examples of These Types of Energy?

Kinetic Energy

Kinetic energy is the energy of motion. All moving objects, from tiny molecules to huge bodies in space, have kinetic energy. The kinetic energy of an object is proportional to its mass and the square of its velocity. Kinetic energy, denoted as *KE*, is described by the following equation:

$$KE = \frac{1}{2}mv^2$$

The most obvious examples of kinetic energy are the motions of bodies through water, on land, or through the sky. This includes fish swimming, boats sailing, people running, cars traveling, fan blades rotating, carousels spinning, birds flying, and balls moving in arced paths through the air.

However, there is kinetic energy in many objects that is not as easily noticeable. When a guitar string is plucked, it vibrates very quickly. The guitar string has kinetic energy, and it transfers this energy to the molecules of air around it, causing them to vibrate as well. In fact, all sound is the result of vibrating molecules in the air. This form of kinetic energy is sound energy.

Consider the motion of atoms in a person's hands. When people rub their hands together, their skin feels warm as a result of the increase in kinetic energy. The rubbing motion causes atoms at the surface of the skin to move faster, increasing their kinetic energy. This increase in kinetic energy causes an increase in the temperature of the skin. In fact, temperature is a measure of how quickly atoms in a substance are moving. This form of kinetic energy is called thermal energy.

Elastic Potential Energy

Potential energy is the energy of an object due to its location, shape, or position. There are many forms of potential energy. Perhaps the easiest to understand intuitively is elastic potential energy. When a rubber band is stretched, it has elastic potential energy. Think of the rubber band as having the "potential" to snap into motion if it were suddenly released from this position. Furthermore, the degree to which the rubber band is stretched will determine the intensity with which the band will snap when released. The farther the rubber band is stretched, the more potential energy it has. In much the same way, springs store potential energy. When a spring is stretched or compressed, it tries to restore itself to its equilibrium position.

The elastic potential energy of a spring is given by the equation:

$$PE_{elastic} = \frac{1}{2}kx^2$$

where k is the spring constant, which depends on the particular properties of the spring, and x is the distance the spring is compressed or stretched.

Gravitational Potential Energy

Take a pencil and lift it as high up in the air as possible. When the pencil is released, it will fall toward the ground. Think of lifting the pencil up into the air in the same way as stretching a rubber band. The lifted pencil has the potential to begin moving once released, similar to a rubber band that has been stretched. When an object is lifted, it is moved against the force of gravity, which acts to pull objects downward toward Earth's surface. Similarly, when a rubber band is stretched, it is moved against the elastic force that acts to keep the rubber band at its equilibrium, or rest, position. The more an object is lifted against the force of gravity, the more energy it will have once it is released. Gravitational potential energy is said to be "stored" in the object when the object is lifted.

The gravitational potential energy of an object is typically denoted by PE, although to avoid confusion, it is sometimes written as PE_{grav}. The gravitational potential energy is proportional to the height h an object is raised above the ground. Suppose the object has a mass m. Then, the value of PE is given by the following equation:

$$PE = mgh$$

In this equation, g is the acceleration due to gravity, which is approximately 9.8 m/s^2 downward.

Chemical and Electric Potential Energy

Two additional forms of potential energy are chemical potential energy and electric potential energy.

Chemical potential energy is stored in the bonds between atoms and molecules. These bonds can be thought of as attractive forces between different atoms and molecules due to their structural properties. Chemical reactions release the chemical potential energy stored in the bonds. When substances such as coal are burned, they release chemical potential energy in the form of thermal energy and light energy. When our bodies digest food, the chemical potential energy in the food is released to fuel many of our vital biological processes.

Electric potential energy is stored in electric fields between charged particles. We can use the principles of chemical and electric potential energy to discuss the function of a battery. The electrochemical structure inside a battery establishes an electric field between the positive and negative terminals. When negative charges (electrons) move within the battery from the positive terminal to the negative terminal (that is, against the field), their potential energy increases.

When a wire connects the positive and negative terminals, it releases the electric potential energy inside the battery. The electrons move through the wire as electric current from the negative terminal to the positive terminal. The energy of the electric current can be converted into other forms of energy, such as light energy, thermal energy, or mechanical energy.

The electric potential energy between two charges, q_1 and q_2, is given by the equation:

$$PE_{elastic} = \frac{kq_1q_2}{r}$$

where k is Coulomb's constant and r is the distance between the charges.

Energy Stored in Fields

All electromagnetic waves, including visible light waves, are made up of electric and magnetic fields. These fields lie in planes that are perpendicular to one another, and they oscillate in phase with each other. Like mechanical waves, such as sound or water waves, electromagnetic waves carry energy from one location to another. However, unlike mechanical waves, electromagnetic waves do not need a medium in order to travel; they can move through a vacuum. Electromagnetic waves from the sun travel over 90 million miles through empty space in approximately 8 minutes as they transmit solar energy to Earth. This energy provides the heat and light that is necessary to sustain life on the planet.

Because electric and magnetic fields both possess energy, the total energy in an electromagnetic wave is the sum of the energy in both of these field components. Both components of the electromagnetic field make an equal contribution to its total energy. The electromagnetic energy density, u, is the amount of total energy per unit volume. In free space, it can be stated in terms of either the amplitude of the electric field, E, or the magnetic field, B:

$$u = \frac{1}{2}\varepsilon_0 E^2 = \frac{1}{2\mu_0}B^2$$

where $\varepsilon_0 = 8.85 \times 10^{-12}\ C^2/N \cdot m^2$ is the permittivity of free space, and $\mu_0 = 4\pi \times 10^{-7}\ N \cdot s^2/C^2$ is the permeability of free space.

In the particle model of light, electromagnetic radiation is made up of small packets of electromagnetic energy called photons. In free space, these photons travel at the speed of light. The energy and wave frequency of the photons is related through the following equation, where f is the wave frequency, and h is a value known as Planck's constant, $6.63 \times 10^{-34}\ m^2 \cdot kg/s$:

$$energy_{photon} = hf$$

As the wave frequency of the photon increases, the amount of energy it contains also increases. Photons with high frequencies, such as X-rays or gamma rays, contain more energy than low-frequency photons such as radio waves. This is why exposure to gamma rays and X-rays is dangerous to humans while radio waves are harmless. The higher energy forms of radiation contain so much energy, they can knock electrons loose from atoms, causing chemical damage to cells.

An electromagnetic beam consists of photons. Therefore, its total energy density is proportionate to the density of photons per unit area within the beam, *n*. As the number of photons increases, the beam energy density also increases. However, the energy density is also proportionate to the square of the electric field strength. This means that the photon density must also be proportional to the square of the magnitude of the electric field:

$$n \propto E^2$$

Energy in Particles

Atoms and molecules have energy because they move and interact with one another. This energy can be transferred between individual particles and, therefore, between larger macroscopic systems. Heat is the energy that is transferred between systems. There are three ways energy can be transferred: conduction, convection, and radiation.

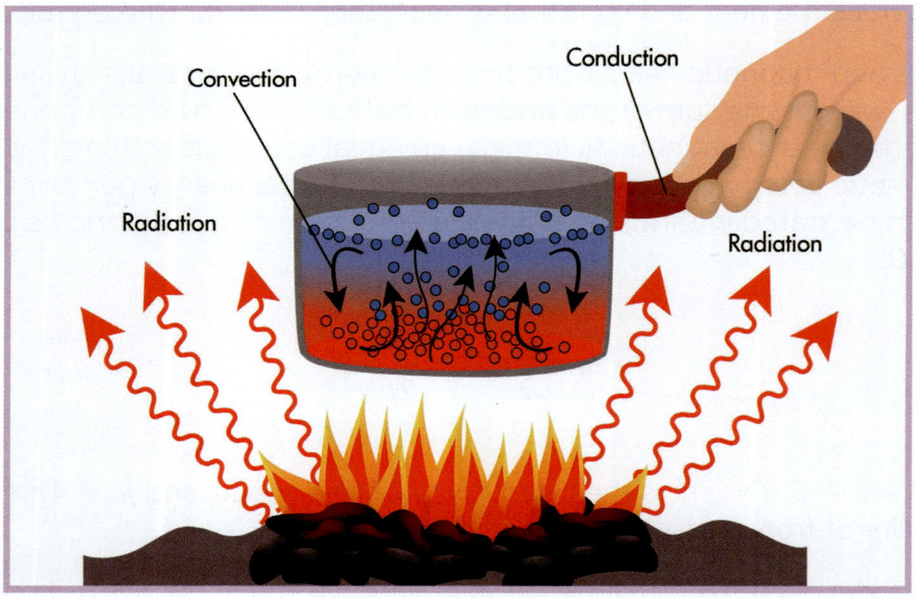

Particles interact in a number of ways. How do particles transfer energy from one to another in each form of heat transfer?

Temperature is a measure of the average kinetic energy of a multitude of particles in a substance. Particles in a higher temperature system move faster than those in a lower temperature system. When the two systems come into contact with one another, the moving particles will collide with each other. When a fast-moving particle collides with a slow-moving particle, some of its kinetic energy is transferred to the less energetic particle.

As a result, the slow particle speeds up. In turn, it can collide with its slow-moving neighbors and transfer some energy to them. This process, called conduction, continues until all of the particles have the same average kinetic energy and all parts of the system are in thermal equilibrium. Thermal conduction only occurs if there is a temperature difference between two systems, and the heat is always transferred from hotter regions to colder regions.

Heat transfer can also occur when particles move from one location to another. This heat transfer process is called convection. Forced convection occurs when an outside agent, such as a fan or pump, pushes air or another fluid to another location. Natural convection is the result of differences in density within a fluid. When a substance such as air or water is heated, the particles gain energy and spread apart. Because hotter fluids typically have fewer particles per volume than their cooler counterparts, they have a lower density. Less dense fluids will rise above the more dense fluids. Therefore, hot fluids rise while cooler fluids sink. Convection is the reason why most attics are hotter than basements.

The final method of thermal energy transfer, radiation, occurs because all particles oscillate. As they oscillate, they emit electromagnetic radiation. Energy is radiated, or carried away, from the source by electromagnetic waves. If you place your hands above a fire, your hands will get hot because heat is being transferred upward primarily due to the process of convection. However, if you put your hands next to the fire, they will still feel warm. This heat transfer is due primarily to radiation.

How Does the Kinetic Energy of an Object Relate to Its Mass and Velocity?

Kinetic Energy, Mass, and Velocity

Recall that the kinetic energy of a moving mass is proportional to its mass and the square of the magnitude of its velocity. Consider a mass *m* moving at a velocity *v*. The kinetic energy *KE* of the mass is described by the following equation:

$$KE = \frac{1}{2}mv^2$$

The units of kinetic energy are Joules, abbreviated as J.

It is possible to rearrange the equation for kinetic energy to solve for different variables such as velocity. This will be helpful in cases where the total kinetic energy and the mass are known but the velocity is unknown. Solving the above equation for *v* gives:

$$v = \sqrt{\frac{2KE}{m}}$$

Kinetic Energy, Mass, and Velocity: Sample Problem: Calculating the Kinetic Energy of a Soccer Ball

A soccer ball is kicked horizontally across a field with an initial velocity of magnitude 15 meters per second (m/s). If the soccer ball has a mass of 0.43 kilograms (kg), what is the kinetic energy of the ball just after it is kicked?

Solution

We know that the ball is kicked with a velocity of magnitude 15 m/s and that it has a mass of 0.43 kg. Thus:

$$v = 15 \text{ m/s}$$
$$m = 0.43 \text{ kg}$$

We want to solve for kinetic energy *KE*. We can use the following equation:

$$KE = \frac{1}{2}mv^2$$

Plugging in the known quantities to solve for *KE* gives:

$$KE = \frac{1}{2}(0.43 \text{ kg})(15 \text{ m/s})^2 = 48 \text{ J}$$

Thus, the kinetic energy of the soccer ball just after it is kicked is 48 J.

Kinetic Energy, Mass, and Velocity: Sample Problem: Calculating the Velocity of a Falling Egg

An egg of mass 0.060 kg is dropped from the top of a building. Just before it reaches the ground, it has a total kinetic energy of 5.9 J. What is the velocity of the egg just before it strikes the ground?

Solution

We know that the egg has a mass of 0.060 kg and a kinetic energy of 5.9 J before it strikes the ground. Thus:

$$KE = 5.9 \text{ J}$$
$$m = 0.046 \text{ kg}$$

We want to solve for the velocity *v*. We can write the equation for kinetic energy and rearrange it to solve for *v*:

$$KE = \frac{1}{2}mv^2$$

$$v = \sqrt{\frac{2KE}{m}}$$

We can plug in the known quantities to calculate *v*:

$$\begin{aligned} v &= \sqrt{\frac{2KE}{m}} \\ &= \sqrt{\frac{2(5.9\ \text{J})}{0.060\ \text{kg}}} \\ &= 14\ \text{m/s} \end{aligned}$$

Thus, the egg is traveling at a velocity of 14 m/s downward just before it strikes the ground.

What Are Some Examples of Conversions between Kinetic Energy and Potential Energy in the World around You?

Conversions between Kinetic and Potential Energy

There are two important facts to remember about energy:

1. The total energy in a closed system is always conserved, or constant. We say that a system is closed when only conservative forces, such as gravity, act on it. This is the law of conservation of energy.

2. While the total energy in a closed system is constant, the energy within that system can change from one form to another.

A common example of energy conversion takes place when objects are in free fall. We say that an object is in free fall when the only force acting on the object is the force of gravity pulling the object toward the surface of Earth. In reality, no objects are ever in total free fall because of the force of air resistance. The force of air resistance can be ignored in this lesson.

When objects are lifted up above the ground, they gain gravitational potential energy. When they are released, the force of gravity causes them to accelerate toward the ground. As the object falls, it loses gravitational potential energy because it decreases in height above the ground, but it gains kinetic energy because its speed increases. The amount of gravitational potential energy that an object loses as it falls is exactly equal to the amount of kinetic energy that the object gains. This is a consequence of conservation of energy

Let us discuss a specific example. Consider an apple of mass *m* at rest on a branch at a height h_1 above the ground. At this time, because the object is at rest, the total energy of the object is in the form of gravitational potential energy, which is given by the equation:

$$PE_1 = mgh_1$$

Assume that at a later time, a strong gust of wind loosens the apple from its branch, causing it to fall. As the apple falls, it loses gravitational potential energy and gains kinetic energy. For example, when the apple has fallen to a new height, h_2, it has a new potential energy:

$$PE_2 = mgh_2$$

Thus, the loss of gravitational potential energy is given by:

$$\Delta PE = PE_1 - PE_2 = mgh_1 - mgh_2$$

The loss in potential energy, *ΔPE*, is equal to the gain in kinetic energy, *ΔKE*. Thus, the *ΔKE* of the apple is given by:

$$\begin{aligned}\Delta KE &= \Delta PE \\ &= mgh_1 - mgh_2 \\ &= mg(h_1 - h_2)\end{aligned}$$

The apple is initially at rest; so it has no initial kinetic energy. Thus, the change in kinetic energy is given by:

$$\Delta KE = \frac{1}{2}mv^2 = mg(h_1 - h_2)$$

where v is the magnitude of the apple's velocity at the height h_2.

The last equation is very helpful in solving problems of motion. In fact, it applies not only to objects in free fall but also to objects that are traveling up and down inclines, such as roller coasters or skiers.

Conversions between Kinetic and Potential Energy: Sample Problem

A roller-coaster car loaded with passengers has a mass of 1,300 kilograms (kg). The first drop of the roller coaster is 64 meters (m) above the ground. The bottom of the first drop is 15 m above the ground. Assuming that the roller coaster comes to rest as it reaches the top of the first drop, determine its speed at the bottom of the first drop.

Solution

The mass of the car is 1,300 kg. The initial height is 64 m, and the final height is 15 m.

$$m = 1{,}300 \text{ kg}$$
$$h_1 = 64 \text{ m}$$
$$h_2 = 15 \text{ m}$$

The objective is to find the value of *v*. The roller-coaster car is initially at rest, so the following equation can be used:

$$\frac{1}{2}mv^2 = mg(h_1 - h_2)$$

Solving it for *v* gives:

$$v = \sqrt{2g(h_1 - h_2)}$$

Note that the mass *m* cancels out of the equation. We can now plug in the known quantities to solve for *v*:

$$\begin{aligned} v &= \sqrt{2(9.8\ \mathrm{m/s^2})(64\ \mathrm{m} - 15\ \mathrm{m})} \\ &= 31\ \mathrm{m/s} \end{aligned}$$

Thus, the speed of the roller-coaster car at the bottom of the first hill is 31 meters per second, or approximately 110 kilometers per hour. Note that the speed does not depend on the mass of the roller-coaster car.

Other Examples of Energy Conversions

Energy changes form all the time. For example, when people clap their hands, kinetic energy transforms into sound energy and thermal energy. Electrical energy in circuits changes into useful forms of energy such as thermal energy and light energy in light bulbs and mechanical energy in spinning fan blades. Living organisms convert energy from their environment into energy needed for life's functions. For example, plants convert light energy from the sun into chemical potential energy during photosynthesis.

Sometimes, energy changes form many times in a system. For example, when a cannonball is fired, chemical potential energy is released, causing an explosion. This energy is then converted into kinetic energy and gravitational potential energy as the cannonball is launched into the air.

Consider the Explain Question:

What steps are involved in converting potential energy to kinetic energy, or kinetic energy to potential energy?

Go online to complete the scientific explanation.

Check for Understanding:

Can you explain two energy transformations that occur when a cannonball is fired upward from a cannon?

STEM in Action

STEM and Types of Energy

Nuclear fusion powers the sun. If it can be controlled in a power plant on Earth, that power plant could possibly produce limitless power because the fuel, a type of hydrogen, would be cheap.

Fusion scientists use certain magnetic devices to contain hydrogen nuclei in an electrically charged plasma. The most popular of these devices is the tokamak. Heavy magnetic coils surround its donut-shaped inner chamber and produce a magnetic field that holds the plasma and confines it. Fusions within this plasma create particles and radiation, which heat a working fluid outside the tokamak. This heat moves to a secondary fluid to make steam, as in many other power plants.

The International Thermonuclear Energy Reactor (ITER) is being built in the south of France. When completed, it will be the largest tokamak on Earth. Scientists will then conduct experiments to produce energy through nuclear fusion, and learn how fusion power plants might one day be created.

Converting Energy

Imagine the situation shown in the diagram. There is a stack of books on the floor. A piece of cardboard rests against the books, creating an inclined plane to the floor. A toy car is placed on top of the books.

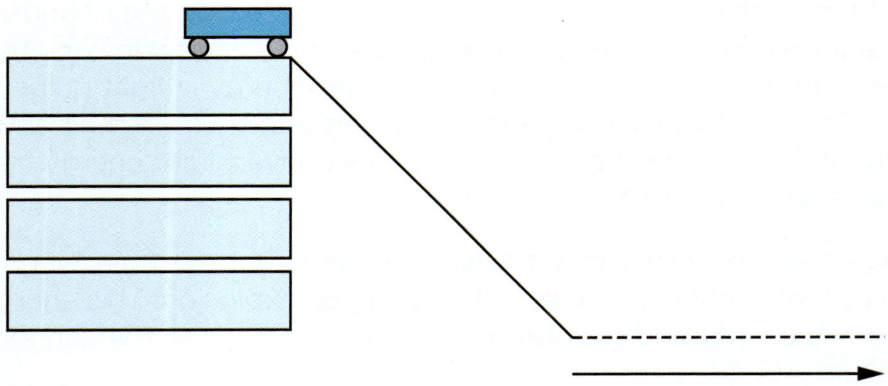

1. How does the height of the stack of books relate to potential energy?
2. In this setup, how can potential energy be converted into kinetic energy?
3. Sketch a labeled potential energy diagram for the cart rolling down the ramp.
4. Describe a technology that relies on the conversion from potential energy to kinetic energy to work.

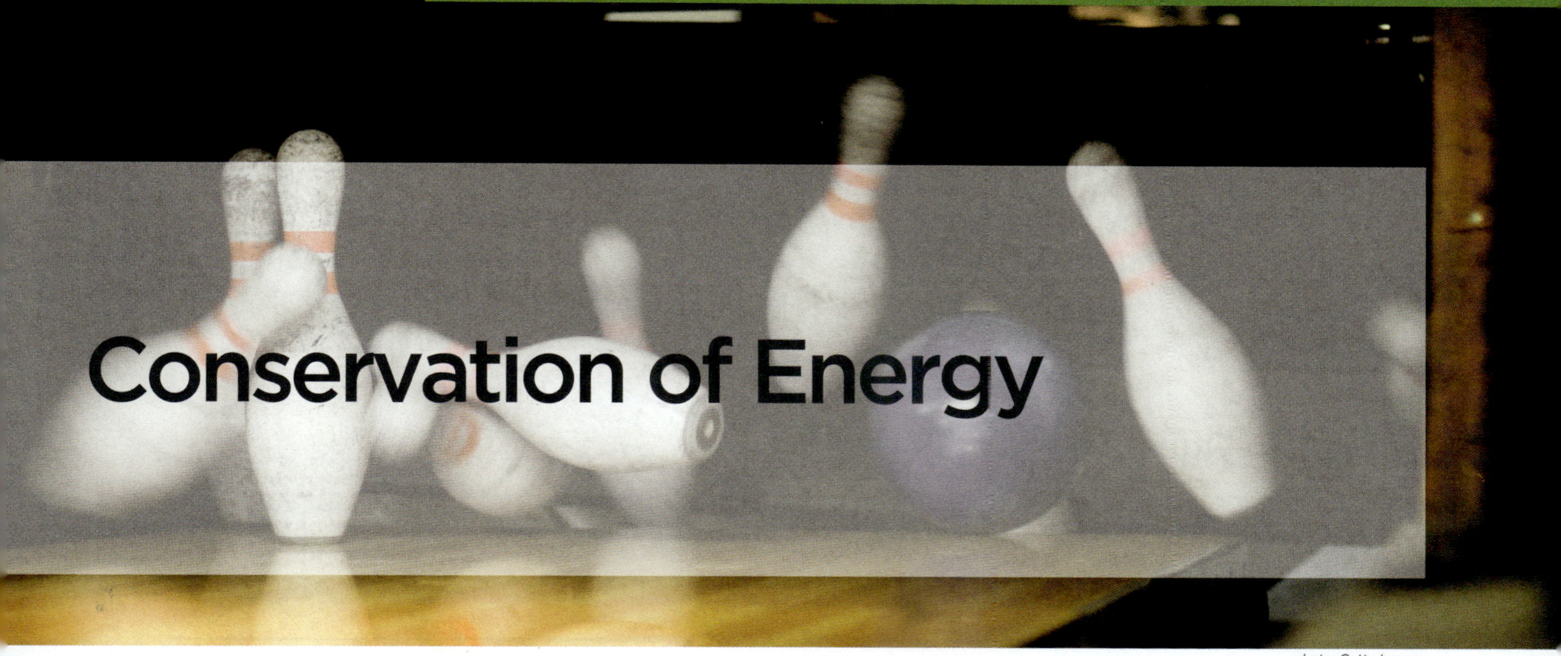

Conservation of Energy

photo: Getty Images

LESSON OVERVIEW

Lesson Questions

- What is the law of conservation of energy, and what are some examples of it?
- How do we calculate changes in gravitational potential energy of objects near Earth's surface?
- How do we investigate the transformation of potential and kinetic energy?
- How can we demonstrate the law of conservation of energy?

Lesson Objectives

By the end of the lesson, you should be able to:

- Explain the law of conservation of energy using examples.
- Calculate changes in gravitational potential energy of objects near Earth's surface using the equation

$$\Delta PE = mg\Delta h.$$

- Solve problems involving potential and kinetic energy.
- Demonstrate the law of conservation of energy.

Key Vocabulary

Which terms do you already know?

- ☐ **chemical potential energy**
- ☐ **conservation of energy**
- ☐ **distance**
- ☐ **elastic collision**
- ☐ **energy**
- ☐ **gravitational potential energy**
- ☐ **inelastic collision**
- ☐ **Joule**
- ☐ **kinetic energy**
- ☐ **potential energy**
- ☐ **thermal energy**

ENGAGE

Where does energy come from? How can we get more energy? Visit the Engage page online to learn more.

Explain Question

How is energy transferred between objects or systems?

EXPLORE

What Is the Law of Conservation of Energy, and What Are Some Examples of It?

Conservation of Energy

The law of conservation of energy says that the total amount of energy in the universe is constant and does not change over time. In a bouncing ball, the energy stored in the ball changes from potential (or stored) energy to kinetic energy (the energy of motion) as the ball falls and bounces; the kinetic energy then changes back to potential energy as the ball returns to the top of its bounce. However, as the ball continues to bounce, some of its kinetic energy transforms into thermal energy and sound. In other words, the ball's energy changes form. Energy cannot be destroyed; it cannot be created; it just exists.

The bouncing basketball shows potential and kinetic energy in action. What would the figure look like if all of the kinetic energy were transformed back into potential energy?

Consider a system where no energy is transformed into thermal energy or sound. The following equation shows that potential and kinetic energy (that is, total mechanical energy) are conserved:

initial mechanical energy = final mechanical energy

$$KE_i + PE_i = KE_f + PE_f$$

(where *i* and *f* are initial and final values)

In other words, the initial sum of kinetic and potential energy is equal to the final sum. This equation can be rewritten to relate the change in kinetic and potential energies:

$$KE_f - KE_i = PE_i - PE_f$$
$$\Delta KE = -\Delta PE$$

The change in kinetic energy is equal to minus the change in potential energy.

Measuring Energy

We measure energy in units called Joules (J). One Joule is equal to 1 Newton-meter.

One thousand Joules is about equal to 1 British thermal unit (Btu). Your kitchen stove's output may be measured in Btu's. One Btu is the amount of thermal energy necessary to raise the temperature of 1 pound of water by 1° Fahrenheit. Thus, about 2,000 Btu's—or about 2,100,000 Joules—are necessary to make a pot of coffee.

Why use Joules rather than Btu's? The reason is simple: The Joule is a metric unit, and scientists use metric units rather than British (or English standard) units. Metric units are related by factors of 10, making conversions easier. For this same reason, scientists measure length in meters rather than feet and measure mass in kilograms rather than pounds. The Joule is the international (SI) unit used for energy.

Measuring Energy: Sample Problem

A student climbs a ladder to free a 0.62-kilogram basketball stuck on a garage roof. The student drops the ball from a height of 4.4 meters. The ball falls to the ground. This is Stage 1. The ball bounces back to a height of 4.4 meters. This is Stage 2. What is the change in kinetic and gravitational potential energy during each stage of the ball's journey?

Solution

During Stage 1, the ball starts from rest at a height of 4.4 meters. Newton's laws for one-dimensional projectile motion will determine the speed of the ball when it hits the ground. If the ball falls from height *h*, the speed *v* is:

$$\begin{aligned} v &= \sqrt{2hg} \\ &= \sqrt{2(4.4\text{ m})(9.8\text{ m/s}^2)} \\ &= \sqrt{86.2(\text{m/s})^2} \\ v &= 9.3\text{ m/s} \end{aligned}$$

The initial speed is 0.0 meters per second and the final speed is 9.3 meters per second. The ball's mass is 0.62 kilograms. The initial kinetic energy is:

$$\begin{aligned} KE_i &= \frac{1}{2}mv_i^2 \\ KE_i &= 0.0\text{ J} \end{aligned}$$

The final kinetic energy is:

$$\begin{aligned} KE_i &= \frac{1}{2}mv_i^2 \\ &= \frac{1}{2}(0.62\text{ kg})(9.3\text{ m/s})^2 \\ &= 27\text{ J} \end{aligned}$$

The change in kinetic energy is:

$$\begin{aligned} \Delta KE &= KE_f - KE_i \\ &= 27\text{ J} - 0\text{ J} \\ \Delta KE &= 27\text{ J} \end{aligned}$$

Mechanical energy is conserved, so the change in potential energy plus the change in kinetic energy is a constant:

initial mechanical energy = final mechanical energy

$$\begin{aligned} \Delta PE &= \Delta KE \\ \Delta PE &= 27\text{ J} \end{aligned}$$

The ball gains 27 Joules of kinetic energy and loses 27 Joules of potential energy.

During Stage 2, the ball reverses its motion. It moves upward with an initial speed of 9.3 meters per second. Its initial kinetic energy is now:

$$\begin{aligned} KE_i &= \frac{1}{2}mv_i^{\,2} \\ &= \frac{1}{2}(0.62\ \text{kg})(9.3\ \text{m/s})^2 \\ &= 27\ \text{J} \end{aligned}$$

The ball comes to rest for an instant at its original height, so the final kinetic energy is 0 Joules. The change in kinetic energy is:

$$\begin{aligned} \Delta KE &= KE_f - KE_i \\ &= 0\ \text{J} - 27\ \text{J} \\ \Delta KE &= 27\ \text{J} \end{aligned}$$

Energy is conserved, so the change in potential energy is minus the change in kinetic energy:

$$\begin{aligned} \Delta PE &= \Delta KE \\ \Delta PE &= 27\ \text{J} \end{aligned}$$

The ball loses 27 Joules of kinetic energy and gains 27 Joules of potential energy. Mechanical energy, which equals 27 Joules, is conserved during both stages.

Gravitational Potential Energy

Picture the bouncing basketball again, only this time, toss it into the air. When the ball is thrown upward, its initial kinetic energy is converted into gravitational potential energy. Although the magnitudes of the ball's kinetic and potential energies change along the way, their sum remains constant. In other words, the law of conservation of energy still applies—energy exists and does not disappear.

As you might guess, sometimes other forces, such as friction, do work on objects in motion. Consider the skiers in the video "Backflipping Skiers" on the Explore page online. Does friction affect their speed down the mountain? Of course it does. As the skiers move down the mountain, they encounter more friction, which keeps them from gaining more speed during the run and slows them down near the end of the run by converting kinetic energy into thermal energy. When an object sliding across a surface slows down because of friction, some of its kinetic energy is converted into thermal energy.

Chemical Potential Energy

Thermal energy is a form of energy that many of us use every day to cook our food and heat our homes. Fossil fuels such as coal and oil are used to convert chemical potential energy into thermal energy. As more coal and oil are used, more thermal energy is produced. However, unlimited supplies of coal and other fossil fuels do not exist—once fossil fuels are converted to other forms of energy, they are gone. They are not a renewable form of energy.

photo: USDA Forest Service

As wood burns, chemical potential energy within the wood is transformed into heat energy. Into what other forms of energy is the chemical potential energy converted?

How Do We Calculate Changes in Gravitational Potential Energy of Objects near Earth's Surface?

Solving Problems Involving Gravitational Potential Energy

Potential energy is the stored energy all objects possess. Potential energy can be gravitational potential energy, which is the stored energy of an object based on its vertical position in a gravitational field; elastic potential energy, which is the stored energy of a stressed or compressed object; chemical potential energy; electromagnetic potential energy; or nuclear potential energy.

Gravitational potential energy is the result of gravitational attraction on an object. Near the surface of Earth, the gravitational potential energy depends on the height and mass of an object.

As an object's height increases, so does its gravitational potential energy. Thus, the gravitational potential energy is directly related to the height of the object above the ground. On the ground, the height is $h = 0$.

photo: Discovery Education

The water has gravitational potential energy at the top of the waterfall, which transforms into kinetic energy as it flows over the edge of the falls. Into what other forms of energy does the gravitational potential energy of the falling water transform?

The gravitational potential energy of an object also depends directly on its mass (m) and acceleration due to gravity (g).

These relationships can be summarized in the equation:

$$PE_{grav} = mgh$$

Sometimes, we are interested in the change in gravitational potential energy as an object moves from one height to another, expressed as Δh. In this case, we can modify the formula:

$$\Delta PE_{grav} = mg\Delta h$$

This equation states that the change in gravitational potential energy (ΔPE_{grav}) is equal to the mass (m) of the object times the acceleration of gravity on Earth (g) times the change in height (Δh) of the object.

Gravitational Potential Energy: Sample Problem

At a fall festival, a crane raises a giant pumpkin 20.0 meters into the air before dropping it on a wrecked car. The pumpkin weighs 450 kilograms, and the acceleration of gravity due to Earth is 9.8 meters per second squared. What is the gravitational potential energy of the pumpkin relative to the ground when the crane has raised it?

Solution

The change in the gravitational potential energy of the pumpkin is the product of the mass of the pumpkin, the height that the crane raises the pumpkin above the ground, and the gravitational acceleration:

$$\begin{aligned}\Delta PE_{grav} &= mg\Delta h\\ &= (450\ \text{kg})(9.8\ \text{m/s}^2)(20.0\ \text{m})\\ \Delta PE_{grav} &= 8.8\times 10^4\ \text{J}\end{aligned}$$

The gravitational potential energy relative to the ground of the pumpkin when it is raised is 8.8×10^4 Joules.

How Do We Investigate the Transformation of Potential and Kinetic Energy?

Elastic Potential Energy

Elastic potential energy is energy stored in an object due to stress or compression. For example, a drawn bow has elastic potential energy. The amount of energy stored is related to the amount of stretch of the device. The farther back the bowstring is pulled, the more energy is stored.

photo: Getty Images

Elastic potential energy, which is the energy stored in elastic materials, such as springs, is due to stretching or compressing. What kind of energy becomes evident when the spring is stretched or compressed many times in a short time interval?

Springs are special instances of elastic potential energy because they can be compressed or stretched. If a spring is neither stretched nor compressed, it is at a zero potential energy position. However, once force is applied to the spring, the potential energy changes.

The potential energy of the spring is directly related to the square of the distance it is stretched, x, and to the elastic property of the spring, called the spring constant, k.

This relationship can be expressed in the equation:

$$PE_{spring} = \frac{1}{2}kx^2$$

This equation shows that you can find a spring's potential energy by multiplying the spring constant (k) by the amount of compression or stretch squared (x) and dividing by two. Once the spring is released, the potential energy transforms into kinetic energy.

Elastic Potential Energy: Sample Problem: Jack-in-the-Box

Suppose a jack-in-the-box has a spring constant of 80 Newtons per meter. How much elastic potential energy is stored in the jack-in-the-box if its spring is compressed 10 centimeters?

Solution

The elastic potential energy of the jack-in-the-box is found by multiplying the spring constant (80 Newtons per meter) by the amount of compression squared (10 cm = 0.1 m).

$$\begin{aligned} PE_{spring} &= \frac{1}{2}kx^2 \\ &= \frac{1}{2}(80\ \text{N/m})(0.1\ \text{m})^2 \\ PE_{spring} &= 0.4\ \text{J} \end{aligned}$$

The elastic potential energy of the jack-in-the-box is 0.4 Joules.

Electric Potential Energy

Have you experienced the buildup of static electricity in your body? Imagine that you are walking across a carpet in the wintertime, when the humidity is very low. If you touch certain objects, you will notice that a spark jumps between you and the object. You also may have observed the same type of electricity when combing or brushing your hair in a dry atmosphere. This electricity is due to the separation of charged particles by friction.

Such charged particles that experience an electric force have electric potential energy. When a charged particle is subjected to an electric force, the particle moves in a way that lowers its electric potential energy and raises its kinetic energy.

A particle with charge q_1, measured in coulombs (the *SI* unit of charge, expressed as *C*), will experience a force due to a second particle with charge q_2, also measured in coulombs. If the particles are separated by distance r, measured in meters, they will have electric potential energy related to the product of their charges and indirectly to the distance between them. This relationship can be expressed in the equation:

$$PE_{elec} = k\frac{q_1 q_2}{r}$$

where PE_{elec} is the electric potential energy, q_1 and q_2 are the charges, r is the distance between the charges, and k is the coulomb constant associated with the medium in which the charges exist. The value for k in a vacuum is:

$$8.99 \times 10^9 \ \mathrm{N \bullet m^2/C^2}$$

Like any form of energy, electric potential energy is measured in Joules.

Electric Potential Energy: Sample Problem: Charged Particles

photo: Getty Images

Lightning is the release of enormous electric potential energy. What accounts for the initial high levels of potential energy that seeks release?

A particle with a charge of –1.2 nano-coulombs (nC), or 10^{-9} C, is located at a distance of 2.3 centimeters from a particle with charge of 2.8 nano-coulombs (2.8×10^{-9} C). The first particle is released from rest. The second particle is fixed at its location. What is the change in electric potential energy when the first particle has moved 1.0 centimeters?

Solution

The charge of the first particle is –1.2 nC = -1.2×10^{-9} C. The charge of the second particle is 2.8 nC = 2.8×10^{-9} C. The particles are separated initially by 2.3 centimeters = 0.023 meters. The initial electric potential energy of the charges is:

$$\begin{aligned} PE_{elec(i)} &= k\frac{q_1 q_2}{r_f} \\ &= \left(8.99 \times 10^9 \frac{\mathrm{N \bullet m^2}}{\mathrm{C^2}}\right)\frac{(-1.2 \times 10^{-9}\mathrm{C})(2.8 \times 10^{-9}\mathrm{C})}{0.023\ \mathrm{m}} \\ &= -1.3 \times 10^{-6}\mathrm{J} \end{aligned}$$

The particles have opposite charges, so they will be attracted. The first particle moves toward the second particle. When the first particle has moved 1.0 centimeters, the particles are separated by 2.3 cm – 1.0 cm = 1.3 cm = 0.013 m.

The electric potential energy is then:

$$\begin{aligned} PE_{elec(i)} &= k\frac{q_1 q_2}{r_f} \\ &= \left(8.99 \times 10^9 \frac{\text{N}\bullet\text{m}^2}{\text{C}^2}\right)\frac{(-1.2 \times 10^{-9}\text{C})(2.8 \times 10^{-9}\text{C})}{0.013\text{ m}} \\ &= -2.3 \times 10^{-6}\text{J} \end{aligned}$$

The change in electric potential energy is therefore:

$$\Delta PE_{elec} = PE_{elec,f} - PE_{elec,I} = -2.3 \times 10^{-6}\text{ J} - \left(-1.3 \times 10^{-6}\text{ J}\right) = -1.0 \times 10^{-6}\text{ J}$$

The electric potential energy has decreased by 1.0×10^{-6} J. Energy is conserved, so the particle's kinetic energy must increase by the same amount. Thus, the particle's speed increases as it moves toward the second particle.

Kinetic Energy

Kinetic energy, the energy of motion, depends on the mass of an object as well as its speed. Remember the bowling ball spinning down the lane to hit the pins as its potential energy transformed into various kinds of kinetic energy, including sound and thermal energy? The mass of the bowling ball and the speed with which it is thrown determine the amount of kinetic energy it possesses, and thus releases. For example, a heavier ball will release more energy than a lighter ball if both are thrown at the same speed.

The kinetic energy (*KE*) of an object in motion is equal to one-half the mass (*m*) of the object times the square of its velocity:

$$KE = \frac{1}{2}mv^2$$

This equation shows you that kinetic energy is directly proportional to the square of its velocity. In other words, if an object's velocity doubles, its kinetic energy increases by a factor of four.

Kinetic Energy: Sample Problem: Roller Coaster

Suppose you are exploring an amusement park that has just opened a new roller coaster. What is the kinetic energy of a 630-kilogram roller-coaster car that is moving with a speed of 20.5 meters per second?

photo: Discovery Education

Roller coasters are designed to convert potential to kinetic energy—for a thrill! How does the kinetic energy change if more people fill the car, or if the initial velocity is 1 meter per second higher?

Solution

Multiply the mass of the car by the square of its speed and divide by two.

$$\begin{aligned} KE &= \frac{1}{2}mv^2 \\ &= \frac{1}{2}(630\text{ kg})(20.5\text{ m/s})^2 \\ KE &= 1.3 \times 10^5\text{ J} \end{aligned}$$

To two significant figures, the kinetic energy of the car is 1.3×10^5 Joules. If you double the speed, the kinetic energy is quadrupled.

How Can We Demonstrate the Law of Conservation of Energy?

Demonstrating with a Falling Object

According to the law of conservation of energy, no matter which form of energy we study, the energy will remain constant in an isolated system. Potential energy transforms into kinetic energy, or thermal energy, or sound, but the total amount always remains the same.

The law of conservation of energy can be demonstrated by measuring the speed of an object after it falls from a certain height. Imagine a 21-kilogram piece of space junk dropped at rest from a rocket 440 kilometers above Earth's surface. What is the change in gravitational potential energy relative to the ground of the space junk at this height?

$$\begin{aligned} \Delta PE_{grav} &= mg\Delta h \\ &= (21\text{ kg})(9.8\text{ m/s}^2)(4.4 \times 10^5\text{ m}) \\ \Delta PE_{grav} &= 9.1 \times 10^7\text{ J} \end{aligned}$$

The gravitational potential energy of the space junk relative to the ground is 9.1×10^7 Joules. If the space junk falls 220 kilometers, its potential energy with respect to its initial position changes by the following:

$$\begin{aligned} \Delta PE_{grav} &= mg\Delta h \\ &= (21\ \text{kg})(9.8\ \text{m/s}^2)(-2.2 \times 10^5\ \text{m}) \\ \Delta PE_{grav} &= -4.5 \times 10^6\ \text{J} \end{aligned}$$

In the real world, the space junk will lose energy to friction because of the air rushing past as it falls. If we ignore this friction for this example, then the decrease in gravitational potential energy should equal the increase in kinetic energy:

$$\begin{aligned} \Delta KE &= \Delta PE \\ \Delta KE &= -4.5 \times 10^7\ \text{J} \end{aligned}$$

The space junk fell from rest, so the initial kinetic energy was 0 Joules. The final kinetic energy is then 4.5×10^7 Joules.

We can check conservation of energy by measuring the speed of the space junk. If energy is conserved, the space junk's speed will be:

$$\begin{aligned} v &= \sqrt{\frac{2KE}{m}} \\ &= \sqrt{\frac{2(4.5 \times 10^7\ \text{J})}{21\ \text{kg}}} \\ v &= 2.1\ \text{km/s} \end{aligned}$$

In the real world, the speed will be less than 2,100 meters per second due to energy lost to friction.

photo: Getty Images

The simple harmonic motion of a pendulum demonstrates the law of conservation of energy. What kind of energy keeps the pendulum in motion? What provides this energy?

Demonstrating with a Spring

To find the elastic potential energy of a spring, measure how much the spring is compressed. Compress the spring with a mass. Then, release the mass upward. The initial stored elastic potential energy equals the change in gravitational potential energy of the mass when it reaches its highest point.

The speed of the mass as it leaves the spring gives another way to check conservation of energy. A motion detector measures the speed of the mass. Using the speed, we can determine the kinetic energy.

Imagine a spring with a spring constant of 2,500 Newtons per meter. A 15-kilogram block is placed on the spring. A student compresses the spring by 12 centimeters below the spring's natural resting point. At this point, the elastic potential energy of the system is as follows:

$$\begin{aligned} PE_{spring} &= \frac{1}{2}kx^2 \\ &= \frac{1}{2}(2500\ \text{N/m})(0.12\ \text{m})^2 \\ PE_{spring} &= 18\ \text{J} \end{aligned}$$

The student then releases the spring. A motion detector measures the speed of the mass as it leaves the spring. The elastic potential energy of the spring is transformed to kinetic energy of the mass. The speed of the mass as it leaves the spring will be:

$$\begin{aligned} v &= \sqrt{\frac{2KE}{m}} \\ &= \sqrt{\frac{2PE_{spring}}{m}} \\ &= \sqrt{\frac{2(18\ \text{J})}{25\text{kg}}} \\ &= \sqrt{2.4(\text{m/s})^2} \\ v &= 1.5\ \text{m/s} \end{aligned}$$

In the real world, the speed of the mass will be less than this amount. A real spring will continue to vibrate after it is released. This fact means that some of the initial elastic potential energy will convert to kinetic energy of the spring.

Demonstrating with a Paddle Wheel

Nineteenth-century British physicist James Joule devised another way to test conservation of energy. (James Joule is the scientist for whom the unit of energy is named.) Joule used a paddle wheel in a tank of water. He connected the wheel to a set of masses. Joule let the masses fall a certain height. As the masses fell, they caused the wheel to turn. Friction between the paddle wheel and the water caused the water's temperature to rise.

Joule's experiment showed that the initial gravitational potential energy of the masses changed to the kinetic energy of the paddle wheel, and then to the thermal energy of the water.

Consider the Explain Question:

How is energy transferred between objects or systems?

Go online to complete the scientific explanation.

Check for Understanding:

Using a bouncing ball as an example, can you describe the conservation of energy?

STEM in Action

STEM and Conservation of Energy

One of the oldest and most commonly used forms of renewable energy in the United States is called hydroelectric energy. It is generated at hydroelectric power plants such as Niagara Falls. In order to create a hydroelectric power plant, a river is dammed to create a reservoir of water that can be released when needed. Although water is a renewable resource, the damming of a river can negatively impact the surrounding environment and can damage the delicate ecosystems that depend on the river.

When the water in a dam is allowed to drop from a high elevation to a lower elevation, its stored gravitational potential energy is converted into kinetic energy. At the bottom, the water flows very quickly because it has a lot of kinetic energy. This energy is used to turn the blades of a turbine. The turbine is connected to an electrical generator that converts the kinetic energy into electrical energy. Do you live near a hydroelectric power plant? If you do, it is likely that some of the electricity that comes into your home was generated from the potential energy stored in water.

The people who design and maintain hydroelectric power plants are called mechanical engineers. In order to build a new power station, an engineer must first assess the terrain and then develop a design that meets the energy needs and that maintains the integrity of the surrounding environment and ecosystem. Engineers frequently used computer-aided design tools to develop prototypes, run simulations, and analyze potential designs. Once a design has been selected, mechanical engineers also oversee the construction of the power plant. Because mechanical engineers are responsible for such a wide variety of tasks, they must have good problem-solving skills, have a solid knowledge of mathematics, and be able to think creatively.

Hydroelectric Power Plant

The following picture shows a schematic of a typical hydroelectric power plant.

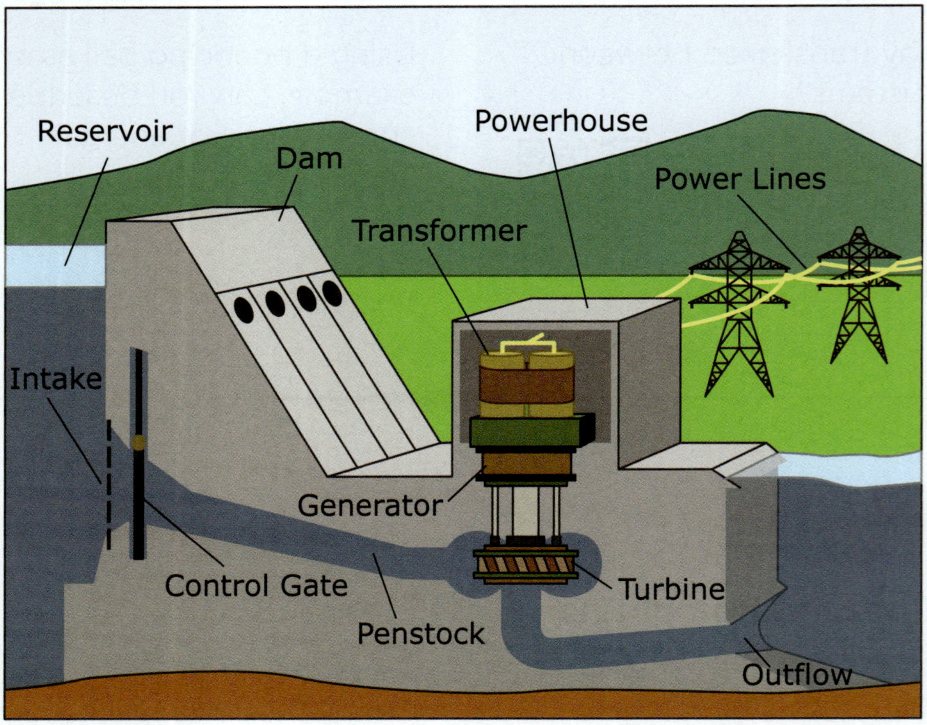

Use this picture to answer the following questions:

1. Describe the energy conversions that take place in a hydroelectric power plant.
2. What would happen if the intake valve was placed lower than the turbine?
3. How can the amount of electricity generated at a hydroelectric power plant be optimized?

Wave Characteristics

photo: Getty Images

LESSON OVERVIEW

Lesson Questions

- What are waves, and how do they transfer energy?
- What are the various characteristics of waves, and how can you describe them?
- What is the relationship among wave speed, wavelength, and frequency? How can you solve problems involving wave speed, wavelength, and frequency?
- How are electromagnetic waves and mechanical waves similar to and different from each other?
- What are the different wavelength bands of the electromagnetic spectrum?
- How can I calculate the energy of electromagnetic radiation?
- How can you describe the energy, power, and intensity of a mechanical wave?
- What is the Doppler effect, and how does it relate to the frequency of a wave?
- How does the speed of a wave relate to the medium through which it travels?
- How can you describe reflection, refraction, polarization, interference, diffraction, and resonance?
- What are some medical or industrial applications of different electromagnetic and mechanical waves?

Key Vocabulary

Which terms do you already know?

- ☐ **amplitude**
- ☐ **diffraction**
- ☐ **Doppler effect**
- ☐ **electromagnetic spectrum**
- ☐ **electromagnetic wave**
- ☐ **frequency**
- ☐ **gamma ray**
- ☐ **Hertz**
- ☐ **intensity**
- ☐ **interference**
- ☐ **light**
- ☐ **longitudinal waves**
- ☐ **photovoltaic effect**
- ☐ **polarization**
- ☐ **power**
- ☐ **radiation**
- ☐ **radio wave**
- ☐ **reduction**
- ☐ **reflection**
- ☐ **refraction**
- ☐ **resonance**
- ☐ **speed**
- ☐ **speed of light (c)**
- ☐ **transverse waves**
- ☐ **wave**
- ☐ **wave speed**
- ☐ **wavelength**
- ☐ **X-ray**

Lesson Objectives

By the end of the lesson, you should be able to:

- Understand the characteristics of waves and how waves transfer energy.
- Describe wave characteristics, including wavelength, wave period, amplitude, frequency, and wave speed.
- Explain the relationships among wave speed, wavelength, and frequency.
- Solve problems involving wave speed, frequency, and wavelength.
- Compare electromagnetic waves and mechanical waves.
- Describe how the speed of a wave varies, depending on the medium through which it passes.
- Model the different wavelength bands within the electromagnetic spectrum.
- Explain the Doppler effect in terms of apparent frequency.
- Describe wave behaviors, including reflection, refraction, polarization, interference, diffraction, and resonance.
- Investigate medical or industrial applications that rely on the characteristics of different types of waves.

ENGAGE

What do you think of when you hear the term *wave*? What are the characteristics of waves? Visit the Engage page online to learn more.

Explain Question

How do you know that the waves sent from the sun to Earth are not mechanical waves?

EXPLORE

What Are Waves, and How Do They Transfer Energy?

Definition of a Wave

Crowds of people do "the wave." What are some properties of this phenomenon? It starts when people in one location stand up and sit down in their seats. When they do this, the people next to them then move up and down, and so on. Through this process, the "up-and-down" motion of people appears to be transmitted across a vast region in an arena. This wave motion is actually quite similar to waves in physics.

Some waves move through matter. The matter that a wave moves through is called the medium of the wave. A traveling wave is a disturbance that moves through space and time, transferring energy without transferring matter. Think of a disturbance as the up-and-down motion of a particle, similar to the up-and-down motion of people in the seats of an arena. The motion and characteristics of a wave are described according to many properties, as will be discussed in this lesson.

How Waves Transfer Energy

Waves originate from an initial disturbance, or oscillation, that provides some energy to the system. As the wave moves, it transmits the energy from the initial disturbance across a region of space. How does it transmit this energy?

Consider the motion of a water wave. The particles in a body of water such as the ocean oscillate in a circular motion near the surface of the water. This motion is typically caused by wind moving across the ocean surface, causing the water particles to move both left and right across the surface as well as up and down.

As the particles oscillate, they disturb the particles around them, causing those particles to oscillate also. To an observer standing on the beach, it appears that a large body of water known as an ocean wave is moving toward the beach, where it eventually crashes. In actuality, only the energy of water particle oscillations is transmitted toward the beach. Said differently, the water particles are not actually moved toward the beach as ocean waves crash; only the energy of the particles is transferred.

In general, as particles in a medium oscillate or vibrate, they transfer the energy of this oscillation to particles around them, and this process continues to transmit the energy of an oscillation to particles across a distance. Through this process, a wave "propagates" through a medium. In this way, waves transfer energy without transferring matter.

When particles oscillate in a direction perpendicular to the direction that the wave propagates, the wave is a transverse wave. This is the most familiar type of wave. However, sometimes the particles in a wave oscillate in the same direction that the wave propagates. These waves are called longitudinal waves. Sound waves and seismic "P-waves" are examples of longitudinal waves. They are occasionally known as pressure waves or compression waves.

What Are the Various Characteristics of Waves, and How Can You Describe Them?

Characteristics of Waves

A periodic wave is a wave that follows a repeating pattern as it propagates through space. That is, the oscillating pattern repeats at regular intervals of time. This repeating motion is also known as periodic motion.

A good way to understand periodic waves is to consider the motion of a spring that is moved up and down. When the spring is at rest, it is at an equilibrium position. When it is moved up and down, it gains some energy. If this motion continues, the spring will oscillate back and forth about the equilibrium point. If the oscillation is plotted against time, the result is the graph of a periodic wave. The maximum position of the spring as it moves up is called the peak and the minimum position is called the trough.

One characteristic of the periodic wave is the wavelength. This is the distance over which the wave completes one full oscillation. To determine this distance, choose a location of the wave at some point in its cycle. For example, choose the peak value of the wave. Then, observe the next oscillation of the wave and find the same location of the wave in this cycle. For example, observe the next peak value of the wave. The distance between these two identical parts of the wave in consecutive cycles is the wavelength. Because wavelength is a measure of distance, it is expressed in units of meters and is usually denoted by the Greek symbol λ.

Parts of a Wave

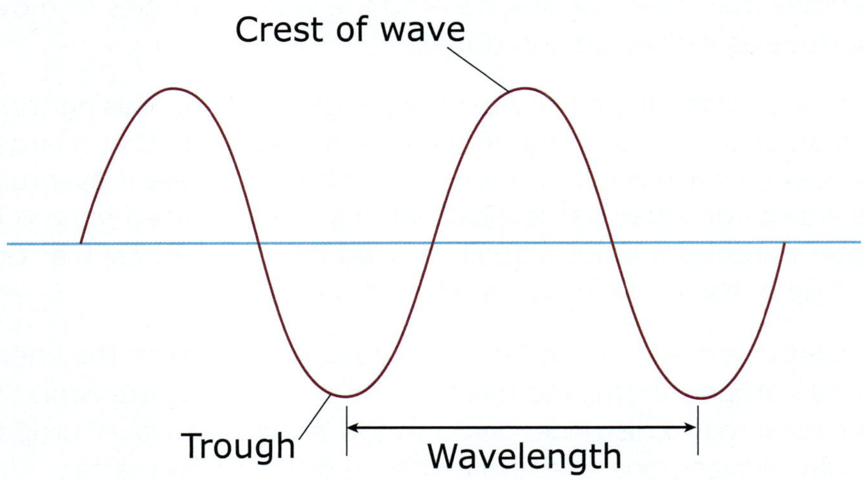

Waves have a crest and a trough; the distance of one full wave cycle is called the wavelength. How can you show the wavelength of a wave?

Similar to the wavelength is the wave period (or just "period") of a wave. The wave period is the time it takes for a wave to complete one full oscillation. The period is a measure in "seconds per cycle," so it is expressed in units of seconds and is represented by the symbol T.

The frequency of a wave is the inverse of the period. While the period is the time it takes for the wave to complete one full oscillation, the frequency is the number of oscillations that a wave makes in 1 second. Frequency is measured in "cycles per second," which has units of 1/s. This unit, 1/s, is referred to as hertz (Hz). Frequency is often denoted by the symbol *f*. The following equations demonstrate the relationship between the period, *T*, and the frequency, *f*, of a wave:

$$f = \frac{1}{T} \text{ and } T = \frac{1}{f}$$

The amplitude of a transverse wave is the maximum distance that the wave travels from its equilibrium position. Think of the amplitude as the "height" of the wave. Because amplitude is a measure of distance, it is expressed in units of meters and is typically denoted by the symbol *A*.

The wave speed is the speed at which the wave propagates through space. Like any other measure of speed, wave speed has units of meters per second, and it is typically expressed by the symbol *v*. The speed at which a wave travels depends upon the nature of the medium that it is passing through. For example, electromagnetic waves travel fastest in a vacuum but slower and sometimes not at all through matter. All mechanical waves require matter for their transmission. The parameters of the medium through which they travel determine their speed.

For example, the speed of sound depends upon the elasticity of the medium. Elastic substances maintain and spring back into shape when they are deformed by a force. For example, iron will rapidly return to its shape. Inelastic substances such as soft clay, on the other hand, deform easily when a force is applied to them and may not return to their original shape. Inelastic or rigid materials have atoms or molecules with strong attractive forces, and as a rule the weaker these forces are in a medium, the slower mechanical waves will travel through them. This means that the phase of a substance can have a big impact on wave speed. You may recall that the forces holding particles together in gases are much weaker than those in liquids, which, in turn, are weaker than those in solids.

If we use sound as an example of a mechanical wave, then we would expect it to travel slower in gases such as air, than in liquid water, or solid iron. At 45 degrees Celsius, the speed of sound is 343.4 meters per second compared with 1,482 m/s in water and 4,910 m/s in iron. Notice that we gave the temperature at which these sounds were measured.

Inertial properties also change the speed of a wave. Inertial properties are the characteristics that make something resistant to change or motion. One inertial property you are already familiar with is the property of density (mass/volume). Generally, the greater the density, the slower the waves will travel. This may seem counterintuitive, but in a single phase of matter, mechanical waves travel slower at high densities and faster at low densities. Normally, density decreases with temperature—the higher the temperature, the lower the density. So, sound, for example, travels slower in cold air than in warm air. This is why the speed of sound in any medium is usually accompanied by a measure of the temperature. The density-wave speed relationship also accounts for the difference in wave speed in different gases. For example sound travels about three times faster in helium because helium is less dense than air.

What Is the Relationship among Wave Speed, Wavelength, and Frequency? How Can You Solve Problems Involving Wave Speed, Wavelength, and Frequency?

Wave Speed, Wavelength, and Frequency

Think back to the definition of speed from kinematics. Recall that average speed is a measure of the distance, *d*, something travels in period of time, *t*: The speed is given by

$$v = \frac{d}{t}$$

How can this principle be applied to waves? The distance a wave travels in one oscillation is the wavelength, λ. The time it takes to complete one full oscillation is the period, *T*. Applying the principle from kinematics:

$$v = \frac{d}{t} = \frac{\lambda}{T}$$

Now, the period *T* is the inverse of frequency. That is:

$$f = \frac{1}{T}$$

Substitute for *f* in the equation relating wavelength and period:

$$v = \frac{\lambda}{T} = \lambda\left(\frac{1}{T}\right) = \lambda f$$

This gives us an important relationship between wave speed, wavelength, and frequency. It shows that the wave speed is equal to the wavelength of the wave multiplied by its frequency, giving the equation:

$$v = \lambda f$$

Problems Involving Wave Speed, Wavelength, and Frequency

It is important not to confuse frequency and wave speed. The wave speed, or speed of a wave, is distance traveled in a given amount of time. In contrast, the frequency is the number of cycles repeated in a given amount of time.

The wave speed, wavelength, and frequency are related by the following equation:

$$v = \lambda f$$

This relationship is used to solve problems of wave motion. For example, if the wavelength and frequency of the wave are known, wave speed can be calculated.

Alternatively, if wavelength and wave speed are known, the frequency can be calculated by rearranging the equation:

$$\lambda f = v$$

to get

$$f = \frac{v}{\lambda}$$

by dividing through by λ.

Similarly, if wave speed and frequency are known, the wavelength can be calculated by rearranging the equation:

$$\lambda f = v$$

to get

$$\lambda = \frac{v}{f}$$

by dividing through by f.

Problems Involving Wave Speed, Wavelength, and Frequency: Sample Problem

A sound wave traveling through a body of freshwater is measured to have a frequency of 440.0 hertz (corresponding to the musical note: "A") and a wavelength of 3.368 meters.

(a) What is its speed in the water?

(b) If the water conditions are held constant, all sound waves will travel at the same speed through water. Use this fact to calculate the wavelength of a 240.0-Hz sound wave (corresponding to "middle C" on a keyboard) traveling through water.

Solution

(a) The following information is given:

$$f = 440.0 \text{ Hz}$$
$$\lambda = 3.368 \text{ m}$$

Use the relationship $v = \lambda f$ to solve for wave speed v. Substituting the known quantities yields:

$$\begin{aligned} v &= \lambda f \\ &= (3.368\ \text{m})(440.0\ \text{Hz}) \\ &= 1{,}482\ \text{m/s} \end{aligned}$$

Thus, the speed of the sound wave in water is 1,482 meters per second.

(b) Assume that another sound wave will travel at the same speed through freshwater as the wave analyzed in part (a). Thus, the following information is known about the next wave:

$$\begin{aligned} f &= 240.0\ \text{Hz} \\ v &= 1{,}482\ \text{m/s} \end{aligned}$$

Solve for wavelength, λ. Because $v = \lambda f$, it follows that:

$$\lambda = \frac{v}{f}$$

Plugging in the known quantities gives:

$$\begin{aligned} \lambda &= \frac{v}{f} \\ &= \frac{1{,}482\ \text{m/s}}{240.0\ \text{Hz}} \\ &= 6.175\ \text{m} \end{aligned}$$

Thus, the wavelength of the 240.0-Hz wave is 6.175 m.

Problems Involving Wave Speed, Wavelength, and Frequency: Sample Problem

The speed of a light wave in air is approximately 2.99×10^8 meters per second (m/s). If a light wave is measured to have a wavelength of 657 nanometers (nm), what is its frequency?

Solution

The wavelength of light is given as 657 nm. First, convert this quantity to meters. Recall that 1 m = 10^9 nm.

$$657\ \text{nm}\left(\frac{1\ \text{m}}{10^9\ \text{nm}}\right) = 6.57 \times 10^{-7}\ \text{m}$$

Thus, the following information about the light wave is known:

$$\lambda = 6.57 \times 10^{-7}\,\text{m}$$
$$v = 2.99 \times 10^{8}\,\text{m/s}$$

We want to solve for frequency, *f*. We can use the following equation $v = f\lambda$. Divide through by λ to get:

$$f = \frac{v}{\lambda}$$

We can plug in the known quantities to solve for *f*:

$$\begin{aligned} f &= \frac{v}{\lambda} \\ &= \frac{(2.99 \times 10^{8}\ \text{m/s})}{(6.57 \times 10^{-7}\ \text{m})} \\ &= 4.55 \times 10^{14}\ \text{Hz} \end{aligned}$$

Thus, the frequency of the light wave is 4.55×10^{14} hertz (Hz), or 455 terahertz (THz). This frequency of light falls within the range of red (visible) light waves.

How Are Electromagnetic Waves and Mechanical Waves Similar to and Different from Each Other?

Mechanical Waves and Electromagnetic Waves

Mechanical waves are propagated by the oscillations of particles in a medium. For example, when a guitar string is plucked, it disturbs the particles in the air around it, causing them to vibrate. These vibrating air particles then cause the particles around them to vibrate and so on. Sound waves can be transmitted by the particles in different media such as air molecules, water, and solid objects.

Sound waves, ocean waves, and seismic waves are all examples of mechanical waves propagated by vibrating particles.

Mechanical waves may be transverse, longitudinal, or a combination of both. Transverse waves, such as tension waves or shear waves, move particles in directions perpendicular to the wave motion. Longitudinal waves, such as sound waves, move particles in directions parallel to the wave motion. Deep-water ocean waves (as opposed to ripples, which are transverse) are waves affected by gravity and the weather. If they reach the surface, they act as circular waves, a combination of both mechanical wave types. All mechanical waves move at speeds that are a function of the properties of the medium through which they are moving.

Electromagnetic waves are transmitted by oscillations within photons, which are particles of light. A photon is composed of an electric field and a magnetic field. As the photon propagates through space, the electric and magnetic fields of the photon oscillate. Photons are massless particles. That is, they are not matter.

Electromagnetic waves are propagated through fields and, therefore, don't require a medium. They can travel in either a vacuum or through matter. Electromagnetic waves are transverse waves and travel at the speed of light in a vacuum.

Comparing Electromagnetic Waves and Mechanical Waves

Electromagnetic and mechanical waves are similar in some ways. Both types of waves propagate with periodic motion, meaning they can be characterized by the properties of wavelength, frequency, period, wave speed, and amplitude. However, these two types of waves are different in important ways.

First, electromagnetic waves are not oscillations of matter; they are oscillations of photons, which are massless particles that are carriers of the electromagnetic force. Because of this, electromagnetic waves do not require a medium through which to travel. In fact, they can travel through a vacuum, such as outer space. Mechanical waves, on the other hand, can only be transmitted through matter. If there is no matter, these waves cannot propagate. For this reason, sound waves and other mechanical waves cannot travel through a vacuum, such as outer space.

Second, electromagnetic waves are transverse waves, while mechanical waves can be either transverse or longitudinal waves. For example, a sound wave is a longitudinal wave because the particles that transmit a sound wave oscillate back and forth in the same direction that the wave propagates. In contrast to this, an electromagnetic wave is transverse because electric and magnetic fields oscillate perpendicular to the direction in which the wave propagates.

Wave-Particle Duality

Electromagnetic radiation such as light can appear to act like waves, like particles, or both. For example, light can interfere and diffract, both of which are properties of all waves. On the other hand, light can also act like a stream of particles or photons, ejecting single electrons from target atoms, an interaction known as the photoelectric effect. This ambiguity is known as the "wave-particle duality" of light.

Both models are actually correct. Light acts like both a wave and a stream of particles. Depending on how an experiment is set up, it can reveal light's wavelike nature or its particle-like nature and sometimes both. In fact, the duality goes both ways. Solid matter can act like a wave! Electrons and other "particles" (such as atoms) have associated wavelengths and can exhibit wavelike behaviors. However, on the scale of everyday objects, these wavelengths are insignificant.

One of the ways scientists resolve the wave-particle duality is to view wavelike properties in terms of statistical effects. A diffraction pattern, for example, can emerge over time as a cumulative effect of multiple single-photon events.

What Are the Different Wavelength Bands of the Electromagnetic Spectrum?

The Electromagnetic Spectrum

Electromagnetic radiation occurs when electric and magnetic fields interact.

Even when electric and magnetic fields are static, they create observable effects. These might include the static electricity you sometimes feel when taking a shirt out of the dryer or the magnetic force holding a magnetized bumper sticker to a car. However, when electric and magnetic fields move near or through one another, that interact on creates electromagnetic waves.

Because electric and magnetic forces do not need a medium in which to propagate, electromagnetic waves can travel through a vacuum. No medium is necessary. Remember, this is a crucial difference between electromagnetic waves and mechanical waves. Mechanical waves can't propagate without a medium.

Electromagnetic energy can be described by frequency, wavelength, or energy, which are all interrelated. Because the speed of light in a vacuum, c, is constant, the wavelength and the frequency of electromagnetic waves have an inverse relationship, $c = \lambda f$. As the wavelength increases, frequency decreases, and vice versa. Low-energy electromagnetic radiation has longer wavelengths, corresponding to lower frequencies. High-energy electromagnetic radiation has shorter wavelengths, corresponding to higher frequencies.

Different Wavelength Bands of the Electromagnetic Spectrum

From lowest energy (longest wavelength) to highest energy (shortest wavelength), the seven groupings along the spectrum are radio waves, microwaves, infrared, visible light, ultraviolet, X-rays, and gamma rays. The visible range of the spectrum is just a tiny portion of the entire electromagnetic spectrum.

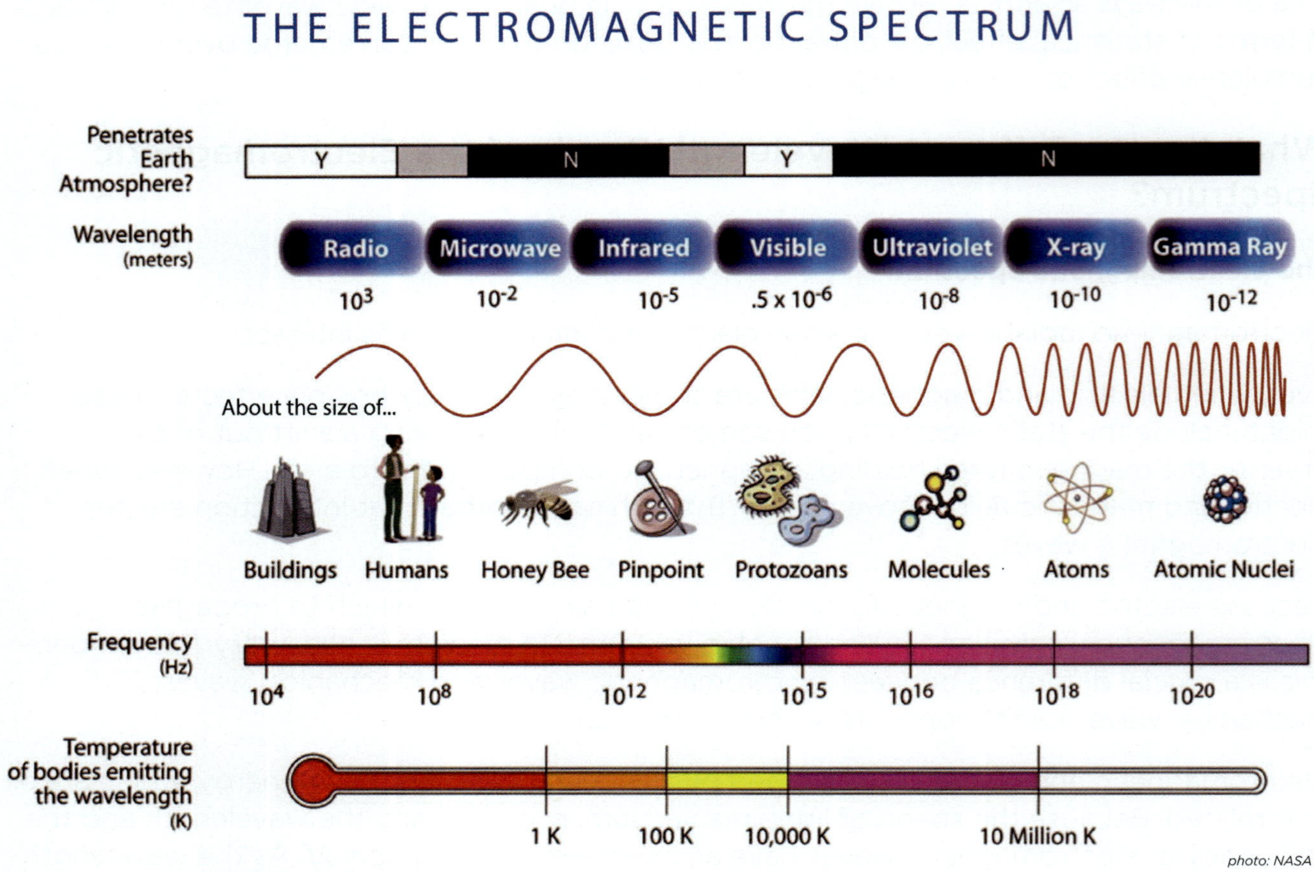

photo: NASA

Observe the many traits of the different wavelengths. Why do you think some waves can penetrate Earth's atmosphere while others cannot?

Radio and microwaves are usually described by frequency (units of hertz); ultraviolet, infrared, and visible light by wavelength (units of meters); and X-rays and gamma rays by energy (units of electron-volts). Though referred to by different names—light, electromagnetic radiation, or rays—all electromagnetic energy is made up of the same kind of waves. The convention of using different units for different parts of the spectrum is simply a convenience that has to do with using numbers that aren't too big or too small. The spectrum doesn't actually have distinct breaks.

Radio waves have the longest wavelength and lowest frequency (and thus lowest energy) of all electromagnetic waves. The wavelength range of radio waves is from thousands of kilometers to meters. Wavelengths in this range are used primarily to transmit radio, television, mobile phone, and other satellite signals (such as GPS navigational information) through the air. Because they have the lowest energy of all electromagnetic waves, radio waves can pass through the human body without harm.

Microwaves have the next shortest wavelength, ranging from meters to millimeters in length. Most people have heard of microwave ovens, which emit microwaves inside a closed chamber. The microwaves pass back and forth through the food millions of times. The microwaves cause water molecules in the food to rotate faster and faster and transfer this rotational energy to nearby molecules. The energy from the microwaves gets transferred to the molecules in the food and is converted into thermal energy, or heat.

This increases the temperature of the food. Microwaves are constantly bombarding Earth, but because they only pass through us once instead of millions of times as in a microwave oven, they do not cause harm to living things.

Infrared rays are next in the electromagnetic spectrum. Their wavelengths range from about 1 millimeter (1 mm) to around 740 nanometers (740 nm). These rays are typically associated with radiated thermal energy such as thermal energy from the sun or a space heater. These rays are mostly associated with thermal energy because their particular wavelengths cause most particles around us to vibrate. These vibrations arise because molecular bonds aren't rigid. Instead, the chemical bonds act like springs, reacting to energy inputs. Energy from the infrared rays increases the intensity of vibration or rotation of molecules in various materials. The resultant molecular movements indicate an increased level of thermal energy, which, in most cases, corresponds to a higher temperature.

Visible light is the next shortest wavelength, ranging from 380 nm to 740 nm. Although it comprises a relatively small portion of the total electromagnetic spectrum, visible light is probably the most familiar type of electromagnetic wave, for this is the range of wavelength the human eye can detect. The wavelengths of visible light are absorbed and reflected by different materials, which we perceive as different colors. From longest to shortest wavelengths, these colors are classified as red, orange, yellow, green, blue, indigo, and violet. (One acronym for remembering this order is ROY G. BIV.) Red waves have the lowest energy, while blue, indigo, and violet waves have the highest energies in this range. When materials absorb visible light, they can heat up as electromagnetic energy is transferred to the material. However, the energy levels in visible light generally aren't high enough to ionize atoms or strip away electrons. Ionization can be harmful for living tissue.

Ultraviolet (UV) rays have shorter wavelengths than does visible light, ranging from 380 nm to 10 nm. Ultraviolet rays are the highest energy waves radiated in large quantities from the sun. Near-UV radiation, with energies just higher than visible light, is strongly absorbed by the surface layer of human skin. These absorbed rays can cause sunburn and, over time, skin cancer. This is why people use sunglasses and sunscreen as protection from these damaging rays.

X-rays have even shorter wavelengths and higher energies than do UV rays. X-rays range from 10 nm to 0.01 nm in wavelength. The energy of X-rays is too high to be absorbed by most atoms. Instead, X-rays tend to knock electrons out of atoms, which can cause mutations and cancer in body cells. The common phrase "X-ray vision" refers to the ability to see through objects, such as walls or humans. This is related to the fact that X-rays can pass through many materials, including most of the tissues in the human body. When X-rays interact with materials, including the human body, they create ionizing effects, meaning the X-rays strip away electrons. Repeated exposure to ionizing radiation can change the chemical balance of cells, in some cases causing cancer. By the same token, if doctors can concentrate radiation on a few cancerous cells, they can help people get better.

Gamma rays have the shortest wavelengths (less than 10 picometers or 0.01 nm) and highest energies of the entire electromagnetic spectrum. These high-energy waves radiate from the nucleus of an atom during radioactive decay. Like X-rays, gamma rays are ionizing, meaning they knock electrons out of atoms. The high energies of gamma rays can cause serious damage to the human body.

When matter absorbs electromagnetic radiation, the energy from the waves affects the material at a molecular level. Lower energy waves, such as microwaves, induce molecules to rotate. Slightly higher energy rays, such as infrared, cause molecules to vibrate. Both rotation and vibration of molecules result in the transfer of thermal energy, or heating. Higher energy radiation, such as some visible light and UV, causes energy transitions within molecules and even greater heating effects. The highest energy radiation, such as X-rays and gamma rays, immediately strip away electrons, changing the chemical nature of matter. For healthy living tissue, absorbing high-energy radiation can be very dangerous, even fatal.

How Can I Calculate the Energy of Electromagnetic Radiation?

Speed of Electromagnetic Waves

In a vacuum, the speed of light is assigned the symbol of *c*. This speed has the value of 2.9979×10^8 meters per second (m/s). All wavelengths of light, from radio waves to gamma rays, travel at the same speed. Even though these forms of radiation have the same speed, they don't have the same energy.

Waves have peaks and valleys or troughs, and the distance from peak to peak or valley to valley is the wavelength. The wavelength is abbreviated as the symbol λ.

The number of peaks (or valleys) that pass by a given point in a certain time period is the wave's frequency. The frequency is abbreviated as the symbol f.

Multiplying wavelength, which has distance units such as meters, by frequency, which has the units such as cycles per second, or Hertz (Hz), results in the wave's speed.

$$\text{Speed} = \lambda f$$

For an electromagnetic wave in a vacuum, this equals *c*, or 2.9979×10^8 m/s, so:

$$c = f\lambda.$$

Notice the units for c are distance per time, which is similar to miles per hour that we might use to describe how fast a car is going. The speed of light has similar units—except light travels at about 700 hundred million miles per hour!

Calculating the Energy of a Wave

A wave's energy is described by the formula:

$$E = hf$$

where E is the amount of energy, f is the frequency, and h is a constant called Planck's constant that describes the linear relationship between energy and frequency. In a mathematical sense, Plank's constant is the slope of the line formed when the relationship between energy and frequency is plotted.

As the formula shows, the energy of an electromagnetic wave depends on its frequency. Energy is directly proportional to frequency. As frequency increases, energy increases. As frequency decreases, energy decreases. Lower frequency radio waves have less energy than higher frequency gamma rays.

Frequency and wavelength are linked, though, so that means energy and wavelength are also linked. However, whereas energy is directly proportional to frequency, energy is inversely proportional to wavelength. This means it has an inverse relationship. As wavelength increases, energy decreases. As wavelength decreases, energy increases.

This relationship is reflected mathematically as:

$$E = hf = \frac{hc}{\lambda}$$

To calculate the energy of a wave, you need to know either its wavelength or its frequency. (To ponder: Why is knowing the speed not enough information?)

For electromagnetic waves, high-energy gamma rays can have 10 million trillion times the energy of low-energy radio waves.

How Can You Describe the Energy, Power, and Intensity of a Mechanical Wave?

Energy, Power, and Intensity of Mechanical Waves

What does it mean for waves to transmit energy? Recall that energy is defined as the ability to do work. Recall also that work is the force applied to an object through a distance. When thinking about the energy of mechanical waves, such as ocean waves, think about how much work the wave could do on an object. That is, how much force would the wave exert on an object to move it through a particular distance?

Remember that the "height" of a wave is known as the amplitude of the wave. It turns out that the energy of a mechanical wave is proportional to the square of the amplitude of the wave. Consider an ocean wave. It is intuitively clear that the bigger the wave, the more force it exerts on objects through a distance. In other words, the bigger the wave (that is, the greater the amplitude of the wave), the more energy it has. We can express this mathematically in the following way:

$$E \propto A^2$$

Here, E is the energy of the wave, A is the amplitude of the wave, and

$$E \propto A^2$$

indicates that E is proportional to A^2.

Recall that power is the amount of energy delivered to, or work done on, an object in a particular amount of time.

Similarly, the power of a wave is defined as the rate at which the wave transmits energy in an interval of time:

$$P = \frac{\Delta E}{\Delta t}$$

Another measure related to the energy and power of a wave is called the intensity of the wave. Intensity, *I*, is defined as the amount of energy that is transmitted through an area of space in a period of time. In other words, it is the power delivered by a wave through an area:

$$I = \frac{\Delta E}{(\text{area}) \times \Delta t} = \frac{P}{\text{area}}$$

Because power is given in units of watts and area has units of square meters, intensity has units of watts per square meter, or W/m^2.

The intensity of a sound wave is related to the loudness of a sound that we hear. There is an inverse square relationship between intensity and distance from the source. The intensity decreases as the wave spreads out from its source. This is why music will be very loud directly in front of a speaker at a concert, but it will sound much quieter standing in the back of an auditorium.

Energy, Power, and Intensity of Mechanical Waves: Sample Problem 1

An ocean wave has an amplitude of 1.0 meter (m).

photo: Paul Fuqua

Imagine that the winds, tides, and waves are not as strong as shown in this image. How would the waves hit the shore?

(a) What happens to the energy of the wave if the amplitude doubles to 2.0 m?

(b) What happens to the energy of the wave if the amplitude decreases to 0.5 m?

Solution

(a) Call the original amplitude of the wave A_1 and the original energy of the wave E_1. Then:

$$E_1 \propto A_1^2$$

Now, let A_2 be the amplitude of the wave after the energy has doubled. Let E_2 be the corresponding energy. Then:

$$E_2 \propto A_2^2$$
$$E_2 \propto (2A_1)^2$$
$$E_2 \propto 4A_1^2$$
$$E_2 \propto 4E_1$$

When the amplitude doubles, the energy quadruples.

(b) The amplitude of the wave, 0.5 m, is half the original amplitude of 1.0 m. Let A_3 be the amplitude of the wave after the energy has halved. Let E_3 be the corresponding energy.

$$E_3 \propto A_3^2$$
$$E_3 \propto \left(\frac{1}{2}A_1\right)^2$$
$$E_3 \propto \frac{1}{4}A_1^2$$
$$E_3 \propto \frac{1}{4}E_1$$

When the amplitude is halved, the energy is quartered.

Energy, Power, and Intensity of Mechanical Waves: Sample Problem 2

A sound originates from a single speaker near the center of a stage. The speaker produces a 60.0-watt sound wave. What happens to the intensity of the sound if the distance of a listener doubles from 10.0 to 20.0 m?

photo: Pixabay

Imagine yourself at a concert. What happens to the intensity of the sound waves as you change your position?

Solution

Call the distance from the source d and the intensity of the sound wave I. The intensity of the sound waves satisfies an inverse square relation:

$$I \propto \frac{1}{d^2}$$

Now, let d1 be the original distance 10.0 m and d2 the doubled distance 20.0 m. Let the original intensity of the sound wave be I_1 and let the intensity at distance d_2 be I_2. Then:

$$I_2 \propto \frac{1}{d_2^2}$$

$$I_2 \propto \frac{1}{(2d_1)^2}$$

$$I_2 \propto \frac{1}{4d_1^2}$$

$$I_2 \propto \frac{1}{4} I_1$$

When the distance doubles, the intensity quarters.

What Is the Doppler Effect, and How Does It Relate to the Frequency of a Wave?

The Doppler Effect

An ambulance siren will rise in pitch as it moves closer, and then as it passes its pitch will lower. This is a common example of the Doppler effect. The Doppler effect is a phenomenon observed when waves originate from a source that is in motion relative to the observer/listener of the wave. This effect results in a change in the perceived frequency of the wave.

To understand how the Doppler effect occurs, consider again the example of the ambulance. If the ambulance were at rest, it would emit sound waves at a specific frequency. Recall that the frequency of the sound wave is the number of oscillations that the sound wave completes in 1 second. The pitch of the sound wave is related to the frequency of the siren that a person hears.

Now, consider the case where the ambulance begins moving toward a person standing beside the street. The siren's sound waves are still emitted at their original frequency, but the person on the street will hear more oscillations per second because the sound waves are compressed closer to one another as the sound source moves forward. This makes the apparent wavelength of the sound waves shorter and, because the speed of sound remains constant for the medium (air) and its conditions, the apparent frequency must be higher. Thus, the pitch of a sound is heard to increase when the sound source moves toward an observer.

Next, consider the case where the ambulance moves away from the person on the street. Again, the sound waves are emitted at their original frequency, but as the ambulance moves away from the observer, the sound waves traveling behind the ambulance are spread farther apart. This makes the apparent wavelength of the sound waves longer and, consequently, the apparent frequency must be lower. Thus, the pitch of the sound is heard to decrease when the sound source moves away from an observer.

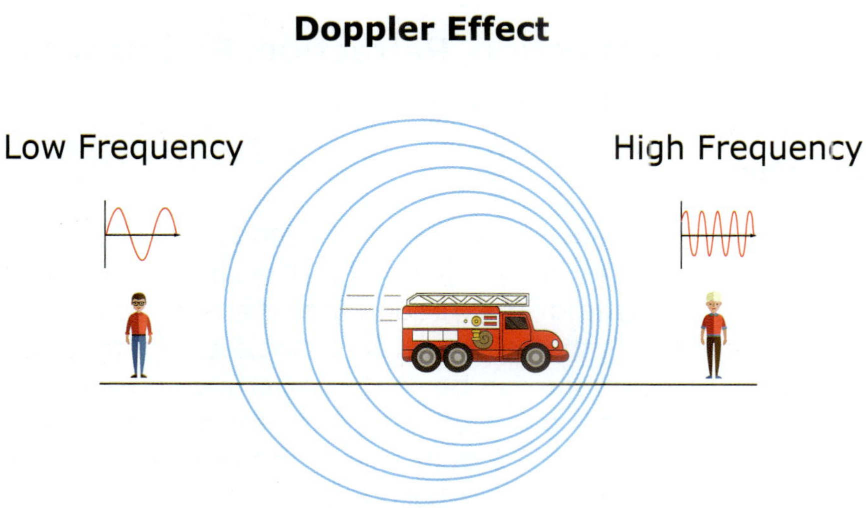

The Doppler effect occurs in moving sources of sound. Under what conditions does the observer hear a sound of shorter wavelength?

Just as the Doppler effect causes a change in the apparent pitch of sound waves, a similar mechanism causes a change in the apparent color of light waves. Light waves emitted by moving objects in outer space exhibit the same behaviors as the sound waves in the ambulance scenarios. When objects in outer space emit light and move toward Earth, the light appears bluer because the frequency of the light wave appears to increase. When objects move away from Earth, the light appears to be redder because the frequency of the light wave appears to decrease. This phenomenon is called, respectively, the "blueshift" and "redshift" of light.

How Does the Speed of a Wave Relate to the Medium through Which It Travels?

Wave Speed through Different Media

Mechanical and electromagnetic waves travel at different speeds when they pass through different media. For mechanical waves, the speed difference is related to the distance between particles in the medium. For example, the particles in a gas, such as air, are spread far apart from one another. As a result, it takes longer for a vibrating air molecule to transmit a vibration to the molecule nearest to it. In contrast, the particles in liquids, such as water, and solids, such as the ground, are packed more tightly together. Therefore, these particles can transmit vibrations to one another more easily, and thus more rapidly, allowing the mechanical wave to travel at a faster speed. In fact, mechanical waves, such as sound waves, travel the fastest through solids. This is why people can hear a train approaching when they press their ear to a solid railroad track, even though the sound of the train cannot yet be heard through the air.

Electromagnetic waves travel at the speed of light through a vacuum. This is the fastest speed that a light wave can travel (and, according to modern physics, it is faster than any object with mass can travel). However, when electromagnetic waves travel through a medium such as air, water, or glass, the waves will change speed because of various properties particular to the medium. This phenomenon is actually the reason light waves refract, or bend, when they pass from one medium to another. Light always travels more slowly through a material than it does through a vacuum.

How Can You Describe Reflection, Refraction, Polarization, Interference, Diffraction, and Resonance?

Reflection and Refraction

Reflection occurs when a wave strikes a boundary between two media and bounces back into the medium from which it came. When light strikes the surface of an object, some light is absorbed and some light is reflected. The frequency of the reflected light is the same as the incident light, which represents the color that we associate with that object. When light reflects off a mirror, a duplicated image can be seen. Similarly, when sound waves reflect off a wall, an echo can be heard. Sound reflects efficiently from hard surfaces such as tiled floors and walls. This is why the sound of footsteps and voices is quite reverberant, or well reflected, in tiled rooms, such as museums or bathrooms.

Refraction is the bending of a wave as the wave changes speed. This typically occurs when the wave passes from one medium to another. The bending of the wave occurs when the wave passes into a new medium at an angle. When this happens, the front edge of the wave will begin to travel at a new speed while the back edge is still traveling at the old speed. This causes the wave to bend, changing directions. Light, for example, will refract when it passes through a prism. During this process, light passes from air to glass and then back to air again. It refracts because light travels at different speeds through air and through glass. White light consists of different colors of light. Each frequency of light will refract a different amount when it passes through glass. Thus, white light will separate into component colors when it refracts through a glass prism.

Polarization

The polarization of a wave refers to the orientation of the oscillations of a transverse wave with respect to the direction in which the wave travels. For example, light is a transverse wave. The polarization of a light wave refers to the direction in which the electric and magnetic fields oscillate. Transverse waves can be linearly polarized, which means that the waves oscillate in a single plane as they travel. Transverse waves can also be circularly or elliptically polarized, which means that the wave oscillations rotate as the waves travel.

Polarization is an important concept in the description of light waves because light can be filtered according to its polarization. Light that travels to Earth from the sun is not polarized, which means that the electric and magnetic fields oscillate in many directions. Light can be filtered when it passes through a material that permits only certain polarizations of light to pass through. For example, certain sunglasses are made of materials that transmit only vertically polarized light. Thus, when randomly polarized light from the sun passes through a polarizing sunglass lens, only a fraction of the light can pass through. This helps reduce the brightness of the sun's rays, thereby reducing damage to a person's eyes.

Interference and Diffraction

When two waves traveling through a medium intersect with one another, the amplitudes of each wave add together. This is called wave interference. It may also be referred to as superposition. An example of this is when an incoming wave encounters its reflection. The interference between the two waves occurs consistently at the same points, which makes those points appear to be standing still. This pattern is called standing-wave pattern.

To understand the effects of interference, it is important to define the phase of a wave. The phase refers to the fraction of a wave cycle that has elapsed relative to the start of the wave. Some waves are said to be "out of phase," such as when one wave starts and a second wave begins after the first wave has already completed some portion of its oscillation. That is, the two waves are out of alignment by some fraction of a wave cycle, and the amount of this difference in this alignment is called the phase.

The phase of waves is important when two or more waves interfere with one another. For example, two waves that start from the same point in their wave cycle are "in phase." If these waves interfere with one another, the amplitudes of each wave will add together fully, forming a wave that has twice the amplitude of each original wave. However, if two identical waves are "out of phase" by a half period, the waves will perfectly cancel each other out when they interfere with one another.

Diffraction occurs when a wave encounters an obstacle or passes through a small opening or slit. As waves pass through the slit, they bend and spread out radially. This often occurs when water waves with large wavelengths strike a barrier with a narrow opening in it. The emerging waves form a semicircular ripple pattern. If the opening is large compared with the wavelength, little or no diffraction occurs.

If a wave encounters a barrier with multiple narrow openings, it will diffract at each of those openings. These diffracted waves will then interfere with one another. For example, if light strikes a barrier with two small slits, the light will diffract through the slits and then interfere on the other side of the slits. This will form what is known as an interference pattern beyond the slits. The pattern will have bright and dark spots or lines corresponding to the places where the diffracted waves interfered.

Resonance

Resonance is the tendency of an object or system to vibrate at a natural frequency called the resonant frequency. This frequency is determined by the particular characteristics of that object. When objects begin to vibrate or oscillate at their resonant frequency, they will vibrate at a greater amplitude than they would at other frequencies.

Resonance explains how many musical instruments produce sound. Consider a trombone, which is a brass instrument with many curved chambers filled with air. A musician vibrates his or her lips into the trombone's mouthpiece, producing many different frequencies of sound waves. When one of these frequencies is a resonant frequency of the trombone, the air in the chambers of the trombone begins to vibrate with a large amplitude. This produces a loud sound. Most musical instruments have numerous resonant frequencies at which they will vibrate with large amplitude. These are known as the "harmonics" of the instrument, and they produce different pitches.

What Are Some Medical or Industrial Applications of Different Electromagnetic and Mechanical Waves?

Applications of Electromagnetic Waves: Medical Uses

There are many applications of waves in medicine. Ultrasound refers to the range of high-frequency sound waves that humans cannot hear. These mechanical waves are higher than 20,000 hertz (Hz). Ultrasound frequencies higher than 10 megahertz (MHz) are typically used in medical applications. In these applications, sound waves are transmitted through a person's body, where they reflect off of internal organs. Technological devices can determine the time it takes for the ultrasound waves to reflect back to the surface of a person's body. This information is then used to construct a detailed image of things inside the human body such as a developing fetus or the arteries around the heart.

The energy of X-rays, which are high-energy electromagnetic waves, is also useful in medical applications. As discussed earlier, X-rays pass through most of the tissues in the human body. However, X-rays are absorbed by the atoms in bones, meaning they do not transmit through bones. This principle allows doctors to produce "negative" images of the human skeleton on X-ray film. An X-ray image will appear black in all the places where X-rays passed through the body, and it will appear white in the places where X-rays were absorbed by bones. The result is a white image depicting the structure of the bones. Doctors can use the X-ray image to diagnose skeletal problems such as broken bones.

Gamma rays are also used in medicine. Gamma rays are the highest energy waves in the electromagnetic spectrum. The high energy of gamma rays can be quite hazardous. However, when focused carefully, gamma rays can help destroy defective cells in human bodies. For example, gamma rays are used to destroy cancer cells and tumors in sick patients.

Applications of Electromagnetic Waves: Digitization

Every day, we use devices that depend on electromagnetic waves—cell phones, TV remotes, computers, light bulbs, GPS units, and many other devices. We use electromagnetic radiation to keep tabs on the planet, too. Satellites take pictures of Earth from space at different wavelengths and store, send, or analyze images. Digitizing information, whether it's an image of the early universe taken by the Hubble Space Telescope or a word processing file on your computer, depends on the application of electromagnetic waves.

Digitization converts information into discrete data "bits" that computers and other processors can interpret and manipulate. Almost any kind of input can be digitized, including text, pictures, sound, music, and video. Converting from analog to digital formats involves sampling and "quantizing" a continuous variable signal so it becomes a fixed series of digital inputs instead.

Digital data is easy to store, organize, and transfer. When you send digital files, the information is being converted into electromagnetic wave pulses and transmitted anywhere in the world—and even beyond. Scientists exchange digital data from satellites, the International Space Station, and even the rovers on Mars.

Applications of Electromagnetic Waves: Photovoltaics

When an electron in a material is hit by a photon, it can absorb the photon's energy. For certain materials, especially metals, the incoming photons cause electrons to jump to a higher energy state where they flow more easily. When electrons move, they create a current. Therefore, incoming light can cause electricity to flow within and across these materials. This is called the photovoltaic effect.

photo: Pixabay

Photovoltaics absorb light energy from the sun that can then be used in many applications. What qualities must light have for photovoltaics to work?

The photovoltaic effect is related to the photoelectric effect. In the photoelectric effect, when a material absorbs very high-energy electromagnetic radiation, the energy of the photons can knock electrons totally away from their atoms. This effect usually occurs only with higher energy radiation such as ultraviolet, X-rays, and gamma rays. The difference is that, for the photovoltaic effect, the electrons get "energized" to a higher level, but they aren't completely stripped away from their atoms.

The discovery of the photovoltaic effect was a big surprise. Scientists found that only certain photon energy levels would have an effect. If the photons were too low energy, nothing would happen, no matter the intensity or number of photons. The photon energies didn't seem to "add." Instead, at a certain energy level of incoming photons, called a threshold energy, electrons would absorb the energy and be ejected. Even one photon at or above the threshold energy level would have an impact, while a million, billion, or trillion photons just below the threshold would have no effect at all.

This surprising result led scientists to a new understanding of light waves as having particle-like qualities and being quantized into discrete chunks, or photons. It was Albert Einstein who helped resolve the mystery by using Planck's constant and proposing that a photon's energy is related to its frequency by the equation:

$$E = hv$$

When the material being hit by incoming radiation can conduct a controlled current, the material can be used to create a PV cell, or a photovoltaic cell. This is what makes it possible to turn sunlight, or solar power, into electricity. So-called solar cells are used to power a wide range of objects and systems, including calculators, cars, orbiting satellites, and even electrical grids in places without generators.

Consider the Explain Question:

How do you know that the waves sent from the sun to Earth are not mechanical waves?

Go online to complete the scientific explanation.

Check for Understanding:

Two students are holding a spring. One student shakes the spring up and down. A wave travels along the spring to the second student. What is the type of wave modeled?

STEM in Action

STEM and Wave Characteristics

Many earthquakes occur daily, and most of them cause little, if any, damage. However, strong earthquakes cause considerable loss of life and property, so the study of earthquakes is an important area of science. The study of earthquakes is called seismology, and a seismologist is a scientist who studies seismology. One of the most important tools of a seismologist is a seismometer. A seismometer is a device engineered to measure the motion of seismic waves that result from a disturbance in the ground, such as an earthquake or large explosion. These devices gather information about the motion of seismic waves, such as their wave speed and frequency. When many seismometers are used in series, they can gather large-scale information about an earthquake. Seismologists can take information from an array of seismometers and can determine the precise location of an earthquake.

Seismologists also use seismometers to gather information about the internal structure of Earth. This process is called reflection seismology. In this process, an explosion is detonated at Earth's surface, causing seismic waves to travel downward toward the internal layers of Earth. When these waves strike boundaries or obstructions, they reflect back toward Earth's surface. Seismometers can detect the time it takes these waves to reflect. Seismologists then use this information to predict the existence of structures or resources underground, such as coal, water, or layers of bedrock.

Just like seismologists study the movement of waves through solid parts of Earth, certain oceanographers study the movement of waves through bodies of water by using sonar. Sonar is the process by which sound waves are emitted, reflected from a surface, and detected by a device. Submarines often use sonar to navigate through hazardous areas and to communicate with other underwater vessels. Sonar can also be used to determine the depth of a region of water in the ocean. Because it is difficult for scientists to gain direct access to the ocean floor, sonar can be used to map its topography, or shape. Sound waves are emitted toward the ocean floor. The waves reflect off the floor and return to the surface. Because oceanographers know the speed at which sound waves travel in water, they can determine the distance that the wave traveled. This allows them to generate detailed maps of the shape of the ocean floor.

Complete the activities "Wave Characteristics" and "Oscillation Procedure" on the STEM in Action page online.

Once you have completed the activities online, use the following to help you analyze your data and draw conclusions. Create a report detailing your procedures and findings.

1. Graph a sinusoidal plot of the vertical displacement of the mass as a function of time. Plot the vertical displacement of the mass on the *y*-axis and time on the *x*-axis.
2. How could you use the table you have drawn to calculate the average amplitude, wave period, and frequency of the wave?
3. What is the average amplitude of this wave? What is the average period? What is the average frequency?
4. Can you determine the wavelength and wave speed of this curve? Why or why not?
5. How is the motion of the mass-spring system similar to and different from an electromagnetic wave propagating through space?
6. Compare and contrast the mass-spring system's motion to the rhythmic motion of a transverse mechanical wave, such as a tension wave in a waving flag or clouds oscillating in altitude due to changing buoyancy. What do these waves have in common? How are they different?
7. Develop a conclusion for your investigation. The conclusion should contain clear thoughts and proper vocabulary. It should support or refute your hypothesis by using logical reasoning to link the hypothesis to the data.

Energy for Life

photo: Getty Images

LESSON OVERVIEW

Lesson Questions

- What is energy, and how does it contribute to maintaining life on Earth?
- What are kinetic and potential energy?
- What are the laws of thermodynamics?
- What is the difference between endergonic and exergonic reactions?
- What is the difference between oxidation and reduction reactions?
- What is the role of photosynthesis and respiration in the carbon cycle?

Key Vocabulary

Which terms do you already know?

- ☐ **aerobic respiration**
- ☐ **anaerobic respiration**
- ☐ **cellular respiration**
- ☐ **chemical energy**
- ☐ **electron transport chain**
- ☐ **endergonic reaction**
- ☐ **energy**
- ☐ **exergonic reaction**
- ☐ **glycolysis**
- ☐ **oxidation**
- ☐ **photosynthesis**
- ☐ **potential energy**
- ☐ **redox reaction**
- ☐ **reduction**
- ☐ **thermal energy**
- ☐ **thermodynamics**

Lesson Objectives

By the end of the lesson, you should be able to:

- Describe the concept of energy and its contributions to maintaining life on Earth.
- Distinguish among the different forms of energy.
- Describe the laws of thermodynamics.
- Distinguish between endergonic and exergonic reactions.
- Distinguish between oxidation and reduction reactions.

ENGAGE

How does your body get energy from the food you eat? How do plants get their energy? Visit the Engage page online to learn more.

Explain Question

Explain how and why energy is important to living things.

EXPLORE

What Is Energy, and How Does It Contribute to Maintaining Life on Earth?

Living Organisms and Energy

All living organisms require energy to perform essential cellular processes. Growth and reproduction are complex processes that need energy. Building larger molecules from smaller ones requires the input of energy; these larger molecules can also be used to store energy for later use. Moving molecules from low to high concentrations requires the input of energy. Organisms also need energy for a variety of other tasks such as movement and sending electrical signals in the nervous system. Without energy, none of the processes of life are possible.

Ultimately, the energy for most organisms on Earth comes from the sun. Photosynthetic organisms such as plants, algae, and some bacteria and archaea can convert solar energy into chemical energy, which is energy stored in chemical bonds. Other organisms such as animals and fungi must acquire their energy from the photosynthetic organisms. When an animal ingests food, the food molecules are broken down, and the energy within is released. Some single-celled organisms are able to use inorganic molecules such as compounds containing sulfur as an energy source.

Defining Energy

All organisms require a source of energy, but what is energy? Energy is the ability to do work or cause change. The processes of living organisms such as moving and building molecules are examples of work that can be accomplished using energy. To do this work, cells convert energy from one form to another. Energy cannot be created or destroyed, but it can be transformed into different forms. Kinetic energy is the energy of motion, thermal energy is heat generated by the kinetic energy of atoms or molecules, and potential energy is stored energy. Chemical potential energy, commonly called chemical energy, is a form of potential energy that is stored in chemical bonds. It is the conversion of energy among these forms that keeps living organisms alive.

Measuring Energy

Energy can be measured using a number of different techniques. Energy is defined as the ability to do work, and work is defined as the force used to move an object for a certain distance (work = force × distance). The standard unit of measurement for energy is the Joule (J), which is defined as 1 Newton of force used to move an object 1 meter:

$$1 \text{ joule} = 1 \text{ Newton} \cdot \text{meter}$$

A watt is defined as the use of 1 J/second. This measurement is often used to measure power expenditure in electrical equipment.

The energy in food is often measured in calories. A calorie (cal) is the amount of thermal energy needed to raise the temperature of 1 gram (g) of water by 1°C. The calories listed on food labels are kilocalories (1,000 calories). On food labels, kilocalories are designated as Calories with a capital letter C.

Energy can also be measured in ergs, which are defined as the kinetic energy of 2 g moving at a velocity of 1 centimeter per second. These measures of energy relate to each other with the following conversions:

$$1\,\text{cal} = 4.184\,\text{J}$$

$$1\,\text{J} = 0.239\,\text{cal}$$

$$1\,\text{erg} = 10^{-7}\,\text{J}$$

In biology, the calorie is the most commonly used unit of energy.

What Are Kinetic and Potential Energy?

Kinetic and Potential Energy

Energy exists in a variety of forms of kinetic and potential energy. Organisms survive by converting energy from one form to another.

Kinetic energy is the energy any object has because it is moving. Objects can have kinetic energy regardless of their mass. Vibrating atoms have kinetic energy because they are moving. A running elephant also has kinetic energy. Energy from food is converted by the elephant's muscles to make it move and give it kinetic energy. Kinetic energy is defined as the work needed to move a body of a given mass to its final velocity, or speed in a given direction.

Thermal energy is a term used to refer to heat generated by the kinetic energy of atoms and molecules. Atoms and molecules are in random constant motion. Temperature is a measure of this motion. Molecules moving more quickly have greater thermal energy than molecules moving more slowly.

Potential energy is the energy in a body or a system based on its position or configuration—it is stored energy. Potential energy exists in a number of forms. These include chemical potential energy, gravitational potential energy, and mechanical potential energy.

Chemical potential energy or chemical energy is the energy that is stored in the bonds between atoms. Chemical energy is released or absorbed during chemical reactions. When bonds are broken in chemical reactions, energy is absorbed. When bonds are made in a chemical reaction, energy is released. The balance of energy between bonds being broken and bonds being formed determines whether a chemical reaction releases or absorbs energy. Organisms store energy as chemical energy. Food compounds, such as glucose, starch, or fats, contain a lot of chemical energy in their molecules. When cells metabolize food during cellular respiration, they react these compounds in the presence of oxygen. This releases energy. Initially, they may store some of the released energy as chemical energy in other "high energy" molecules. One of these molecules is adenosine triphosphate (ATP). This stored chemical energy is then readily available to do work within the cell. Chemical energy is required for all life processes.

Energy can be stored in other ways. A ball at the top of a hill has potential energy because of its position in relation to the center of the Earth. If the ball rolls down the hill, this gravitational potential energy is converted to kinetic energy. Stretched or compressed objects, such as springs or rubber bands, can also store energy. This mechanical potential energy is converted to kinetic energy when the force holding them in position is released.

What Are the Laws of Thermodynamics?

The Laws of Thermodynamics

The study of energy is called thermodynamics. A scientific understanding of thermodynamics has resulted in three basic laws.

The first law of thermodynamics is related to the conservation of energy and is often called the law of conservation of energy. The law states that energy cannot be created or destroyed. The total energy in the universe remains constant. Energy can, however, be transformed from one form into another.

For example, when an organism breaks down glucose molecules during cellular respiration, some of the chemical potential energy of the glucose molecules is transformed into the chemical potential energy of other molecules such as ATP, some into other compounds, and some into thermal energy. No energy disappears. The amount of energy at the beginning is the same as at the end.

The second law of thermodynamics states that the entropy of a system cannot decrease. Entropy is the measure of disorder, or chaos, in a system. Systems increase in disorder until equilibrium is reached. This law of thermodynamics explains why energy cannot be transferred from a colder body to a hotter body and why energy conversions are never 100 percent efficient. Most of the energy taken in from food is lost as thermal energy. Only a small percent of the energy taken in is used to do work. This thermal energy is why people feel hot when exercising. The conversion of potential energy to kinetic energy releases a lot of thermal energy.

The third law of thermodynamics defines the point at which the thermal energy of molecules is zero. This point is called absolute zero and is defined as 0K (Kelvin); this occurs at –273°C. Absolute zero is the lowest possible temperature because atoms have ceased moving.

It is important to understand that the laws of thermodynamics apply to closed systems. A closed system is one in which no energy can enter or leave. The ultimate closed system is the universe. In nature, all systems are open—energy can get in or out. Organisms are open systems within the bigger closed system of the universe. The laws of thermodynamics apply to them just as they do to the rest of the universe.

What Is the Difference between Endergonic and Exergonic Reactions?

Endergonic and Exergonic Reactions

The laws of thermodynamics can be used to classify chemical reactions into two categories: reactions that take in energy from their surroundings and reactions that release energy into their surroundings.

Endergonic reactions require a net input of energy from the surroundings. In a biological system, free energy (G) is energy that can do work if temperature and pressure are held constant. The change in free energy (ΔG) of these reactions is positive because the products of the reaction contain more energy than the reactants. Because these reactions have a positive ΔG, they are not spontaneous; that is, they are not energetically favorable and require the input of energy to proceed.

Exergonic reactions release energy. The ΔG of these reactions is negative because the products of the reaction contain less energy than the reactants. This energy is released into the surroundings and can be used to power endergonic reactions. Because these reactions have a negative ΔG, they are spontaneous; that is, they are energetically favorable and occur without the input of energy.

In cellular processes, endergonic reactions are often linked to exergonic reactions. A common exergonic reaction that is used to power a variety of cellular processes is the breakdown of ATP into ADP (adenosine diphosphate). This breaking down of larger molecules into smaller molecules is exergonic and can be used to power endergonic reactions.

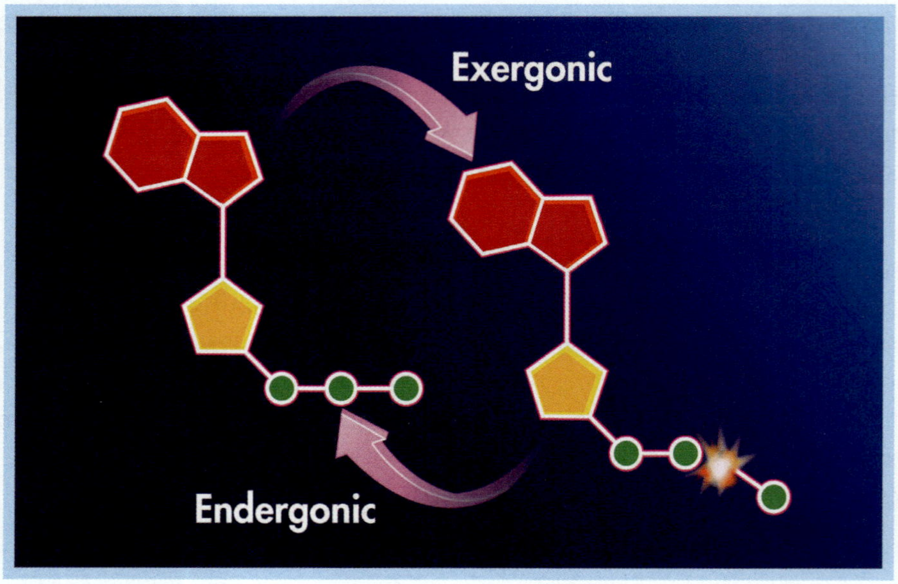

Exergonic reactions such as the breakdown of ATP to ADP are used by cells to power endergonic reactions. Why is this pairing necessary?

Cells regenerate their ATP store by adding a third phosphate to ADP; synthesizing larger molecules from smaller subunits is endergonic. Cells can break down food molecules (an exergonic process) to release the needed energy to build ATP (an endergonic process). ATP is then stored until its chemical energy is needed to do work.

What Is the Difference between Oxidation and Reduction Reactions?

Redox Reactions

Redox reactions are used to store and release energy in biological systems. Electrons can be moved from a lower energy state in one atom to a higher energy state in another atom; this serves to store energy. Electrons can also move from higher energy states in one atom to lower energy states in another, thus releasing energy. The movement of electrons from one atom to another changes the oxidation number of both atoms. Any chemical reaction that changes the oxidation number of atoms is called a redox reaction.

The term *redox* is a combination of the terms reduction and oxidation. Reduction is the gain of electrons, which decreases the oxidation state of the atom or ion. In reduction, the potential energy of electrons is increased; this stores energy. Oxidation is the loss of electrons, which increases the oxidation state of the atom or ion. In oxidation, the potential energy of the electrons is decreased, which releases energy. Oxidation and reduction reactions often occur together as one atom gains an electron from another atom. This reaction is called a redox reaction because one atom or ion has been reduced from the oxidation of another atom or ion.

In living things, energy is usually stored and released through redox reactions. For example, in photosynthesis carbon dioxide is reduced to make sugars, and water is oxidized to form oxygen. Chemical reactions like this occur in many small steps. Each step is a small redox reaction.

What Is the Role of Photosynthesis and Respiration in the Carbon Cycle?

The Carbon Cycle

Carbon is constantly circulating between living things and the environment. As it circulates, it changes form—from gas to solid and back again. There are several ways in which carbon is stored on Earth. A large quantity of stored carbon is called a reservoir. Carbon reservoirs include fossil fuel deposits, biomass (living plants and animals), oceans, and rocks. Carbon is essential to life on Earth, and the carbon cycle helps keep the amount of carbon in the atmosphere relatively constant.

The exchange of carbon between biomass and the atmosphere is also referred to as the fast carbon cycle because this cycle occurs relatively quickly. Carbon is taken from the atmosphere when plants use it to produce food. Then, animals consume the food and use its energy for the animal's life processes. When the animal dies, it decomposes and releases the carbon into the atmosphere or into the soil.

Photosynthesis and Respiration

Photosynthesis and respiration are two processes that exchange carbon between the atmosphere and living things in the environment. They are essential to the functioning of the fast carbon cycle.

Photosynthesis is the biochemical process by which plants use energy from the sun to produce food from carbon dioxide and water. In this way, carbon moves from the atmospheric reservoir to the biomass reservoir.

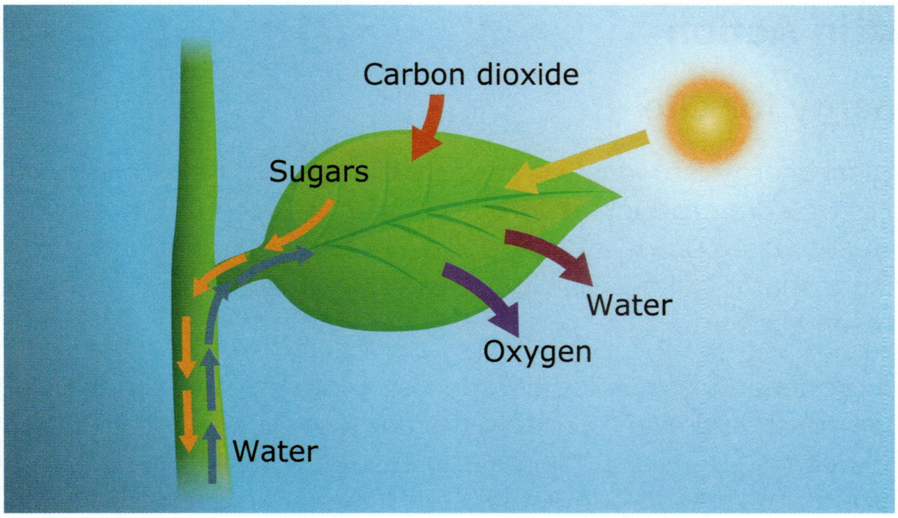

Plants use the process of photosynthesis to produce food from carbon dioxide and water. What role does photosynthesis play in the carbon cycle?

Cellular respiration is the set of metabolic processes by which plants and animals convert the chemical potential energy in glucose and other nutrients into adenosine triphosphate (ATP), which is then used to fuel other biochemical processes. The overall reaction for cellular respiration converts glucose into carbon dioxide (which is exhaled) and water. Thus, carbon moves from biomass back into the atmosphere.

Photosynthesis and cellular respiration do not only cycle matter in the form of carbon through the carbon cycle, they also cycle energy through the carbon cycle. Photosynthesis converts solar energy into the chemical potential energy contained in glucose and other carbohydrates. Then, cellular respiration converts the chemical potential energy into ATP, which fuels other biochemical processes.water.

Consider the Explain Question:

Explain how and why energy is important to living things.

Go online to complete the scientific explanation.

Check for Understanding:

After eating a healthy breakfast, a student starts a five-mile hike. How does this scenario illustrate kinetic and potential energy?

STEM in Action

STEM and Energy for Life

Converting chemical energy into kinetic energy is the basis of how cars move. The potential energy stored in petroleum products is used to power the automobile engine. This combustion produces waste products that pollute the environment and is a nonrenewable resource. This means that it will eventually get used up. Alternative fuel sources for cars are currently being researched.

Biofuels are fuels produced from biological sources. Biofuels can include using agricultural waste products, sawdust, and even corn to generate fuel such as ethanol. Corn is of particular interest because of its abundance and ease of converting into ethanol. The corn is cooked with different enzymes to convert its starch into simple sugars.

Corn production has reached all-time highs in the United States. Of the 1.5 billion bushels of corn produced in 2011, 600 million were used for biofuel. This is good news for the environment.

To meet the ever-growing demand for biofuels, crop science has become more and more popular as a course of study. Crop scientists conduct research on crops and other agricultural products to find new and improved ways to use these for fuel. A crop scientist may test several types of perennial grasses to see which can be most efficiently broken down into simple sugars. Crop scientists also work to improve crop yields by using techniques that could enhance feedstock production efforts. A career in this field would require a strong biology background, as well as a lot of field training.

1. The following graph shows the basal metabolic rate of several different types of organisms. This is the number of calories they burn due to their overall body mass. Describe the trend in basal metabolic rate that is present in this graph.

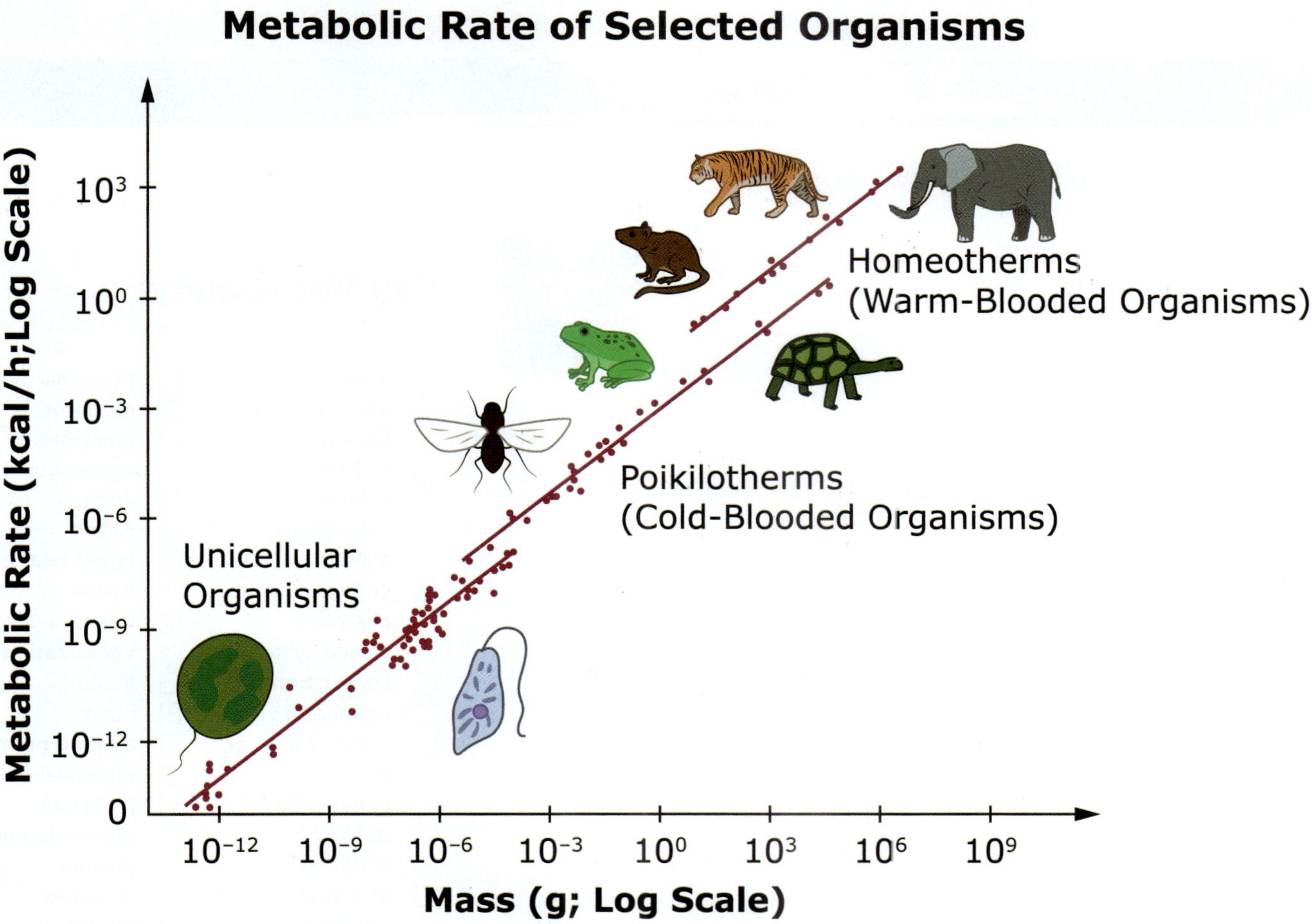

1 kcal/h = 1.162 watts

2. What type of animal would represent the shift from negative log metabolic rate and mass to positive log metabolic rate and mass? Why do you think they exist here?

3. Where would humans fall on this graph? Why would they go there?

Heat

photo: Getty Images

LESSON OVERVIEW

Lesson Questions

- What is heat?
- What is the difference between thermal energy and temperature?
- How can problems involving conversions between Joules, calories, and kilocalories be solved?
- How does thermal-energy transfer (heat) cause substances to change state?
- How can two objects of different masses have the same temperature but different amounts of thermal energy?
- What are some examples that show how thermal energy is transferred by means of conduction, convection, and radiation?
- How can applying the law of conservation of energy help solve calorimetry problems?

Key Vocabulary

Which terms do you already know?

- ☐ atom
- ☐ average kinetic energy
- ☐ boiling point
- ☐ calorie
- ☐ calorimetry
- ☐ Celsius
- ☐ change of state
- ☐ chemical
- ☐ closed system
- ☐ conduction
- ☐ conductor
- ☐ conservation of energy
- ☐ convection
- ☐ density
- ☐ distance
- ☐ electromagnetic spectrum
- ☐ electromagnetic wave
- ☐ electron
- ☐ Fahrenheit
- ☐ force
- ☐ freezing point
- ☐ freezing point depression
- ☐ frequency
- ☐ gamma ray
- ☐ gas
- ☐ heat
- ☐ heat energy
- ☐ insulator
- ☐ inversely proportional
- ☐ Joule
- ☐ Kelvin
- ☐ latent heat of fusion
- ☐ latent heat of vaporization
- ☐ liquid
- ☐ mass
- ☐ melting point
- ☐ microwave
- ☐ molecule
- ☐ phase change
- ☐ plasma
- ☐ pressure
- ☐ radiation
- ☐ rotation
- ☐ solid
- ☐ specific heat
- ☐ temperature
- ☐ thermal energy
- ☐ thermal equilibrium
- ☐ transport
- ☐ visible light
- ☐ wavelength
- ☐ work
- ☐ X-ray

Lesson Objectives

By the end of the lesson, you should be able to:

- Model that heat is the transfer of thermal energy from an object of higher temperature to an object of lower temperature until an equilibrium temperature is reached.
- Explain the difference between thermal energy and temperature.
- Provide evidence of how two objects of different masses can have the same temperature but different amounts of thermal energy.
- Describe the energy transfer that occurs when substances change state.
- Use examples to explain how thermal energy is transferred via conduction, convection, and radiation.
- Solve problems involving conversions between Joules, calories, and kilocalories.
- Solve calorimetry problems by applying the law of conservation of energy.

ENGAGE

Why can sand burn your feet on a sunny day? What is heat, and how does it behave? Visit the Engage page online to learn more.

Explain Question

What is heat, how is it related to temperature and thermal energy, and how does the transfer of thermal energy affect matter?

EXPLORE

What Is Heat?

Heat is the transfer of thermal energy from a high-temperature object to a low-temperature object that are in *thermal contact*. Matter does not contain heat. Heat is energy in transit. Heat transfer will continue until the two objects are at the same temperature. When the objects reach the same temperature, they are said to be in thermal equilibrium.

In spontaneous processes, thermal energy moves (as heat) in only one direction, from higher-temperature areas to lower-temperature areas. Sometimes, people speak of hot and cold, but in truth, these words are relative terms used to describe only the temperature of a substance, not the total amount of thermal energy of a substance. There is no physical process in which thermal energy is transferred from a lower-temperature object to a higher-temperature object unless outside influences, such as work, force it to do so.

People sometimes express the movement of thermal energy by heat incorrectly. When they are talking about cold weather, they may say that their house needs better insulation "to keep the cold out." What they mean is that they want to keep the thermal energy in. Thermal energy always flows from regions of higher temperature to regions of lower temperature. If you hold a frozen snowball in your hand, heat flows from your hand to the snowball. The snowball melts because it gains thermal energy; your hand gets cold because it loses thermal energy.

Heat and Conservation of Energy

The law of conservation of energy applies to isolated systems. An isolated system is a distinct region that is closed off from its surroundings from which no matter or energy can be added or removed. Objects within the system may gain or lose thermal energy, but the system's total energy is constant. Any thermal energy lost by one object must be gained by another object.

The total energy of an isolated system is conserved. One way to express this principle is to state that energy can neither be created nor destroyed, but it can be transformed to a different form of energy or transported somewhere else. During the manufacturing process, power plants and factories generate a lot of extra thermal energy. This "waste heat" can be recycled in one of two ways. It can be transferred to another type of useful energy such as electrical energy. A common method for converting thermal energy into electrical energy is to use the energy to make the steam that drives a steam turbine. A steam turbine is a machine that generates electricity. Extra thermal energy can also be transported away from the factory or power plant and used somewhere else. The thermal energy is used to heat up water. The hot water is then pumped into nearby buildings where it is used as a source of heat.

What Is the Difference between Thermal Energy and Temperature?

Thermal Energy vs. Temperature

Thermal energy and *temperature* are terms that have specific definitions in physics. The thermal energy of an object is the sum of all the kinetic energy of its molecules. Its temperature is a measure of their average kinetic energy. Different objects with the same temperature can have very different thermal energy. For example, seawater in the ocean and seawater in a bucket can have the same temperature. However, the ocean will have more thermal energy because it has more molecules.

Molecules of water in that bucket have kinetic energy. This would be low if that water were frozen. This would be higher if it were a liquid, and still higher if it was a vapor. For the same mass of a substance, molecules of vapor have more thermal energy and higher temperature than those of a liquid. Molecules of a liquid have more thermal energy and higher temperature than those of a solid.

Units for Temperature

Temperature is typically measured using one of three common scales:

Fahrenheit scale: This scale is in everyday use in the United States. It uses the freezing point (32°F) and the boiling point (212°F) of water as reference points. It divides the range between them into 180 equal degrees. Although almost all of the rest of the world has abandoned it, the Fahrenheit scale does have the advantage of having divisions that are smaller compared with those in the other two major scales. This allows for greater resolution of temperature. The scale is specified by using either the notation °F or the phrase "degrees Fahrenheit."

Celsius scale: This scale is in everyday use throughout most of the world. It is used in science and increasingly in industry in the United States. It also uses the freezing and boiling points of water as reference points. It sets them at convenient temperatures of 0 and 100°C, respectively, dividing the range into 100 equal parts. Temperatures below the freezing point of water are negative. The scale is specified by using either the notation °C or the phrase "degrees Celsius."

Kelvin scale: This scale is in universal scientific use. Its divisions are the same size as the degrees in the Celsius scale. The lowest possible temperature is set to absolute zero. At absolute zero, there is no energy from molecular motion available for transfer to other objects. Absolute zero on the Kelvin scale (0 K) is equivalent to –273.15°C. It is very difficult to produce the conditions required to observe matter at this temperature, even if it is possible. A laboratory in Finland has been able to supercool a piece of rhodium to 0.0000000001 K, or 100 pK above absolute zero. In contrast with the other two scales, the word *degrees* and the symbol° are not used with the Kelvin scale. Examples are 2,300 K and 273.15 kelvin.

How Can Problems Involving Conversions between Joules, Calories, and Kilocalories Be Solved?

Units for Thermal Energy

Thermal energy and temperature are related, but they have different units. Thermal energy uses the energy units Joules (J) or calories (cal). It can also be expressed using kilojoules (kJ) and kilocalories (kcal).

A Joule (J) is an amount of energy equal to the work done by a force of 1 Newton (N) acting through a distance of 1 meter (m).

One calorie (cal) is the energy required to raise the temperature of 1 gram of water by 1°C.

The prefix "kilo" means 1,000, so one kilojoule (kJ) is equal to 1,000 Joule, and one kilocalorie (kcal) is equal to 1,000 cal.

Another word for a kcal is Calorie, denoted by a capital C. This unit is used on food labels to indicate the amount of energy available in the food.

Units of Joules can be converted to calories, and vice versa, by using the following conversion factors:

- 1 cal = 0.001 kcal = 4.186 J
- 1 kcal = 1,000 cal = 4186 J
- 1 J = 0.239 cal = 0.000239 kcal

Units for Thermal Energy: Sample Problem

Convert 516 calories (cal) into Joules (J) and kilojoules (kJ).

Solution

$$(516 \text{ cal})(4.186 \text{ J/cal}) = 2{,}160 \text{ J}$$

To convert to kilojoules, divide by 1,000:

$$(2{,}160 \text{ J})(1 \text{ kJ}/1{,}000 \text{ J}) = 2.16 \text{ kJ}$$

How Does Thermal-Energy Transfer (Heat) Cause Substances to Change State?

Effect of Thermal Energy

When thermal energy is added to or removed from a substance, the effect is either to change the substance's temperature or to change its state of matter. A change of state is also called a phase change. It occurs when a substance changes among its solid, liquid, gas, and plasma states.

If thermal energy is transferred to a substance, under most circumstances, its atoms and molecules gain kinetic energy. This increase in average kinetic energy increases the temperature of the substance. When the substance's melting or boiling point is reached, however, adding more thermal energy does not immediately result in a further temperature increase. It causes the substance to change state instead.

In a substance, the electrical potential energy between atoms and molecules is stored in the form of intermolecular forces. During a change of state, the additional thermal energy allows particles in the substance to overcome the intermolecular forces holding them together. The average kinetic energy of the particles does not change. The temperature of the substance remains constant. Only when the phase change is complete does additional thermal energy cause the temperature increase to resume.

The principle is also true in reverse. As thermal energy is removed from a substance, the average kinetic energy of its particles decreases. Therefore, the temperature also decreases until the substance reaches a phase-change point. Then, the temperature remains constant while more thermal energy is removed. Only when the phase change is complete will further removal of thermal energy cause the temperature to resume its decrease.

For rising temperatures, the temperatures at which phase changes take place are called the substance's melting point and boiling point. For falling temperatures, these same temperatures are known as the substance's freezing point and condensation point.

Phase change temperatures are not the same for all substances. For example, while the solid/liquid and liquid/vapor phase-change temperatures for water are 0°C and 100°C, respectively, those for silver are approximately 962°C and 2,212°C.

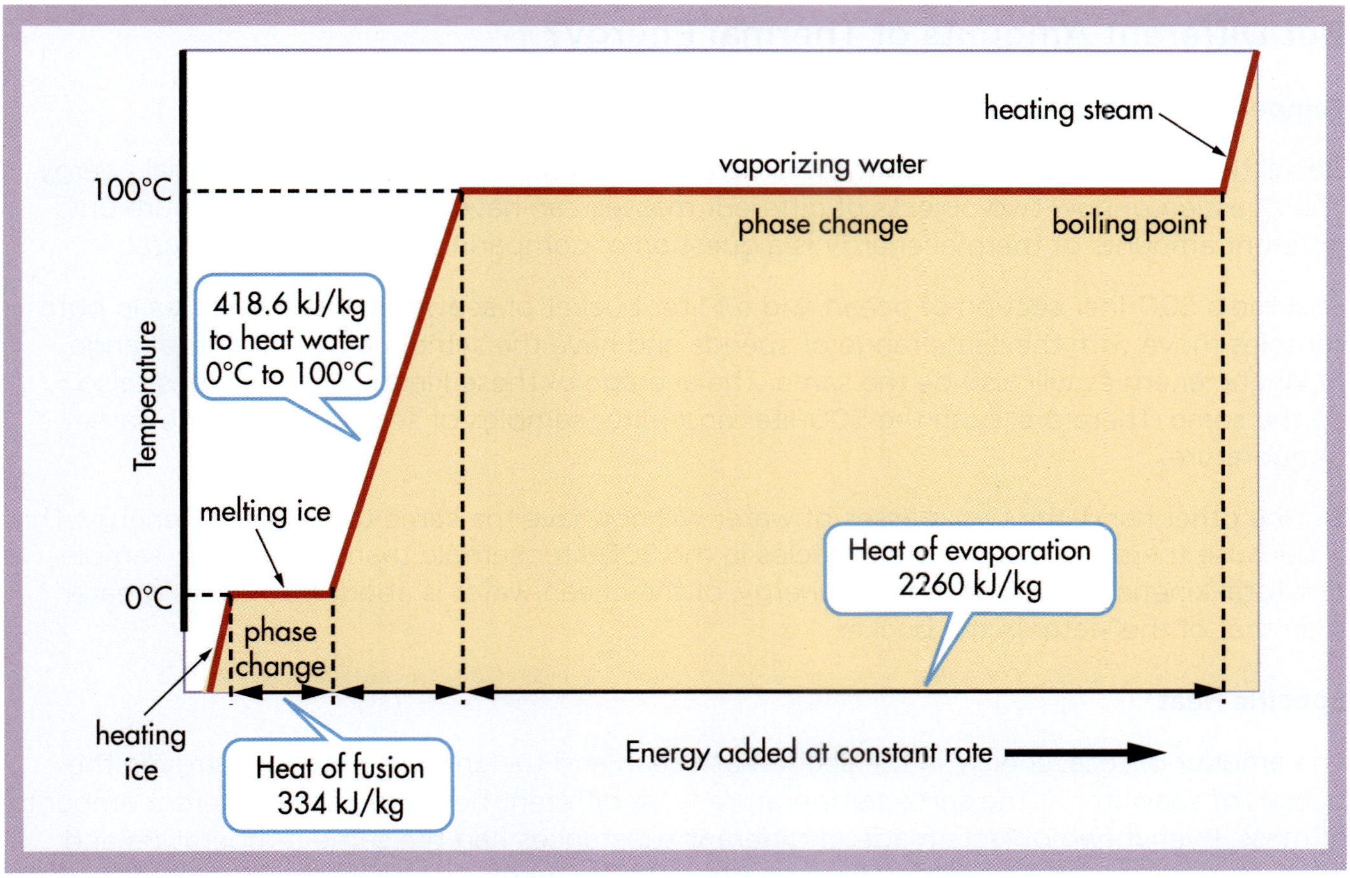

As energy is added to water, the temperature does not increase at a constant rate. What happens to the relationship between the energy added and the temperature during a phase change?

Not only must a substance be at a certain temperature in order to change phase, it must also gain or lose a certain amount of thermal energy. The heat needed to change a certain amount of a substance from the solid to the liquid phase without changing its temperature is its latent heat of fusion. The heat required to change a certain amount of a substance from the liquid to the vapor (gas) phase without changing its temperature is the latent heat of vaporization.

Water, for example, requires about 334 kilojoules per kilogram of thermal energy to transition between solid and liquid form. It needs 2,260 kJ/kg of thermal energy to transition between liquid and gaseous states.

The amount of thermal energy a substance gains in a phase change is the same amount of energy it loses in the reverse phase change. For example, the same amount of thermal energy added to a solid to melt it completely must be removed from the liquid to freeze it to the solid phase.

Whether a substance changes state at a given temperature is affected by various factors. Under special conditions, thermal energy can be added or removed from a substance such that it does not undergo a phase change as it normally would. The temperature continues to increase or decrease past the phase-change points. The substance is then said to be superheated or supercooled, and it is unstable. The phase change is likely to occur suddenly when conditions shift slightly.

How Can Two Objects of Different Masses Have the Same Temperature But Different Amounts of Thermal Energy?

Temperature Is Average Kinetic Energy

Recall the earlier discussion about the difference between temperature and thermal energy. The question of how two objects of different masses can have the same temperature but different amounts of thermal energy is a question of comparing an average to a total.

Assume a 500-liter section of ocean and a 1-liter bucket of seawater. If the particles in both samples move with the same range of speeds and have the same masses, then the range of kinetic energies will also be the same. The average of these kinetic energies must also be the same. Therefore, both the 500-liter and 1-liter samples of seawater have the same temperature.

On the other hand, the two masses of water will not have the same total thermal energy. This is because there are many more particles in the 500-liter sample than in the 1-liter sample. The total kinetic energy, or thermal energy, of the ocean water is about 500 times greater than that of the water in the bucket.

Specific Heat

The amount of total energy in the section of ocean and the amount of total energy in the bucket of seawater at the same temperature were different because of the different amounts of mass. Even if two objects made of different substances had the same temperature and mass, they still might have different amounts of total energy.

The abilities of materials to absorb and release thermal energy vary. The amount of energy required to change the temperature of 1 gram of a given substance by 1°C (or 1 K) is its specific heat. Specific heats are typically given in units of J/(g × K) or kJ/(kg × K).

For equal masses, different materials require different amounts of energy to achieve the same temperature change. Conversely, given the same amount of thermal energy, equal masses may achieve different temperature changes if the masses are not made of the same substance.

In the example of playground slides, one was made from steel and the other from plastic. Steel has a specific heat of about 0.5 kJ/(kg × K), while plastic has a specific heat of about 2.0 kJ/(kg × K). The steel slide's significantly lower specific heat causes it to achieve a higher temperature than does the plastic slide.

If a substance has a low specific heat, it takes relatively little energy to change its temperature. If it has a high specific heat, it takes a lot of energy to change its temperature. A substance with high specific heat resists changes in its temperature.

Among common substances, water is very unusual in that it has a high specific heat of 4.186 kJ/(kg × K). In other words, it would take 4,186 J to raise the temperature of 1 kg of water by 1°C. Changing the temperature of water involves a lot of energy.

What Are Some Examples That Show How Thermal Energy Is Transferred By Means of Conduction, Convection, and Radiation?

Three Methods of Heat Transfer

Heat transfer occurs in three main ways: conduction, convection, and radiation.

In conduction, thermal energy is transferred between different areas of the same substance. Thermal energy applied in one part of the substance causes atoms and molecules in the substance to move more rapidly. This causes collisions with neighboring particles, causing them to move faster as well. Further collisions transfer kinetic energy to particles farther away, and so forth. Thus, thermal energy spreads throughout the substance.

An example of conduction is the heating of one end of an iron bar. Conduction causes thermal energy to travel from the area of higher temperature to the area of lower temperature at the other end of the bar. Heat transfers energy until thermal equilibrium is reached.

Conduction works best in solids because the atoms and molecules are close together. Metal solids make particularly good conductors of thermal energy for the same reason that they are good conductors of electricity. Their atoms have free electrons that can move easily through the material and transmit the energy to other particles through collisions. Materials with tightly held electrons, such as wood and paper, make poor conductors. These materials are called insulators. A poor conductor is a good insulator, and vice versa.

Conduction is not possible in a vacuum. There are no particles in a vacuum to transfer energy. Air, as a gas, has particles, but they are spread farther apart than in liquids or solids. Materials that have air spaces in them, such as fiberglass insulation, plastic foam, and the feathers in down comforters, effectively slow the transfer of thermal energy.

Convection is the transfer of thermal energy by way of moving molecules in fluids. Fluids include liquids and gases.

Convection works because of differences in fluid density. Less dense fluids tend to rise, while more dense fluids sink. When thermal energy is applied to one part of a mass of fluid, nearby particles become more energetic and spread apart. The fluid in that area becomes less dense. Less dense fluid rises away from the heat source, while more dense fluid sinks to take its place. Eventually, the heated fluid cools, becomes denser again, and sinks. The resulting cycle of moving fluid transports thermal energy.

The upward flow of hot air above a room heater is an example of convection. Hot air moves away from the heater, and cool air moves to replace it. The air in the room circulates. Similarly, convection circulates air in the atmosphere, affecting the weather. The sun transfers energy to Earth's surface. Air near the surface warms, becomes less dense, and rises. Cooler air rushes in underneath, causing wind.

Convection is also important in the world's ocean currents. Cold, dense water near Earth's poles sinks and moves along the bottom of the seafloor. Water warmed near the equator and pushed by wind and Earth's rotation moves along the surface toward the poles. Warm water passes near landmasses, warming them and significantly affecting their climate.

Radiation is the transfer of energy by electromagnetic waves. Energy from the sun arrives on Earth by radiation. Unlike conduction and convection, it can travel in a vacuum. Thermal interaction by radiation does not require actual physical contact.

Radiant energy includes radiation from across the electromagnetic spectrum, from the short wavelengths and high frequencies of gamma rays and X-rays, through visible light, to the long wavelengths and low frequencies of microwave and radio energy. All objects are at a temperature above absolute zero and, therefore, they emit some form of electromagnetic radiation. The radiation wavelength an object emits most (its peak wavelength) is inversely proportional to its absolute temperature in kelvins.

Some objects, such as the sun and incandescent light bulbs, emit radiation that includes the visible portion of the electromagnetic spectrum. Moving from the blue-violet end of the visible light spectrum toward the red end, the wavelengths of electromagnetic radiation get longer and the frequencies get lower. At frequencies just below those for red light, in the infrared range, our eyes no longer perceive the radiation as light. However, we can feel it as heat. A hot fire emits some of its radiant energy as the visible flames and glowing embers and some as invisible infrared energy that is felt as warmth.

How Can Applying the Law of Conservation of Energy Help Solve Calorimetry Problems?

Calorimetry

The total energy in an isolated system is conserved. This makes it possible to solve problems in which an amount of thermal energy is transferred from one substance to another with known thermal properties. Calorimetry is the study of thermal energy changes that occur in a nearly isolated system.

The heat energy transferred by an object or substance going through a temperature change is expressed by this equation:

$$Q = mc\Delta T$$

where Q is the amount of heat, m is the mass of the substance, c is its specific heat, and ΔT is the temperature change.

This equation, combined with the understanding that energy is conserved, can be used to solve calorimetry problems.

Calorimetry: Sample Problem

Consider 2.50 kilograms of hot copper at 98.0°C, which is dropped into a tank containing 10.0 kg of water at 15.0°C. As thermal energy transfers from copper to water, the metal's temperature decreases while the water's temperature increases. Because energy is conserved, the thermal energy removed by heat from the copper will equal the heat transfer to the water. Both copper and water will arrive at the same final temperature—the temperature of thermal equilibrium: T_{eq}. Find T_{eq}.

Solution

Using the formula following, and recalling that the specific heat of water is 4,186 J/(kg × °C), the amount of heat energy received by water is:

$$\begin{aligned} Q_{water} &= (10.0\text{ kg})(4{,}186\text{ J}/(\text{kg}\times{}^\circ\text{C}))(T_{eq} - 15.0\,{}^\circ\text{C}) \\ &= (41{,}860\text{ J}/{}^\circ\text{C})T_{eq} - 627{,}900\text{ J} \end{aligned}$$

Meanwhile, copper, which has a specific heat of 386 J/(kg × °C), has cooled from 98.0°C to T_{eq}:

$$\begin{aligned} Q_{copper} &= (2.50\text{ kg})(386\text{ J}/(\text{kg}\times{}^\circ\text{C}))(98.0\,{}^\circ\text{C} - T_{eq}) \\ &= 94{,}570\text{ J} - (965\text{ J}/{}^\circ\text{C})T_{eq} \end{aligned}$$

Because of the conservation of energy, $Q_{water} = Q_{copper}$. Thus:

$$(41{,}860\text{ J}/{}^\circ\text{C})T_{eq} - 627{,}900\text{ J} = 94{,}570\text{ J} - (965\text{ J}/{}^\circ\text{C})T_{eq}$$

Solving algebraically for T_{eq} and rounding for significant digits: $T_{eq} = 16.9$°C.

Intuitively, this makes sense, because T_{eq} should be a temperature that is between the initial temperatures of copper and water. It is closer to the initial temperature of water, which has the higher specific heat and greater mass.

Consider the Explain Question:

What is heat, how is it related to temperature and thermal energy, and how does the transfer of thermal energy affect matter?

Go online to complete the scientific explanation.

Check for Understanding:

What is the difference between thermal energy and temperature?

STEM in Action

STEM and Heat

When civil engineers develop designs for products and projects, they often must consider the thermal properties of the materials used in their manufacturing and construction. They often work with engineers who develop new materials or new applications for old materials. Materials engineers might specialize in include metals, ceramics, plastics, or other innovative materials.

When objects are heated, the transferred thermal energy causes the molecules of the object's material to move faster and to move farther apart. As temperatures rise, most materials expand. As temperatures fall, most contract. Depending on the material and the shape of the object, the change can be significant.

Water represents a special case, especially if the temperature is near one of its phase-change points. For example, at around 4°C, a given volume of water is at its most compact. As the temperature drops just below the freezing point (0°C), the water (ice) expands to take up more space than at any other temperature below the boiling point. In areas that experience freezing temperatures, water commonly damages concrete and masonry structures as it infiltrates, changes to ice, and expands.

photo: U.S. Department of Transportation

Plastic sheeting allows the concrete to slide easily over the ground as it expands and contracts. This prevents uneven stress in the concrete that could cause cracks.

Civil engineers are particularly concerned with designing roads and bridges that accommodate the thermal expansion and contraction of concrete pavements. The design must include features that allow the pavement to remain level, aligned, and crack-free as it changes size.

To allow the pavement to slide easily over the ground as it changes size, workers can install polyethylene sheeting on a carefully leveled asphalt base before the concrete is placed on top.

photo: Getty Images

What features of particular metals do you think are important to civil engineers when they design expansion joints?

The decks of bridges expand and contract more than other pavements because of the open space underneath the bridge. Why is thermal expansion and contraction not as much of a problem for pavement that lies over the ground?

In addition, road designs typically include gaps between pavement slabs to allow the concrete to shrink or grow without cracking or buckling. In some places, such as bridges, the pavement is expected to move so much that the gap must be 75–100 millimeters wide. In this case, an interlocking "armored" metal joint protects the gap. For smaller expected movements, a doweled expansion joint is sufficient.

When the concrete slabs are poured, steel dowel rods implanted along their edges connect them to neighboring slabs across a gap. The rods keep the slabs aligned vertically and horizontally as they change size. They also help to smoothly distribute the load from one slab to the next as heavy trucks roll over them. On one or both ends of the rods inside the concrete, there is a cap with a space in the end so that dowels can move in and out of their holes slightly.

The designed size of the gap depends partly on the extremes of the temperatures the road will be experiencing and partly on the temperatures on the day the road is constructed. Normal shrinkage of the concrete as it cures is also accounted for. Finally, the gap is sealed at the top with polyurethane and silicon to keep water out. As the road ages and the gap seals deteriorate, maintenance crews can remove and replace them. This is much easier than replacing concrete that has been damaged by thermal expansion and contraction.

photo: U.S. Department of Transportation

The finished joint allows the concrete slabs to expand and contract while supporting traffic and excluding potentially troublesome water. What might happen if the expansion joint were to fail?

Predicting an Outcome

Why is it that some materials take longer to heat up than others?

photo: Getty Images

How long does it take to heat certain metals?

Many physical constants, including thermal properties, can be calculated from experimental data. This is how those properties were discovered and confirmed in the first place: scientists conducted experiments, took careful measurements, and analyzed the data, looking for patterns in the results.

Complete the "Specific Heat" activity on the STEM in Action page online. You will calculate the specific heat of aluminum, predict the outcome of an experiment based on your calculation, and then test your prediction.

1. Refer to your completed data table in which you measured the temperature change of 100 milliliters of water in a coffee cup calorimeter after a mass of metal (aluminum) at 100°C was immersed in it.

2. Use the data from your experiment, the specific heat of water: 4,186 J/(kg × °C), and the equation for thermal energy ($Q = mc\Delta T$) to calculate the specific heat of aluminum in J/(kg × °C). Show your calculations.

3. What if you doubled the mass of aluminum (at the same temperature) but kept the mass and temperature of the water the same? Use the value of specific heat for aluminum that you just found to calculate a predicted thermal equilibrium, T_{eq}, for the water and aluminum in the calorimeter. Indicate the equation you used for your calculation.

Write up a plan for the experimental testing of the value you predicted, based on your work in the activity "Specific Heat." Include a title, objective, hypothesis, materials list, procedure, and safety precautions. After your teacher has approved your experiment plan, conduct the experiment, record your data, and write the analysis and conclusion.

Your final report should have these parts:

- **Title:** Record the title of your experiment.
- **Objective:** Describe what you are trying to find out.
- **Hypothesis:** Record your proposed explanation.
- **Materials:** Provide a detailed list of the materials you will need.
- **Safety:** Provide a description of the safety precautions you will take.
- **Procedure:** Provide a detailed description of the procedure you will use. Be sure to describe the data you intend to collect.
- **Results:** Provide your data. Be sure to show your calculations.
- **Analysis and Conclusion:** Provide an analysis of your results and any conclusion you have reached. Include possible sources of error for your experiment in your conclusion.

Fundamental Forces

photo: Getty Images

LESSON OVERVIEW

Lesson Questions

- What are the ways in which matter and energy can interact?
- What is plasma, and how does it compare to other states of matter?
- How were the four fundamental forces discovered?
- What factors determine the magnitude of the gravitational force between two objects?
- How can atoms and molecules be identified by how they absorb and emit light?

Lesson Objectives

By the end of the lesson, you should be able to:

- Demonstrate the various ways in which matter and energy can interact.
- Describe plasmas and compare them with other states of matter.
- Describe the discovery of the four fundamental forces: gravity, electromagnetism, weak nuclear force, and strong nuclear force.
- Solve problems involving the gravitational force between two objects.
- Explain how atoms or molecules can be identified by how they absorb and emit light.

Key Vocabulary

Which terms do you already know?

- ☐ absorption
- ☐ absorption spectra
- ☐ atomic nucleus
- ☐ atomic spectroscopy
- ☐ electron
- ☐ emission
- ☐ gravitational force
- ☐ gravity
- ☐ light
- ☐ mass-energy equivalence
- ☐ matter
- ☐ neutrino
- ☐ neutron
- ☐ Newton's laws
- ☐ nucleus (atom)
- ☐ photoelectric effect
- ☐ quantum
- ☐ quantum mechanics
- ☐ quark
- ☐ solar system
- ☐ star
- ☐ state of matter
- ☐ strong nuclear force
- ☐ subatomic
- ☐ thermal energy

ENGAGE

What forces are at work on Earth and in the universe? How are these forces alike and different? Visit the Engage page online to learn more.

Explain Question

How do electromagnetism and gravitation differ from the strong and weak nuclear forces?

EXPLORE

What Are the Ways in Which Matter and Energy Can Interact?

Four Forces

Newton's laws describe the way objects react to forces. Many types of forces come up in physics problems. In physics, matter and energy interact through just four fundamental forces. The four fundamental forces are the gravitational force, the electromagnetic force, the strong nuclear force, and the weak nuclear force. All other forces and interactions can be explained in terms of these four fundamental forces.

For example, the friction between two objects, such as a helium-filled balloon resting against a ceiling, is actually caused by the electromagnetic forces between atoms in the balloon and the ceiling. Energy produced in a nuclear reactor is due to the weak nuclear force.

Particles, Forces, and Quanta

The four fundamental forces explain how different types of matter interact. Matter makes up the physical world and includes solids such as sandstone, liquids such as water, and gases such as carbon dioxide. Matter also includes plasma in stars and the clouds of dust and gas out in space.

In the 1800s, many scientists thought that the smallest particles were atoms. Near the turn of the 20th century, atoms were discovered to have "subatomic" parts that were named electrons, protons, and neutrons. Investigations of electromagnetic forces by Max Planck and others starting in 1900 showed that energy in light and other radiation is emitted in packets Planck called "quanta." In the mid-20th century, protons and neutrons were proven to consist of even smaller particles that were named quarks and neutrinos, identified in part by their quanta. These subatomic particles interact with each other through the fundamental forces.

Not all fundamental particles can interact through all of the fundamental forces. For example, only particles that have an electric charge will experience an electromagnetic force. However, all particles interact through the gravitational force. The study of the behavior and relationships of subatomic particles is called quantum mechanics, or quantum physics.

The standard model of particle physics is the theory that describes interactions of three of the four fundamental forces: electromagnetism, the strong nuclear force, and the weak nuclear force. Gravity is described by the theory of general relativity. Physicists today are working to find an expected, but currently unknown, connection between the standard model and the theory of general relativity.

What Is Plasma, and How Does It Compare to Other States of Matter?

Plasmas

Subatomic particles interact through the four fundamental forces. These forces allow particles to combine into atoms and molecules, and also to form larger objects such as desks and planets. The familiar matter surrounding us is made of combinations of countless subatomic particles.

Gases, liquids, and solids are familiar states of matter, but particles can also combine as a fourth state of matter called plasma. Plasma is a state of matter similar to a gas. Particles in plasma can move much more freely than in a solid or liquid. In a gas, particles form electrically neutral combinations of atoms or molecules. For example, gaseous carbon dioxide molecules do not have net electric charges.

The difference is that atoms in plasma do have a net electric charge. An atom can pick up an electric charge if it loses an electron. An ion is an electrically charged particle. Plasma is like a gas of charged particles: positively charged ions and negatively charged electrons.

Plasmas are not as common on Earth as the other matter states, but plasmas are by far the most common state of matter in the universe as a whole. In fact, the sun and other stars are giant balls of plasma.

Creating Plasmas

Plasmas can be created by heating a gas, which gives thermal energy to the gas. The added energy allows electrons to escape from the atoms. The atoms become ionized when the atoms gain enough energy and lose electrons. This process means that the temperature of plasmas is often much higher than the temperatures of gases, liquids, or solids.

Lightning is plasma created by friction between moving particles within a cloud. Friction between ice particles and super-cooled liquid causes the ice to become positively charged and the liquid to become negatively charged. This is similar to the charge separation created when a person rubs their socks across a carpet in the wintertime. Plasmas are used inside neon lights and CFLs. “Plasma” TV screens and computer monitors also contain plasma.

How Were the Four Fundamental Forces Discovered?

Gravitational Force

Gravity was the first fundamental force to be well understood. Physicist Sir Isaac Newton realized that the force that causes an apple to fall to the ground is the same force that keeps the planets in orbit around the sun. In fact, this force of gravity explains the motions of planets, stars, galaxies, and the universe as a whole. Newton's law of universal gravitation described this force. Gravity is a long-range force. Even objects that are very far apart still interact through gravity. Gravity is an attractive force. Masses, even separated by huge distances, feel a force pulling them toward each other. Earth feels a pull toward the sun, and it also feels a pull toward stars on the other side of the Milky Way galaxy. It even feels a pull toward stars across the universe.

Newton explained how objects can be attracted by gravity, but not fall into each other. As a free body, the moon would move in a straight line unless acted upon by a force. Earth's gravity pulls the moon toward Earth, and this changes the direction the moon moves. It doesn't change how fast the moon is moving. The moon is thus always falling toward Earth, but missing it and staying in its orbit.

Ancient thinkers believed the orbits of the moon and the planets were circular. The astronomer Johannes Kepler showed that they are slightly elliptical. Kepler introduced the first two of his laws of planetary motion in 1609, about 80 years before Newton presented his theory of universal gravitation in 1687. Newton showed how he could derive Kepler's laws by using this theory. This helped scientists of the time understand and accept it.

Physicists now describe gravity with the theory called general relativity. Albert Einstein developed general relativity in the early part of the 20th century. Newton's theory is still a useful approximation for many situations, such as describing the orbit of Earth around the sun.

Electromagnetic Force

Researchers once thought electrical discharges and magnetic fields were separate. The theory of electromagnetism linked the two. Electric currents flow and magnetic fields arise because of this common cause.

The electromagnetic force is between particles that have an electric charge. Charged particles also interact with electromagnetic fields. The electromagnetic force explains electricity, magnetism, light, and other electromagnetic waves.

In 1795, Charles-Augustin de Coulomb studied the force between two charged objects. He noticed that the closer the charges are, the stronger the force. His measurements showed that the force is inversely proportional to the square of the distance between charges. Coulomb's law was an early statement about the electromagnetic force.

In 1831, Michael Faraday showed an electric current can create a magnetic field. Also, a moving magnetic field can induce an electric current. This led to the invention of motors and generators. Both devices use the relationships between electricity and magnetism to transform energy of one type into another.

In 1864, Scottish physicist James Clerk Maxwell realized the connection between electricity, magnetism, and light. He drew upon the work of Coulomb, Faraday, and others to help him understand it. In 1873, he published a set of four equations that came to bear his name. These give a way to use math to think about and predict these forces.

Today, physicists think of electromagnetism as one of the forces explained by the standard model. This more general theory includes the strong and weak nuclear forces as well.

Nuclear Forces

The nucleus of an atom contains a mix of positively charged protons and uncharged neutrons. The protons in a nucleus repel each other due to the electromagnetic force and their positive charge. The neutrons do not have a charge, so they do not experience an electromagnetic force. The attractive force due to gravity between protons and neutrons is very weak because of their small masses. This fact means gravity cannot hold the nucleus together. What does hold the nucleus together?

The answer is a third fundamental force called the strong nuclear force. The strong nuclear force involves gluons that act to hold protons and neutrons together in an atomic nucleus. Gluons are thought to be massless particles that can bind quarks together.

The fourth fundamental force is called the weak nuclear force. The weak nuclear force is an interaction between subatomic particles that is responsible for radioactivity. The weak force allows protons to transform into neutrons and neutrons to transform into protons. These changes allow an atom's nucleus to change its combination of protons and neutrons. The nucleus emits other types of subatomic particles when the protons and neutrons transform through the weak force. The generation of these emitted particles is called radioactive decay.

Standard Model

The strong, weak, and electromagnetic forces are described by the theory called the standard model. The standard model was developed during the 20th century after researchers had discovered quantum mechanics. The standard model and quantum mechanics describe the interactions of subatomic particles through three of the fundamental forces. The fourth force, gravity, does not yet fit into the standard model. Researchers are still searching for a theory that will unify the four fundamental forces.

What Factors Determine the Magnitude of the Gravitational Force between Two Objects?

Gravitational Forces

Researchers currently describe gravity using the theory of general relativity developed by Einstein. Newton's law of universal gravitation is still a useful rule though, particularly in typical day-to-day situations, where Newton's theory is often a very good approximation.

The law of universal gravitation is useful for describing the gravitational force attracting two large, slow-moving objects such as Earth and the sun. Newton's law gives the force F_g attracting two masses m_1 and m_2 separated by a distance r. The gravitational force between them has a magnitude of:

$$F_g = \frac{Gm_1m_2}{r^2}$$

The gravitational constant G is about 6.67×10^{-11} N × m²/kg².

Each mass exerts a force on every other mass, meaning that the forces occur as pairs. The force on object 1 points toward object 2, and vice versa.

Gravitational Forces: Sample Problem

Earth has a mass of 6.0×10^{24} kilograms (kg), and the sun has a mass of 2.0×10^{30} kg. Earth orbits the sun at an average distance of 1.5×10^{8} kilometers. Calculate the force of gravity on Earth due to the sun.

Solution

The force of gravity F_g attracting two masses m_1 and m_2 separated by a distance r has a magnitude of:

$$F_g = \frac{Gm_1m_2}{r^2}$$

G has a value of 6.67×10^{-11} N × m²/kg². If m_1 is 6.0×10^{24} kg, m_2 is 2.0×10^{30} kg, and r is 1.5×10^{8} km (or equivalently 1.5×10^{11} m) then F_g is:

$$\begin{aligned} F_g &= \frac{Gm_1m_2}{r^2} \\ &= \frac{\left(6.67\times10^{-11}\,\frac{\text{N}\times\text{m}^2}{\text{kg}^2}\right)\left(6.0\times10^{24}\,\text{kg}\right)\left(2.0\times10^{30}\,\text{kg}\right)}{\left(1.5\times10^{11}\,\text{m}\right)^2} \\ &= 3.6\times10^{22}\,\text{N} \end{aligned}$$

The force of gravity on Earth due to the sun is 3.6×10^{22} Newtons directed from Earth toward the sun.

How Can Atoms and Molecules Be Identified by How They Absorb and Emit Light?

Identifying Atoms

Subatomic particles combine into atoms, and atoms combine into molecules. Particles in atoms and molecules still interact through the fundamental forces. The interactions are very complicated and difficult to calculate. Fortunately, the complicated effects have at least one simple result. Each atom or molecule can only absorb and emit particular wavelengths of light.

The specific wavelengths depend on the type of atom or molecule. The particular pattern of wavelengths that an atom or molecule absorbs or emits can be used to identify the atom or molecule. This pattern is called the atom's spectrum. For example, carbon atoms absorb different wavelengths than oxygen atoms, and carbon dioxide molecules emit different wavelengths than methane molecules.

The standard model predicts that electrons in an atom can only have certain values of energy called energy levels. The particular values are different for different atoms. An electron in an atom can change energy level, but it can only change to one of the allowed energy levels. An electron raises its energy level by absorbing the energy of a photon. It lowers its energy level by emitting a photon that carries away energy.

Atoms only absorb or emit photons of light with energies that match the energy gap between their electron energy levels. The energy of a photon is directly linked to the wavelength of the photon. The result is that atoms can only absorb or emit photons with particular wavelengths.

Spectroscopy

Atomic spectroscopy is the technique that allows atoms to be identified by the light they absorb or emit. The pattern of wavelengths is called the atomic spectrum, so spectroscopy is the method of measuring the spectrum of an atom.

Researchers have developed many techniques for measuring atomic spectra. Researchers can, for instance, put a gas in a closed container and then shine white light on the gas. White light contains light of all visible wavelengths. The gas will absorb only the photons with the particular wavelengths associated with its spectrum. Researchers measure the light that passes through the gas and look for the spectrum of wavelengths that were absorbed and did not pass through. This technique is called absorption spectroscopy. The pattern produced is the absorption spectrum.

Another technique is called emission spectroscopy. Researchers excite the atoms in a gas, usually by heating them. The excited atoms release the energy they have absorbed by emitting photons. The emitted photons can only have energies equal to the differences in energy levels. By measuring the wavelengths of the emitted light, researchers measure the spectrum of the gas. The pattern is the emission spectrum.

Consider the Explain Question:

How do electromagnetism and gravitation differ from the strong and weak nuclear forces?

Go online to complete the scientific explanation.

Check for Understanding:

Can you give an example of a phenomenon that illustrates each fundamental force?

STEM in Action

STEM and Fundamental Forces

Physicists use their knowledge of the fundamental forces to explain physical phenomena and solve applied problems. For example, plasmas can be found in stars, but they also exist on Earth in common devices, such as fluorescent lamps and some television monitors. Research into the behavior of plasmas involves advanced knowledge about the electromagnetic force, and in subjects such as nuclear fusion physics, physicists study nuclear forces as well, so as to provide insights into the process that powers stars and may provide clean energy in the future.

Another pursuit in physics is to understand why the four fundamental forces are different, and under which conditions they become the same. This is part of an effort to unify the four forces so that all physics may be explained with just one theory. While advances in this field have been made (such as the unification of electromagnetism with the weak nuclear force), a complete unified theory will take a long time, advanced engineering, and a lot of hard work to realize. Researchers will have to be very creative to design experiments that will produce data linking the four fundamental forces.

photo: Getty Images

A fluorescent lamp contains a noble gas and a tiny amount of mercury vapor. When you turn the lamp on, plasma is produced. What is plasma?

Gravity

photo: Getty Images

LESSON OVERVIEW

Lesson Questions

- How is the magnitude of the gravitational force between two objects related to their masses and the distance between them?
- What is Newton's law of universal gravitation?
- How does Newton's law of gravitation relate the distance between two objects, their masses, and the force of attraction between them?
- What is the difference between center of mass and center of gravity?
- How does the law of universal gravitation apply to objects near Earth's surface?
- How do we solve problems that involve universal gravitation?
- How do we use Newton's second law of motion and the law of universal gravitation to solve problems that involve the orbital motion of satellites?

Key Vocabulary

Which terms do you already know?

- ☐ distance
- ☐ force
- ☐ gravitational force
- ☐ gravitational potential energy
- ☐ gravity
- ☐ mass
- ☐ Newton's law of universal gravitation
- ☐ Newton's laws
- ☐ orbit

Lesson Objectives

By the end of the lesson, you should be able to:

- Show how the magnitude of gravitational force between two objects is proportional to their masses and inversely proportional to the square of the distance between them.
- Identify each term in the equation for Newton's law of universal gravitation and summarize what they mean.
- Cite evidence that relates Newton's law of gravitation to the distance between two objects, their masses, and the force of attraction between them.
- Differentiate between center of mass and center of gravity.
- Interpret how the law of universal gravitation applies to objects near Earth's surface.
- Construct solutions to problems that involve universal gravitation.
- Construct solutions to problems involving the orbital motion of satellites using the equations for Newton's second law of motion and the law of universal gravitation.

ENGAGE

How do you think the force of gravity affects the motion of our solar system through the Milky Way galaxy? Visit the Engage page online to learn more.

Explain Question

What evidence is there that gravity affects Earth and the sun?

EXPLORE

How Is the Magnitude of the Gravitational Force between Two Objects Related to Their Masses and the Distance between Them?

Gravitational Force

There is a force that pulls the moon toward the center of planet Earth. This force acts on the moon during every part of its orbit. The force is proportional to the masses of both the moon and Earth. This force is gravity.

The gravitational force is an interaction between any two objects in the universe that have mass. The magnitude of the force of gravity is directly proportional to the masses of the two objects. Therefore, the force increases as the mass of either object increases. The magnitude of the force of gravity between Earth and the moon is large, because the masses of the two objects are large.

photo: Getty Images

Even though the sun and Earth are far apart, they are very massive. Therefore, they exert a significant gravitational pull on each other. What is the result of this attractive force?

The force of gravity is smaller when two objects are far apart. Earth is fairly close to the moon, so the force of gravity between them is strong. The moon, Io, which orbits Jupiter, has a mass roughly similar to that of the moon. Io is far from Earth, however, so the force of gravity between Io and Earth is weaker.

Gravitational force is inversely proportional to the square of the distance between the two objects. As the distance increases, the force becomes weaker. As the distance decreases, the force becomes stronger.

What Is Newton's Law of Universal Gravitation?

Newton's Law of Universal Gravitation

The sun exerts a gravitational force on Earth. Earth exerts an equal but opposite force on the sun. Gravity and Earth's momentum keep Earth and the sun in orbit around each other. The gravitational force from Earth also holds you to the ground and causes baseballs, rocks, and apples to fall downward.

The English physicist Isaac Newton discovered a simple law that describes the force of gravity between any two objects.

photo: Library of Congress

Sir Isaac Newton discovered the law of universal gravitation. How did he represent gravitation mathematically?

Newton's law of universal gravitation states that the gravitational force between two objects has a magnitude of:

$$F_g = \frac{Gm_1m_2}{r^2}$$

where F_g is the magnitude of the force, m_1 and m_2 are the objects' masses, and r is the distance between the two objects. The constant G is the universal gravitational constant.

$$G = 6.67 \times 10^{-11}\,\mathrm{N} \times \mathrm{m}^2 / \mathrm{kg}^2$$

Gravitational Fields

The law of universal gravitation describes the gravitational field of every object. The gravitational field of an object, such as Earth, represents the force that another object, such as the moon or a rock, will feel due to Earth, divided by that object's mass:

$$g_{\text{Earth}} = \frac{F_g}{m_{\text{Moon}}}$$

$$g_{\text{Earth}} = \frac{1}{m_{\text{Moon}}}\frac{Gm_{\text{Earth}}m_{\text{Moon}}}{r^2}$$

$$g_{\text{Earth}} = \frac{Gm_{\text{Earth}}}{r^2}$$

If another object, such as the moon, is at a distance *r* from Earth, it feels a force equal to the moon's mass times Earth's gravitational field:

$$F_g = m_{\text{Moon}}g_{\text{Earth}}$$

$$F_g = \frac{Gm_{\text{Earth}}m_{\text{Moon}}}{r^2}$$

If another object, such as the moon, is at a distance *r* from Earth, it feels a force equal to the moon's mass times Earth's gravitational field at that distance.

Gravitational Fields: Sample Problem

A sphere that has a mass of 1,400 kilograms (kg) sits at one location. At another location 130 kilometers (km) due south, sits a sphere with a mass of 5,800 kg.

How strong is the force of gravity exerted by the second sphere on the first one? Compare this with the force exerted by the first sphere on the second.

Solution

The law of universal gravitation states that the magnitude of the force is:

$$F_g = \frac{Gm_1m_2}{r^2}$$

We know the mass of the first sphere, m_1, is 1,400 kg. The mass of the second sphere, m_2, is 5,800 kg. The distance, *r*, is 130 km, or 1.3×10^5 m. Newton's constant, *G*, is 6.67×10^{-11} N · m²/kg².

Insert these values and calculate:

$$F_g = \frac{Gm_1m_2}{r^2}$$
$$= \frac{(6.67\times10^{-11}\,\text{N}\times\text{m}^2/\text{kg}^2)(1{,}400\text{ kg})(5{,}800\text{ kg})}{(1.3\times10^5\text{ m})^2}$$
$$F_g = 3.2\times10^{-14}\,\text{N}$$

The first sphere exerts a force of magnitude 3.2×10^{-14} Newtons on the second sphere.

The second sphere exerts a force of magnitude 3.2×10^{-14} Newtons on the first sphere.

How Does Newton's Law of Gravitation Relate the Distance between Two Objects, Their Masses, and the Force of Attraction between Them?

Forces between Objects

It is important to remember that the gravitational force is an interaction between the two objects. The same amount of force is acting on both of them. Earth exerts a force on the moon, but the moon also exerts a force on Earth. The strengths of these forces are the same in both directions. We can describe these forces as "equal but opposite."

Newton's law of universal gravitation describes the force between any two objects: Mars and the sun, two distant galaxies, or even you and Earth. The law states that the gravitational force between two objects has a magnitude of:

$$F_g = \frac{Gm_1m_2}{r^2}$$

We can say that:

$$F_{g1} = F_{g2}$$

The magnitude of the force is the same between the two objects.

The force of gravity is always attractive. This means that the direction of the force on each object points toward the other object. The gravitational force on Earth due to the sun points from Earth's center toward the sun's center. The gravitational force on the sun due to Earth has the same magnitude but the opposite direction.

The force of gravity is considered to come from the center of each object. When drawing force diagrams related to gravity, the force vector points from the center of one object to the center of the other.

The gravitational field of an object is represented by a vector at every point in space. This vector points in the same direction as the gravitational force experienced by a second object at that point. The field of Earth points toward Earth's center because an astronaut at any point in space feels a force due to Earth that points toward Earth's center.

Forces between Objects: Sample Problem

A 1,600-kilogram object feels a gravitational force of 24 Newtons due to a 2,300-kilogram object. What is the distance between the two objects?

Solution

The law of universal gravitation states:

$$F_g = \frac{Gm_1m_2}{r^2}$$

This equation can be solved for the distance *r*:

$$r^2 = \frac{Gm_1m_2}{F_g}$$

$$r = \sqrt{\frac{Gm_1m_2}{F_g}}$$

The mass m_1 has the value 1,600 kg, and m_2 has the value 2,300 kg. The force has a magnitude F_g of 24 N. Newton's constant G has the value 6.67×10^{-11} N × m²/kg². Inserting these values gives the distance:

$$r = \sqrt{\frac{Gm_1m_2}{F_g}}$$

$$= \sqrt{\frac{(6.67 \times 10^{-11}\,\mathrm{N} \times \mathrm{m}^2/\mathrm{kg}^2)(1{,}600\,\mathrm{kg})(2{,}300\,\mathrm{kg})}{24\,\mathrm{N}}}$$

$$r = 3.2 \times 10^{-2}\,\mathrm{m}$$

The two objects are separated by a distance of 3.2×10^{-2} m.

What Is the Difference between Center of Mass and Center of Gravity?

Center of Mass

The sun, eight planets, and other orbiting bodies make up the solar system. Any two bodies in the system attract each other according to the law of universal gravitation. The sun and all orbiting bodies have mass.

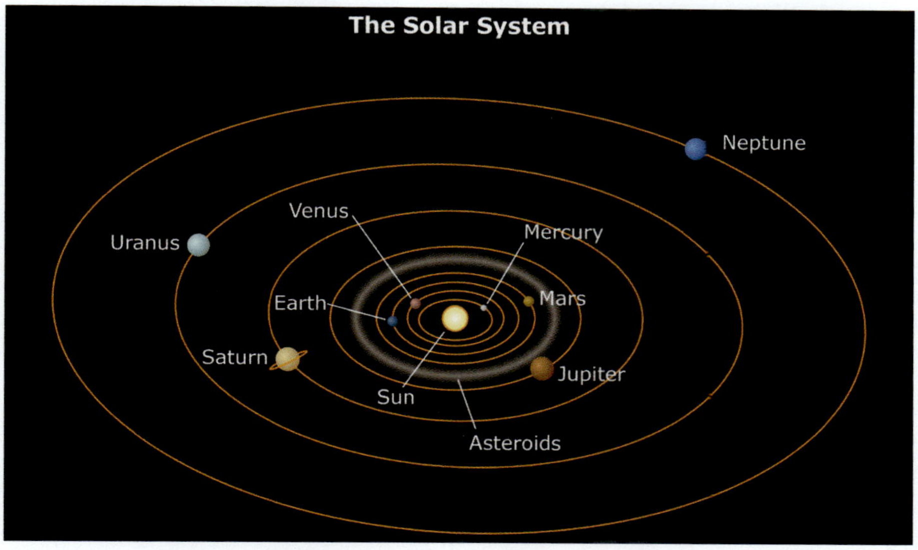

The center of mass in the solar system is very close to the center of the sun. How can we mathematically compute this?

The center of mass of a system such as the solar system is the average location of the mass in the system. The location of the center of mass is described by a vector $\vec{r}_c$ that is defined by the equation:

$$\vec{r}_c = \frac{\sum_i m_i \vec{r}_i}{\sum_i m_i}$$

where the sum is the number of individual objects of the system, the m_i are the masses, and the $\vec{r}_i$ are the location vectors of the objects.

You can calculate the center of mass of the solar system by taking the sum of the values for the sun and each planet. In fact, the sun has much more mass than the rest of the solar system combined. This means that:

$$\sum_i m_i \approx m_{\text{Sun}}$$

and

$$\sum_i m_i \vec{r}_i \approx m_{\text{Sun}} \vec{r}_{\text{Sun}}$$

Putting these facts together gives:

$$\vec{r}_c = \frac{\sum_i m_i \vec{r}_i}{\sum_i m_i}$$

$$\approx \frac{m_{\text{Sun}} \vec{r}_{\text{Sun}}}{m_{\text{Sun}}}$$

$$\vec{r}_c \approx \vec{r}_{\text{Sun}}$$

The center of mass of the solar system is very close to the center of the sun.

Center of Gravity

The center of gravity of a system, such as the solar system, is the average location of the gravitational force on the system. Physicists use different ways to define the center of gravity. The center of gravity does not always equal the center of mass.

One definition of center of gravity uses the average of the weights of the objects. The weight of an object is simply the magnitude of gravitational force on the object. A book's weight on Earth is equal to Earth's gravitational pull on the book. The weight of the book on Mars would equal Mars' gravitational pull on the book.

Imagine a system of objects such as the solar system. Object 1 (the sun) has a location vector $\vec{r}_1$. A gravitational force of magnitude w_1 acts on it. This is also true for object 2, object 3, and so on. The center of gravity is:

$$\vec{r}_g = \frac{\sum_i w_i \vec{r}_i}{\sum_i w_i}$$

Just like the center of mass, the center of gravity of the solar system is very close to the center of the sun.

Sample Problem

What is the acceleration due to gravity felt by the International Space Station? The ISS has a mass of 4.5×10^5 kg and is in an orbit of about 380 km; Earth has a mass of 6.0×10^{24} kg and a radius of 6,400 km.

Solution

The distance between the ISS and the center of Earth is the sum of Earth's radius and the altitude of the ISS:

$$380 + 6{,}400 = 6{,}780 \text{ km} = 6.78 \times 10^6 \text{ m}$$

Earth has a mass of 6.0×10^{24} kg. Newton's constant *G* has the value 6.67×10^{-11} N × m²/kg².

The acceleration's magnitude is:

$$\frac{(6.0\times10^{24})\times(6\times10^{-11})}{(6.78\times10^{6})\times(6.78\times10^{6})} = 8.7\ \mathrm{m/s^{-2}}$$

The ISS has an acceleration of 8.7 m/s² downward toward the center of Earth.

Comparing Center of Mass and Center of Gravity

The center of mass of a system and the center of gravity are not always equal. Mass and weight are different things.

The center of mass and the center of gravity are equal when the gravitational field is uniform. The field is uniform if it points in the same direction and has the same strength at every point in the system. The weight of an object is its mass multiplied by the strength of the gravitational field. The field strength divides out when the field is uniform:

$$\begin{aligned}\vec{r}_g &= \frac{\sum_i w_i \vec{r}_i}{\sum_i w_i} \\ &= \frac{\sum_i m_i g \vec{r}_i}{\sum_i m_i g} \\ &= \frac{\sum_i m_i \vec{r}_i}{\sum_i m_i} \\ \vec{r}_g &= \vec{r}_c\end{aligned}$$

In the case of a uniform field, the center of mass equals the center of gravity.

The center of mass and the center of gravity of a body may actually lie outside the body. These locations depend on the body's shape and the way that the mass is distributed throughout the body. The center of mass and the center of gravity of the solar system are inside the sun. In many systems, however, these points are not inside any of the bodies present.

Comparing Center of Mass and Center of Gravity: Sample Problem

A 6.0-kilogram object sits 2.0 meters west of a 3.0-kilogram object. A 4.0-kilogram object sits 1.5 meters east of the 3.0-kilogram object. Find the center of mass of this system.

Solution

The position of the center of mass is a convenient choice for the origin of the frame of reference. The line connecting the three objects is a convenient choice for the *x*-axis. The direction east is a convenient choice for the positive-*x* direction.

The center of mass $\vec{r}_c$ is given by:

$$\vec{r}_c = \frac{\sum_i m_i \vec{r}_i}{\sum_i m_i}$$

The position vectors of the three objects only have *x*-components, so $\vec{r}_c$ will lie along the *x*-axis.

The sum of the masses is:

$$\sum_i m_i = 6.0 \text{ kg} + 3.0 \text{ kg} + 4.0 \text{ kg} = 13 \text{ kg}$$

The center of mass location is given by:

$$\begin{aligned} r_c &= \frac{\sum_i m_i r_i}{\sum_i m_i} \\ &= \frac{1}{(13 \text{ kg})}\big((6.0 \text{ kg})(-2.0 \text{ m}) + (3.0 \text{ kg})(0.0 \text{ m}) + (4.0 \text{ kg})(1.5 \text{ m})\big) \\ &= \frac{-6.0 \text{ m} \times \text{kg}}{13 \text{ kg}} \\ r_c &= -0.46 \text{ m} \end{aligned}$$

The negative sign in the answer indicates that the center of mass sits 0.46 m west of the origin.

How Does the Law of Universal Gravitation Apply to Objects near Earth's Surface?

Objects near Earth's Surface

The law of universal gravitation states that spherical objects act as though their masses are concentrated at the center. Because Earth is spherical, objects on its surface are pulled by gravity toward its center.

Remember that the strength of gravity is lower when objects are far apart. The change in distance must be relatively large to see this effect. If you ride an elevator, your distance from Earth's center does not change much. Even if you climb a high mountain or ride in an airplane, the change in gravity is not significant enough to affect everyday life.

For objects near Earth's surface:

$$F_g = mg$$

Here, *m* is the object's mass, and *g* is the acceleration due to gravity. For objects near Earth's surface, *g* is a constant with approximate value

$$g = 9.8 \text{ m/s}^2$$

The value of *g* decreases with increasing altitude, so there is in fact a small change in gravitational force between Denver, Colorado, and New Orleans, Louisiana, for example.

Newton's second law of motion states that the acceleration of an object is proportional to the force on it:

$$F = ma$$

Objects near Earth's surface that experience only the force of gravity have an acceleration equal to *g*:

$$F = F_g$$
$$ma = mg$$
$$a = g$$

Because of this connection, *g* is the acceleration due to gravity near Earth's surface.

Objects near Earth's Surface: Sample Problem

What is the force of gravity due to Earth on a 65-kilogram student standing on Earth's surface? Compare this with the force of gravity on the student due to an 18,000-kg house that is 2.5 meters away.

Solution

The force of gravity on the student due to Earth has a magnitude of:

$$F_{\text{g}} = m_{\text{student}} g$$

The student has a mass of 65 kg, and *g* has the value 9.8 m/s^2. Inserting these values into the equation gives the force magnitude:

$$F_g = m_{\text{student}} g$$
$$= (65 \text{ kg})(9.8 \text{ m/s}^2)$$
$$F_g = 640 \text{ N}$$

The force of gravity on the student has a magnitude of 640 Newtons and is directed downward toward Earth's center.

The law of universal gravitation says that the force of gravity on the student due to the house is:

$$F_g = \frac{Gm_{\text{student}} m_{\text{house}}}{r^2}$$

The student has a mass of 65 kg, and the house has a mass of 18,000 kg. The distance *r* between them is 2.5 m. Newton's constant *G* has the value 6.67×10^{-11} N × m^2/kg^2. Inserting these values gives the magnitude of force:

$$F_g = \frac{Gm_{\text{student}} m_{\text{house}}}{r^2}$$
$$= \frac{(6.67\times10^{-11} \text{ N}\times\text{m}^2/\text{kg}^2)(65 \text{ kg})(18{,}000 \text{ kg})}{(2.5 \text{ m})^2}$$
$$F_g = 1.2\times10^{-5} \text{ N}$$

The force of gravity on the student due to the house has a magnitude of 1.2×10^{-5} N. This force is more than 10^{-8} times smaller than the force due to Earth.

How Do We Solve Problems That Involve Universal Gravitation?

Using Newton's Law

Newton's law of universal gravitation describes the effect of gravity on all objects in the universe. This law, when used together with Newton's laws of motion, can describe the motions of all objects: from pebbles to people to galaxies.

Problems involving gravitation can be solved using Newton's law of universal gravitation.

$$F_g = \frac{Gm_1 m_2}{r^2}$$

The law can be used to find the force between two objects. For a given force, Newton's law of universal gravitation can be combined with Newton's laws of motion to describe the motion of an object.

Newton's second law states that the force *F* on an object is proportional to the object's acceleration *a*:

$$F = ma$$

where *m* is the object's mass.

We can rearrange Newton's second law to solve for *a*. Next, substitute Newton's universal law for *F*.

$$\begin{aligned} a &= \frac{1}{m}F \\ &= \frac{1}{m}F_g \\ &= \frac{1}{m}\frac{Gmm_2}{r^2} \\ a &= \frac{Gm_2}{r^2} \end{aligned}$$

The direction of the acceleration is toward the center of the second object.

Using Newton's Law: Sample Problem

A spacecraft is orbiting Mars at an altitude of 110 kilometers above the surface. The spacecraft releases a probe that falls from rest toward the surface. Mars has a mass of 6.4×10^{23} kilograms and a radius of 3,400 kilometers. What is the acceleration of the spacecraft when the probe is released?

photo: NASA

In 2013, NASA launched MAVEN, a satellite that orbits Mars, to study the planet's atmosphere. How would releasing a probe affect MAVEN's acceleration?

Solution

First, find the distance between the satellite and the center of Mars. This distance equals the sum of the radius of Mars and the altitude of the satellite:

$$\begin{aligned} r &= r_{\text{Mars}} + r_{\text{altitude}} \\ &= 3.4\times10^{6}\,\text{m} + 1.1\times10^{5}\,\text{m} \\ r &= 3.5\times10^{6}\,\text{m} \end{aligned}$$

Mars has a mass of 6.4×10^{23} kg. Newton's constant *G* has the value 6.67×10^{-11} N × m^2/kg^2.

The acceleration's magnitude is:

$$\begin{aligned} a &= \frac{Gm_{\text{Mars}}}{r^2} \\ &= \frac{(6.67\times10^{-11}\,\text{N}\times\text{m}^2/\text{kg}^2)(6.4\times10^{23}\,\text{kg})}{(3.5\times10^{6}\,\text{m})^2} \\ a &= 3.5\,m/s^2 \end{aligned}$$

The spacecraft has an acceleration of 3.5 m/s^2 downward toward the center of Mars.

How Do We Use Newton's Second Law of Motion and the Law of Universal Gravitation to Solve Problems That Involve the Orbital Motion of Satellites?

Satellites

The gravitational force between a satellite and Earth keeps the satellite in orbit around Earth. If a satellite were moving around Earth and the force of gravity suddenly stopped, Newton's first law of motion states that the satellite would continue in a straight line toward the outer reaches of the solar system. Satellites remain in orbit because the force of gravity pulls them toward Earth.

The momentum of satellites keeps them from crashing to Earth's surface. Gravity exerts just enough force to change the direction of a satellite's velocity. This keeps the satellite on its circular path.

If the gravitational force around any planet, star, or other object is known, Newton's second law can be used to describe the satellite's orbit.

Satellites: Sample Problem

What is the acceleration due to gravity felt by the International Space Station? The ISS has a mass of 4.5×10^5 kilograms (kg) and is in an orbit of about 380 kilometers (km). Earth has a mass of 6.0×10^{24} kg and a radius of 6,400 km.

Solution

The distance between the ISS and the center of Earth is the sum of Earth's radius and the altitude of the ISS:

$$\begin{aligned} r &= r_{\text{Earth}} + r_{\text{altitude}} \\ &= 6.4 \times 10^6 \text{ m} + 3.8 \times 10^5 \text{ m} \\ r &= 6.8 \times 10^6 \text{ m} \end{aligned}$$

Earth has a mass of 6.0×10^{24} kg. The universal gravitational constant has the value 6.67×10^{-11} N×m²/kg².

The acceleration's magnitude is:

$$\begin{aligned} a &= \frac{Gm_{\text{Earth}}}{r^2} \\ &= \frac{(6.67 \times 10^{-11} \text{N} \times \text{m}^2/\text{kg}^2)(6.0 \times 10^{24} \text{kg})}{(6.8 \times 10^6 \text{m})^2} \\ a &= 8.7\ m/s^2 \end{aligned}$$

The ISS has an acceleration of 8.7 m/s² downward toward the center of Earth.

Consider the Explain Question:

What evidence is there that gravity affects Earth and the sun?

Go online to complete the scientific explanation.

Check for Understanding:

Can you describe how the law of universal gravitation can be seen at work during a baseball game?

STEM in Action

STEM and Gravity

Gravitational waves cannot be detected with ordinary tools, such as light or radio telescopes. Instead, scientists and engineers have developed a new technology, called the Laser Interferometer Gravitational-Wave Observatory (LIGO), which is sensitive enough to detect these waves. Currently, there are two LIGO facilities, one in Louisiana and one in Washington. Each detector forms the shape of an L with arms that are approximately 4 kilometers long. LIGO detectors have been in operation since 2002. Their goal is to find direct evidence for the existence of gravitational waves.

On February 11, 2016, members of the LIGO team announced that they had detected a gravitational wave emitted from the collision of two distant black holes. This monumental scientific discovery was only possible because of the collaborative efforts of thousands of scientists. This is an exciting time to be an astronomer working with the LIGO scientific community!

Newton and Einstein gave two different explanations for the same phenomenon—gravity. Answer these questions about the two theories:

1. What are the similarities between Newton's and Einstein's theories?
2. What are the differences between Newton's and Einstein's theories?
3. What technologies are needed to verify each theory?

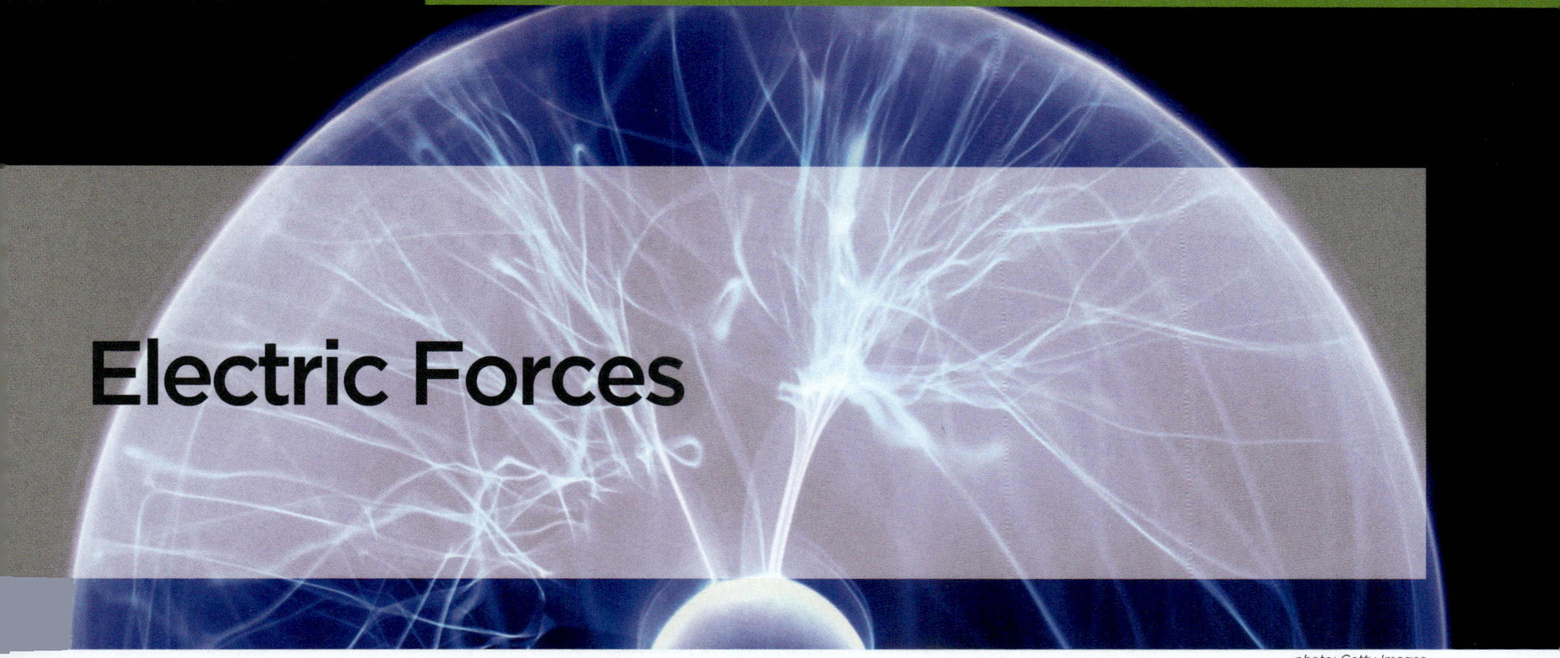

Electric Forces

photo: Getty Images

LESSON OVERVIEW

Lesson Questions

- How does the electric force relate to the common contact forces people experience each day?
- What determines the strength of the electric force?

Lesson Objectives

By the end of the lesson, you should be able to:

- Describe how most observable forces ultimately result from electric forces that act between atoms and molecules.
- Show how the strength of the electric force between two charged objects is proportional to the charges and inversely proportional to the square of the distance between them.

Key Vocabulary

Which terms do you already know?

- ☐ atom
- ☐ Coulomb's law
- ☐ electron
- ☐ force
- ☐ molecule
- ☐ negative charge
- ☐ nucleus (atom)
- ☐ positive charge
- ☐ proton

ENGAGE

What do many of the technologies you use every day have in common? How do electric forces work? Visit the Engage page online to learn more.

Explain Question

What is Coulomb's law?

EXPLORE

How Does the Electric Force Relate to the Common Contact Forces People Experience Each Day?

The Electric Force

The ancient Greeks noticed that when a piece of amber is rubbed on the fur of an animal, it starts to exhibit strange properties. If brought near small objects such as bits of wool, it attracts the wool. It can also lift the wool off the ground. The Greeks were intrigued by this behavior and offered theories to explain it. They thought a mysterious fluid flowed from the fur into the amber, causing the unusual properties. Their theories were not completely accurate, but they did offer some interesting insights. Today, scientists know that the electric force is behind this phenomenon.

photo: Paul Fuqua

An electrically charged piece of amber exerts an electric force on small bits of wool. How can amber become electrically charged?

Atoms are made of subatomic particles with different electric charges. The electron has a negative charge while the proton, which is inside the nucleus, has a positive charge. Therefore, people do not usually notice the electrical forces between objects. The fluid that the Greeks described is actually a flow of electrons from one object to another. Normally, molecules of a substance are electrically neutral because they contain the same number of electrons and protons.

However, this balance of positive and negative charges is sometimes thrown off. Electrons may flow away from certain atoms (leaving an excess of positive charge) and flow toward other atoms (creating an excess of negative charge). When this happens on a large scale, two objects can become oppositely charged. That produces an electric force that is attractive. It pulls the two charged objects or particles together. By contrast, two like charges such as two negative charges or two positive charges produce a repulsive force. Both are examples of the electric force.

Another interesting demonstration of the electric force arises when an acrylic rod is given a negative charge and brought near a stream of flowing water. The stream of water is attracted to the negatively charged rod. The electric force deflects it. Even a fast-moving stream of water can be deflected significantly by a moderate amount of electric charge which shows that the electric force is quite strong.

Common Contact Forces

In the study of physics, common forces such as a person's push are often called applied forces. Applied forces and other common forces such as the normal force are often known as contact forces because two objects appear to be in contact with each other. However, when one examines what is happening on the atomic level between two objects that appear to be in contact, it might be surprising to know that there is never any contact at all. Instead, atoms of one object are simply coming very close to atoms of the other object. Once this happens, the strength of the electric force becomes very strong between the two objects. Therefore, they are repelled away from each other. The electric force can act at a distance and therefore does not require that the atoms ever make contact at all for this repulsive force to exist.

For example, when a person pushes on a wall, there is no actual contact between the person and the wall. The electrons in the person's hand and the electrons in the wall repel one another. The sensation of this repulsive force makes people feel like they are in contact with another object. However, the repulsion from the electric force is so strong that contact cannot happen.

What Determines the Strength of the Electric Force?

Coulomb's Law

It is clear that two charged objects exert forces on each other. Like charges repel one another, and unlike charges attract. How can the strength of these two charges be determined? Coulomb discovered this relationship in the late 1780s. He discovered that the strength of the electric force, F_e, between two charged objects depends on two factors: the amount of charge on the two objects and the distance between them.

The equation for Coulomb's law is:

$$F_e = \frac{kq_1q_2}{r^2}$$

In this equation, q_1 and q_2 represent the amount of charge on objects 1 and 2, respectively. Charge is measured in units of coulombs (C). The lowercase r represents the distance between the two charges measured in meters. Finally, $k = 8.99 \times 10^9 \text{ N} \times \text{m}^2/\text{C}^2$ is a constant of proportionality.

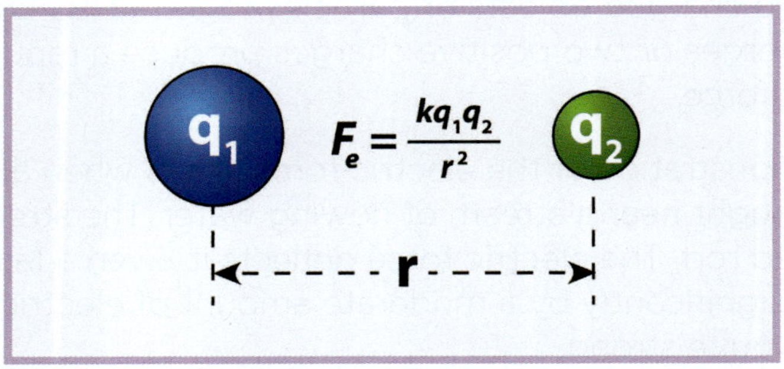

Coulomb's law states that the electric force between two objects increases as the charge on the two objects increases and it decreases as the distance between the objects increases. When does Coulomb's law apply?

Examining Coulomb's law, it should be clear that the greater the magnitudes of the two charges, the stronger the electric force. In other words, the magnitude of the electric force is directly proportional to the product of the magnitudes of the charges. However, the reverse is true of the distance between the two charges. As the distance increases, the magnitude of the electric force decreases. Therefore, the electric force is inversely proportional to the square of the distance between the two objects. The equation for Coulomb's law follows what is called an inverse square law. Because force is inversely proportional to the square of the distance rather than just the distance, the strength of the force decreases rapidly as the distance increases.

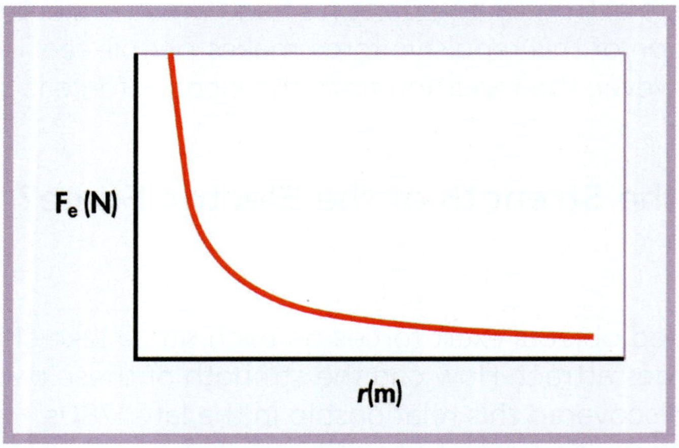

Because the electric force is inversely related to the square of distance, it decreases very rapidly as the distance between the two charges increases. How could this relationship be represented mathematically?

Coulomb's Law: Sample Problem

Two large metal spheres are given charges. The first sphere has a charge of $q_1 = 2.0 \times 10^{-4}$ C. However, q_2 is unknown. A scientist measures that the distance between the centers of two charges as 0.20 meters and the force between the two spheres as 450 Newtons. What is the charge on the second object?

Solution

Apply Coulomb's law:

$$F_e = \frac{kq_1q_2}{r^2}$$

Every quantity in the law is known except for one of the charges. Thus, rearranging the equation and solving for the second charge yields:

$$q_2 = \frac{F_e r^2}{kq_1}$$

Substituting the known quantities leads to:

$$\begin{aligned} q_2 &= \frac{F_e r^2}{kq_1} \\ &= \frac{(450\ \mathrm{N})(0.20\ \mathrm{m})^2}{(8.99 \times 10^9\ \mathrm{N \cdot m^2 / C^2})(2.0 \times 10^{-4}\ \mathrm{C})} \\ &= 1.0 \times 10^{-5}\ \mathrm{C} \end{aligned}$$

Consider the Explain Question:

What is Coulomb's Law?

Go online to complete the scientific explanation.

Check for Understanding:

Without using an equation, how would you describe the magnitude of the electric force between two charges?

STEM in Action

STEM and Electric Forces

When a net charge builds up on an object, it is known as an electrostatic charge. Static in this case means stationary, so the word electrostatic can be thought of as electric charge at rest. Like a comb on a dry winter day, a lot of objects can build up electric charge under the right circumstances. Too much built-up charge can lead to electric forces that could be dangerous to electronics and other technological equipment.

Engineers need to take this danger into account when designing systems. For instance, airplane designers must account for the varying amounts of charge that different materials on the outside surface of an airplane pick up from the air. Then, they must figure out how to conduct any excess charge away from areas where they could be dangerous.

A computer repair technician is someone who is trained to fix and maintain computers and computer servers. The education required is normally a technical school, and a computer repair technician needs to know the ins and outs of all aspects related to computers.

Computer repair technicians also need to worry about electric charge. Too much static electricity can damage hard drives and other sensitive electronic equipment. Therefore, technicians often wear armbands that ground their hands to an object that can accept electric charge. In this way, they ensure that their hands never build up too much charge as they work on sensitive electronic equipment.

photo: Getty Images

How might static electricity damage a computer's motherboard?

In chemistry, lattice energy is the energy required to break a bond between two ions. In other words, it describes the strength of the bond.

Analyze the chart of lattice energies between some alkali metals (Li^+, Na^+, K^+ and halide ions F^- Cl^-, Br^-, I^-). Then, answer the questions that follow.

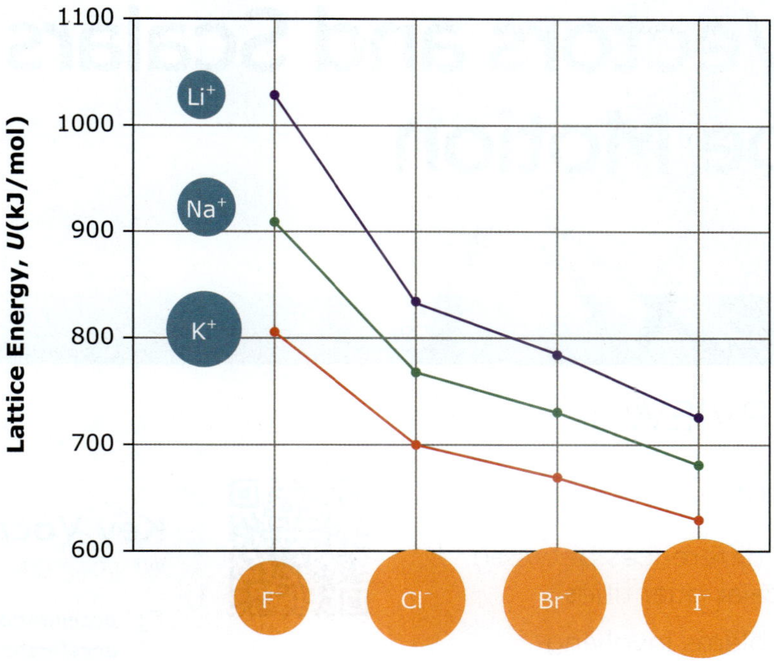

1. Because the charge on both the alkali metal and halide are +1 and –1 respectively, what property of the ions is influential in the lattice energy? Explain.
2. Does your answer make sense according to Coulomb's law?
3. Research the lattice energy of Magnesium chloride, $MgCl_2$ (which consists of Mg^{+2} and Cl^- ions).
4. How does its lattice energy compare with sodium chloride?
5. What do you think has a greater influence in the lattice energy: charge or size?

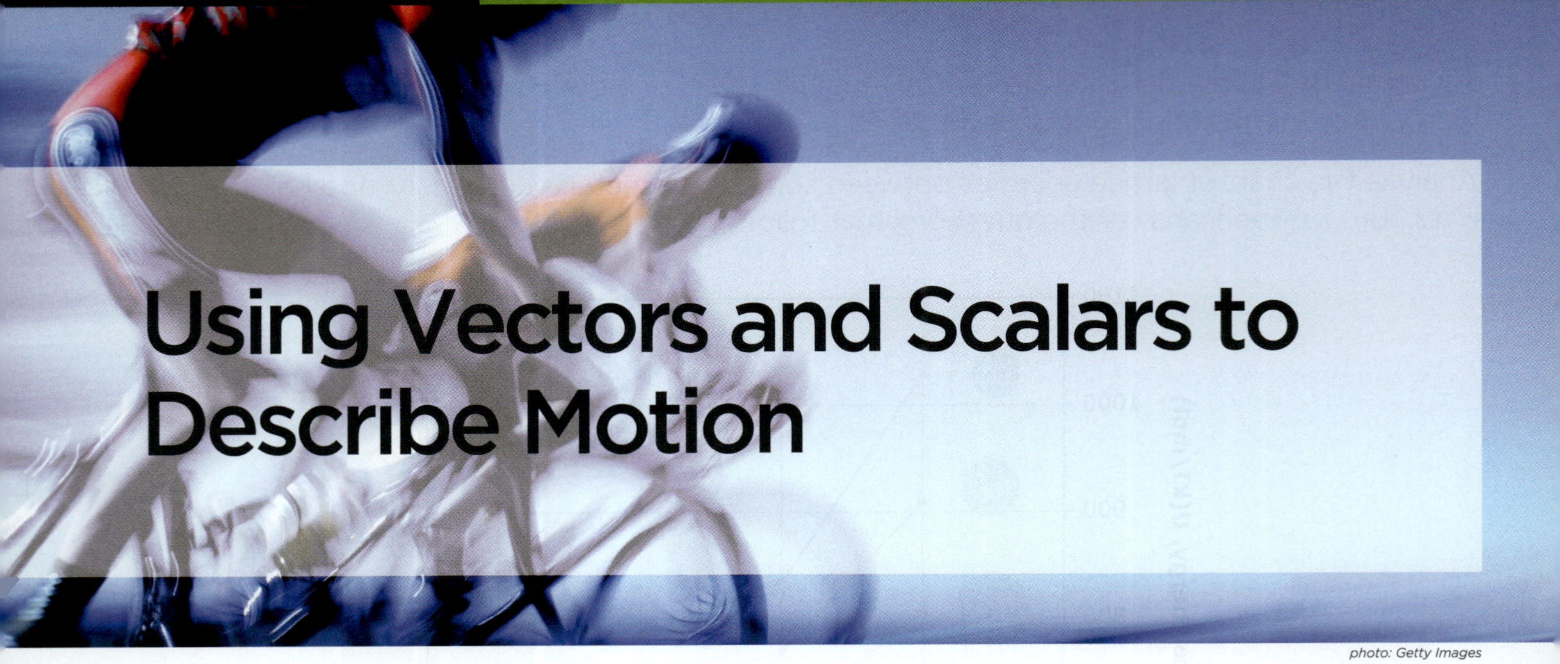

Using Vectors and Scalars to Describe Motion

photo: Getty Images

LESSON OVERVIEW

Lesson Questions

- What are the differences between vector and scalar quantities?
- How may problems involving two-dimensional vectors and their components be solved?
- When a force is applied to an object in a direction perpendicular to its motion, what are the effects on its speed and direction?

Lesson Objectives

By the end of the lesson, you should be able to:

- Distinguish between vector and scalar quantities and cite examples.
- Construct solutions to problems involving two-dimensional vectors and their components.
- Predict the result of applying a force to an object perpendicular to its direction of motion.
- Show how physical concepts can be applied to design a device that minimizes the force on an object resulting from a collision.

Key Vocabulary

Which terms do you already know?

- ☐ **acceleration**
- ☐ **acceleration due to gravity**
- ☐ **average speed**
- ☐ **centripetal force**
- ☐ **displacement (vector)**
- ☐ **distance**
- ☐ **energy**
- ☐ **force**
- ☐ **free body diagram**
- ☐ **gravitational force**
- ☐ **gravity**
- ☐ **mass**
- ☐ **measurement**
- ☐ **momentum**
- ☐ **power**
- ☐ **scalar**
- ☐ **speed**
- ☐ **uniform circular motion**
- ☐ **vector**
- ☐ **velocity**
- ☐ **work**

ENGAGE

How could you use velocity and mathematics to find a missing person? Visit the Engage page online to learn more.

Explain Question

Why is the total velocity of a roller coaster zero at the end of the ride, and how does its speed vary with its location?

EXPLORE

What Are the Differences between Vector and Scalar Quantities?

Scalar Quantities

In physics, speed is one of many scalar quantities. A scalar quantity is a value or measurement that has magnitude, or size, only. It has units, but does not include information about direction. Other examples of scalar quantities include:

- mass
- time
- distance
- electric charge
- length
- area
- volume
- density
- pressure
- temperature
- energy
- work
- power

Scalar quantities are expressed with a number and the relevant units, for example:

- 31 kg
- 52 s
- −24°C

Vector Quantities

Many other quantities in physics are vectors. These quantities have magnitude and direction. The direction information can be just as important as the magnitude information.

Physicists use the vector term *velocity* more often than they use the scalar term *speed*. A velocity has both a magnitude and a direction. For physicists, speed and velocity are not the same thing, and they cannot be used interchangeably.

Other common vector quantities in physics include:

- displacement
- acceleration
- momentum
- force

Sometimes, vectors are indicated in text in boldface or italic type, and vectors are often written with an arrow along the top. The direction and units are included. Here are some examples of vectors:

$$\vec{v} = 67 \text{ m/s east}$$

$$\vec{F} = 13 \text{ N downward}$$

$$\vec{a} = 45 \text{ m/s}^2 \text{ to the right}$$

It is often useful to draw vectors as part of a diagram of an object that is moving or that is subject to forces. The vectors are shown as arrows. The length of the arrow indicates the magnitude of the vector, and the direction the arrow shows the direction of the vector. For this reason, vectors are referred to as directed line segments. Vectors are often simply labeled v (v for vector). If you want to refer to the magnitude of a vector, then put it inside a double absolute value: ||v||. However, for most purposes, it is easier to use arrow length to represent magnitude, or the arrow above the vector letter representation.

Vector quantities from the real world can point in any direction in any of the three dimensions, so their arrows can point in any direction. Physics problems often show simpler situations, with vector quantities acting in only two dimensions.

Two or more vectors of the same type can be combined. Vectors pointing in the same direction are added. If the vectors point in opposite directions, one is subtracted from the other.

A vector cannot, however, be added to a different type of vector. For example, an object's velocity cannot be added to its acceleration in a meaningful way.

Vector quantities cannot be added to scalar quantities. For example, the acceleration of an object cannot be added to its mass in a meaningful way.

How May Problems Involving Two-Dimensional Vectors and Their Components Be Solved?

Manipulating Vectors

The first step in solving problems involving vector quantities is usually to make a diagram. Sometimes, the arrows representing vectors are drawn so that their lengths are proportional to the magnitude of the vectors they represent. In the diagram, a scale is chosen such that the vectors can be drawn with lengths directly proportional to their magnitudes.

For example, in the following illustration, Car 1 has velocity to the right of 10 meters per second. Car 2 has velocity to the right of 20 m/s, so its arrow is twice as long. A person making the drawing might draw Car 1's arrow 10 centimeters long and Car 2's arrow 20 cm long.

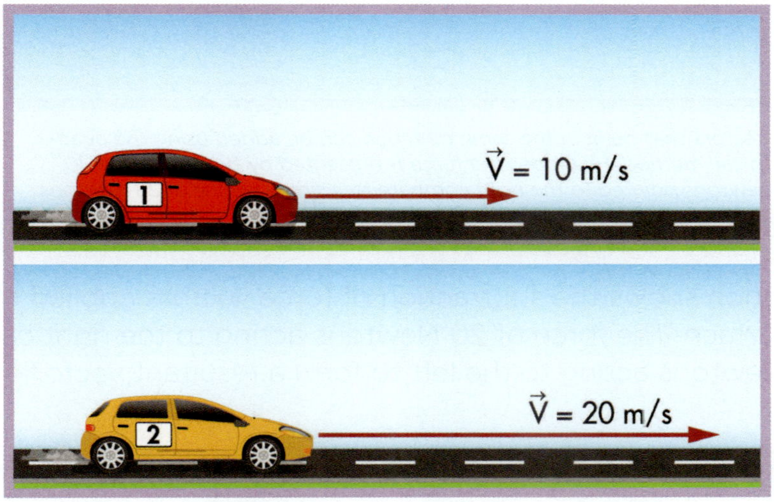

Vectors are drawn with lengths proportional to their magnitudes. What is the relationship between the velocities and vectors for the yellow car and the red car?

Vectors can be combined when they represent the same units and they act on the same object. If they act in the same direction, they are added. If they act in opposite directions, they are subtracted.

Suppose that two forces of 10 Newtons act on the same object in the same direction. The two vectors of 10 Newtons each would be added as in the following illustration by using the head-to-tail method to form a combined resultant vector of 20 Newtons.

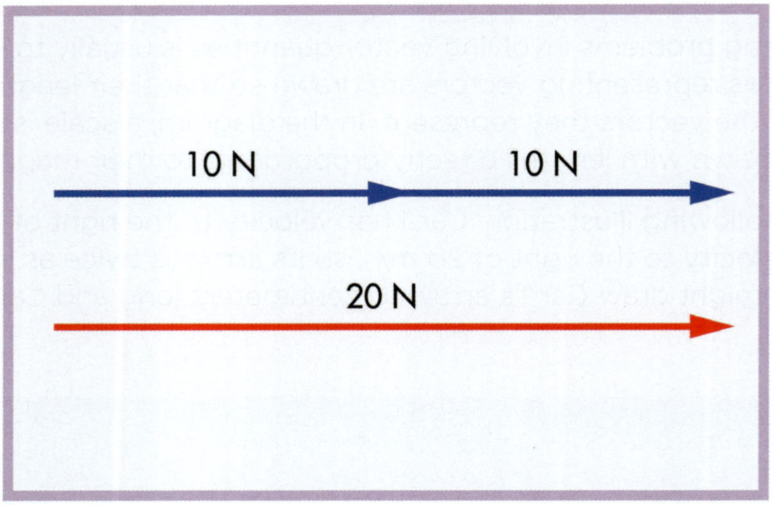

Vectors that point in the same direction can be added using the head-to-tail method. How does the force represented by two vectors of 10 newtons pointed to the right compare to a single vector of 20 newtons vector pointed to the right?

The following illustration shows the subtraction of force vectors applied to the same box on a level, frictionless surface. The force of 20 Newtons acting to the right can be combined with the force of 5 Newtons acting to the left to form a resultant vector of 15 Newtons acting to the right.

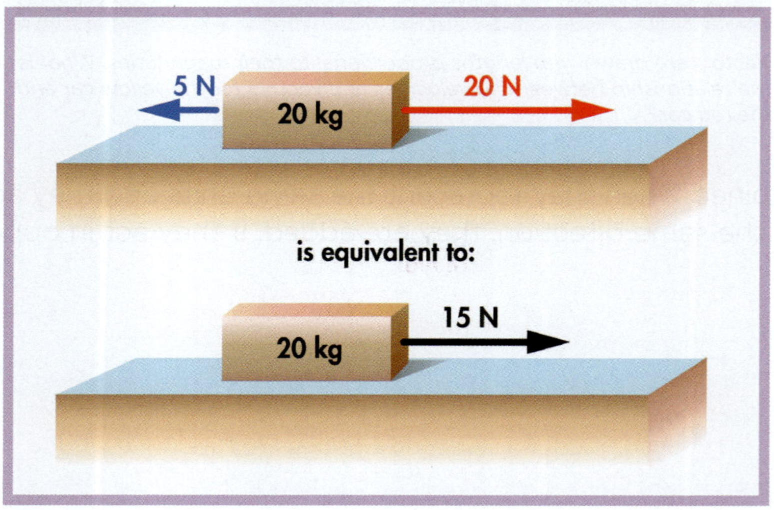

Vectors that act in opposite directions can be subtracted from each other. What is the relationship between the initial vectors and the resultant vector?

Drawing a diagram of the forces acting on an object can be very useful in analyzing motion problems.

When two-dimensional vectors of any type are to be added, they might not act in the same or exactly opposite directions. In this case, they can be combined using a diagram of a parallelogram, as shown in the following illustration. The two vectors are drawn as adjacent sides of a parallelogram, with their lengths proportional to their magnitudes. Sides added parallel to these vectors complete the parallelogram shape. Finally, drawing the diagonal of the parallelogram that starts at the base of the two original vectors gives the resultant vector. Its length gives the combined vector's magnitude, and its direction gives the combined direction.

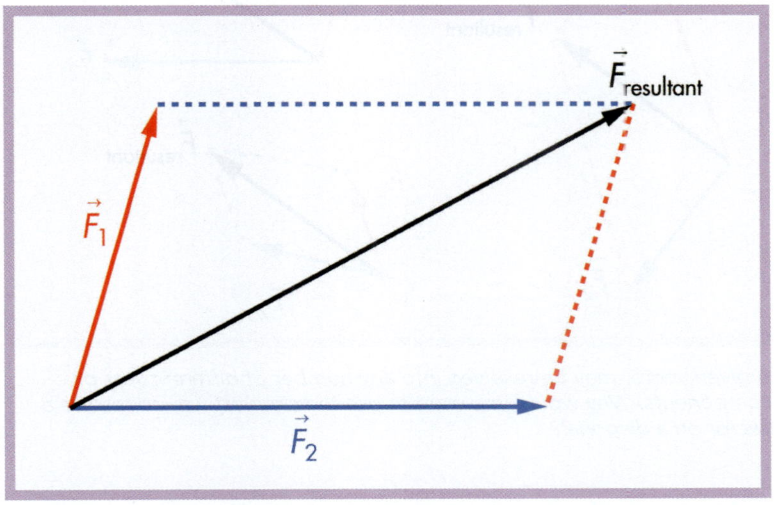

Forces acting in different directions can be depicted as vectors in a parallelogram, using the diagonal to find the resultant force. How does the resultant vector compare with the two initial vectors?

In the special case of the two vectors being at right angles to each other, the Pythagorean theorem can be used to find the length of the resultant vector.

If there are more than two vectors to add, draw the vectors in a chain by linking them head to tail in sequence. The resultant vector can then be drawn from the first vector's tail to the last vector's head, as in the following illustration. The order of the combinations does not matter.

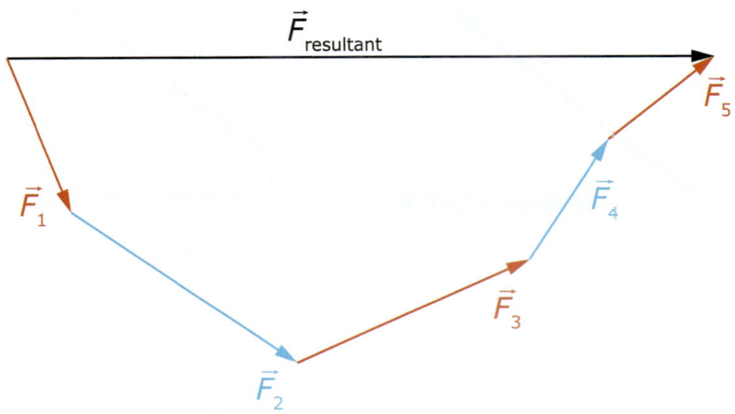

Vectors are placed head to tail in any order, and the resultant goes from the tail of the first vector to the head of the last. What is the resultant vector if all of the vectors combine so that the head of the final vector is at the same point as the tail of the first vector?

The process works the other way, too. Instead of combining vectors into a single resultant, a vector can be separated into two or more components. There are many ways to do this. The following illustration shows three of the ways $\vec{F}_{resultant}$ could be broken down into two component vectors $\vec{F}_1$ and $\vec{F}_2$.

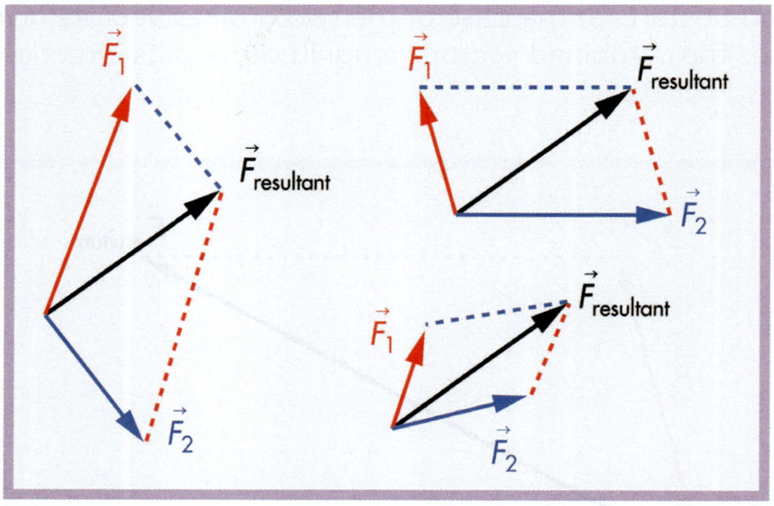

A given vector may be resolved into any number of different sets of components. Why would you want to use the simplest way to resolve a vector on a diagonal?

In physics, when a vector quantity is in two dimensions, it is often useful to break it into two components that are at right angles to each other. This results in a parallelogram that is a rectangle. Often, the alignment of the component vectors is chosen in a way that is convenient for analysis. For example, one component might be parallel to the angled surface of an inclined plane. The most common choice is to choose components so that they are parallel to the horizontal *x*-axis and vertical *y*-axis. The following illustration shows an example.

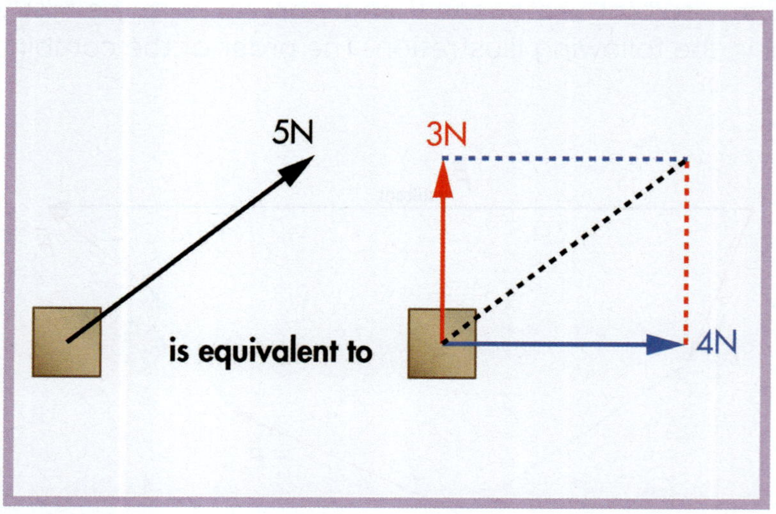

A vector may be resolved into component vectors that are at right angles to each other and can be aligned with the x- *and* y-*axes of a diagram. What is the advantage of aligning the vector components along the* x- *and* y-*axes?*

Instead of a drawing, trigonometry can be used to determine the components of a vector in two perpendicular directions.

Manipulating Vectors: Sample Problem

A force of 5 Newtons acts on a box at an angle to the horizonta . Use trigonometry to find the horizontal and vertical components of this force.

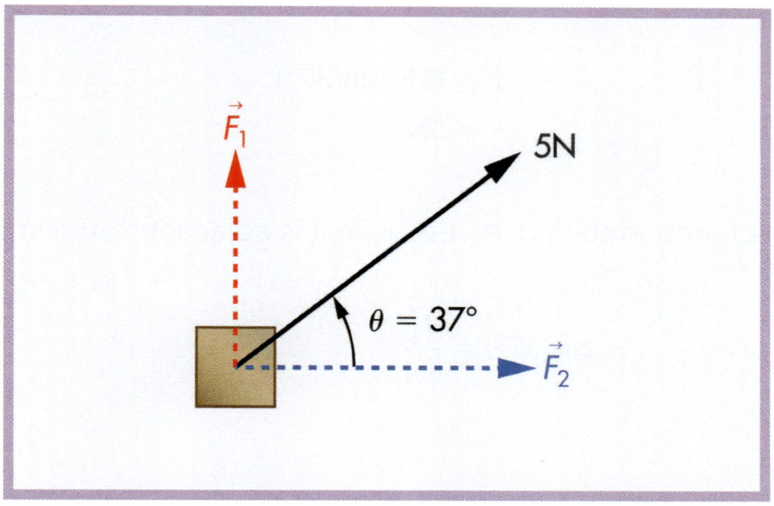

A force of 5 Newtons acts on a box at a 37° angle. What are the vertical and horizontal components of this vector?

Solution

The lengths of all the vectors are proportional to their magnitudes, so trigonometric definitions can be used, just as if the force diagram were a plane figure of a triangle. Redrawing the diagram makes this clearer.

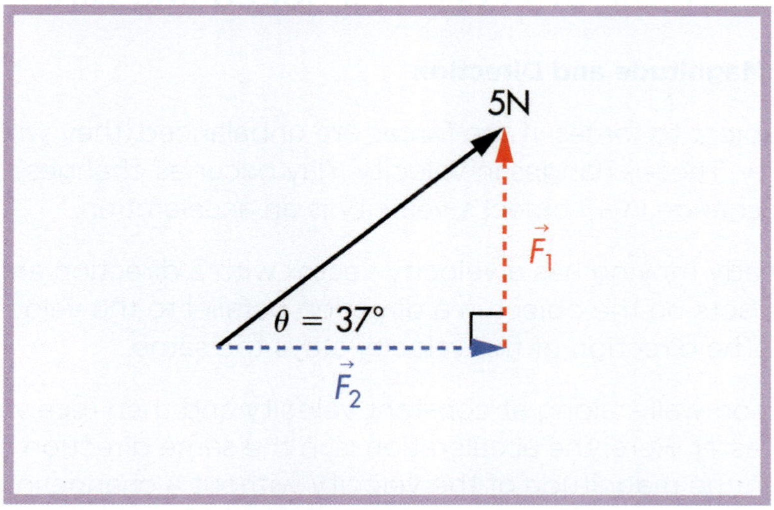

The vector can be represented in the form of a right triangle. How does this representation enable the vertical and horizontal components of this vector to be determined?

Also, using the sine function to find $\vec{F}_2$ because it is opposite to the known angle:

$$\sin(37°) = \frac{\vec{F}_1}{5\text{N}}$$

Rearranging,

$$\begin{aligned}\vec{F} &= (5\text{N})\sin(37°)\\ &= 3\text{N}\end{aligned}$$

Also, using the cosine function to find $\vec{F}_2$ because it is adjacent to the known angle:

$$\cos(37°) = \frac{\vec{F}_2}{5\text{N}}$$

Rearranging,

$$\begin{aligned}\vec{F}_2 &= (5\text{N})\cos(37°)\\ &= 4\text{N}\end{aligned}$$

The horizontal component vector is 4 N, and the vertical component vector is 3 N.

When a Force Is Applied to an Object in a Direction Perpendicular to Its Motion, What Are the Effects on Its Speed and Direction?

Effects of Forces on Magnitude and Direction

An object may be subject to forces. If the forces are unbalanced, they will cause a change in the object's velocity. These changes in velocity may occur as changes in the magnitude, direction, or both. A change in an object's velocity is an acceleration.

An object that is already moving has a velocity vector with a direction and a nonzero magnitude. If a force acts on the object in a direction parallel to the velocity, the velocity's magnitude changes. The direction of the velocity stays the same.

For example, if a person walks along at constant velocity and then receives a push from behind, he will walk faster. Here, the acceleration is in the same direction as the motion. The result is an increase in the magnitude of the velocity without a change in direction.

Next, think about a ball rolling in a straight line experiencing friction along the surface. The friction acts parallel and opposite to the ball's velocity. The magnitude of the ball's velocity changes by slowing down, and the direction does not change.

What happens if two billiard balls, both in motion, collide with each other at an angle? The likely result is that each ball will have a change in both the magnitudes and the directions of their respective velocities.

photo: Pixabay

Pocket billiards, also called pool, is a game based on understanding vectors. How can a player use knowledge of vectors to ensure that a ball ends up in the pocket?

On the other hand, if the force on the moving object acts in a direction perpendicular to the velocity, the velocity's direction changes, and the magnitude of the velocity stays the same. A common example is uniform circular motion. Objects travel along circular paths at constant speed. Their direction is always changing, and thus their velocity is always changing, and therefore they are accelerating. Whenever there is an acceleration, there must be an unbalanced force acting on the object. In uniform circular motion, this is the centripetal force, which acts to accelerate the object toward the center of the circle of its motion. In the case of orbiting objects, the centripetal force is provided by the force of gravity between the orbiting object and a massive object in the center.

Consider the Explain Question:

Why is the total velocity of a roller coaster zero at the end of the ride? How does its speed vary with its location?

Go online to complete the scientific explanation.

Check for Understanding:

Using specific examples, what is the difference between a vector and scalar quantity?

STEM in Action

STEM and Using Vectors and Scalars to Describe Motion

Aeronautics engineers analyze aircraft flight by separating it into four force vectors: lift, gravity, thrust, and drag. Because they are vectors, each of the four forces has both magnitude and direction. And, these aircraft flight force vectors act in opposing pairs: lift versus gravity and thrust versus drag. In order for the airplane to fly, lift must equal or exceed gravity and thrust must exceed drag.

When the force of gravity is greater than the force of lift created by air moving over the wings, then the aircraft will not rise from the ground. If the drag force is greater than the thrust provided by the engines, the aircraft cannot move forward. To reduce drag, airplane wings are designed and shaped to let air flow easily and smoothly over them with a minimum of resistance. Wings are also designed to maximize the lift created by air flowing over the wings as the plane moves forward.

Aloft, when the airplane is in what engineers and pilots call "straight and level flight," the thrust vector, which is the vector propelling the plane forward, exceeds the drag vector and the airplane maintains a constant velocity. Meanwhile, all the upward lift force is balanced by the downward weight of the airplane due to gravity and the plane remains at a steady altitude. The plane and its wings are also engineered so that the lift acts perpendicular to the wings.

When the airplane turns, the lift force vector continues to act perpendicularly to the wings. The lift force now has both a vertical and a horizontal component. The horizontal component causes the plane to turn. This is similar to a ball whirling on the end of a string that turns in a circle when the string provides a force toward the circle's center.

In turning the plane, however, less of the original lift vector is available to offset gravity's weight vector. To remain at the same altitude, the pilot takes action to increase the wings' lift during a turn. The pilot may achieve this by pointing the nose slightly upward or by lowering wing flaps to increase the wings' surface area parallel to the ground.

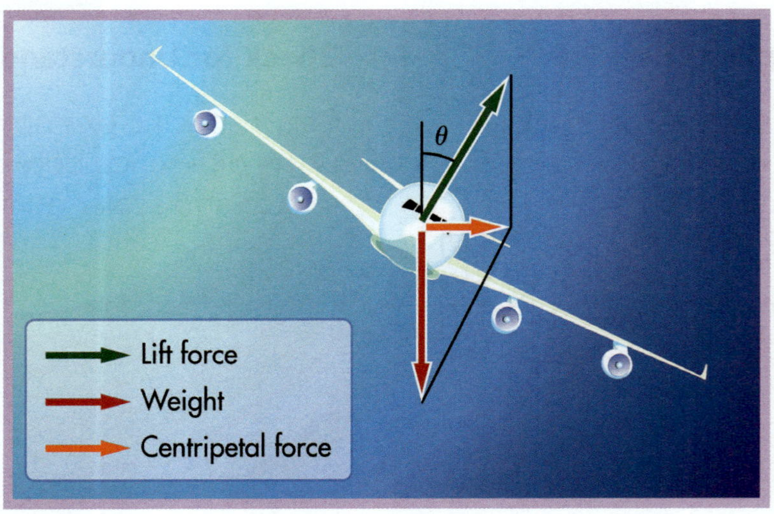

This diagram shows the vector forces acting on a banking airplane. How are vectors used to model airplane flight?

You walk from your house to the library as shown in the image. Calculate your displacement. Express all answers to three significant figures.

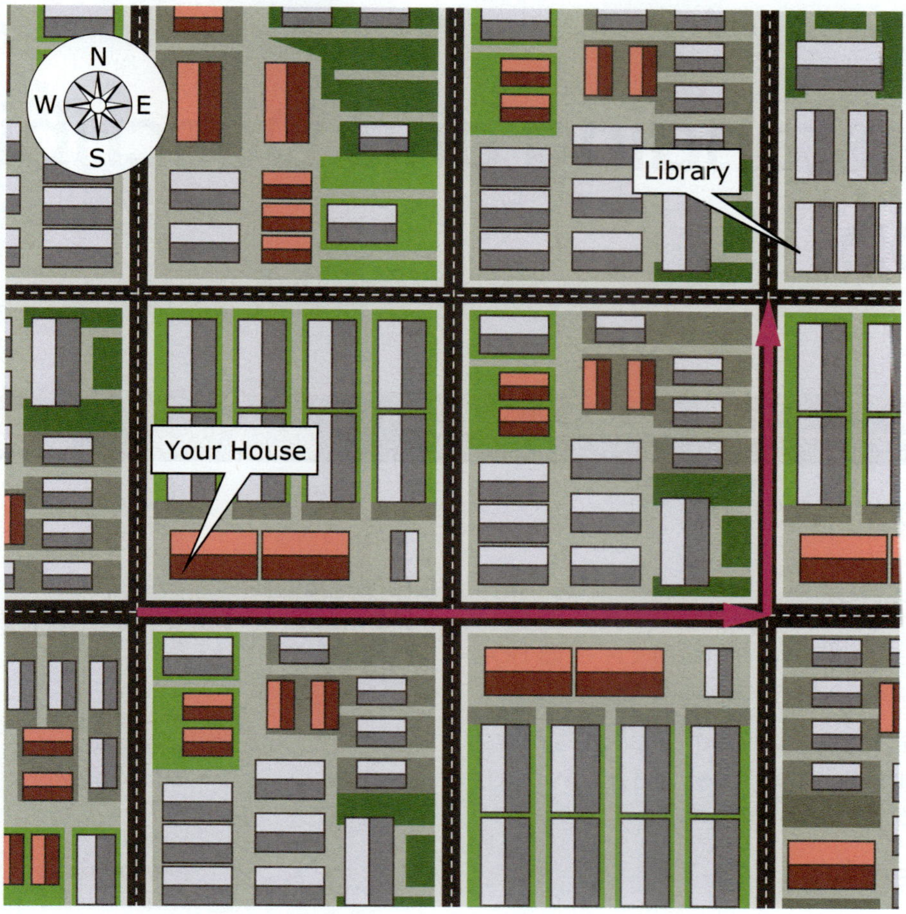

Understanding and Describing Motion

photo: Getty Images

LESSON OVERVIEW

Lesson Questions

- What are frames of reference, and how do they relate to motion?
- How can physicists use Newton's laws of motion and universal gravitation to predict objects' motions?

Lesson Objectives

By the end of the lesson, you should be able to:

- Use frames of reference to describe motion.
- Model Newton's laws of motion and universal gravitation to predict the motion of most objects.

Key Vocabulary

Which terms do you already know?

- ☐ acceleration
- ☐ acceleration due to gravity
- ☐ data
- ☐ directly proportional
- ☐ displacement (vector)
- ☐ distance
- ☐ force
- ☐ frame of reference
- ☐ gravitational force
- ☐ gravity
- ☐ inertial frame of reference
- ☐ inversely proportional
- ☐ mass
- ☐ measurement
- ☐ motion
- ☐ Newton's law of universal gravitation
- ☐ Newton's laws
- ☐ non-inertial frame of reference
- ☐ observation
- ☐ speed
- ☐ velocity

ENGAGE

What do you think of when you picture a moving object? How do you describe motion? Visit the Engage page online to learn more.

Explain Question

How can Newton's laws of motion and the law of gravitation predict the motion of an object, and how can frames of reference be used to describe that motion?

EXPLORE

What Are Frames of Reference, and How Do They Relate to Motion?

Frames of Reference: Coordinate Systems

Physicists describe the positions of objects with respect to frames of reference. A frame of reference provides a system of coordinates for identifying an object's location.

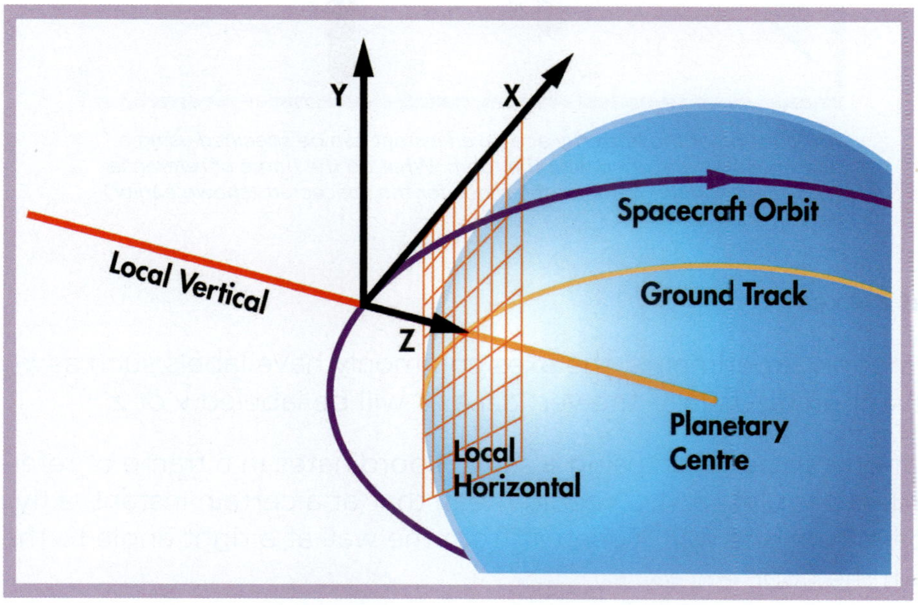

NASA uses a frame of reference that moves with the spacecraft for spacecraft orbiting bodies in space. What are the advantages to using this type of frame of reference?

For example, suppose a student drops a ball straight down from the top of a tall building. A vertical axis along the ball's path could be chosen as a frame of reference. The ball's changing positions can be measured relative to an origin at ground level.

It would have been entirely reasonable to choose another frame of reference. For example, the release point at the top of the building could be chosen as the origin.

Reference frames can also be moving. In the example of the falling ball, a frame of reference could be chosen to be at the ball's center. For observers within this frame of reference, the ball would remain stationary while the surroundings would move. For outside observers, the frame itself would be accelerating toward the ground.

Using Rooms as Frames of Reference

In the lab, physics motion studies might use the room itself as a convenient reference frame. One corner at the floor or a particular spot on a piece of equipment can act as a reference frame's origin. The three axes of the frame can lie parallel to the room's three dimensions of length, width, and height. These three axes are typically at 90° to one another.

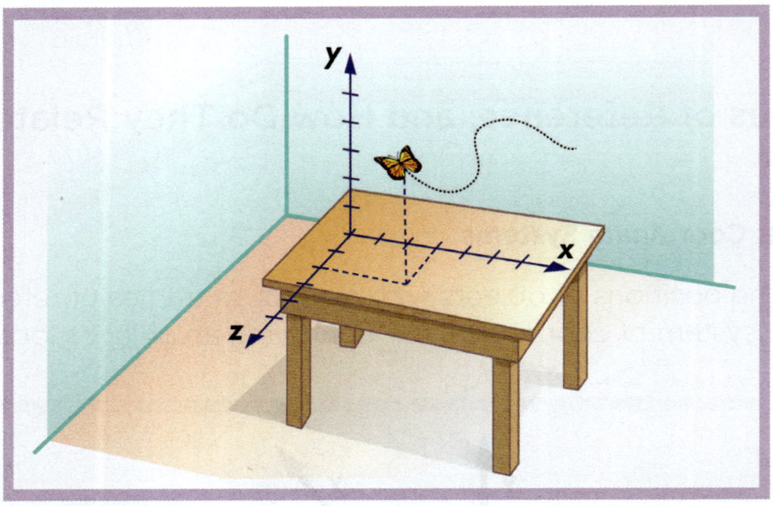

The position of the butterfly at a given instant can be specified using a reference frame aligned with the room. What do the frame of reference for a room and the frame of reference for the spacecraft (shown earlier) have in common?

Using *x*-, *y*-, and *z*-Axes

In diagrams of motion experiments, the axes commonly have labels such as *x*, *y*, and *z* and include the relevant units. Usually, the vertical axis will be labeled *y* or *z*.

An object's location can be given using a set of coordinates in a frame of reference. For example, the ordered triplet (3, 4, 2) could mean that at a certain instant, a fly in the room might be 3 meters from one wall, 4 meters from the wall at a right angle to the first, and 2 meters up from the floor.

Objects sometimes move along only one or two axes. For example, students might observe an object falling straight down along a vertical axis. Neither horizontal coordinate changes so the students might only record the change in the object's vertical coordinate. Physicists describe this kind of motion as one-dimensional.

As another example, a toy cannon might launch a projectile at an angle. The projectile would move along both vertical and horizontal axes in two-dimensional motion. Two coordinates would be necessary to describe its position at a given time, but the third might be omitted because it is constant.

Inertial and Non-inertial Frames of Reference

Physicists distinguish between two types of frames of reference: inertial and non-inertial. The difference has to do with the first of Newton's laws of motion, which are discussed in detail in another section. The first law applies in the case that there are no unbalanced forces acting on an object. It says that objects remain motionless if they are at rest, and the motion of an object will continue in a straight line at a constant speed if it is already moving.

An inertial frame of reference is one that acts in accordance with the first law. It has a constant velocity, which could have zero magnitude. Its velocity does not change in magnitude or direction. Objects moving within the inertial reference frame may accelerate, but the frame itself does not.

For example, imagine a train traveling in a straight line on a smooth track. It has a constant nonzero velocity. The train acts as an inertial frame of reference. People inside place drinks on tables in the dining car. Snack carts roll down the aisles only when pushed. Falling objects accelerate straight down toward Earth just as they do when the train is sitting still at the station. In fact, if the window shades were down, there would be no motion experiment that passengers could perform inside the train car that could prove whether the train was moving (if the tracks were perfectly smooth).

photo: Getty Images

Trains constantly enter and leave the railway terminal at different speeds. How does the frame of reference of the observer affect the apparent speed of the train?

However, if the speed or direction of the train changed, the frame of reference would become non-inertial. Suddenly, it might seem as if outside forces were pushing drinks onto the floor and rolling carts around the car.

Physicists often consider Earth to be an inertial frame of reference. However, Earth itself moves with nonconstant velocity because the direction of its motion is always changing. It rotates on its axis. It moves along its orbit around the sun. It is a part of a moving solar system and galaxy. Newton's first law does not precisely apply, so Earth is not truly an inertial reference frame. Fortunately, the effects of its changing motion on the human-scale experiments on Earth are usually small. The results from using Newton's laws are usually very good approximations.

How Can Physicists Use Newton's Laws of Motion and Universal Gravitation to Predict Objects' Motions?

Forces on Objects

Using a given frame of reference, physicists can analyze and predict changes in objects' positions over time—their motion. Under most ordinary circumstances, they can use laws of motion developed by the brilliant scientist, Sir Isaac Newton, in the 1600s. These laws describe effects of forces on objects' motions.

photo: Library of Congress

Sir Isaac Newton (1642-1727) published his ideas about motion in his 1687 book Philosophiæ Naturalis Principia Mathematica, and he also developed practical telescopes and calculus. How does his work allow us to predict how objects will move?

Newton's Law of Universal Gravitation

Additionally, Newton developed an elegant explanation for the way gravity works. In Newton's law of universal gravitation, he showed that the force, due to gravity, between two bodies is directly proportional to their masses. Therefore, it is stronger between more massive objects. At the same time, gravity is also inversely proportional to the square of the distance between the bodies. So, it is stronger between objects that are closer together.

As an example, suppose a student stands on Earth. The distance between the student's and Earth's centers of mass is very nearly the radius of Earth. The student's (relatively small) mass and Earth's (relatively large) mass pull on each other with substantial gravitational force.

The measure of this force is the student's weight on Earth. It is important to remember the difference between mass and weight. Mass is the amount of material that makes up an object. An object's weight is the force that mass exerts in a gravitational field. Scientists use units of kilograms when discussing mass, and units of Newtons when describing weight.

If the same student were standing on the moon, his mass would be the same as it would be anywhere, but his weight on the moon would be only one-sixth of his weight on Earth. Although he would be closer to the center of the moon than he would be to the center of Earth, the moon's mass is much less than Earth's.

The section on gravity gives more details of Newton's law of universal gravitation.

Newton's Law of Universal Gravitation: Sample Problem

Newton's second law of motion says that any force exerted on an object will cause that object to accelerate. Any two objects with mass will exert a gravitational force on each other. This force will then create an acceleration that can be used to describe the motion of the two objects. Before the motion can be described, the force of gravity between the two objects must be found. Newton's law of universal gravitation says that the force of gravity can be described by the following equation:

$$F_g = \frac{Gm_1m_2}{r^2}$$

where G is Newton's gravitational constant, 6.67×10^{-11} N × m²/kg², m_1 and m_2 are the masses of the two objects, and r is the distance between them.

The moon and Earth have masses of 7.35×10^{22} kilograms and 5.98×10^{24} kilograms, respectively. The distance between these bodies is about 3.84×10^{8} meters. To calculate the force of gravity between them, use Newton's law of universal gravitation:

$$F_g = \frac{(6.67 \times 10^{-11}\,\mathrm{N} \times \mathrm{m}^2/\mathrm{kg}^2)(7.35 \times 10^{22}\,\mathrm{kg})(5.98 \times 10^{24}\,\mathrm{kg})}{(3.84 \times 10^{8}\,\mathrm{m})^2}$$

$$F_g = 1.99 \times 10^{20}\,\mathrm{N}$$

The force of gravity that Earth exerts on the moon is about 2×10^{20} Newtons.

Validity and Usefulness of Newton's Laws

Newton's laws are most useful only within inertial frames of reference. They can be carefully modified for use in non-inertial frames.

They work best in everyday situations, with everyday-sized objects traveling at everyday speeds. When objects are very tiny, such as atoms and molecules, they do not follow Newton's laws of motion. Another set of physics laws, quantum mechanics, works better at that scale. When objects travel very fast, more than about one-tenth of the speed of light, Newton's laws do not work well either. The theory of relativity better describes objects' behavior at those speeds.

Even with these restrictions, Newton's laws have many applications. If physicists know what forces act on everyday objects, the laws allow practical predictions of their motions.

Consider the Explain Question:

How can Newton's laws of motion and the law of gravitation predict the motion of an object, and how can frames of reference be used to describe that motion?

Go online to complete the scientific explanation.

Check for Understanding:

What type and axes of a reference frame would be used in a high school physics classroom for the motion of an object in three dimensions?

STEM in Action

STEM and Understanding and Describing Motion

When scientists launch rockets from Earth, they carefully set up the launch conditions to make sure that the rocket travels along a prescribed trajectory that avoids unwanted collisions with planetary bodies and space debris. From a frame of reference based out in space, Earth is traveling along its orbital path around the sun, and its direction in relationship to the other planets varies depending on the date. Because Earth makes one full rotation about its axis every 24 hours, locations on the surface of Earth face in different directions throughout the day. Also, objects near Earth's equator travel through space faster than those at its poles. Scientists must carefully choose the launch time and date to ensure the spacecraft travels an exact trajectory.

Many of the scientists who calculate the launch times and trajectories needed for spaceflight work for the National Aeronautics and Space Administration (NASA), the federal agency that oversees this country's space program. Its mission is the peaceful exploration of space in order to advance human understanding of science. In addition to scientists, NASA employs engineers, pilots, technicians, mathematicians, and engineering designers.

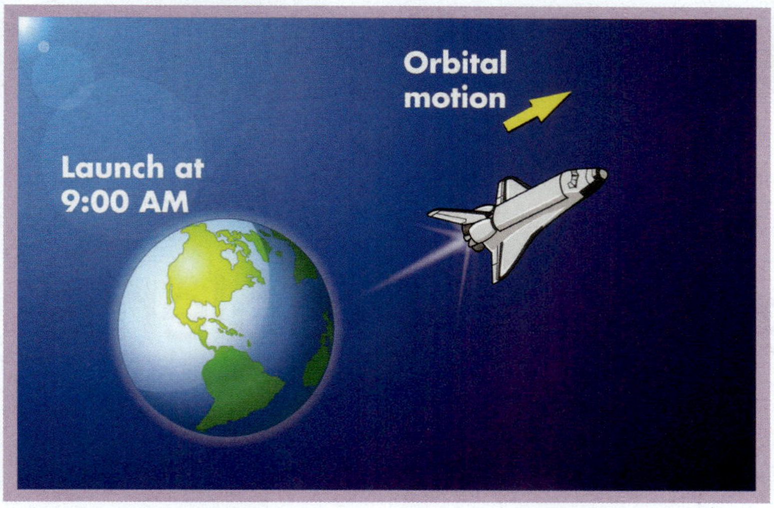

Earth rotates about its axis as it orbits the sun. How does this rotational motion affect the location and timing of spacecraft launches?

The graph shows the orbital speed of a satellite as a function of its distance from the surface of Earth. Use this graph to answer the questions.

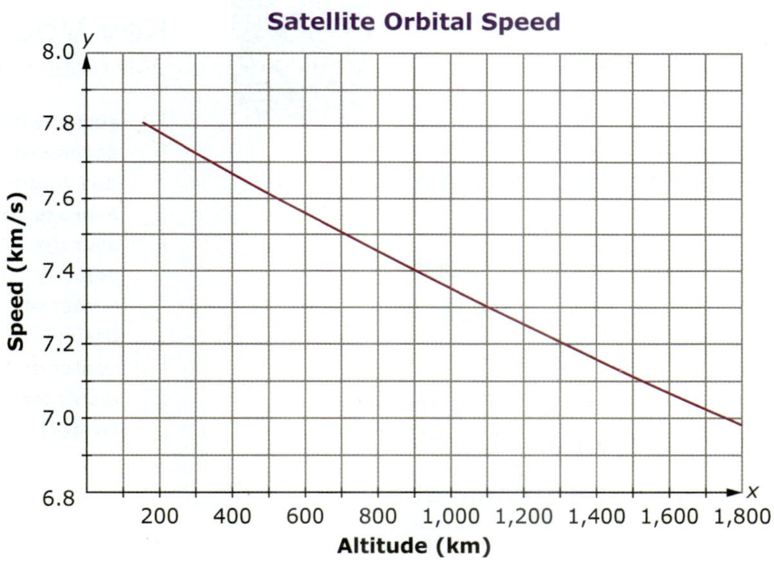

1. Describe the relationship between the orbital height and the speed of the satellite. Suggest an explanation for this behavior.
2. Describe how the acceleration of the satellite changes as the distance from Earth changes.

Solving Motion Problems

photo: Getty Images

LESSON OVERVIEW

Lesson Questions

- How does an object's speed affect the distance it travels in time?
- How can position, velocity, and acceleration be used to describe the motion of an object?
- How can motion be represented graphically?
- How do objects, such as a basketball, move in the air?
- How can motion sensors be used to record and describe the motion of an object?
- How does friction affect the motion of an object?

Key Vocabulary

Which terms do you already know?

- ☐ acceleration
- ☐ acceleration due to gravity
- ☐ average speed
- ☐ average velocity
- ☐ center of gravity
- ☐ center of mass
- ☐ displacement (water)
- ☐ distance
- ☐ force
- ☐ instantaneous speed
- ☐ instantaneous velocity
- ☐ kinetic friction
- ☐ scalar
- ☐ speed
- ☐ static friction

Lesson Objectives

By the end of the lesson, you should be able to:

- Solve problems that involve both average and constant velocity.
- Distinguish between equations that use acceleration, average velocity, instantaneous velocity, velocity, distance, and displacement to describe motion in one dimension.
- Create graphs to illustrate and analyze motion.
- Apply an equation to describe the motion of a projectile in two dimensions.
- Demonstrate use of technology, such as photo gates or motion detectors, to measure motion.
- Distinguish between kinetic friction and static friction, and differentiate their effects on motion.

ENGAGE

When you think about motion, what objects come to mind? What do you think is the difference between speed and acceleration? Visit the Engage page online to learn more.

Explain Question

How do the displacement, velocity, and acceleration of a runner change as he races from the starting line toward the finish line?

EXPLORE

How Does an Object's Speed Affect the Distance It Travels in Time?

What Is Speed?

Speed is a measure of how fast an object is moving. Faster-moving objects cover a greater distance than do slower-moving objects in the same amount of time. Speed is the rate at which distance is traveled.

photo: Pixabay

A speedometer measures only the instantaneous speed of an object and cannot measure average speed because it has no record of speed over time. What features would a speedometer need to measure the average speed?

What Are Average Speed and Constant Speed?

A car driving down a road does not usually maintain the same speed as it travels. It may travel faster at certain times but then slow down because of a slower-moving vehicle in front. It may even come to a stop for a stoplight. The car's actual speed is likely to change over time. The speed at a particular instant in time is the car's instantaneous speed. A more accurate picture of how fast the car is going over a period of time can be found using the average speed during that period of time.

Sometimes, objects travel at the same speed over a period of time. In this case, the speed is constant. For instance, on the highway, drivers often set the cruise control of their cars in an attempt to maintain a constant speed. Then, the car maintains that speed for a certain period of time. Spacecraft speed up as they leave Earth but then maintain a constant speed as they escape Earth's orbit in space.

When an object is moving at a constant speed, its instantaneous speed at any moment is the same as its constant speed. Likewise, its average speed is equal to the constant speed and, therefore, the instantaneous speed at any moment.

How Is Average Speed Calculated?

Speed is simply the rate at which an object is covering distance. Therefore, the average speed of an object can be determined by finding the total distance traveled in a particular amount of time. For instance, if a car traveled 200.0 kilometers in 8.0 hours, then the average speed of the car is:

$$s = \frac{d}{t} = \frac{200.0\ \text{km}}{8.0\ \text{h}} = 25\ \text{km/h}$$

In physics, SI units are typically used. The SI unit for distance is meters, and the SI unit for time is seconds. It would be typical to express a speed in meters per second (m/s). For instance, if a sprinter in the Olympics travels 100.0 meters in 10.4 seconds, then his or her average speed is:

$$s = \frac{d}{t} = \frac{100.0\text{ m}}{10.4\text{ s}} = 9.61\text{ m/s}$$

Distance Traveled for an Object Moving at Constant Speed

To find the distance traveled by an object moving at constant speed, rearrange the equation for speed and solve for the distance. Multiplying both sides by *t* gives the equation:

$$d = st$$

If both the speed of an object and the distance it traveled are known, the time can be found by solving the equation for time algebraically. Multiplying both sides by *t* and dividing both sides by *s* gives the equation:

$$t = d/s$$

Distance Traveled for an Object Moving at Constant Speed: Sample Problem

How long would it take a long-distance runner to run a distance of 1,100 m at a constant speed of 6.0 m/s?

Solution

The distance traveled and the speed are known, so use the equation for speed that has been rearranged and solved for time:

$$t = d/s = (1{,}100\text{ m})/(6.0\text{ m/s}) = 183\text{ s}$$

How Can Position, Velocity, and Acceleration Be Used to Describe the Motion of an Object?

How Is the Displacement of an Object Determined?

The displacement of an object is defined as the vector between an object's starting point and ending point. The starting point is called the initial position and the ending point is called the final position.

For instance, the movement of an ant can be measured with a one-dimensional number line. Suppose an ant starts at x = 4 m and walks to x = 8 m on the number line.

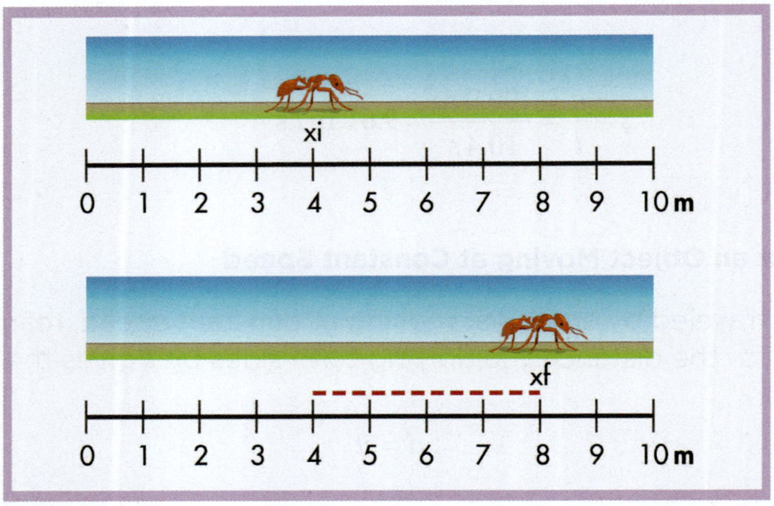

The ant moves from its starting position of x_i to its final position x_f. How far did it travel?

The displacement is calculated as follows:

$$
\begin{aligned}
x &= x_f - x_i \\
&= 8\text{m} - 4\text{m} \\
&= 4\text{m}
\end{aligned}
$$

This equation describes motion in one dimension with the concept of displacement. The change in position or displacement of the ant is 4 m. However, what if the ant started walking backward from where it came? Suppose that after reaching 8 m, the ant walked backward 6 m and arrived at a position of 2 m on the number line.

What is the new displacement of the ant?

$$
\begin{aligned}
\Delta x &= x_f - x_i \\
\Delta x &= 2\text{m} - 4\text{m} \\
&= -2\text{m}
\end{aligned}
$$

Note that only the final position of 2 m and the initial position of 4 m matter. The fact that the ant walked all the way to the 8-m mark and back does not factor into the calculation. Displacement considers only the starting and ending points of the motion. The negative sign indicates the direction of overall motion. The displacement is negative because the net effect of this motion was that the ant walked in the negative direction (from 4 m back to 2 m).

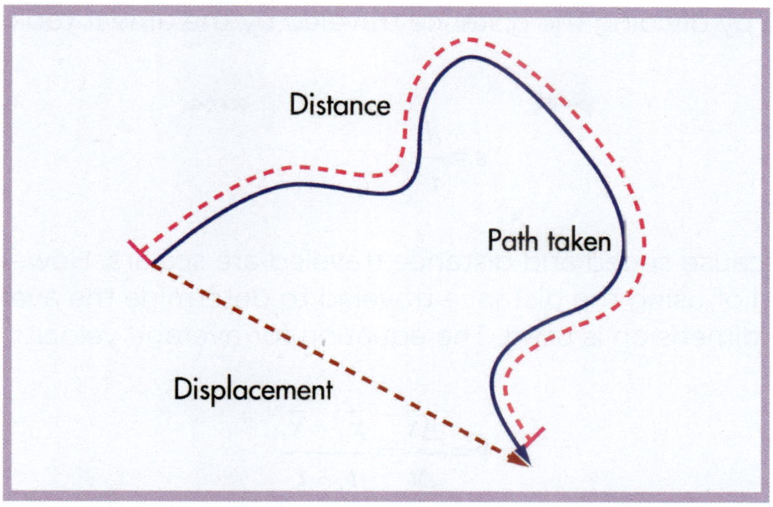

Displacement takes only an object's starting (initial) point and ending (final) point into consideration, while distance measures the total distance an object travels along its path. Are there any situations in which the displacement will be greater than the distance traveled?

What Is the Difference between Displacement and Distance Traveled?

In the previous example, the displacement did not take into account the ant's total movement from 4 m to 8 m and back to 2 m. It only took into account the change in position from a starting point to an ending point. In contrast, the distance traveled measures the total movement. In the previous example of the ant, the ant moved from 4 to 8 m covering a distance of 4 m. Then, it moved from 8 to 2 m covering a distance of 6 m. The total distance traveled is just the sum of the absolute values of these distances:

$$\text{Distance traveled} = 4\text{ m} + 6\text{ m} = 10\text{ m}$$

Note that, unlike the displacement, the distance traveled ignores the direction of motion. Moving from the 8-m mark to the 2-m mark is just considered moving 6 m. The fact that the ant was moving in the negative direction is irrelevant.

Distance traveled is a scalar quantity because the direction of each individual motion is not considered in the summation.

What Is the Difference between Speed and Velocity?

In everyday life, the words speed and velocity are sometimes used interchangeably. However, in physics, there is a difference. Velocity is displacement over time, and speed is distance over time. To specify the velocity of an object, one needs to state both how fast it is going (its speed) and the direction in which it is moving. For instance, the speed of a car might be 45 miles per hour (mph), but the velocity of the same car might be 45 mph south.

Velocity requires a direction, and, therefore, it is a vector quantity. Speed takes into account only the magnitude and not the direction of the object, and, therefore, it is a scalar quantity. For the velocity of an object to change, either the speed of the object will change, or the direction the object is traveling will change.

Speed is determined by dividing the distance traveled by the time it took the object to travel that distance:

$$s = \frac{d}{t}$$

This makes sense because speed and distance traveled are scalars. However, velocity is a vector, so instead of using the distance traveled to determine the average velocity, displacement in one dimension is used. The equation for average velocity is:

$$v = \frac{\Delta\vec{x}}{\Delta t} = \frac{\vec{x}_f - \vec{x}_i}{t_f - t_i}$$

The velocity of the object at any particular moment during this time interval is known as the instantaneous velocity, and it may be different from the average velocity. To understand instantaneous velocity $\vec{v}_{inst}$, from a mathematical viewpoint, think about how displacement in average velocity changes as the difference between t_f and t_i becomes smaller. At some very small difference in time, the displacement likewise becomes smaller, so that the ratio of $\Delta\vec{x}$ to Δt is still a finite value. This change to an infinitesimally small time interval is indicated by the expression "$\lim\limits_{\Delta t \to 0}$," which stands for "in the limit as the difference in time approaches zero." The concept of instantaneous velocity for motion in one dimension is described by the following equation:

$$\vec{v}_{\text{inst}} = \lim_{\Delta t \to 0} \vec{v} = \lim_{\Delta t \to 0} \frac{\Delta\vec{x}}{\Delta t}$$

Just as average speed "v" is the scalar of average velocity, so the scalar "v_{inst}" indicates "instantaneous speed," or the magnitude of instantaneous velocity.

In motion that takes place in more than one dimension, the difference between velocity and speed becomes more evident. Objects can have a constant speed but a varying velocity if their direction is changing. Distance increases for all motion because distance is the sum of displacement magnitudes. If this increase is uniform with the change in time, the speed will be constant.

However, if the direction of motion changes, the velocity is not constant, even if the speed is. For instance, a car traveling around a bend in the road may keep its speed constant but as it turns, its velocity is continuously changing because its direction is changing. Similarly, a satellite might have a nearly constant speed in orbit, but because it travels in a circle around Earth, its direction of motion is constantly changing and, therefore, its velocity is changing.

Speed and velocity can also differ if the motion is not in a straight line. Because velocity is displacement over time, there can be a dramatic difference between the speed and velocity of an object. For example, suppose a jogger runs 400.0 m around an oval track in 72.2 s, returning precisely to her starting point. Her average speed is (400.0 m)/(72.2 s) = 5.54 m/s. However, because her displacement is zero, her average velocity is zero.

The Difference between Speed and Velocity: Sample Problem

A driver travels 6,240 m north on a straight stretch of road for 4.0 minutes. Analyzing the motion in one dimension, what is his average velocity?

Solution

First, convert 4.0 minutes to seconds, using the conversion factor of 60 seconds per minute:

$$4.0 \text{ minutes} = 4.0 \text{ minutes} \times \frac{60\text{s}}{1 \text{ minute}} = 240\text{s}$$

Next, use the equation for average velocity, substituting 6,240 m north as the change in displacement, and 240 s for the change in time:

$$\vec{v} = \frac{\Delta \vec{x}}{\Delta t} = \frac{6{,}240 \text{ m, north}}{240\text{s}} = 26 \text{ m/s, north}$$

The Average Acceleration of an Object

Acceleration is defined as the rate of change in the velocity of an object per unit time. For instance, when a stoplight turns green, the cars waiting at the intersection begin to accelerate to increase their speed in a particular direction. The equation used for determining acceleration is:

$$\vec{a} = \frac{\Delta \vec{v}}{t} = \frac{\vec{v}_f - \vec{v}_i}{t}$$

This equation describes motion in one dimension with the concept of acceleration. Suppose an electric car that is initially traveling at a velocity of 12 m/s forward increases its speed to 34 m/s in 5.0 s without changing its direction. What is the acceleration?

$$\begin{aligned}\vec{a} &= \frac{\vec{v}_f - \vec{v}_i}{t} \\ &= \frac{34 \text{ m/s} - 12 \text{ m/s}}{5.0 \text{ s}} \\ &= \frac{22 \text{ m/s}}{5.0 \text{ s}} \\ &= 4.4 \text{ m/s}^2\end{aligned}$$

The acceleration, in this case, is in the direction of motion. An acceleration of 4.4 m/s^2 means that for every second, the electric car gained 4.4 m/s of velocity. The velocity of the car increased by 4.4 m/s each second. Phrasing the acceleration in this manner helps one to understand why the unit has seconds squared in the denominator.

The Average Acceleration of an Object: Sample Problem

A race car traveling forward at a velocity of +75 m/s crosses the finish line. After 9.0 s, the car has a velocity of +23 m/s. Analyzing the motion in one dimension, what is the car's acceleration?

Solution

Note that $\vec{v}_f = +23\text{m/s}, \vec{v}_i = +75\text{m/s}$ and t = 9.0 s. Substitute these values into the equation for average acceleration:

$$\vec{a} = \frac{\vec{v}_f - \vec{v}_i}{\Delta t} = \frac{(+23\text{m/s}) - (+75\text{m/s})}{9.0\text{s}} = \frac{-52\text{m/s}}{9.0\text{s}} = -5.8\text{m/s}^2$$

The negative sign indicates that the acceleration is in the opposite direction to the motion of the car, and, therefore, the car is slowing down.

Distance Traveled under Uniform Acceleration

When people speak of "uniform acceleration," they mean that the acceleration is constant. When the acceleration of an object is uniform, the velocity of an object changes at a constant rate. The equation used to determine the distance, or magnitude of the total displacement traveled, Δx, by an object starting at rest that is accelerating uniformly with acceleration a is:

$$\Delta x = \left(\frac{1}{2}\right)at^2$$

Consider the previous case, but now suppose the object was not initially at rest. If the magnitude of the object's initial velocity is v_i, the total distance traveled will increase. In this situation, the equation for the distance traveled is:

$$\Delta x = v_i t + \left(\frac{1}{2}\right)at^2$$

How Can Motion Be Represented Graphically?

Graphing Position vs. Time

The one-dimensional motion of an object can be described by a graph of its position over time, where position is plotted as the dependent, y-axis variable and time is plotted as the independent, x-axis variable.

For instance, consider the graph:

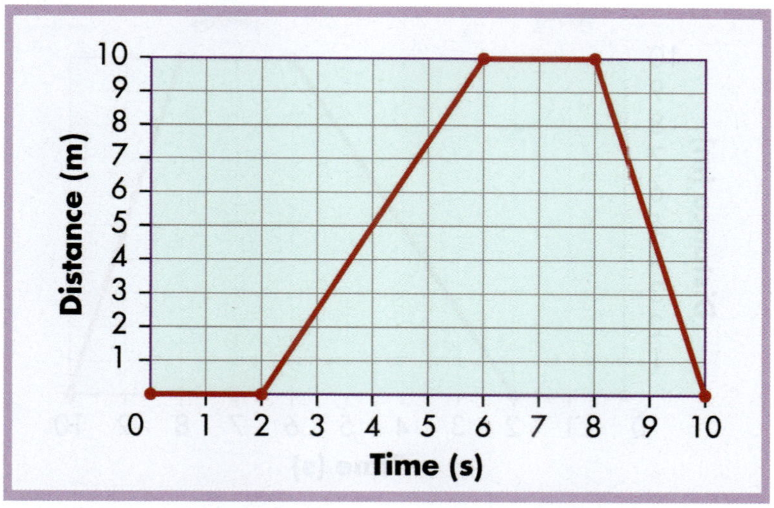

In this graph, an object's position is graphed versus time. How does the object's position change over time?

In the first 2 s, the object's position is not changing. This means that the object is not moving, or as physicists often say, the object is "at rest." After 2 s, the object begins moving because the position begins changing in one dimension over time. The speed of the object can be determined by using the data available on the graph. The object starts at a position of 0 meters at the 2 s mark and travels to a position of 10 m at the 6-s mark. Using this information about position and time, the magnitude of the object's average velocity is:

$$v = \frac{\Delta x}{\Delta t} = \frac{x_f - x_i}{t_f - t_i} = \frac{10\text{m} - 2\text{m}}{6\text{s} - 2\text{s}} = \frac{8\text{m}}{4\text{s}} = 2\text{m/s}$$

Note that because the change in position is equal to the rise of the line between 2 s and 6 s and the time is equal to the run of the line, the velocity is the slope of the line during that time period. Remember that the slope of a line is the rise divided by the run. This is true at every moment in time along the position-time graph and is a useful insight that allows physicists to calculate the average velocities of objects given information about their positions over time. After 6 s, the object stops moving because the position is not changing in the 6-s to 8-s interval. Then, at the 8 s mark, the position begins to decrease. The change in position is negative, so the object is moving backward toward its starting position.

Graphing Position vs. Time: Sample Problem

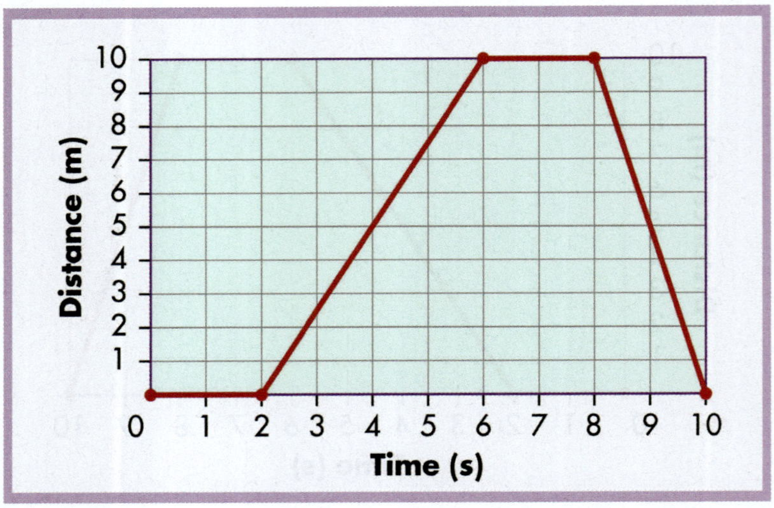

Use the preceding graph to answer the following:

(a) What is the object's average velocity in the time interval between 8 s and 10 s?

(b) What is the object's instantaneous velocity at 9 s?

(c) Do the answers for (a) and (b) differ? Why or why not?

Solution

(a) The average velocity of the object is equal to the slope of the position-time graph:

$$v = \frac{x_f - x_i}{t_f - t_i}$$
$$= \frac{0\text{m} - 10\text{m}}{10\text{s} - 8\text{s}}$$
$$= -5\text{m/s}$$

Note that the average velocity during this time interval equals the slope, which is -5 m/s. The minus sign indicates that the object is moving backward, not that it is slowing down. In physics, the negative sign for displacement or velocity in one dimension indicates direction.

(b) The instantaneous velocity at 9 s is equal to the slope of the line at t = 9 s. This is the same as the slope over the interval between 8 s and 10 s, so the instantaneous velocity is -5 m/s.

(c) The instantaneous velocity is the same as the average velocity in this case. This is because there is no change in the slope of the line in the interval between 8 s and 10 s. The slope at the instant of 9 s is no different from the overall slope during the 2-s interval.

Graphing Velocity vs. Time

The motion of an object can also be described by a graph of its velocity over time. The velocity is plotted as the dependent, *y*-axis variable and time is plotted as the independent, *x*-axis variable. For instance, consider the velocity vs. time graph shown for the same motion described in the preceding position vs. time graph:

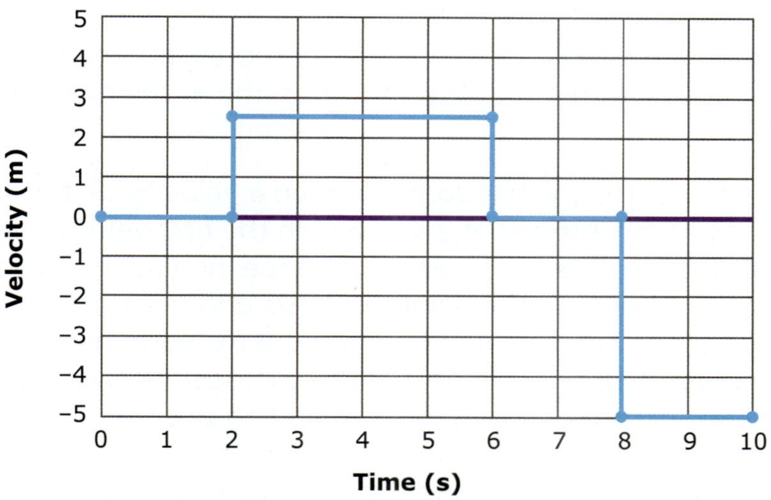

This graph shows changes in distance over time. How does this graph differ from one that shows position vs. time?

In the first 2 s, the object's velocity is zero. This means that the object is not moving. From 2 s to 6 s, the object's velocity is 2.5 m/s. From 6 s to 8 s, the velocity is 0 m/s indicating that the object is again at rest. Finally, the object moves backward with a negative velocity of −5 m/s in the time interval 8 s to 10 s.

Consider a different situation now. In the following velocity vs. time graph, the velocity increases gradually from 0 m/s to 20 m/s over 5 s.

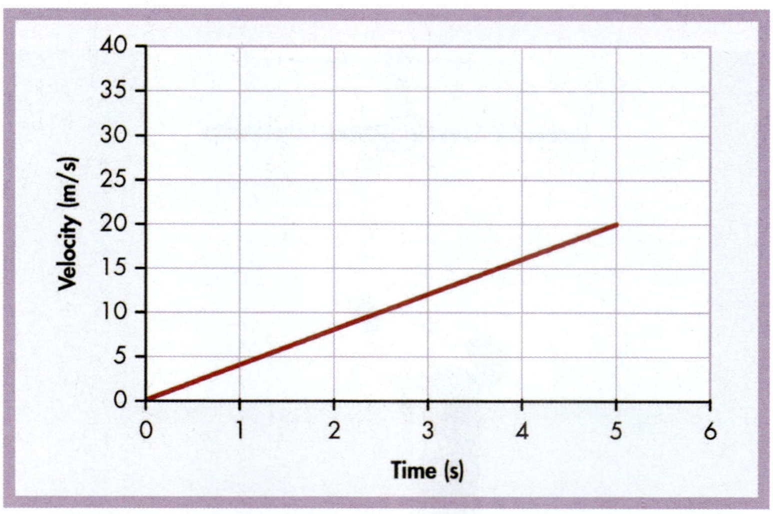

This graph shows changes in velocity over time. How does this graph differ from one that shows distance vs. time?

Just as the slope of the position vs. time graph has significance (it represents the velocity), so has the slope of the velocity vs. time graph. The slope of the velocity vs. time graph is the acceleration. Remember that acceleration is:

$$\vec{a}=\frac{\Delta\vec{v}}{\Delta t}=\frac{\vec{v}_f-\vec{v}_i}{t}$$

On the velocity vs. time graph, the change in velocity represents the rise of the line and the change in time represents the run of the line. Thus, the acceleration is the slope of the velocity vs. time graph.

The area between the curve and the horizontal axis on a velocity vs. time graph also has significance. Consider the time interval of 2 s to 6 s on the first velocity vs. time graph shown. The area under the curve is the width of the time interval multiplied by the height of the graph during that time. Because the height is 2 m/s and the width of the time interval is 4 s, the area is 8(m/s)s. The seconds cancel, leaving 8 m. Note that this is equal to the displacement of the object during that time. Review the position vs. time graph and observe that it moved from a starting position of 2 m to an ending position of 10 m during the time interval of 2 s to 6 s. This example demonstrates that the area under a velocity vs. time graph between two moments in time is the displacement of the object during that time. In fact, this is always true for every time interval on a velocity vs. time graph.

How Do Objects, Such as a Basketball, Move in the Air?

Introduction to Projectile Motion

When a basketball player lines up at the free throw line and shoots, the basketball has what in physics is known as projectile motion. In fact, any object that is launched into the air is known as a projectile and has projectile motion. A projectile is any object launched into the air, with gravity being the only force acting on it. Air resistance is typically ignored because it greatly simplifies the system, and it has only a very small impact on the path of the projectile.

photo: Pixabay

When the player throws the basketball, it follows a trajectory. How can mathematics describe the ball's motions?

As a basketball travels toward the hoop, it follows a curved trajectory. The trajectory is parabolic. It is the same shape that is traced out by a function like $y = -x^2$.

What are the laws of physics that cause projectiles to follow a parabolic path? To understand this, it is important to distinguish between an object's motion in the horizontal (or left-right) direction and the vertical (or up-down) direction. In the horizontal direction, there are no forces pulling on an object once it is in the air (if air resistance is ignored). When there are no forces acting on an object, Newton's second law says that there is no acceleration. When there is no acceleration, the velocity is constant. Therefore, the horizontal component of a projectile's velocity is constant.

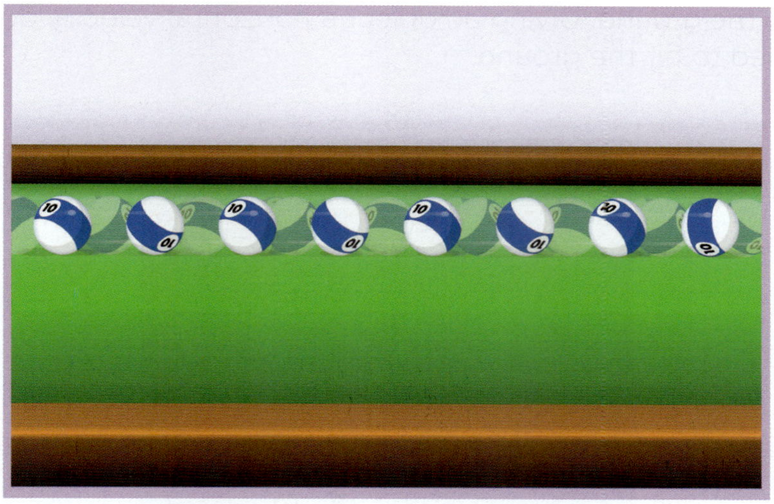

For an object moving at constant speed in the horizontal direction, the velocity is constant. How does the velocity of a projectile compare with an object like this rolling ball?

In the vertical direction, the story is different. The force of gravity pulls the object downward, and, therefore, in accordance with Newton's second law, the object experiences an acceleration that is directed downward. Furthermore, because the force of gravity is constant, so is the acceleration. This causes the object to accelerate in the vertical direction.

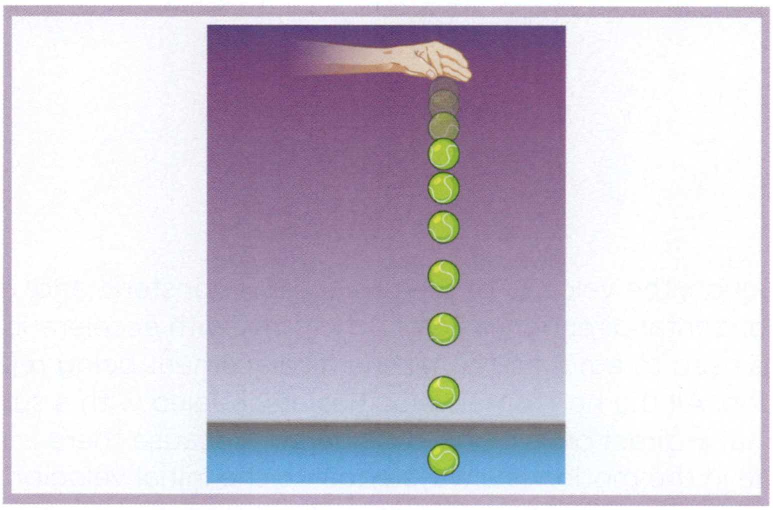

Projectiles experience acceleration in the vertical direction and, therefore, fall a greater and greater distance in each time interval. How could measurement using this time-lapse image demonstrate that?

The parabolic or curved shape of projectile motion is caused by the combination of horizontal and vertical motion. Because the object is traveling at a constant velocity in the horizontal direction but is accelerating in the vertical direction, the distance traveled for each successive time interval is different in the two directions.

Would an object launched sideways (horizontally) fall faster or slower than an object dropped straight down from the same height? (Assume that neither object is given any downward velocity to start with.) The answer is neither. Both objects would hit the ground at exactly the same time. The reason is that the horizontal and vertical components of a projectile's motion are independent of each other. What determines how long an object takes to hit the ground is the rate of acceleration due to gravity, the object's vertical velocity, and its height above the ground. Giving an object a horizontal velocity does not in any way affect the time needed to hit the ground.

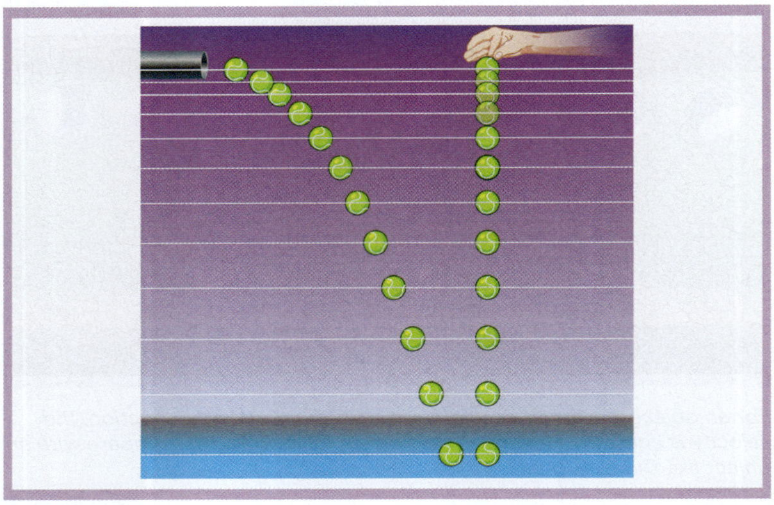

The initial horizontal velocity given to the yellow ball affects the direction of the ball. What explains why both balls hit the ground at the same time?

Equations of Projectile Motion

To describe projectile motion, apply the existing equations of motion for velocity and position to a two-dimensional situation. As a reminder, these equations are:

$$\Delta x = x_0 + v_i t + \frac{1}{2} a t^2$$

$$v_f = v_i + at$$

In the horizontal direction, the velocity of the projectile is constant, and, therefore, the acceleration in the horizontal direction is zero. Any terms with acceleration are therefore zero. The variable x is used to emphasize that the displacement being referenced is only in the horizontal direction. All the horizontal velocities are labeled with a subscript "x" to show that only the horizontal x-direction is being considered. Because there is no acceleration, the velocity v_x at any time in the motion is always equal to the initial velocity v_{ix}. This gives the equation $v_x = v_{ix}$. Because the velocity is constant, the distance in the horizontal direction is just $x = v_x t$.

In the vertical direction, there is an acceleration. That acceleration is caused by the force of gravity and is therefore denoted as *g*. It is vital to use –9.8 m/s², a negative value, for the acceleration due to gravity. The direction of the acceleration must be included. The subscript "y" is used to show that the equations are considered only for the vertical direction. Displacement is now *y*, to emphasize that the direction being referenced is only in the vertical direction.

$$y = y_0 + v_{iy}t + \frac{1}{2}gt^2$$

$$v_y = v_{iy} + gt$$

When you encounter projectile problems within this course, assume the motion starts from the origin of the coordinate system $x_0 = 0$, $y_0 = 0$

Equations of Projectile Motion: Sample Problem

A ball rolls off a 30-m high cliff with a horizontal velocity of 3 m/s. (a) How long does it take to reach the ground below? (b) How far has the ball traveled horizontally when it reaches the base of the cliff?

Solution

(a) To determine how long the ball takes to reach the ground below, use the equations for vertical motion. The height of the cliff is known and can be used for the vertical distance *y* that the ball must travel to reach the ground.

The equation:

$$y = y_0 + v_{iy}t + \frac{1}{2}gt^2$$

can be simplified in this case because the initial velocity in the *y*-direction is zero. The resulting equation is:

$$y = \frac{1}{2}gt^2$$

Rearrange this equation and solve for time:

$$t = \sqrt{\frac{2y}{g}}$$

When the ball hits the ground, it will have a vertical displacement of –30.0 m. Plugging in the known values yields:

$$t = \sqrt{\frac{2(-30.0\text{ m})}{-9.8\text{ m/s}^2}}$$
$$= 2.5\text{ s}$$

(b) To determine how far the ball traveled horizontally before reaching the ground, use the equations for motion in the horizontal direction. The time is known, so apply the equation for horizontal position:

$$x = x_0 + v_x t = (3.0\ \mathrm{m/s})(2.5\ \mathrm{s}) = 7.5\ \mathrm{m}$$

How Can Motion Sensors Be Used to Record and Describe the Motion of an Object?

Measuring Motion

The human eye in combination with the visual cortex (the part of the human brain responsible for vision) has evolved to detect motion in the environment. It is an important survival mechanism. This ability relies on the use of sophisticated biological sensors in our eyes.

Electronics can also play the role of observing and recording motion. In a physics laboratory, stopwatches are used to record how long it takes for something to happen. However, a far more accurate way to measure the motion of an object is to use a device called a photogate or a motion sensor.

A photogate measures the time at which an object passes between two probes on the photogate. When an object blocks an optical or ultrasonic signal being sent between these two probes, the photogate senses that an object is present. This technology can be used to very accurately determine the time at which something happens, such as when a lab cart reaches a certain position on a track.

A motion sensor can record the position of an object over time. A sequence of motion sensors can be used to determine exactly how far away an object is at any point in time by sending ultrasonic sound pulses. The sensor records how long it takes for these pulses to bounce off an object and return to the sensor. Then, a computer interprets the information and determines the distance of the object from the sensor.

Motion sensors and photogates are very useful to students and scientists in the laboratory. They provide much more accurate data about the position of objects over time than more primitive methods such as a stopwatch.

How Does Friction Affect the Motion of an Object?

What Is Friction?

Friction is a force that resists motion. Friction can be felt all around. For instance, when a person slides his or her foot against a rough carpet, the friction created between shoes and carpet requires effort to overcome. Ice, for example, has very low levels of friction, and so is very slippery. However, even ice has some friction.

Friction is caused by the microscopic roughness between two surfaces. Even surfaces that look very smooth to the naked eye are actually quite rough when seen through a microscope. It is this roughness that makes it hard for two objects to slide past one another when a force is applied.

What Is the Difference Between Static and Kinetic Friction?

Static friction is the name given to the type of friction that exists between two objects when the objects are at rest with respect to one another. (*Static* means "stationary.") For instance, if a person tries to push a sofa that is at rest across a carpeted floor, the person is trying to overcome static friction. However, once the sofa is moving, the force of friction that is opposing motion is called kinetic friction. (*Kinetic* means "in motion.")

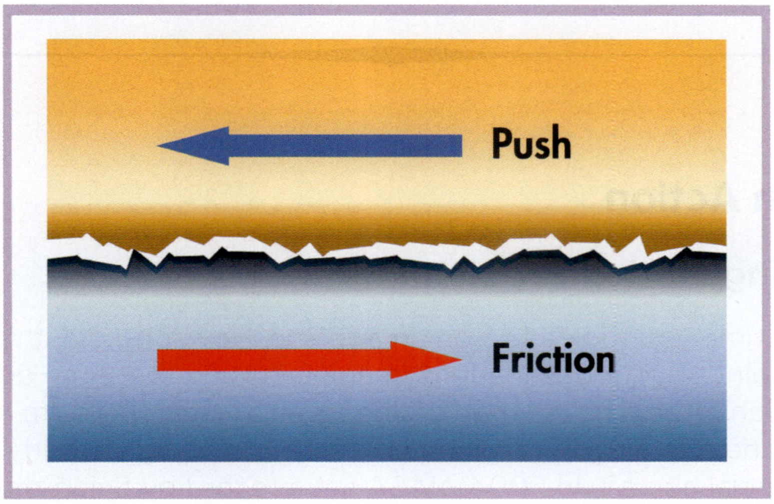

Static friction keeps objects from moving, whereas kinetic friction is the force of friction experienced by objects in motion. How is the motion of an object determined by the two kinds of friction?

Friction Occurs in Fluids

Friction exists not only between solid objects, but also between objects in liquid or gaseous states, including between a fluid and a solid. For instance, boats moving through the water experience a frictional force known as "drag" that is caused by the boat bumping into the water molecules and trying to move past them.

Skydivers experience another type of friction known as air resistance. Skydivers accelerate rapidly, and once they reach high velocities, they experience a significant amount of air resistance caused by the interaction between their bodies and the air molecules around them. It is this type of drag that allows skydivers to slow down enough for a safe landing after their parachute has opened.

The magnitude of the force due to air resistance experienced by a skydiver increases as his or her speed increases. If a skydiver is falling fast enough, air resistance will balance the force of gravity. When this happens, skydivers stop accelerating and they reach what is called "terminal velocity."

Consider the Explain Question:

How do the displacement, velocity, and acceleration of a runner change as he races from the starting line toward the finish line?

Go online to complete the scientific explanation.

Check for Understanding:

A woman pushes a couch, from rest position, across a carpeted floor. How does this example illustrate static and kinetic friction?

STEM in Action

STEM and Solving Motion Problems

Think about the last time you took a trip on an airplane. How long did it take you to travel from your starting point to your destination? Imagine cutting that time down by half, three-quarters, or even more. This is what engineers and researchers are attempting to do. They hope that one day, airplanes will be able to travel at significantly higher speeds than today's commercial aircraft. In 2004, NASA set a record by flying the hypersonic, rocket-powered X-43 A at almost 7,000 miles per hour. That is just over nine times the speed of sound!

How would the use of superfast commercial aircrafts change our world? Imagine flying from New York to Tokyo in about two hours; that is better than the 15 to 17 hours currently required for a flight on today's commercial aircraft. This ability would revolutionize the air travel industry!

However, traveling at extremely fast speeds presents many engineering design challenges. An engineer who designs and builds these types of vehicles is called an aerospace engineer. The drag, or air resistance, that acts on an aircraft increases with increasing speed. Therefore, a superfast aircraft would experience an enormous drag force, which would generate a lot of heat.

When designing these aircrafts, engineers must develop a mechanism for dissipating this thermal energy quickly and efficiently. In spite of these and other design challenges, aerospace engineers are working hard to solve these problems.

These graphs show the horizontal and vertical positions of a ball that is thrown off of a cliff.

Horizontal Position

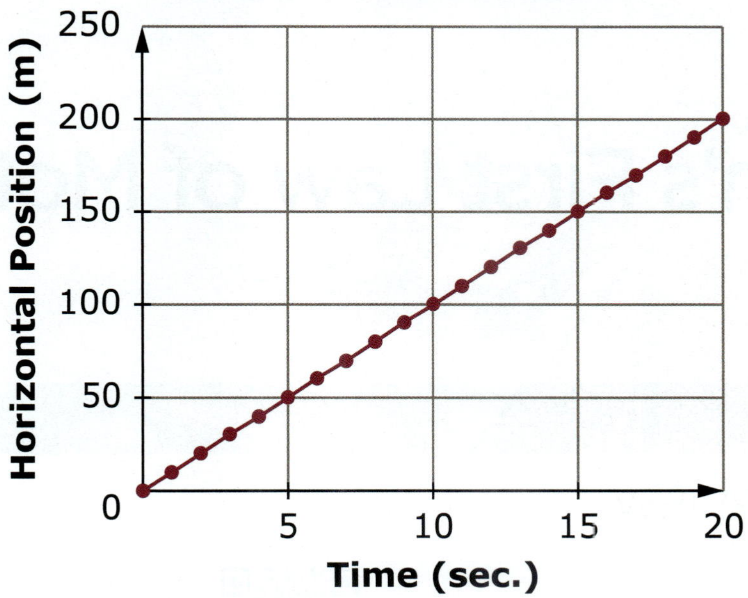

Vertical Position

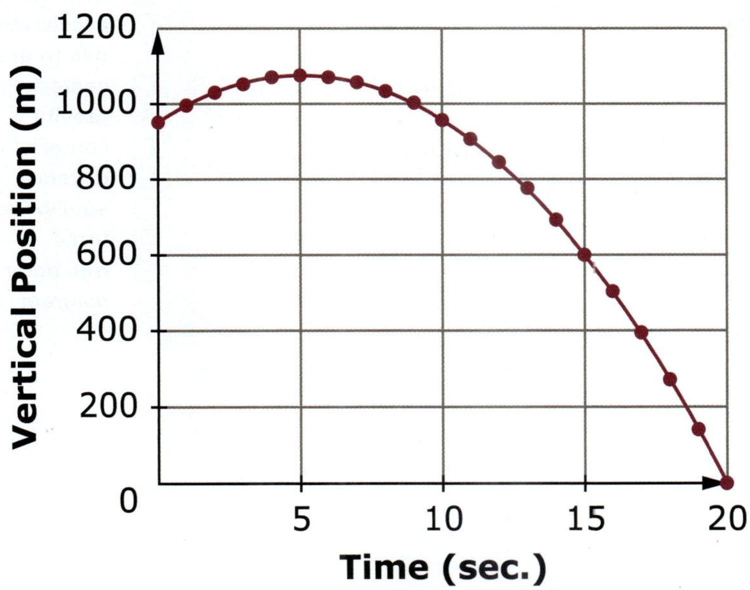

1. Describe the trajectory of the ball as it travels through the air.
2. Describe the horizontal and vertical components of the velocity of the ball as it travels through the air.
3. Describe the horizontal and vertical components of the acceleration of the ball as it travels through the air.

Newton's First Law of Motion

photo: Getty Images

LESSON OVERVIEW

Lesson Question

- What is Newton's first law of motion?

Lesson Objective

By the end of the lesson, you should be able to:

- Analyze data and construct evidence of Newton's first law of motion.

Key Vocabulary

Which terms do you already know?

- ☐ acceleration
- ☐ acceleration due to gravity
- ☐ center of gravity
- ☐ center of mass
- ☐ distance
- ☐ equilibrium
- ☐ force
- ☐ free body diagram
- ☐ gravitational force
- ☐ gravity
- ☐ inertia
- ☐ mass
- ☐ observation
- ☐ speed
- ☐ velocity

ENGAGE

What causes objects to stop moving or change the way they are moving? Visit the Engage page online to learn more.

Explain Question

How can the motion of objects be explained in terms of the balanced and unbalanced forces acting upon them?

EXPLORE

What Is Newton's First Law of Motion?

Newton's First Law of Motion

Aristotle tried to imagine motion in a "void" or "vacuum" and concluded that it would not be possible because there would be no medium, such as air or water, through which an object could travel. Aristotle would have been surprised to learn that someday people would move and work in space. His explanations of motion mostly made sense on the basis of everyday observation, but they were sometimes incorrect. Because of his influential achievements in other areas and because the medieval Church in Europe used his ideas about motion to support its theology, these ideas were accepted for about 2,000 years. Then, in the 1600s, the Italian scientist and astronomer Galileo Galilei performed experiments that led to an improved way of viewing the nature of motion.

One set of Galileo's experiments involved inclined planes. An inclined plane is a flat surface placed at an angle to the direction of the force of gravity. A ramp is an example of an inclined plane.

Many physics motion experiments involve moving objects sliding or rolling up or down inclined planes. In his experiments, Galileo observed that balls rolling from rest down inclined planes accelerate, or increase in velocity. He also saw that balls rolling up inclined planes decelerate until they come to a stop.

Next, Galileo put two inclined planes together. Balls released from rest at the top of one slope rolled down it and up to a stop on the other one. He noticed that a ball would roll up the second plane almost to the same height as its starting point on the first.

The smoother the surfaces, the closer the ball could come to its starting height.

This was true regardless of the angles of the planes, although the distances the balls traveled along them varied. If the second plane was at a steep angle, the ball lost its speed quickly and came to a stop before it had traveled very far. If the second plane was at a shallow angle, the ball lost its speed slowly as it climbed, and it rolled a long distance before stopping. The flatter the angle of the second board, the more slowly the ball lost its speed.

Galileo reasoned that if there were no friction, a rolling ball would lose no speed at all, and it would continue rolling along in a straight line forever.

Aristotle did not see friction as separate from motion and did not imagine what motion would be like without it. He believed that without a force pushing a moving object along, it would slow and stop. Galileo's motion experiments showed that it is not the absence of a force that causes moving objects to slow and stop, but the presence of the force of friction.

Galileo died in 1642. In the same year, Isaac Newton was born. In the late 1600s, Newton developed ideas about motion that still are useful today. He wrote about these ideas in one of the most important scientific books ever published, the *Philosophiæ Naturalis Principia Mathematica*, sometimes simply called the Principia.

In the Principia, Newton explained his ideas about gravity and laid out three "laws" governing the motion of objects. These three laws are now known as Newton's laws of motion. The first of these three laws clarified the concept of inertia. Translating from Latin and expressing the ideas in modern terms, the first law says: "An object will remain at rest or continue moving in a straight line with a constant speed unless acted on by an unbalanced force." In other words, an object that is not moving will tend to stay motionless. An object that is moving will tend to continue moving with the same speed and in the same direction. Objects tend to continue doing whatever they were already doing. The only thing that will change that is an unbalanced force.

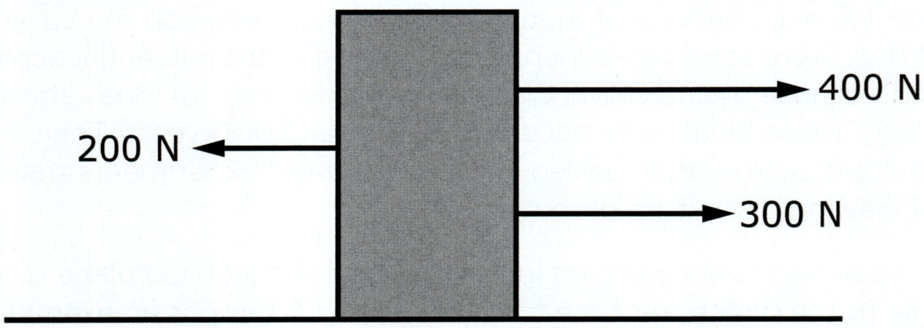

Three forces are acting on this box, with different Newtons (N) of force at different directions. Are the forces on this box balanced?

Usually, several forces act on an object at the same time. A force is a push or a pull that acts on objects. For example, a gravitational force may pull a book down toward Earth's center. A table surface may exert a normal force on the same book, pushing it upward so that it remains at the same height. A student might push on the book along the table to the left. A frictional force to the right might resist the book's sliding motion. How the book actually moves depends on how the forces add up in magnitude and in direction. The student will only be able to move the book across the table if his pushing force overcomes the opposing frictional force. That is, the student has to apply an unbalanced force. On the other hand, the vertical forces might become unbalanced if the weight of the book overcomes the table's ability to provide upward support. If that happened, the gravitational force between the book and Earth would break the table and the book would fall downward.

If all the forces acting on an object exactly cancel out, they are said to be balanced, and the system is said to be in equilibrium. An object will not change its motion if the forces on it are balanced. However, an unbalanced force on an object will cause it to accelerate.

It may seem obvious that an object at rest in a state of equilibrium will not start moving. An unbalanced force will be needed to cause it to move. It is important to remember that Newton's first law also applies to objects that already are moving. An object that already is in motion at a constant velocity is also in a state of equilibrium. The forces on it are balanced. An unbalanced force will be needed to cause a change in its motion as well.

Newton's First Law of Motion: Sample Problem

A car is traveling along a flat, straight highway with a speed of 25 meters per second. If the car's engine provides a forward force of 2,250 Newtons, calculate the force of friction acting on the car.

photo: Pixabay

When a car maintains a steady speed, there are still forces at work. What forces balance to maintain the car's speed?

Solution

Because the car is moving with a constant velocity (speed and direction are unchanged), according to Newton's first law, all forces acting on it must be balanced. The force of friction acting opposite to the direction of motion must therefore be equal to the force of the engine acting in the direction of motion. The force of friction on the car also must be 2,250 Newtons.

Objects in Space

There are only a very few atoms in space, so there is little friction. Because there is less friction, Newton's first law of motion can be demonstrated clearly. In the example of the astronaut's wrench, once nothing is touching the wrench and it is not subject to any unbalanced force, its velocity does not change. It continues indefinitely in the same direction at the same speed.

Inertia

Inertia is a Greek word that means "lazy." The motions of the astronaut's wrench, Galileo's rolling balls, and books on tables all illustrate the property of inertia. Inertia means that objects resist changes in velocity. Their velocities tend to stay constant. Inertia is not a force. Instead, it is a property of matter. Newton's first law of motion is also known as his law of inertia.

Consider the Explain Question:

How can the motion of objects be explained in terms of the balanced and unbalanced forces acting upon them?

Go online to complete the scientific explanation.

Check for Understanding:

Can you describe an example of Newton's first law?

STEM in Action

STEM and Newton's First Law of Motion

How are elevators of today different from those made 40 or 50 years ago? A mechanical engineer might apply Newton's first law of motion to revise elevator design to meet criteria for today's requirements of efficiency and smoothness.

Newton's First Law of Motion

Every time an elevator starts and stops, it experiences a change in velocity. An engineer is required to incorporate application of forces to facilitate these velocity changes.

Use the Exploration "Newton's First Law of Motion" on the STEM in Action page online to identify patterns of velocity of a sled as different forces are applied in opposite directions. Notice how the sled's speed changes when different net forces are applied.

1. Imagine you are riding in a car traveling at a constant velocity holding an open glass of water. The driver stops suddenly and the water sloshes over the glass. Which direction will the water spill from the glass? How can you explain the spill using Newton's first law of motion?

2. Why is Newton's first law of motion sometimes called the law of inertia? Explain using scientific terms.

3. You are at a party and one of the guests performs a demonstration in which he pulls the tablecloth out from under a glass of water on a small table. The glass remains on the table and the water is not spilled. Explain how the glass remains on the table using Newton's first law of motion.

4. What is the net force acting on the table shown? What will happen to the table?

Newton's Second Law of Motion

photo: Getty Images

LESSON OVERVIEW

Lesson Question

- How do forces affect the motion of objects?

Lesson Objective

By the end of the lesson, you should be able to:

- Analyze data and construct evidence of Newton's second law of motion.

Key Vocabulary

Which terms do you already know?

- ☐ acceleration
- ☐ acceleration due to gravity
- ☐ center of gravity
- ☐ center of mass
- ☐ distance
- ☐ force
- ☐ mass
- ☐ net force

ENGAGE

What factors affect the relationship between force and motion? Visit the Engage page online to learn more.

Explain Question

What is the relationship between force and motion?

EXPLORE

How Do Forces Affect the Motion of Objects?

What Causes Motion?

Motion is a central part of existence. Think about all the things that are moved on a daily basis. To get to a class, students pick up their backpacks. To write, someone moves a pencil or pen. To walk, a person must move her legs. What causes all this motion to happen?

The answer is force, and force causes motion. Specifically, an unbalanced force causes a change in an object's motion. For instance, if someone wanted to pick up a backpack that is initially at rest, he would apply a force with his hand. That force causes the backpack to begin moving.

Newton discovered an important relationship between force and motion. He discovered that force causes acceleration, and the more force that is applied, the greater the acceleration. Remember that acceleration is a change in an object's velocity over time, which is a change in the object's motion. The direction of the acceleration is always the same as the direction of the net force.

Think about a bobsled at the start of a race. One of the keys to winning a bobsled race is to give the sled a really strong push at the beginning of the race. The greater the initial force that is applied, the greater the acceleration of the bobsled. The greater the acceleration, the greater the velocity will be as it travels down the track.

Keep in mind that often there are many forces acting on objects. For instance, in the case of the bobsled, there is a small amount of friction from the ice acting on the bobsled. In the case of lifting a backpack, the force of gravity acts against the upward pull from the lifting hand. The acceleration of an object therefore depends not on a single force but on the combination of all forces acting on the object. This is called the net force.

The acceleration increases as the applied force increases, and an object's acceleration is directly proportional to the net force acting on it. That means that if the force doubles, the acceleration doubles. If the force triples, the acceleration triples, and so on.

The mathematical expression of this concept is $\vec{F}_{Net} = m\vec{a}$ where $\vec{F}_{Net}$ is the net force, m is the mass, and $\vec{a}$ is the acceleration. The metric unit for force is Newtons (N). Mass is measured in kilograms (kg), and acceleration is measured in meters per second squared (m/s^2). In the English system, the unit for force is pound-force (lbf) while mass is measured in slugs or stones.

An Object's Mass Matters

Newton discovered that acceleration depends not only on the net force applied but also on the object's mass. In the World's Strongest Man competition, competitors sometimes pull massive 18-wheel trucks by themselves, with nothing more than a harness. While they are able to move these trucks, the acceleration they produce is very small.

Return to the example of the bobsled team. If the team pushed with the same force as before, but now had to push a bobsled loaded with passengers, they would have a much more difficult time. If the mass of the bobsled doubled, the acceleration would be halved. If the mass was tripled, the acceleration would be cut by a factor of 3 (to one-third of the original acceleration). When the force is constant, the acceleration is inversely proportional to the mass of the object.

Because of the truck's large mass, the strongman's force is able to produce only a very small acceleration. How is the truck's acceleration related to its physical properties?

Putting It All Together—Newton's Second Law of Motion

Given that an object's acceleration depends on both the net force and the mass, Newton's second law can be written as:

$$a = \frac{\vec{F}_{\text{Net}}}{m}$$

where *a* represents the acceleration of the object, $\vec{F}_{\text{Net}}$ represents the net force acting on the object, and *m* represents the mass of the object.

Putting It All Together—Newton's Second Law of Motion: Sample Problem 1

Two men push a bobsled with a total force of 250 Newtons. The combined mass of the men and the bobsled is 230 kilograms. Ignoring friction, what is the acceleration of the bobsled?

photo: Pixabay

A women's bobsledding team might have less total mass than a men's team. How would the lower mass of the women's team affect their sled's acceleration compared with the men's?

Solution

Apply Newton's second law. The mass and the force are known, and the acceleration is the unknown. The magnitude of the acceleration is:

$$a = \frac{F_{\text{Net}}}{m} = \frac{250\text{N}}{230\text{kg}} = 1.1\text{m}/\text{s}^2$$

The acceleration is in the direction of the force.

Putting It All Together—Newton's Second Law of Motion: Sample Problem 2

Two men push a bobsled with a total force of 250 Newtons. The combined mass of the men and the bobsled is 230 kilograms. Suppose that there is a constant 30-Newton force due to friction. What is the acceleration of the bobsled?

Solution

Remember that it is the net force that is important when calculating the acceleration. The force of friction opposes motion and therefore acts against the pushing of the sled. Thus, the net force will be the difference of the forces. The magnitude of the acceleration is therefore:

$$a = \frac{F_{\text{Net}}}{m} = \frac{250\text{N} - 30\text{N}}{230\text{kg}} = 0.96\ \text{m}/\text{s}^2$$

It makes sense that the acceleration is slightly smaller when there is an opposing force due to friction.

Consider the Explain Question:

What is the relationship between force and motion?

Go online to complete the scientific explanation.

Check for Understanding:

A 1,382 kg car is at rest. The car begins to accelerate at a rate of 2.5 m/s^2. What is the force that the engine is applying to the car?

STEM in Action

STEM and Newton's Second Law of Motion

Aircraft engines are essentially giant turbines. The engines pull air in the front and push it out the other side. In order to propel the large mass of an airplane forward, the engines need to be extremely powerful. A single engine on a 747 aircraft can produce as much as 282,000 Newtons of thrust. A 747 aircraft has four such engines that, when combined, can produce more than 1 million Newtons of thrust. If the mass of a 747 is about 350,000 kilograms, the magnitude of the maximum acceleration of the aircraft as it takes off down the runway can be computed as follows:

$$a = \frac{F_{\text{Net}}}{m} = \frac{4(282{,}000)\,\text{N}}{350{,}000\,\text{kg}} = 3.2\frac{\text{m}}{\text{s}^2}$$

The acceleration is in the direction of the net force.

photo: Getty Images

In this image, two astronautical engineers prepare a spacecraft destined to explore the atmosphere of Mars. Why do you think the engineers are fully covered by special clothing for this job?

Aerospace engineers design aircraft and spacecraft. Those that specifically design spacecraft are known as astronautical engineers. Aerospace engineers who focus on airplanes and other aircraft are known as aeronautical engineers.

Aeronautical engineers design all parts of an aircraft, including the engines that provide force in the form of thrust. They also design or select the materials used to build the plane, making sure these materials are both lightweight and strong. They run tests on aircraft fuselages with different shapes to determine which design makes the best cargo plane, suitable to haul heavy loads, and which is built for speed and maneuverability. They design wings to provide the plane with maximum lift, making it more efficient and reliable.

Usually, each aeronautical engineer has his or her own specialty and works with other engineers to build planes that have the most acceleration for the force produced by their engines. How do you think the job of an astronautical engineer differs from that of an aeronautical engineer?

photo: Getty Images

Engineers have designed this plane that could be driven on the road or flown. Which variable—force, mass, or acceleration—is constant, no matter where the plane is used?

If you have not already done so, please complete the Hands-On Activity "Exploring the Relationship between Force and Motion" on the STEM in Action page online. Then, answer the following questions.

1. In the diagram of the experimental setup, which portion represents the force?
2. How could you slow the acceleration of the lab cart without altering either the mass on the cart or the mass attached to the end of the string?

Newton's Third Law of Motion

photo: Getty Images

LESSON OVERVIEW

Lesson Question

- When you push on a wall, what does the wall do to you?

Lesson Objective

By the end of the lesson, you should be able to:

- Describe and give an example of Newton's third law of motion.

Key Vocabulary

Which terms do you already know?

- ☐ acceleration
- ☐ acceleration due to gravity
- ☐ center of gravity
- ☐ center of mass
- ☐ distance
- ☐ force
- ☐ mass

ENGAGE

What are the forces involved when a tennis ball is hit with a racket, or a baseball is hit with a bat? Visit the Engage page online to learn more.

Explain Question

When a baseball bat hits a baseball, the force of the bat causes the ball to change velocity, but what happens to the bat?

EXPLORE

When You Push on a Wall, What Does the Wall Do to You?

Newton's Third Law of Motion Defined

One of Newton's most important observations is that a force cannot exist by itself. Rather, forces always come in pairs known as an action and reaction pair of forces.

For example, when a hammer strikes a nail, the hammer exerts a large force on the nail. However, the nail also exerts a force back onto the hammer. We know this has to be the case because the hammer stops moving and the only thing that can cause a change in an object's motion is a force. Put another way, if there was no force, then the hammer would continue moving by Newton's first law of motion.

Similarly, when a child kicks a football, his foot exerts a force on the ball. However, the ball also exerts a force back on the child's foot. Anyone who kicks a football can feel the force of the ball on their foot.

This idea brought Newton to his third and final law of motion: For every action force, there is a reaction force that is equal in magnitude and opposite in direction.

photo: Getty Images

A hammer exerts a force on a nail as it strikes the nail. However, the nail also exerts a force back on the hammer. So, do forces always exist in pairs?

Examples of Newton's third law of motion are literally everywhere. When a book is placed on a desk, the weight of the book exerts a force down on the desk. In response, the desk exerts a force back upward on the book. The upward response force prevents the book from falling through the desk. If someone leans against a wall, they exert a force on the wall. However, the wall also exerts a force back on the person. Without the response force, the person would go crashing through the wall.

Newton's third law of motion can also explain why birds fly and how rockets take off. When a bird flaps its wings in the air, the wings exert a downward force against the air. In return, the air exerts an upward reaction force (or "lift") on the bird, which is what allows the bird to stay in the air. Rockets rely on a similar principle. The rocket pushes downward on the gases that exit the rocket. In return, the gases push upward on the rocket, causing the rocket to take off.

Identifying the Action and Reaction Forces

One of the easiest ways to identify the action and reaction forces is to identify one of the forces and then switch the language in the sentence to identify the other. For instance, in the case of a bird flapping its wings, we could say the action is "the bird's wings pushing down on the air." If that is the case, then to identify the reaction force, simply switch the order of the objects in the sentence and reverse the direction of the force. For instance, the reaction force is "the air pushing back upward on the bird's wings."

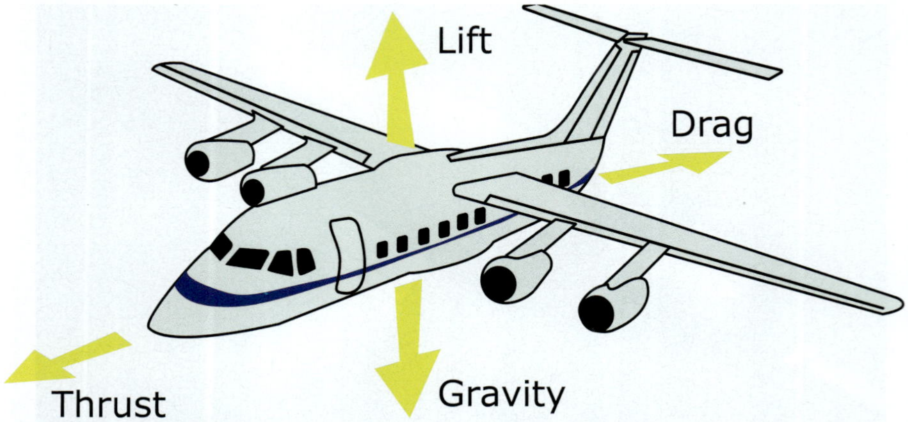

Accomplishing flight meant finding ways to fight the forces of gravity and drag. What forces were used to balance these, allowing human flight?

The following question is often asked: How can the action force be recognized apart from the reaction force? There is no need to worry about this. It does not matter which is which. Either force can be labeled as the action and the other as the reaction. What is important is not which is which, but that there are always two forces and these two forces must always be equal in magnitude and opposite in direction.

The Importance of Mass

Newton's second law says that forces affect the motion of objects. It would make sense to think that because action and reaction forces are equal in magnitude, they must affect the motion of the two objects involved in the same way. However, this is not always true because force is not the only factor involved. Mass is important, too.

The Importance of Mass: Sample Problem

Two boys are floating in a swimming pool on their own inner tubes. What will happen when the lighter boy pushes the heavier boy away?

A 40.0-kilogram boy is lying in an inner tube in a swimming pool when a larger boy with a mass of 80.0-kilogram floats by in his own inner tube. The first boy pushes the second with a force of 60.0 Newtons. Assuming that the mass of each inner tube can be ignored, what is the acceleration of each boy and his inner tube?

Solution

If the first boy pushes the other boy's inner tube (the action force), then the reaction force is the other boy's inner tube pushing back on the first boy's hand. We know these forces are equal in magnitude but opposite in direction. Thus:

$$\begin{aligned} F_{action} &= -F_{reaction} \\ F_{action} &= 60\text{N} \\ F_{reaction} &= -60\text{N} \end{aligned}$$

Note that the negative sign for the reaction force indicates that the direction is opposite. We also know that the action force causes the second boy's inner tube to accelerate in the water, and the reaction force causes the first boy's inner tube to accelerate in the opposite direction (because the reaction force is in the opposite direction). From Newton's second law:

$$a = \frac{F}{m}$$

Use the forces and masses to determine the accelerations:

$$\begin{aligned} a_{1stBoy} &= \frac{F_{1stBoy}}{m_{1stBoy}} \\ &= \frac{-60\text{N}}{40.0\text{kg}} \\ &= -1.50\ \text{m}/\text{s}^2 \end{aligned}$$

$$\begin{aligned} a_{2ndBoy} &= \frac{F_{2ndBoy}}{m_{2ndBoy}} \\ &= \frac{60\text{N}}{80.0\text{kg}} \\ &= 0.750\text{m}/\text{s}^2 \end{aligned}$$

Note that the first boy accelerates away at twice the rate of the second boy. Newton's second law:

$$F = ma,$$

indicates that if the force is constant, then mass and acceleration are inversely proportional. If the mass is doubled, the acceleration must be halved. The accelerations have opposite signs because they are in opposite directions.

Do Action and Reaction Forces Cancel Out?

Because action and reaction forces are equal in magnitude and opposite in direction, it is tempting to think that they cancel out. However, they do not. Consider the diagram and note which objects the action and reaction forces act on. The action force acts on the nail, while the reaction force acts on the hammer. Because action and reaction forces act on different objects, they do not cancel out. Equal and opposite forces cancel only when they are both acting on the same object, but this never happens with an action and reaction pair.

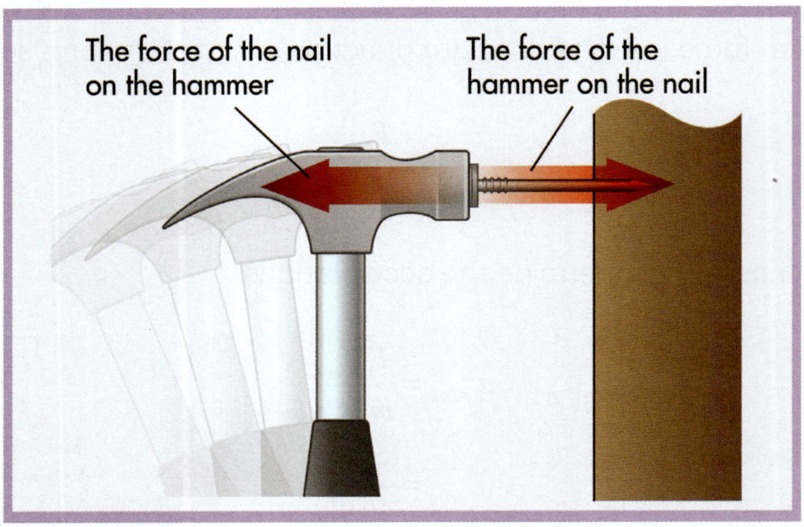

The action and reaction forces between a hammer and a nail do not cancel out because the forces act on different objects. What is the action force and what is the reaction force?

Examples of Newton's Third Law at Work

Newton's third law of motion is universally true for all forces and in all circumstances. For example, gravity is an attractive force between any two objects that have mass. Objects on Earth tend to fall toward the ground because the objects experience an attractive gravitational force toward the center of Earth. When a person drops a pen, Earth exerts a gravitational force on the pen. Because of Newton's third law of motion, we know that there must also be a force equal in magnitude but opposite in direction. What is this force? It is the gravitational force that the pen has on Earth. Just as the pen accelerates toward the center of Earth, the center of Earth accelerates toward the pen. We do not notice the acceleration of Earth because it is extremely small when compared with that of the falling pen.

Newton's third law of motion also explains why the pen stops moving when it reaches the ground. If gravity exerts a force between any two objects, why does the pen not accelerate all the way to the center of Earth? That is due to the normal force that is equal in magnitude but opposite in direction to the weight of the object. The normal force is exerted by the ground on the pen. Because of the normal force, the pen stops accelerating toward Earth, and Earth stops accelerating toward the pen when the pen is resting on the ground.

Just as Newton's third law of motion applies to gravity, it is also applicable to electrostatic forces. When two point charges are placed by each other, they experience an attractive or repulsive force that is proportional to the magnitude of their charges, depending on whether they bear the same charge or opposite charge. If two positive charges or two negative charges are placed by each other, they each will repel the other so that both point charges accelerate away from each other. If a positive and a negative charge are placed by each other, they both will accelerate toward each other. In all these cases, both point charges experience forces of the same magnitude but opposite in direction.

Newton's third law of motion can also explain why dragging objects across the floor leaves scratches. Friction is the force that opposes the momentum of an object as it moves. It is proportional to the normal force and is exerted in the direction opposite to the direction of motion. Friction is caused by the rubbing of two surfaces. When a heavy box is dragged across the floor, the floor exerts a frictional force on the box. Because the floor exerts a frictional force on the box, the box must also exert a force on the floor. This force on the floor can sometimes displace the surface of the floor, leaving scratches.

Consider the Explain Question:

When a baseball bat hits a baseball, the force of the bat causes the ball to change velocity, but what happens to the bat?

Go online to complete the scientific explanation.

Check for Understanding:

Can you explain why a cannonball exiting a cannon moves more than the cannon?

STEM in Action

STEM and Newton's Third Law of Motion

Newton's third law of motion states that for each force acting on a body, the body exerts another force that is equal in magnitude and opposite in direction. Engineers use their understanding of this law to design machines, infrastructures for our communities, and processes for the production of goods that apply concepts of action and reaction forces.

Aerospace engineers design rockets based on Newton's laws. All rocket engines apply force that produces reaction force. This reaction force is greater than the force of gravity, so the craft rises from the ground. A rocket must maintain this force against gravity for several minutes, using a limited amount of fuel in the process. Rockets also use small quantities of fuel to move in space, where gravitational force is small. Because the fuel makes up a part of the rocket's overall mass, the rocket's design must take the mass of the fuel into account. Here again, engineers use Newton's third law of motion as the guiding principle for the spacecraft's design.

Aerospace engineering is just one of many careers where forces are manipulated using Newton's third law of motion. This type of engineering specifically involves the design of rockets and spacecraft, with particular attention to the propulsion systems, so they can exert the necessary forces for motion with the most efficient use of fuel. Aerospace engineers also work on jet engines for airplanes flying close to Earth.

Complete the activity "Newton's Third Law of Motion" on the STEM in Action page online. Then, answer the questions:

1. What action-reaction pair caused the balloon to move?
2. Can force be calculated from the information obtained from the Exploration? Why or why not?

Applying Newton's Laws of Motion

photo: Getty Images

LESSON OVERVIEW

Lesson Questions

- In which situations are Newton's laws of motion not applicable?
- How can Newton's laws of motion be applied to practical problems involving forces and motion?

Lesson Objectives

By the end of the lesson, you should be able to:

- Understand that Newton's laws of motion provide good approximations except for situations in which the speed of an object is close to the speed of light or the object is so small that quantum effects are significant.
- Understand that Newton's laws of motion can be applied to practical problems involving forces and motion.

Key Vocabulary

Which terms do you already know?

- ☐ **acceleration**
- ☐ **acceleration due to gravity**
- ☐ **center of gravity**
- ☐ **center of mass**
- ☐ **distance**
- ☐ **electron**
- ☐ **force**
- ☐ **free body diagram**
- ☐ **gravitational force**
- ☐ **gravity**
- ☐ **mass**
- ☐ **net force**
- ☐ **Newton's laws**
- ☐ **quantum mechanics**
- ☐ **speed**
- ☐ **vector**
- ☐ **velocity**
- ☐ **wave**

ENGAGE

If you observe a sports competition, you notice that athletes who are participating in the events are almost continuously moving. How can you describe the forces involved in this motion? Visit the Engage page online to learn more.

Explain Question

How do Newton's three laws of motion explain the movement of people and objects around you?

EXPLORE

In Which Situations Are Newton's Laws of Motion Not Applicable?

Motion at High Speeds

Newton's laws of motion use the concepts of mass, inertia, force, velocity, acceleration, and time. These ideas have fairly straightforward definitions in physics. However, at very high speeds, things change in odd and interesting ways. For example, physicists working with high-speed objects use two terms for mass: rest mass and relativistic mass. They are the same for an object at rest, and very nearly the same for objects moving at everyday speeds. Curiously, at speeds approaching the speed of light, the measurement of an object's mass increases sharply and is said to become "relativistic." The faster an object moves, the greater its relativistic mass. In addition, as an object's speed increases, the passing of time is perceived more slowly by outside observers.

Newton's laws do not work well at very high speeds because mass and time behave differently. The theory of relativity is needed to accurately explain and predict the behavior of objects at speeds approaching the speed of light. The particular speed at which relativistic effects become important varies with the application.

Motion at Small Scales

Some models of atoms describe electrons whizzing around the nuclei like tiny satellites. One might expect from these models that the motion of an electron would be similar to the motion of a satellite. However, the motion of subatomic particles cannot be analyzed by using Newton's laws.

For example, a satellite in space can occupy any of a range of stable orbits determined by its distance from Earth and its speed. However, electrons may occupy only certain orbits around a nucleus. This is because matter at the atomic scale can be considered not only as particles but also as waves. Matter waves are like other waves in that they can interfere constructively and destructively. Waves of electrons must constructively interfere to form closed standing orbits, which is only possible at certain radii from the nucleus, and at certain energies. These restrictions complicate their motion.

Also, the wave nature of matter means that the position of an electron at any given moment is not absolutely determined. Newton's laws assume that the position of the moving body is known, but scientists can only discuss an electron's probable positions. In 1925, the scientist Erwin Schrödinger developed an equation that describes the behavior of matter over time at very small scales, much as Newton's laws and equations describe it at ordinary human scales. It is the basis of the theory of quantum mechanics.

How Can Newton's Laws of Motion Be Applied to Practical Problems Involving Forces and Motion?

Known Forces Affecting Motion

Within their range of applicability, Newton's laws of motion provide excellent explanations and predictions of the motion of objects.

The equation forms of Newton's second law express the relationship between force, mass, and acceleration. The equation can be rearranged algebraically to solve for any of the three quantities:

$$\vec{F} = m\vec{a},$$

$$m = \frac{F}{a}, \text{ and}$$

$$\vec{a} = \frac{\vec{F}}{m}.$$

Known Forces Affecting Motion: Sample Problem

Find a force that could accelerate a 0.5-kilogram toy car by 2 meters per second squared.

Solution

Use the $\vec{F} = m\vec{a}$ form of Newton's second law and substitute the known values:

$$\begin{aligned}\vec{F} &= m\vec{a} \\ &= (0.5\text{kg})\,(2\text{m/s}^2) \\ &= 1\text{N}\end{aligned}$$

Known Forces Affecting Motion: Sample Problem

A 2.00-kilogram mass is given a 30.0 Newton push. Find the acceleration of the mass.

Solution

Use the $\vec{a} = \frac{\vec{F}}{m}$ form of Newton's second law and substitute the known values.

$$\vec{a} = \frac{\vec{F}}{m} = \frac{30.0\ \text{N}}{2.00\ \text{kg}} = 15.0\ \text{m/s}^2$$

Net Force and Acceleration

Sometimes, there are forces on an object that do not result in an acceleration. For example, gravity is pulling down on a person sitting at rest in a chair. If the person does not accelerate toward the ground, the upward force provided by the chair must be balancing the downward gravitational force. The total or net force is zero, so the acceleration is zero. The key point is that Newton's second law is about the net force.

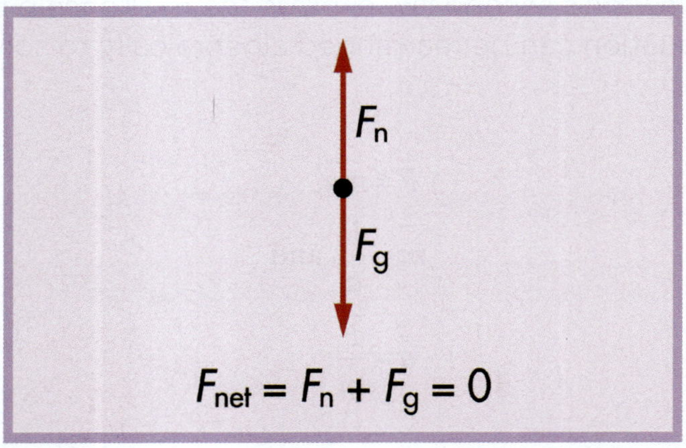

The downward force of gravity on the person sitting on the chair, F_g, is balanced by the upward normal force of the chair on the person, F_n. What would happen if the two forces were not balanced?

Objects can be moving without accelerating. For example, a child dragging a toy across the floor may be pulling it along at a constant velocity. Again, if there is no acceleration, the child's pulling force must be balancing the frictional forces. Again, the net force is zero, so the acceleration is zero.

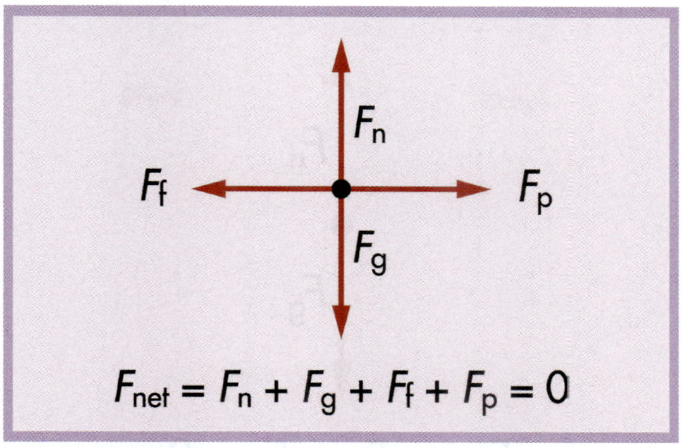

The downward force of gravity on the toy, F_g, is balanced by the upward normal force of the floor on the toy, F_n. The force of the child pulling to the right, F_p, is balanced by the force of friction, F_f, acting to the left. What would happen if the left-right forces were unbalanced, but the up-and-down forces remained balanced?

In each case of zero acceleration, the net force is also zero. Sometimes, people say the forces "cancel each other out." A word of caution: The person sitting in the chair feels the chair pushing up on him, and the chair bears the weight of the person pushing down. Forces are present and acting; they are not canceled, but balanced.

Common Applications

Newton's laws of motion apply in most situations where objects are subject to forces, including gravity. Some common classes of physics problems include those involving inclined planes (ramps), circular or rotational motion, pulleys, falling objects, and projectiles.

A projectile is an object that is fired or launched and then flies freely. Unlike a rocket, it has no means of propulsion of its own. After launch, typically it is considered to have only the force of gravity acting on it, and no air resistance.

Projectile motion can illustrate all three of Newton's motion laws. The velocity vector of a projectile typically has horizontal and vertical components. The horizontal component at each point along the projectile's path demonstrates Newton's first law: If no net force acts on an object, it will continue in straight-line motion at constant speed. The vertical components demonstrate Newton's second law: A gravitational force acts on the projectile's mass and accelerates it straight down toward the ground. As a result, the vertical velocity decreases at a constant rate. This means that initially, it slows its upward motion and stops, and then increases its downward motion. At the moment of launch, the cannon, the person, or whatever is providing the launching force recoils back in accordance with Newton's third law.

Total Force and Acceleration

Sometimes, there are forces on an object that do not result in an acceleration. For example, gravity is pulling down on a person sitting at rest in a chair. If the person does not accelerate toward the ground, the upward force provided by the chair must be balancing the downward gravitational force. The total force is zero, so the acceleration is zero. The key point is that Newton's second law is about the net force.

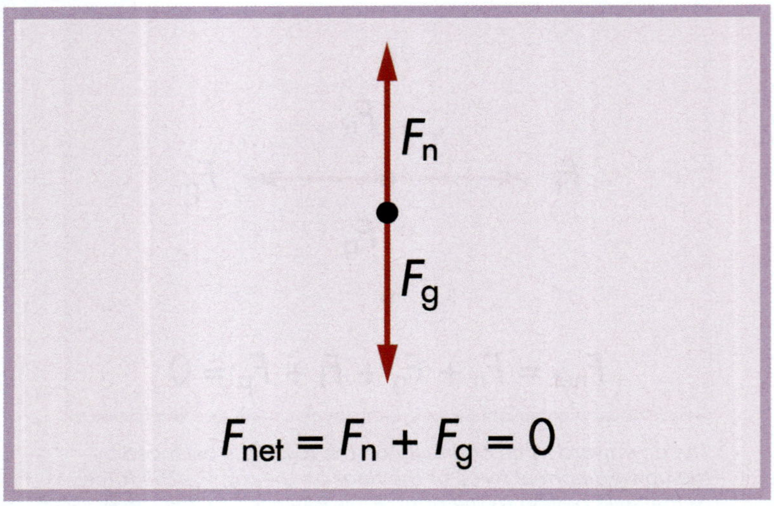

Free Body Diagram of a Chair. The downward force of gravity on the chair, F_g, is balanced by the upward normal force of the floor on the chair, F_n. The net force is zero.

Objects can be moving without accelerating. For example, a child dragging a toy across the floor may be pulling it along at a constant velocity. Again, if there is no acceleration, the child's pulling force must be balancing the frictional forces. Again, the total force is zero, so the acceleration is zero.

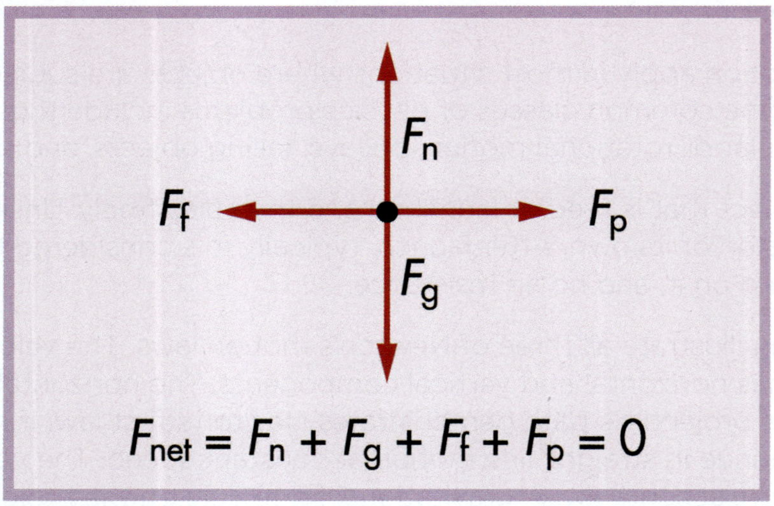

Free Body Diagram of a Child Dragging a Toy. The downward force of gravity on the toy, F_g, is balanced by the upward normal force of the floor on the toy, F_n. The force of the child pulling to the right, F_p, is balanced by the force of friction, F_n, acting to the left. The net force is zero.

In each case of zero acceleration, the total force is also zero. Sometimes people say the forces "cancel each other out." A word of caution: The person sitting in the chair feels the chair pushing up on him, and the chair bears the weight of the person pushing down. Forces are present and acting; they are not canceled, but balanced.

Common Applications

Newton's laws of motion apply in most situations where objects are subject to forces, including gravity. Some common classes of physics problems include those involving inclined planes (ramps), circular or rotational motion, pulleys, falling objects, and projectiles.

A projectile is an object that is fired or launched and then flies freely. Unlike a rocket, it has no means of propulsion of its own. After launch, typically it is considered to have only the force of gravity acting on it, and no air resistance.

Projectile motion can illustrate all three of Newton's motion laws. The velocity vector of a projectile typically has horizontal and vertical components. The horizontal component at each point along the projectile's path demonstrates Newton's first law: If no net force acts on an object, it will continue in straight-line motion at constant speed. The vertical components demonstrate Newton's second law: A gravitational force acts on the projectile's mass and accelerates it straight down toward the ground. As a result, the vertical velocity decreases at a constant rate. This means that initially it slows its upward motion and stops, and then increases its downward motion. At the moment of launch, the cannon, the person, or whatever is providing the launching force recoils back in accordance with Newton's third law.

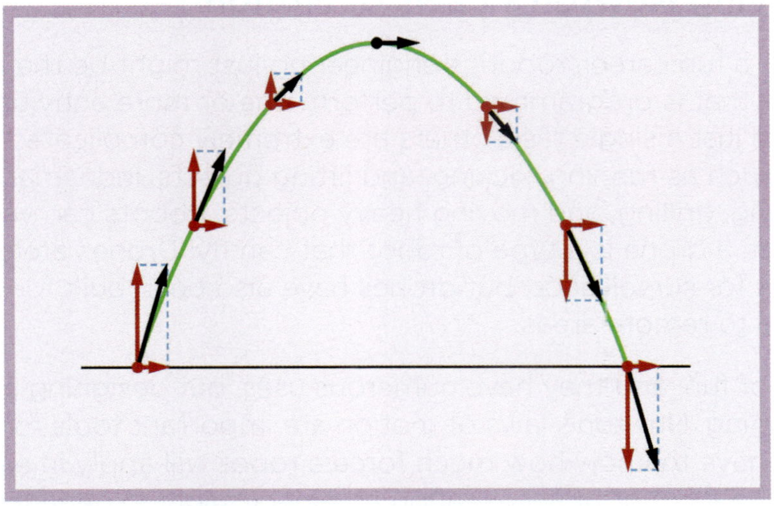

Note that while the horizontal components of the velocity are the same throughout the trajectory, the vertical components are subject to acceleration due to gravity.

More details about projectile and other kinds of motion are given in the concept *Solving Motion Problems*.

Consider the Explain Question:

How do Newton's three laws of motion explain the movement of people and objects around you?

Go online to complete the scientific explanation.

Check for Understanding:

Can you describe two situations in which Newton's laws do not give accurate descriptions of an object's motion?

STEM in Action

STEM and Applying Newton's Laws of Motion

If you are looking for a fun career, robotics engineering just might be the thing for you. A robot is a machine that is programmed to perform one or more activities. Although some robots perform just a single task, others are extremely complicated and can perform sophisticated tasks such as running, kicking, and lifting objects. Industrial applications for robots include welding, drilling, and moving heavy objects. Robots can vacuum your floor or cut your grass. In fact, a drone is a type of robot that can fly! Drones are used by the military and law enforcement for surveillance, but drones have also been built for delivering packages and medical supplies to remote areas.

Robots can be a lot of fun, and they have numerous uses, but designing and programming them can be challenging. Newton's laws of motion are important tools for engineers who design robots. They have to know how much force a robot will apply in each situation and, just as importantly, the amount of acceleration an object might experience as a result of that force. Newton's third law of motion is also a consideration. If a robot is programmed to exert a force on another object, engineers have to make sure the robot can withstand the same amount of force being exerted back on the robot.

Applying Newton's Laws of Motion

Suppose you are an engineer responsible for designing the robot hand in the following picture.

photo: Getty Images

1. Describe factors that you would have to consider in order for the robot to be able to grasp items, pick up loads, and push against objects.
2. Explain an example of how you would apply each of Newton's three laws of motion in your design and construction.
3. Describe mathematical calculations related to Newton's laws of motion that you would have to perform as part of the design and construction.

Work and Power

photo: Getty Images

LESSON OVERVIEW

Lesson Questions

- How are work problems solved, including those in which force and displacement vectors are at an angle?
- How are power problems solved?

Lesson Objectives

By the end of the lesson, you should be able to:

- Analyze and solve problems involving work, including those in which the force and displacement vectors are at an angle.
- Solve problems involving power.

Key Vocabulary

Which terms do you already know?

- ☐ conservation
- ☐ displacement (vector)
- ☐ displacement (water)
- ☐ force
- ☐ gravitational force
- ☐ gravity
- ☐ Joule
- ☐ kinetic energy
- ☐ potential energy
- ☐ power
- ☐ scalar
- ☐ significant digit
- ☐ speed
- ☐ uniform circular motion
- ☐ vector
- ☐ work

ENGAGE

When a martial artist uses his or her hand and arm to transfer force into a stack of blocks, what happens? Visit the Engage page online to learn more.

Explain Question

How do you solve work and power problems?

EXPLORE

How Are Work Problems Solved, Including Those in Which Force and Displacement Vectors Are at an Angle?

Basic Work Definition

Work has a special definition in physics: Work is done when a force acts on an object and causes it to move through a displacement.

Work is "done on objects" by forces, which are vector quantities. In order for the force to perform work, it is not enough for the force to simply act on the object; the object must also move to a new location. The change in the object's location can be described with a displacement vector.

When an object is displaced a distance d by a constant force F applied in the same direction as the displacement, the amount of work done on the object is given by:

$$W = Fd$$

Work is a scalar quantity.

This formula applies only when the displacement and the force are parallel. It turns out that only the force that is parallel to the displacement is considered to be doing work. A more general formula will need to be developed for the case that the displacement and force vectors are not in the same direction.

If the force is applied perpendicular to the displacement, no work is done. For example, if a person walks across a level floor, gravity does not do any work on them because the force of gravity acts at a right angle to the displacement.

The unit of work is Joules, named after the English physicist James Joule. A Joule, which is equivalent to a Newton meter, is the amount of work done in applying a force of 1 Newton through a displacement of 1 meter. The Joule is also the unit for energy.

General Formula

Sometimes, the force vector is not in the same direction as the displacement vector. In this case, there is an angle θ between the force and the displacement. Only the component of the force vector that is parallel to the displacement vector is considered to be doing work.

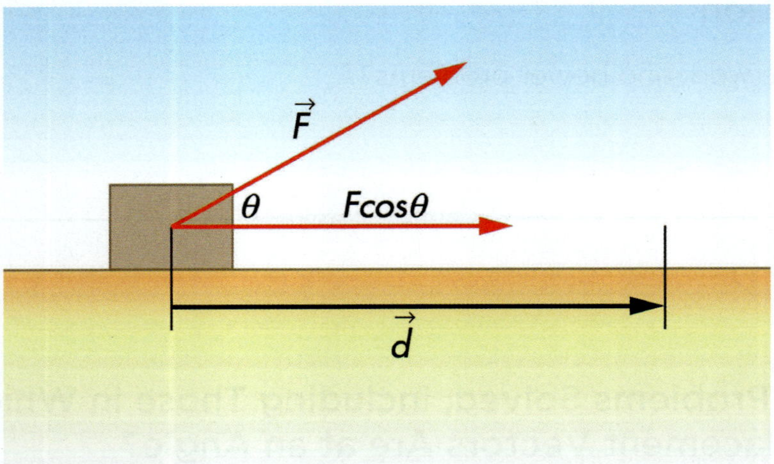

Only the component of the force vector that is parallel to the displacement vector performs work on the object. How can you find this component mathematically?

For example, an object's change in height leads to a change in its gravitational potential energy. This change is the result of work done by gravity. Even if gravity does not cause an object to move straight down, gravity still does work on the object. Only the component of the force of gravity in the same direction as the displacement is considered to have done work.

Suppose that the magnitude of a force is F and the magnitude of a displacement is d, and that θ is the angle between the force and displacement vectors. Using trigonometry, the magnitude of the component of the force in the same direction as the displacement is $F \cos(\theta)$, where θ is the angle between the force and displacement vectors. Thus, the general work formula is:

$$W = Fd\cos(\theta)$$

Depending on the value of $\cos(\theta)$, work can be positive, negative, or zero. When the force and displacement vectors are parallel, θ is equal to 0°, and then $\cos(\theta) = \cos(0°) = 1$, and thus the formula reduces to its basic form:

$$W = Fd$$

It is when $\theta = 0°$ that the work done on an object is at its maximum positive value because at that angle, the cosine function is at its maximum.

When $\theta = 180°$, the value of the cosine function is –1. The work done is at its least (most negative) value. In this case, the work directly opposes the motion. Friction is an example of a force that does negative work on objects.

When the force and displacement vectors are perpendicular to each other, $\theta = 90°$ and then the value of the cosine is $\cos(\theta) = \cos(90°) = 0$. No work is done. In other words, forces (or components of forces) that are perpendicular to the displacement vector cannot do work on the object.

General Formula: Sample Problem 1

An object is moving to the east at 15 meters per second when a pair of balanced forces is applied to the object over 85 meters. The first force is 25 Newtons to the north, and the second force is 25 Newtons to the south. What is the work done by the first force?

Solution

The force is balanced, so it does not result in a change in direction. The 85-m displacement is to the east. Work is calculated by using the formula:

$$W = Fd\cos(\theta)$$

The angle between the displacement vector and the first force is 90°, so the work done is:

$$W = (25\,\mathrm{N})(85\,\mathrm{m})\cos(90°) = 0$$

No work is done by a force applied perpendicular to a displacement.

There are often several angles present in physics work problems, which can cause confusion. For example, students may be asked to find the work done by a force pulling an object up a plane at a given angle. The angle may even be labeled θ because that is a common name for an angle. However, this is not the same angle as is used in the general work formula. In the work formula, θ refers only to the angle between the force vector and the displacement vector.

General Formula: Sample Problem 2

A 40-N force pulls an object at a 60° angle with respect to the horizontal. Due to friction, the object moves at a constant velocity. The object moves 5 m horizontally to the right. How much work is done on the object by the force?

Solution

Only the portion of the force that is parallel to the 5-m displacement of the object does work on it. Substituting it into the general formula for work gives:

$$W = Fd\cos\theta$$
$$W = (40\,\text{N})(5\,\text{m})\cos(60°)$$
$$W = 100\,\text{J}$$

Work and Energy

One definition of energy is that it is the capacity to do work. Some people define work as the transfer of energy from one object or system to another or from one form to another.

Work, with its units of Joules, is equivalent to energy. Work is not exactly energy. While objects can "possess energy," they cannot "possess work." Work is related to energy in that it is the capacity to change the amount of mechanical energy of an object. An object's mechanical energy is the sum of the potential energy associated with its position and the kinetic energy of its motion. Work's effect on the object's mechanical energy depends on the kind of force involved. Forces can be categorized as internal and conservative or external and nonconservative. Both involve a transfer of energy.

Internal, conservative forces include gravity, electric force, magnetic force, and spring forces. For these forces, work changes only the form, not the amount, of the mechanical energy that the object possesses.

For example, when gravity does work on a falling object, it changes the object's potential energy of position into kinetic energy of motion. The object's total amount of mechanical energy is conserved. The work done on the object is equal to the amount of energy that was converted from one form to another.

External, nonconservative forces include applied forces, friction, air resistance, tension, and normal forces. External forces can do work to increase the object's total amount of mechanical energy by transferring some to it or to reduce its energy by transferring some away. In a larger sense, the total amount of energy in the universe is still conserved. However, work by nonconservative forces can move energy from one object or system to another.

For example, a golfer increases a golf ball's total mechanical energy by applying an external force to it with the golf club. The golf ball acquires kinetic and potential energy equal to the amount of work done to transfer energy from the moving club to the ball.

As another example, the work done by a frictional force on a book sliding across a table removes some of the book's kinetic energy. Energy is not "lost." The work done by friction transfers some of the book's kinetic energy by converting it to thermal energy. The sliding book's total mechanical energy is reduced by the amount of work done by the frictional force.

How Are Power Problems Solved?

Definition of Power

The work equations do not use time. The amount of time that a force is applied is not relevant to the amount of work done by the force. A person lifting a 5-kilogram box up 2 meters does the same work whether they do the task in 4 seconds or 12 seconds.

Instead, power is the scalar quantity that considers the rate at which the work is performed. Another way to think about power is that it is the rate at which energy is expended. The formula for power is:

$$P = \frac{W}{t},$$

where *W* is the amount of work done and *t* is the time in which the work is done. The arrangement of terms in the formula shows that an increase in the amount of power can do the same work in less time or more work in the same time.

The usual SI units are Joules for work and seconds for time. The unit for power, the watt (W), is equivalent to one Joule per second. The watt is named for the Scottish engineer James Watt.

Definition of Power: Sample Problem

An 82-kilogram (kg) athlete does a chin-up to lift his body 0.33 meters (m) in 1.5 seconds. What is the power expended by his muscles?

Solution

With work done against gravity, the force required is:

$$F = mg$$

The work done by this force is:

$$W = Fd = (mg)d$$

Power is the amount of work divided by the time over which the work is done:

$$P = \frac{W}{t}$$

Substituting the expression for *W* gives:

$$P = \frac{W}{t}$$

$$P = \frac{mgd}{t}$$

$$P = \frac{(82\ \text{kg})(9.8\ \text{m/s}^2)(0.33\ \text{m})}{1.5\ \text{s}}$$

$$P = 180\ W$$

Consider the Explain Question:

How do you solve work and power problems?

Go online to complete the scientific explanation.

Check for Understanding:

How fast would someone with a mass of 45 kg have to climb an 11-m rope in order to use 38 Joules of power?

STEM in Action

STEM and Work and Power

Flight controllers at NASA's Mission Control Center work with orbiting objects daily, using what they know about vectors, circular motion, energy, work, and power. They track the velocity and path of The International Space Station (ISS) as it orbits Earth. The flight controllers, who have engineering or other technical degrees, decide when to give the ISS an occasional boost to counteract the effects of "atmospheric drag."

When a force is perpendicular to motion, the force does no work. To better understand why, consider the ISS or other satellites orbiting Earth.

In an idealized case, a satellite orbiting Earth has uniform circular motion, which means it travels at constant speed in a circular orbit. In accordance with Newton's first law of motion, the satellite would travel in a straight line forever, but the force of Earth's gravity continually pulls the satellite toward the center of the circle, causing the direction of its velocity to curve into a circular path. The satellite's speed is not affected. The angle between the force of gravity and the satellite's velocity is always 90 degrees, so Earth's gravitational force does no work on the satellite.

But, most satellite orbits are elliptical, not circular. As a result, the gravitational force does not always act at 90 degrees. Sometimes, a component of the force vector acts in the same or opposite direction as the velocity vector. This component does work on a satellite, and that work affects the satellite's energy.

Because gravity is an internal, conservative force, it can only convert a satellite's potential energy to kinetic energy and vice versa. It cannot add or subtract from the satellite's total energy. As the satellite travels in its orbit, the changing forces do work to make it speed up or slow down and to change its distance from Earth, but the total energy remains the same.

photo: NASA

Flight controllers in the Mission Control Center monitor the undocking of the Soyuz TMA-14M spacecraft. What forces must they control to manage the orbit of the ISS?

Satellites might orbit Earth forever if there were no external forces to change their total mechanical energy. But, space is not completely empty. The masses of other planets and the sun affect satellites.

Also, space is not a perfect vacuum; there are a few molecules of atmosphere and widely spaced dust particles everywhere. These materials cause friction opposing the satellites' motion. Eventually, if energy were not added with periodic boosts in speed and altitude, satellites' orbits would degrade as they lose energy to the friction of "atmospheric drag."

Complete the Exploration "Work and Power" on the STEM in Action page online, and record your data and observations. Throughout this exercise, you should keep a table that includes the following information, each with its own column: Type of Object, Mass of Object, Length of Ramp, Height of Ramp, Coefficient of Friction, Applied Force, Force Down the Ramp, Frictional Force, and Resultant Force.

photo: NASA

How does NASA counteract "atmospheric drag" on the International Space Station?

After you have completed the activity online, answer the questions.

1. Draw a free body diagram for the wooden box as it is being pushed up the ramp. Assume that the applied force is provided by a person pushing the box parallel to the ramp.
2. Analyze the data and see if you can make sense of it by utilizing your free body diagram. The resultant force is the sum of all forces parallel to the ramp, with those forces pointing up the ramp being positive and those pointing down the ramp being negative.
3. Refer to your data table and calculate the amount of work required to move the box up the ramp and onto the truck bed. Determine the amount of work you would need to provide in order to move the concrete block up the same ramp and onto the truck bed. Explain your calculations.

Classification of Matter

photo: Getty Images

LESSON OVERVIEW

Lesson Questions

- How are the four states of matter described?
- How are pure substances distinguished from mixtures?
- What is a mixture?
- How are different types of mixtures distinguished from each other?

Lesson Objectives

By the end of the lesson, you should be able to:

- Compare and distinguish between the four states of matter.
- Differentiate between pure substances (i.e., elements and compounds) and mixtures.
- Identify and classify different types of mixtures.
- Describe how different mixtures can be separated.

Key Vocabulary

Which terms do you already know?

- ☐ **alloy**
- ☐ **atomic number**
- ☐ **chemical property**
- ☐ **colloid**
- ☐ **compound**
- ☐ **element**
- ☐ **emulsion**
- ☐ **gas**
- ☐ **heat energy**
- ☐ **heterogeneous**
- ☐ **homogeneous**
- ☐ **liquid**
- ☐ **mixture**
- ☐ **nucleus (atom)**
- ☐ **phase of matter**
- ☐ **plasma**
- ☐ **pure substance**
- ☐ **sedimentation**
- ☐ **solid**
- ☐ **solute**
- ☐ **solution**
- ☐ **solvent**
- ☐ **suspension**

ENGAGE

What properties of snow allow glaciers to form? How are the properties of steam different from the properties of water? Visit the Engage page online to learn more.

Explain Question

Explain how different states of matter are distinguished from each other, and how pure substances are distinguished from mixtures.

EXPLORE

How Are the Four States of Matter Described?

The Four States of Matter

Matter exists in four states. Each is characterized by its physical properties. The four states of matter are solids, liquids, gases, and plasmas.

Solids are materials that have a definite shape and a definite volume. An example would be an ice cube. An ice cube is a six-sided object whose volume can be determined by multiplying its height, depth, and width. The particles in an ice cube, which are molecules of water (H_2O), are densely packed and vibrate in fixed positions. Solids are almost incompressible, which means they resist being packed together more tightly.

Liquids are materials that have an indefinite shape and a definite volume. We say that liquids have an indefinite shape because they take the shape of the container they occupy. However, their volume is definite. They do not expand to fill the container they occupy. For example, liquid water in a graduated cylinder will take the shape of the cylinder. The water has a definite volume, which can be determined by reading the cylinder's scale. The water will fill only a certain space in the cylinder. In other words, 50 milliliters (mL) of water in a 100-milliliter cylinder will not expand to fill the cylinder. The volume will remain 50 mL, no matter what the container looks like. The particles in liquid water are relatively loosely arranged, and are somewhat free to move past one another. Liquids are almost incompressible. Like solids, they resist being packed together more tightly.

Gases are materials that have an indefinite shape and an indefinite volume. Water vapor, which is a gas, will take the shape and volume of the container in which it is enclosed. This means that the volume of a sample of water vapor will vary with the volume of its container.

For example, water vapor contained in a 1-liter bottle will have a volume of 1 liter (L). The same mass of water vapor contained in a 2 L bottle will have a volume of 2 L. The particles of a gas are widely separated and are very free to move from one place to another. Gases are compressible.

Physical Properties of Solids, Liquids, and Gases				
State	**Shape**	**Volume**	**Movement of Particles**	**Compressibility**
Solid	Definite	Definite	Vibrate in place	Very limited
Liquid	Indefinite	Definite	Relatively free to move	Very limited
Gas	Indefinite	Indefinite	Very free to move	Highly compressible

How might cooling affect the properties of the states of matter?

Each of these states of matter can be transformed into another state of matter by adding or subtracting energy. This is commonly accomplished by adding or subtracting heat energy, that is, heating and cooling the matter. For example, heat transforms a solid ice cube to a liquid, and then a gas. Cooling a gas, such as water vapor, transforms the gas to a liquid, and then to a solid.

Plasma is the fourth state of matter. Plasma is similar to a gas in some ways, but particles in the plasma state have much greater kinetic energy. When atoms or molecules absorb extremely high amounts of energy, they break apart into ions (charged particles) and free electrons. The electric charges of these ions and electrons affect the properties of plasma. Usually, the energy absorbed when gas changes to plasma is in the form of heat and involves temperatures greater than 5,000°C.

Conditions on Earth's surface do not naturally sustain matter in the plasma state. While plasma does exist in Earth's atmosphere for short periods of time, it is unstable. Plasma on Earth quickly releases energy and returns to the gas state. Examples include static electricity (which we see as lightning) and the auroras.

Plasma is found in outer space. The atmospheres of stars, for example, are made of plasma. Also, in the space between stars, tiny amounts of matter exist, often in the plasma state.

Scientists on Earth can produce plasma under certain conditions and use them in everyday products. Plasma TVs, welding machines, fluorescent lights, and neon signs are some examples. More complex devices use plasma in the investigation of nuclear energy, especially energy produced by fusion reactions. A fusion reaction occurs when the nuclei of two atoms combine. Fusion reactions produce the energy in stars. If harnessed, fusion might become an unlimited source of energy for humankind.

Because of its extremely high energy, plasma often is confined by magnetic fields that attract and repel the charged particles, keeping it in place.

How Are Pure Substances Distinguished from Mixtures?

Pure Substances

Matter falls into three classes. These are elements, compounds, and mixtures. An element is a form of matter that consists of one kind of atom. A compound consists of the atoms of two or more elements that are chemically bonded to one another.

Elements and compounds are pure substances. A pure substance is a type of matter whose samples, wherever found, have the same composition and chemical properties. Mixtures are combinations of pure substances that are not chemically bonded to one another.

Iron is an element and, therefore, a pure substance. An iron sample found in the United States will have the same physical and chemical properties as an iron sample found anywhere else on Earth, or even on Mars.

For example, all samples of iron will be shiny, silvery-gray, solid at room temperature, have a density of 7.87 grams per cubic centimeter, melt at 1,538°C, boil at 2,861°C, and react chemically with oxygen to form iron oxide or rust. There are over 100 known elements, and each is a pure substance that possesses its own unique set of physical and chemical properties. An element's properties are the same no matter where it is found.

The other type of pure substance is a compound. As is true of elements, a sample of a compound possesses a defined set of specific physical and chemical properties. All other samples of the same compound possess the same composition and the same set of physical and chemical properties, no matter where they are found.

For example, hydrogen chloride is a compound made up of one atom of the element hydrogen bonded to one atom of the element chlorine. Wherever it is found, a sample of hydrogen chloride will be a colorless gas that melts at –114.17°C, boils at –85°C, has the chemical properties of an acid, and will react with certain metals to form a salt and hydrogen gas.

Each of the almost countless other compounds known to science has its own unique set of physical and chemical properties. And each sample of one of these compounds has the same properties as any other sample of the same compound, wherever it is found.

What Is a Mixture?

The Nature of Mixtures and Processes to Separate Them

Mixtures consist of two or more substances. Although mixed together, each substance retains its own physical and chemical properties. This allows a mixture to be separated into its component substances by physical means, such as evaporation, distillation, straining, or other physical processes.

For example, iron is a pure substance. One of iron's physical properties is its attraction to a magnet. Table salt is also a pure substance, and one of its physical properties is its lack of attraction to a magnet.

If iron filings and grains of table salt are parts of a mixture, they can be separated by passing a magnet over the mixture. The iron filings will leave the mixture as they are picked up by the magnet. The grains of salt will remain behind.

photo: Getty Images

Magnets can be used to separate mixtures of ferrous and non-ferrous materials. What applications can you think of for separating mixtures this way?

Other physical properties of iron and table salt also can be used to separate them from a mixture. Table salt is soluble in water. Iron is not. So, if a mixture of iron filings and table salt is stirred in a beaker of water, the salt will dissolve in the water, but the filings will not. The liquid can be poured off, leaving behind the iron filings.

This process, of course, produces another mixture, one that contains table salt dissolved in water. How can the components of this mixture be separated? Again, the physical difference of table salt and water can be used to get the job done. Water evaporates. Table salt does not. If the mixture is allowed to stand in a warm place, the water will evaporate, leaving behind crystals of salt. The process can be sped up by heating the water.

How might a mixture of two liquid substances be separated into its components? The two substances likely will have different boiling points. If so, then by slowly heating the mixture, the substance with the lower boiling point will boil off first, leaving behind the substance with the higher boiling point. The vapors of the substance with the lower boiling point can be condensed and captured by passing them through a long cooling tube. This process is called distillation.

How might a mixture of two or more gases be separated into its components? Air is such a mixture. It consists of about 78 percent nitrogen, 21 percent oxygen, and 1 percent other gases. Leaving aside the 1 percent other gases, how might the nitrogen and oxygen in air be separated? A clue to the answer lies again in the boiling points of nitrogen and oxygen. Nitrogen boils at –196°C. That means it also changes from a gas to a liquid at this temperature. Oxygen boils at –183.0°C. So, if a sample of air is slowly cooled, oxygen will become a liquid first. It can be collected, leaving the nitrogen behind.

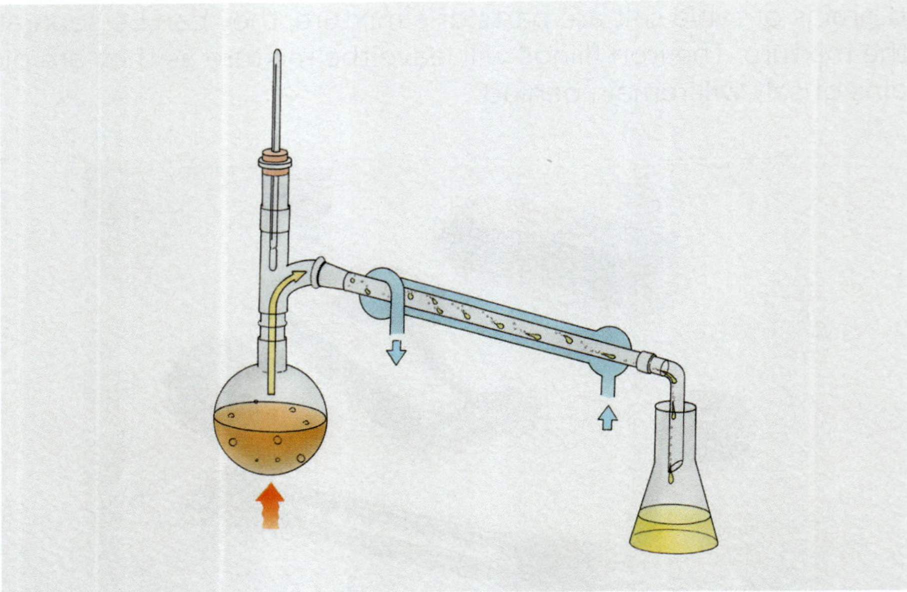

photo: Getty Images

Distillation uses heat to separate liquid mixtures. How is this process used commercially?

How Are Different Types of Mixtures Distinguished from Each Other?

Homogeneous Mixtures

Mixtures may be homogeneous or heterogeneous. In a homogeneous mixture, the components of the mixture are uniform throughout. Any part of the mixture will be identical in composition to any other part of the mixture. The mixture is homogeneous because the components of the mixture are distributed evenly.

A solution is a homogeneous mixture composed of two or more substances, but only exhibiting one phase of matter. Solutions may be in the gaseous, liquid, or solid states of matter. Air is a gaseous solution. Seawater is a liquid solution. Steel is a solid solution.

Different combinations are possible from these three states of matter. A solution can consist of a gas dissolved in another gas, in a liquid, or in a solid. A solution also may consist of a liquid dissolved in a gas, in another liquid, or in a solid. Finally, a solution may consist of a solid dissolved in a gas, in a liquid, or in another solid. The substance that is dissolved is called the solute. The substance that does the dissolving is called the solvent. In a glass of salt water, the solute is salt. The solvent is water.

An alloy is a solid solution composed of two or more elements. Steel is a solid solution made up of iron and other substances, such as carbon, nickel, and chromium. Bronze is also a solid solution. It is an alloy made of copper mixed with tin. Brass is an alloy of copper and zinc. The dime, the U.S. 10-cent coin, is made of an alloy of copper and nickel.

Heterogeneous Mixtures

In a heterogeneous mixture, the components of the mixture are not uniform throughout the mixture. That is, any part of the mixture will not be identical in composition to any other part of the mixture. The mixture is heterogeneous because the components of the mixture are distributed unevenly.

Soil is a solid heterogeneous mixture. It is made up of fine and large particles of decayed plant and animal parts, clay, sand, pebbles, and rocks. A sample of soil in one place may not resemble a sample of soil in another place. For example, beach soil is much different in composition from forest soil.

Salad dressings are also liquid heterogeneous mixtures. They may consist of vinegar, oils, and solid materials, such as spices. One sample of salad dressing will be different from another sample of the same dressing. This is because the materials in the dressing are not uniformly distributed.

Oil and water do not easily dissolve in one another. So, they do not form a homogeneous mixture. If such a mixture is shaken vigorously, it will look as if the oil and water are permanently distributed uniformly. But if this mixture is allowed to stand for a while, the oil will separate and rise to the top of the mixture. This occurs because oil is less dense than water and will float on top of the water. The parts of this mixture are temporarily suspended, or "hanging in place." A mixture in which the parts are temporarily suspended is called a suspension.

Soil mixed with water is another example of a suspension. If this mixture is placed in a beaker, it will look like muddy water at first. But as time passes, soil particles will settle out at the bottom of the beaker, and less murky water will be found at the top.

If the particles in a liquid mixture are very small, they may not spontaneously separate. They will remain suspended. But they do not form a solution. Such a mixture can be thought of as being somewhere between a solution and a suspension. Mixtures like these are called colloids. If a colloid consists of a liquid dispersed in another liquid, the mixture is called an emulsion.

One difference between solutions, colloids, and suspensions is the size of the solid particles in the mixture.

Solutions, Colloids, and Suspensions		
Mixture	**Size of Particle (nm)**	**Behavior**
Solution	Less than 1	Parts do not separate spontaneously.
Colloid	Between 1–100	Parts do not separate spontaneously.
Suspension	More than 100	Parts separate spontaneously.

Particle size is one characteristic of different types of mixtures. What other properties are used to classify mixtures? Note: The symbol nm stands for nanometer. One nanometer equals one billionth of a meter.

How can a person determine whether a mixture is a colloid or a suspension? There are a few methods. One method is to separate the parts of a suspension by passing the mixture through ordinary filter paper, such as a coffee filter. The parts of a solution or colloid will pass right through ordinary filter paper.

Another way to identify a colloid is based on particle size. Many colloids appear homogeneous because the particles are too small to be seen without a microscope. However, when a beam of light passes through a colloid, it is scattered by the colloidal particles and the beam's path can be seen clearly in the colloid. You may have seen a car's headlight beam in the fog at night. This is known as the Tyndall effect. When a light beam shines through a colloidal liquid, the beam's path can be seen clearly in the liquid. Colloids and suspensions can exhibit the Tyndall effect. Because solutions do not show the Tyndall effect, it is helpful to determine if a mixture is a suspension, colloid, or solution.

Examples of colloids include whipped cream (gas in a liquid), mayonnaise (liquid in a liquid), smoke (solid in a gas), butter (liquid in a solid), paint (solid in a liquid), and fog (liquid in a gas).

photo: Getty Images

Light beams shining through the fog illustrate the Tyndall effect. Which types of mixtures exhibit the Tyndall effect?

Consider the Explain Question:

Explain how different states of matter are distinguished from each other, and how pure substances are distinguished from mixtures.

Go online to complete the scientific explanation.

Check for Understanding:

What is the difference between homogeneous mixtures and heterogeneous mixtures?

STEM in Action

STEM and Classification of Matter

The substances in most containers of the milk people purchase today are evenly distributed throughout the liquid, and they do not separate out on standing. In other words, the substances in the container produce a colloid.

However, this description does not apply to the milk that comes directly from a cow. Raw milk, the milk that comes out of a cow, does not produce a colloid. If allowed to stand, the substances in raw milk will separate. Large fat globules, which are the primary components of cream, will float to the top, leaving behind a more watery liquid. This indicates that raw milk is a suspension.

For a number of reasons, milk producers (human farmers and manufacturers, not the cows) and milk consumers prefer milk that does not separate. In other words, they desire milk as a colloid, not a suspension. This desire led to the development of a technology that transforms the milk suspension into milk as a colloid.

A food scientist is someone who uses chemistry to study, analyze, and improve the methods by which food is created and processed. In this situation, the technology that is used is called homogenization, and the product is called homogenized milk. The following description explains how this technology works.

First, remember that an important distinction between a suspension and a colloid is particle size. The particles in a suspension are larger than the particles in a colloid. This difference indicates how raw milk might be homogenized by breaking large fat globules into much smaller fat globules.

photo: Paul Fuqua

When homogenized, milk is a colloid. Why is it important that milk is homogenized?

The original patent for a machine able to homogenize milk was granted to French inventor Auguste Gaulin in 1899. Although the machine has been improved during the past century, its basic principles have remained unchanged. Milk is forced through very narrow tubes at high pressure and high velocity (up to 120 meters/second). During this process, the average diameter of fat globules may be reduced by a factor of about eight. At this reduction in diameter, the fat globules become evenly dispersed and do not separate from the mixture.

Cleaning an Oil Spill

One of the jobs of environmental engineers is to test and evaluate methods of cleaning up oil spills. Part of this job includes cleaning the oil from bird feathers. Work with a partner to research different methods of cleaning up oil from feathers. Methods that you might consider are dispersants (detergents), absorption (cloth), and skimming with a spoon.

Test each method to clean a feather dipped in oily water.

1. First, use a spoon to remove as much of the oil as possible using the skimming method. Based on your observations, rate the effectiveness of this method from 1 (very ineffective) to 5 (very effective). Based on your research, consider the materials that would be required to implement this method in a real-world situation. Based on your observations and research, rate the skimming method's cost from 1 (very expensive) to 5 (very inexpensive).
2. Next, repeat the cleaning activity using new feathers, with the dispersant method and the absorption method. Record your ratings of effectiveness and cost. Create a bar graph that compares both the effectiveness and the cost of cleaning feathers using each method.
3. Briefly describe your observations, and write a conclusion about how the three methods of cleaning oil from bird feathers compare in terms of effectiveness and cost.

Chemical and Physical Properties and Changes

photo: Getty Images

LESSON OVERVIEW

Lesson Questions

- What are the similarities and differences between chemical and physical properties?
- How are chemical and physical changes different?
- What are the observable indicators that can be evidence a chemical change might have occurred?

Lesson Objectives

By the end of the lesson, you should be able to:

- Compare and contrast chemical and physical properties.
- Distinguish between chemical and physical changes.
- Identify the observable indicators that can be evidence a chemical change might have occurred.

Key Vocabulary

Which terms do you already know?

- ☐ chemical change
- ☐ chemical property
- ☐ extensive properties
- ☐ indicator
- ☐ intensive properties
- ☐ phase of matter
- ☐ physical change
- ☐ physical property
- ☐ precipitate
- ☐ sublimation

ENGAGE

What happens to a coffee bean when it is roasted? What happens to ice when it melts? Visit the Engage page online to learn more.

Explain Question

Explain how chemical and physical properties are alike and different, and explain what happens during a chemical change or a physical change.

photo: Getty Images

In what ways does an apple change while it is rotting?

EXPLORE

What Are the Similarities and Differences between Chemical and Physical Properties?

Physical and Chemical Properties

Water naturally exists in three states of matter: solid, liquid, and gas. These are observable physical qualities of matter.

Some elements feel rough or smooth, while others may be shiny or dull in appearance. Chlorine gas has a yellowish-green color, while water vapor is colorless. These are also observable physical qualities of matter.

Physical properties are any properties of matter that can be measured or observed without changing or altering the chemical identity of matter. Physical properties are used to describe matter.

Color, texture, density, boiling point, melting point, freezing point, odor, appearance, solubility, and polarity are just some of the common physical properties of matter. These properties can be determined without changing the composition of the matter being tested. If an apple is cut in half, both pieces are still apple, and both retain the distinct properties of the original apple.

Chemical properties differ from physical properties. To determine chemical properties of a substance, the substance must be altered in some way.

Consider a glass of water and a glass of rubbing alcohol, like the kind used to clean cuts and scrapes. The two liquids look very similar. The water is stable, nonflammable, and nontoxic. The alcohol is also stable, but it is quite flammable and is very toxic if swallowed. Numerous tests on each of these samples are required to determine these chemical properties.

Chemical properties include reactivity, toxicity, flammability, stability, and many others. The way atoms bond and the types of bonds that atoms form determine chemical properties. Many chemical properties can be observed during a chemical reaction.

Extensive and Intensive Properties

Physical properties are classified as extensive or intensive. Extensive properties depend on the amount of matter in a sample. An extensive property will change if matter is added or removed from a sample. Consider a sample of sugar with a mass of 397 grams and a volume of 250 milliliters. The sugar is a clear, odorless solid that tastes sweet. If more sugar is added, the mass and volume will increase. Mass and volume are extensive properties, as are shape and length. However, other properties of the sugar such as color and taste will remain the same no matter how much sugar is present.

Intensive properties do not depend on the amount of matter in a sample. In the previous example of the sugar, the color, taste, odor, and state of the sugar stayed the same when more sugar was added. These are all intensive properties. Other examples of intensive properties include hardness, density, boiling point, melting point, freezing point, and solubility. Intensive properties can be used to help identify an unknown sample. For example, a clear liquid with a melting point of 0°C, a boiling point of 100°C, and a density of 1 gram per milliliter can be identified as water.

Solids, liquids, and gases can be compared in terms of shape and volume, both extensive properties, and compressibility and structure, both intensive properties. (Compressibility refers to how easily a substance can be squeezed into a smaller volume.)

- A solid has a specific shape and volume. Solids cannot be compressed easily into a smaller volume. This lack of compressibility is due to the internal structure of a solid, in which the particles are packed close together in an ordered pattern.

- A liquid has a specific volume, but no definite shape. Instead, a liquid flows to fill the shape of its container. This is because the particles in a liquid are less closely packed and have no pattern. The particles in a liquid are free to move and flow. Liquids are not very compressible, because the particles in a liquid are still close together.
- A gas does not have a specific shape or volume. Unlike solids and liquids, gases can be compressed easily because the particles in a gas are so far apart. The particles in a gas move freely and quickly, allowing the gas to expand or contract to fit its container.

How Are Chemical and Physical Changes Different?

Physical and Chemical Changes

When water changes from liquid to solid form, a physical change called freezing occurs. The ice contains all the same molecules as it did when it was water. The process can also be reversed by melting the ice. During a physical change, the chemical composition of a substance does not change.

Freezing, melting, boiling, condensation, and sublimation are examples of physical changes. If an ice-cold glass of lemonade is set outside on a hot day, water droplets will form on the outside of the glass. That is condensation, the transformation of a gas into a liquid. The water vapor in the air condenses into liquid because the glass is cool. Dry ice is solid carbon dioxide, which transforms into gas without becoming a liquid in a process called sublimation. This is yet another example of a physical change. Physical changes are usually easily reversible.

Unlike a physical change, a chemical change is not easily reversible. A chemical change occurs when substances are changed into new substances with new properties.

For example, when iron (Fe) reacts with oxygen (O_2) and water (H_2O), it becomes a different compound, iron(III) hydroxide ($Fe(OH)_3$). The reaction forms a kind of rust, from which the iron cannot easily be removed and returned to its original chemical state.

What Are the Observable Indicators That Can Be Evidence a Chemical Change Might Have Occurred?

Observable Indicators of a Chemical Change

There are many signs that indicate a chemical change has occurred. The production of heat, light, or other types of energy indicates that a chemical change has occurred. For example, when wood burns, heat and light are produced. This is also seen when fireworks and sparklers are ignited. Fireworks and burning wood also produce sound, another sign of chemical change. Because this is a chemical change, the original substances have changed into different chemical substances.

A chemical indicator will change color in the presence of certain acids or bases. Litmus paper will change from blue to red or red to blue in response to a change in pH. Blue litmus paper changes to red when a solution is acidic, while red litmus paper turning blue indicates an alkaline solution. Another example of a chemical change producing a color change is oxidation. When a cut apple is left to sit on the counter at room temperature, it reacts with oxygen, turning from pale yellow or white to a reddish-brown color.

photo: Getty Images

Oxidation is responsible for color changes in many materials. What color change occurs when blood undergoes oxidation?

Many chemical changes will have a change in odor. Fresh eggs have little, if any, odor, while rotten eggs have a distinct smell of sulfur compounds. Rotting is one example of a chemical change.

Another indicator that a chemical change has occurred is the formation of a precipitate. When some water solutions are mixed together, a solid substance will form at the bottom of the beaker or test tube. This solid substance is called a precipitate. Precipitates are either insoluble in water or are only partially soluble in water. The production of a precipitate indicates that a chemical change has occurred. Kidney stones that form in the body are a precipitate made partly of excess calcium ions.

The release of a gas from a system when heat has not been added is another indicator of a chemical change. When seltzer tablets are added to water, they produce bubbles. These bubbles are actually carbon dioxide being released during the reaction. When some metals are added to hydrochloric acid, hydrogen gas bubbles out. These are reactions that cannot be reversed easily.

Consider the Explain Question:

Explain how chemical and physical properties are alike and different, and explain what happens during a chemical change or a physical change.

Go online to complete the scientific explanation.

Check for Understanding:

You are presented with an unknown liquid. How will you identify the substance?

STEM in Action

STEM and Chemical and Physical Properties and Changes

The Water Cycle

Nature includes a variety of cycles that involve changing physical and chemical properties of matter. The water cycle includes several physical changes. Water accumulates in oceans, lakes, and rivers and then undergoes evaporation. This water vapor condenses in the atmosphere to produce clouds, which then release precipitation, rain or snow, onto Earth's surface. Some water remains in the solid phase as ice or snow in colder regions of Earth. The rest falls back to Earth as a liquid and again becomes part of oceans, lakes, and rivers.

Meteorologists carefully monitor atmospheric segments of the water cycle as well as atmospheric temperature and pressure for the purpose of predicting weather events. Changes in temperature and pressure can alter the rate and nature of physical changes that occur in the water cycle.

The Water Cycle

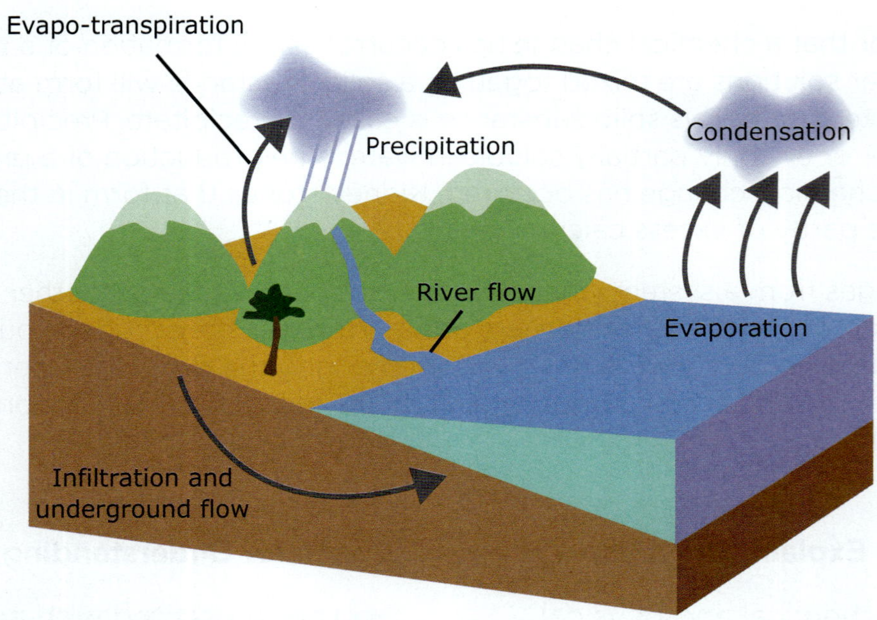

The water cycle involves physical changes of water. What change in water occurs during the process of evapotranspiration?

The Carbon Cycle

The physical and chemical changes that comprise Earth's carbon cycle are somewhat different from those of the water cycle. Solid materials such as wood, soil, plants, and limestone contain the element carbon. Over time, these materials break down and release carbon into the soil. Sometimes, the carbon forms fossil fuels, such as coal or oil.

photo: Getty Images

Some of the carbon in the carbon cycle is stored deep underground. What forms can it take?

Carbon exists in atmospheric gases such as carbon dioxide (CO_2) and a small amount of carbon monoxide (CO). Some of this carbon originates from living organisms, such as humans, who exhale carbon dioxide. Some comes from the oxidation of fossil fuels. Environmental scientists monitor the level of carbon dioxide in the atmosphere because it is one of the contributing factors of global climate change.

Plants use energy from the sun to react carbon dioxide in the atmosphere with water in photosynthesis. The carbon from the gaseous CO_2 is absorbed into the plants. When the plant dies, the leaves, stem, and root break down, releasing carbon into the soil and continuing the carbon cycle.

Study the diagram and answer the questions that follow.

Carbon Cycle

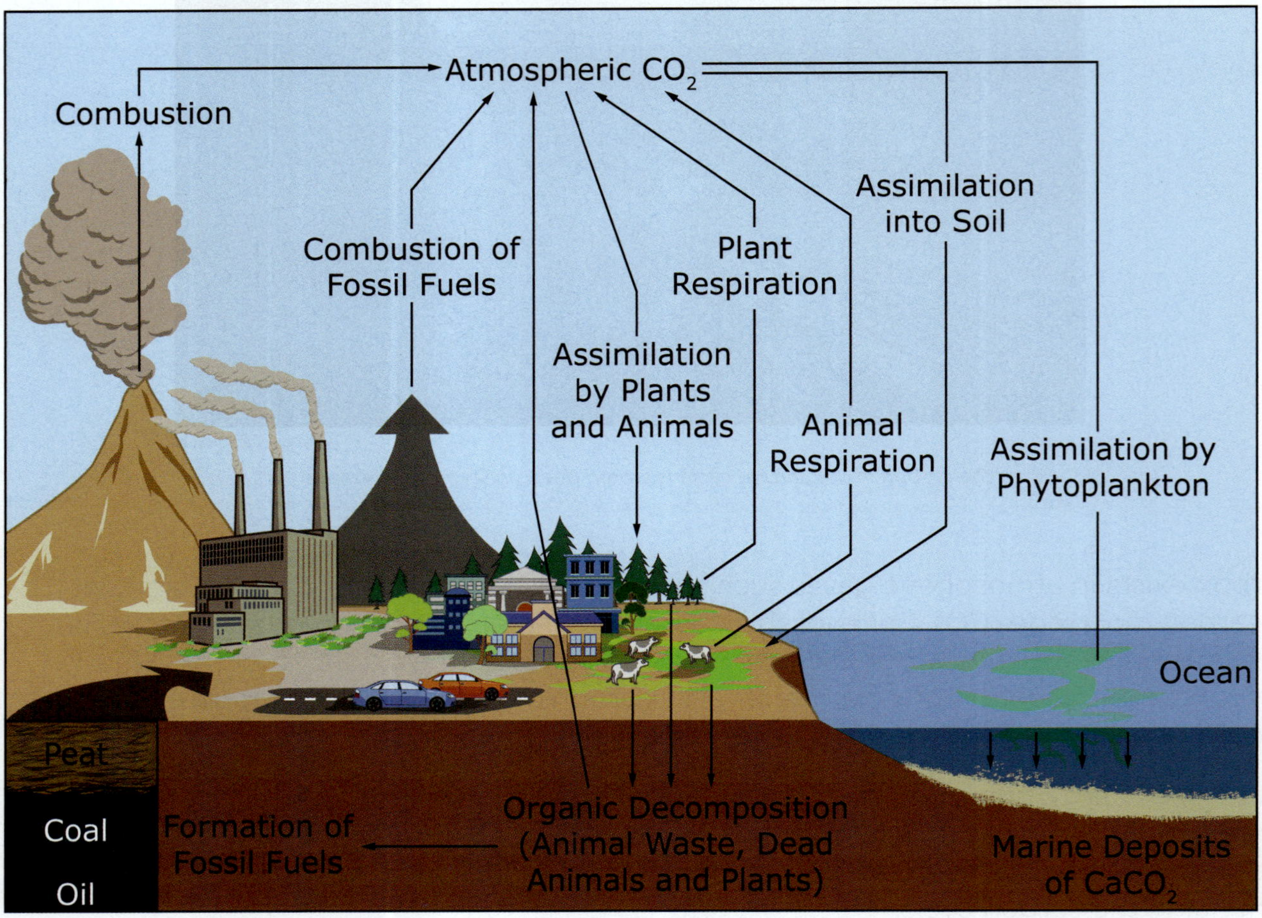

Carbon cycles through Earth's atmosphere, life-forms, and organic matter. What types of chemical changes must occur in order to continue this cycle?

1. Identify and describe two examples of physical changes of matter that are part of the carbon cycle.
2. Identify and describe two examples of chemical changes of matter that are part of the carbon cycle.
3. One method by which excess carbon enters the atmosphere is through emissions by industrial smokestacks. Suggest a way of reducing the carbon contained in these emissions using a physical change and another way using a chemical change.

Physical Behavior of Matter

photo: Getty Images

LESSON OVERVIEW

Lesson Questions

- What is the kinetic molecular theory of matter?
- How does the absorption or release of energy produce a change in state?
- How is a change of state related to changes in kinetic and potential energy?

Lesson Objectives

By the end of the lesson, you should be able to:

- Explain the kinetic molecular theory.
- Explain that a change of state occurs when heat energy is absorbed or released from a system.
- Describe changes of state on the basis of changes in kinetic and potential energy.

Key Vocabulary

Which terms do you already know?

- ☐ boiling point
- ☐ condensation
- ☐ deposition (phase change)
- ☐ evaporation
- ☐ freezing
- ☐ heat energy
- ☐ heating curve
- ☐ intermolecular force
- ☐ kinetic energy
- ☐ kinetic molecular theory
- ☐ liquid/vapor equilibrium
- ☐ phase of matter
- ☐ potential energy
- ☐ state of matter
- ☐ temperature
- ☐ thermal energy
- ☐ vapor pressure
- ☐ vaporize
- ☐ volume

ENGAGE

Have you ever burned your bare feet walking on black asphalt on a hot summer day? Why does this happen? Visit the Engage page online to learn more.

Explain Question

How is energy related to the way particles behave when matter changes to different states?

EXPLORE

What Is the Kinetic Molecular Theory of Matter?

Motion and Molecules

All of the particles that make up matter are in constant motion. The metal ions that make up a locker are in motion, even though humans cannot detect any movement. The molecules that make up ice cream in a bowl are in motion, even before you pick up a spoonful. All particles have kinetic energy, or the energy of motion. When the particles are in the gas state, the postulates of the kinetic molecular theory relate the properties of the gas to the kinetic energy of the molecules. The kinetic molecular theory has been expanded to cover other states of matter as well, so it is also referred to as the kinetic molecular theory of matter.

A molecule is a small particle of matter consisting of two or more atoms bonded together. The atoms may be of the same element or different elements. A molecule of oxygen (O_2) consists of two atoms of oxygen, and a molecule of methane (CH_4) consists of four hydrogen atoms bonded to a carbon atom. The kinetic energy of a single particle, such as an atom or a molecule, cannot be directly determined. However, the average kinetic energy of a group of particles can be measured. That value is defined as temperature. Temperature is a measure of the average kinetic energy of molecules in a sample of matter. As the average kinetic energy of molecules in a sample increases or decreases, the temperature of the sample increases or decreases as well.

Solids, Liquids, and Gases

Matter is most often found in solid, liquid, or gas form. These are the main states of matter. The state at a given temperature depends on the attractive forces between the atoms or molecules of the substance. These intermolecular forces can dominate when kinetic energy (temperature) is low. As kinetic energy (temperature) increases, the intermolecular forces may be overcome and the molecules become separated from each other.

A substance can exist in each of the common states of matter. When a substance is in the solid state, its particles possess the least kinetic energy. These particles are held in place in a rigid structure. The particles that make up solid matter vibrate constantly, but their motion through the sample is very limited.

photo: Getty Images

When ice melts, water changes from a solid to a liquid. How does water behave in its liquid form that is different from its behavior as a solid form?

When a substance is heated, its particles gain kinetic energy. These particles flow past one another, but they do not possess sufficient kinetic energy to fully overcome the strong attractions due to the intermolecular forces between them. Thus, the particles in a liquid tend to stay close together. They move more than particles in the solid state, but this movement is still influenced by the neighboring molecules.

When more heat is added to the substance, its particles eventually gain enough kinetic energy to overcome the intermolecular forces that hold particles close together. The substance enters the gaseous state, and its particles are free to move away from one another. The motion of these particles is limited only by the walls of the container in which they are being held.

Matter, whether in the solid, liquid, or gaseous state, occupies space. The measure of that space is called volume. The volume of a substance is independent of the size of its molecules or atoms. Instead, volume depends on the space between particles. The greater the space between particles, the greater the volume of the substance.

Because of the molecules' close spacing, the volume of a solid or liquid tends to be relatively fixed, as long as temperature and pressure do not change. The volume of a gas, however, will continue to expand until it fills the container in which it is placed. This expansion is caused by the wide spacing and free movement of the gas particles.

When gas particles move throughout a container, they frequently collide with one another. The particles also collide with the walls of the container. During these collisions, the total energy of the system remains constant. That is, energy may be transferred from one particle to another, but no energy is lost from the system.

How Does the Absorption or Release of Energy Produce a Change in State?

Changes in State

Kinetic energy governs the behavior of particles in solids, liquids, and gases. Kinetic energy also affects changes between these states.

One of the most common types of energy that particles can absorb is thermal energy. When thermal energy is transferred into or out of a system, it is referred to as heat. In most conditions, the addition of heat increases the kinetic energy of the molecules in a system, and the removal of heat decreases the kinetic energy of the particles. Remember, this change in kinetic energy is usually illustrated by a change in temperature. Thermal energy is the part of the potential and kinetic energy of an object or system that results in the system's temperature.

As a sample of matter is heated, collisions occur with increasing force. Eventually, this gives the particles sufficient energy to overcome intermolecular forces and enables them to move farther apart. This causes a change in state.

The temperature at which a solid changes state to become a liquid, or melts, is called the melting point. The reverse process occurs when heat is removed from a substance. As kinetic energy decreases, intermolecular forces increase in strength, collisions are minimized, and particles move closer together. The freezing point of a substance is the temperature at which a liquid changes into a solid. The freezing and melting points of a substance are identical.

When additional heat is added to a liquid, the particles gain even more kinetic energy and collisions increase. The molecules begin to overcome intermolecular forces and break away from each other to behave as a gas. Temperature (kinetic energy) remains constant as added heat is used to separate the molecules. As a substance changes to its gas state, it is said to vaporize.

When the change to a gas state occurs at the surface of a liquid, the process is referred to as evaporation. Evaporation of a substance can occur at any temperature above its freezing point. Remember that temperature reflects the average kinetic energy of the molecules that make up matter. Individual molecules might have kinetic energy that is greater or less than this average. If the high-energy molecules move to the surface of a liquid, where intermolecular forces are only applied from three sides, some of them break free and change to the gas state.

When an entire sample of liquid is changing to a gas state, bubbles form as gas molecules group together. The bubbles then rise to the liquid's surface because the gas is less dense than the liquid around it. The bubbles release vapor into the atmosphere in a process referred to as boiling. The temperature at which a liquid becomes a gas is called the boiling point. Again, the process of vaporization is reversible. When gas particles lose energy, they move closer together and form a liquid during a process called condensation.

Air pressure significantly affects the boiling points of a particular substance. For molecules to escape from the surface of a liquid, for example, the molecules must have enough energy to overcome both the attraction between molecules and the pressure exerted on the liquid's surface by the atmosphere. In other words, in order for molecules to leave the liquid, the vapor pressure exerted upward by molecules of the substance must exceed the pressure exerted downward by the gas particles that make up the atmosphere.

The normal boiling point of a substance is the temperature at which its vapor pressure equals the air pressure at sea level. This air pressure is referred to as standard atmospheric pressure and is defined as 1 atmosphere (atm) or 101.325 kilopascals (kPa). The kilopascal is a unit of pressure measurement defined under the International System of Units (SI).

The normal boiling point of water is 100°C. However, the actual temperature at which water boils varies with changes in atmospheric pressure. Water boils at lower temperatures in locations where air pressure is lower, such as the top of a mountain. It will take more energy to raise water to its higher boiling point in locations where air pressure is higher, such as at the bottom of a valley with an altitude below sea level.

Most substances change state from solid to liquid to gas or the reverse. However, some substances skip the intermediate step and change directly between the solid and gaseous states. This change is called sublimation. Solid carbon dioxide (CO_2), or "dry ice," undergoes such a transformation. This process is reversible. In other words, a substance that can change directly from a solid to a gas can also change directly from a gas to a solid in a process called deposition. For carbon dioxide, the sublimation-deposition point occurs at a temperature of –78.5°C.

photo: Getty Images

Solid carbon dioxide, known as dry ice, changes directly from a solid into a gas through a process called sublimation. How does the energy of carbon dioxide molecules in dry ice compare with the energy of carbon dioxide molecules in their gaseous state?

Liquid/Vapor Equilibrium

Imagine a container filled nearly to the top with water, then sealed tightly with a lid. The container holds water molecules. Most of the water inside is in the liquid state. The container also holds molecules in the gas state in the space between the water's surface and the lid. Most of the gas is air, trapped there when the container was closed. Some of the gas is water vapor.

Even if the temperature of the water is below the boiling point, some of the molecules in the liquid have enough kinetic energy to break free from the surface of the liquid. They evaporate and enter the space above the water in the container, moving freely throughout that space as gas particles.

As evaporation continues, more and more molecules change into the gaseous state. Some of those gas molecules come into contact with the surface of the liquid and are pulled back in by intermolecular forces. Other molecules join together when they collide and condense into water droplets.

If the temperature of the container remains constant, the rates of evaporation and condensation eventually become equal. Both processes are still occurring, but molecules are constantly moving from one state to another and back again. The amounts of liquid and vapor in the container remain constant. The system is in dynamic equilibrium because the two ongoing processes are in balance. If energy is added or removed from the system, the rates of evaporation and condensation will change, and the equilibrium will be upset.

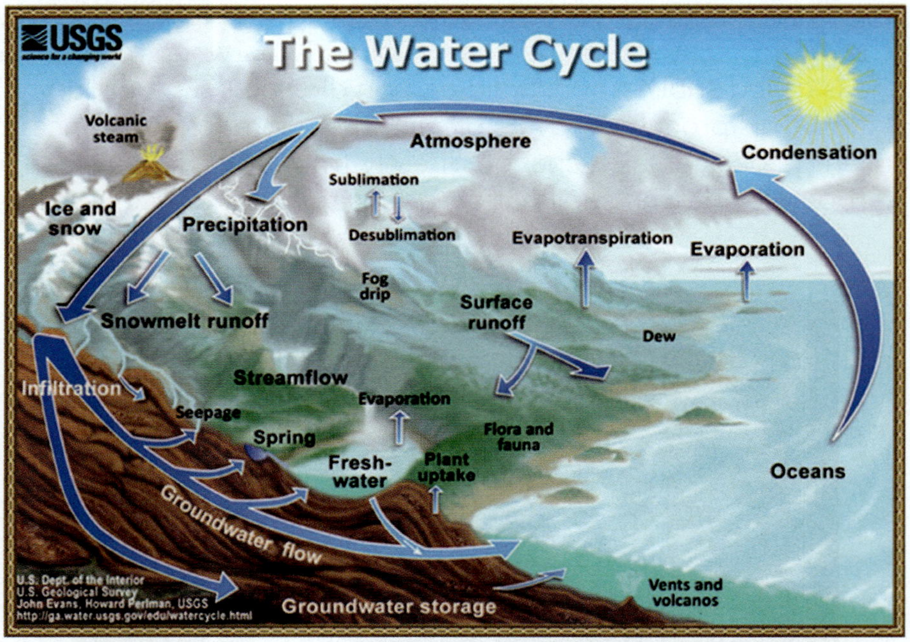

photo: USGS

Evaporation and condensation can occur in dynamic equilibrium. How does sublimation play a part in the water cycle?

How Is a Change of State Related to Changes in Kinetic and Potential Energy?

Kinetic and Potential Energy Changes

Kinetic energy is the energy of motion. In contrast, potential energy is the energy stored in a system. Often, potential energy is related to the position of the system or the position of its parts. Potential energy may also be stored chemically, to be released when a reaction occurs. Potential energy stored in chemical bonds is sometimes called chemical energy.

Kinetic energy can be transformed into potential energy and vice versa. For example, a boulder resting on a mountain side has a great deal of potential energy. If it moves, it will carry a great deal of force. When the boulder begins to roll, some of its potential energy becomes kinetic energy. The farther it rolls down the mountain, the faster it rolls as its kinetic energy increases. At the same time, its potential energy decreases as the distance to the bottom of the mountain decreases. During these transformations, no energy is gained or lost. That is, the total amount of energy remains unchanged.

A heating curve is a graph that relates temperature to the behavior of a substance over time. During a change in state, variations occur in the kinetic and potential energies of the particles that make up the substance. The heating curve helps explain these.

For example, if a sample of ice (pure water) is heated, the following occurs:

- The kinetic energy of the molecules in the sample increases as the temperature increases, until the solid reaches its melting point and begins to change into a liquid.
- The temperature of the solid-liquid system remains constant while the potential energy of the molecules increases. The molecules overcome intermolecular forces and move farther apart. This process continues until all of the solid material has become liquid.
- As the liquid continues to be heated, the kinetic energy of the molecules in the sample increases as the temperature increases until the liquid reaches its boiling point and begins to change into a gas.
- The temperature of the liquid-gas system remains constant while the potential energy of the molecules increases. The molecules overcome intermolecular forces and move independently of one another. This process continues until all of the liquid has become gas
- As the gas is heated, the kinetic energy of the molecules in the sample increases as the temperature continues to increase.

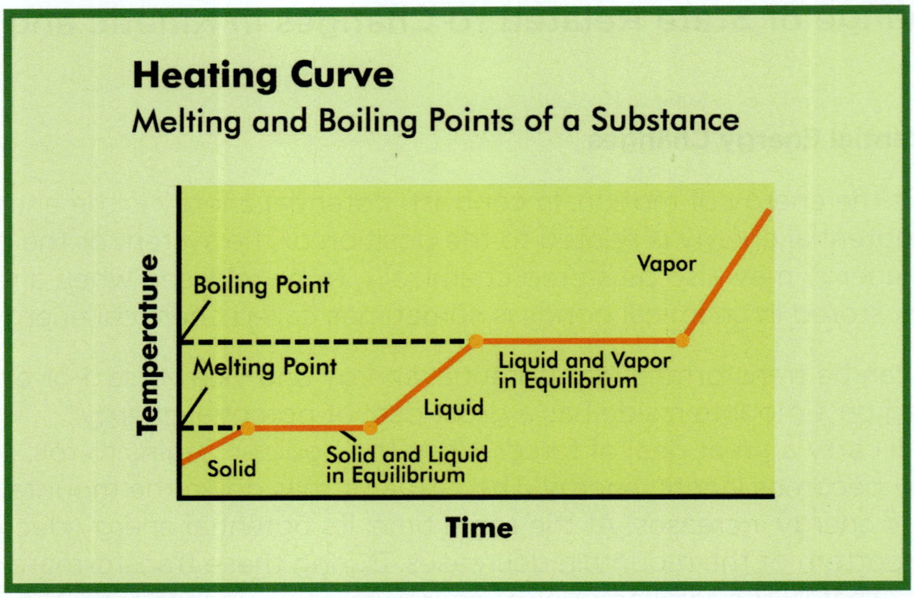

A heating curve shows the changes in temperature that occur as energy is added and a substance changes between states. At what point(s) in this graph does potential energy increase? At which point(s) does kinetic energy increase?

What does the graph reveal about the kinetic and potential energy of the system as energy is added over time? The segments of the graph that have an upward slope show an increase in temperature, which corresponds to an increase in kinetic energy. The segments of the graph that are flat show no increase in temperature and, thus, no increase in kinetic energy. This occurs even though energy (in the form of heat) is being added to the system. What is happening to that energy?

Most of the time, when heat is added to a substance it is converted to kinetic energy. The heat increases the motion of the particles and thus increases the temperature of the substance. During a change of state, however, heat is converted to potential energy instead. The energy is absorbed by the particles, allowing them to overcome intermolecular forces. If more heat is added after that step is completed, then the kinetic energy of the particles again increases.

The energy that was absorbed during the state change is not released. It is stored by the particles as potential energy and can be converted again to kinetic energy if the changes in state are reversed. For example, water vapor releases thermal energy as it condenses into liquid water, releasing heat to the surroundings. Likewise, liquid water releases thermal energy when it forms ice. Again, heat is released to the surroundings and the potential energy of water molecules decreases as they change state to become ice.

Changes of State and Kinetic Molecular Theory

Thermal energy is the energy of moving or vibrating particles in matter. Heat is the thermal energy that flows between matter. Even things that are very cold have thermal energy. Thermal energy will move from a high-temperature object to a low-temperature object. It will end when the objects are at the same temperature. The total energy released by the warmer object will be the same as the total energy absorbed by the cooler one. For example, 1 gram of ice at 0°C has less heat than 100 grams of ice at the same temperature.

Temperature is proportional to the average amount of heat per molecule of matter. Temperature is a measure of the average kinetic energy of the particles in matter. The higher the temperature, the faster the particles move. It makes sense to think that energy moves from warm areas to cool ones. In other words, energy moves from an area where particles are moving quickly to an area where they are moving more slowly.

Energy is measured in Joules (J) or calories (cal). One Joule of energy can also raise the temperature of 1 gram of liquid water by 0.239°C. One calorie is the amount of energy it takes to raise the temperature of 1 gram of liquid water by 1 degree Celsius. With that information, can you calculate how many calories make up 1 Joule?

$$1\text{ J} = 0.239\text{ cal}$$

Context is important when using calories as a unit of energy. There are 1,000 calories in one kilocalorie (kcal). Sometimes, kilocalories are also called Calories (Cal) when referring to the amount of energy in foods. When talking about science, the term calorie will always refer to the small "c" calories.

Conservation of Energy and Energy Transfer

According to the law of conservation, energy can be converted from one form to another but cannot be created or destroyed. This means the total amount of energy in the universe is constant. It also means that every increase in energy needs to be accompanied by an equal decrease in energy somewhere else in the same system. In chemical systems, chemical energy is transformed into heat, light, or sound as the bonds between atoms are made or broken.

Temperature allows us to track the flow of heat in a system. When two objects at different temperatures come into contact with one another, heat flows from the object with a higher temperature into the one with a lower temperature until the temperature of the system reaches equilibrium.

The amount of heat (*q*) gained or lost as the temperature of an object changes depends on its mass and specific heat (*c*):

$$q = m \times c \times (T_f - T_i)$$

Specific heat is the amount of heat that it takes to change the temperature of 1 gram of a substance by 1 degree. For example, to calculate the amount of heat required to warm 100 grams (g) of liquid water from 25°C to 100°C, you need to know the specific heat of water.

$$c(\text{liquid water}) = 4.18\text{ J}/(\text{g} \times {}^\circ\text{C})$$

$$q = 100\text{ g} \times 4.18\text{ J}/(\text{g} \times {}^\circ\text{C}) \times (100^\circ\text{C} - 25^\circ\text{C}) = 31{,}350\text{ J} = 31.35\text{ kJ}$$

Remember, a change of 1 degree Celsius is equal to a change of 1 degree Kelvin so you do not need to convert from Celsius to Kelvin when making these calculations. The value of q is negative when the temperature of the substance has decreased and positive when it has increased.

However, a substance undergoing a change of state will not increase or decrease in temperature. For example, if you put an ice cube in a glass of water, thermal energy flows from the water to the ice cooling the liquid water. Meanwhile, the water molecules in the ice are gaining enough kinetic energy until the intermolecular bonds can no longer hold them in a rigid solid state and they change into liquid water. Only when all of the ice is melted, can the temperature of the water increase further?

The energy that is required to change 1 gram of a substance from a solid to a liquid state is called the heat of fusion (H_f) or the latent heat of fusion (L_f), and the energy that is required to change 1 gram of a substance from a liquid state to a gaseous state is called the heat of vaporization (H_v) or latent heat of vaporization (L_v). The amount of heat transferred when a substance is changing state is given by the following equation:

$$q = m \times L$$

The value of q is negative when the substance is changing from a higher energy state to a lower energy state (freezing or condensing) and positive when the substance is changing from a lower energy state to a higher one (melting or evaporating).

How much energy is required to melt 100 g of ice?

$$L_f(\text{ice}) = 344 \text{ J/g}$$

$$q = 100 \text{ g} \times 344 \text{ J/g} = 34{,}400 \text{ J} = 34.4 \text{ kJ}$$

To calculate the energy required for a sample of ice at –15°C to change to water vapor at 110°C, we need to calculate five separate energies and add them together:

1. The amount of energy needed to increase the temperature of ice from –15°C to the melting point.
2. The amount of energy needed to melt the ice.
3. The amount of energy needed to increase the temperature of liquid water to the boiling point.
4. The amount of energy needed to boil the water.
5. The amount of energy needed to increase the temperature of water vapor from the boiling point to 100°C.

Consider the Explain Question:

How is energy related to the way particles behave when matter changes to different states?

Go online to complete the scientific explanation.

Check for Understanding:

Can you describe how the potential and kinetic energy of a pure substance changes as it is heated and cooled?

STEM in Action

STEM and Physical Behavior of Matter

Pure substances change state at predictable temperatures; people count on their physical behavior when they use the materials. But mixtures have properties that are different from those of the pure substances that make them up. Fortunately, the properties of the mixtures are also understood, predictable, and useful for various purposes.

For example, the functioning of the internal combustion engine depends on the properties of the materials it is made of, such as metal alloys, and of the materials it uses to run, such as gasoline and diesel fuel. Internal combustion engines are the kind of engine that hums under the hoods of most cars, trucks, and buses. The internal combustion engine operates by burning—or combusting—fuel inside of an engine. This process produces a great amount of heat, which must be carried away from the metal engine parts lest they become seriously damaged.

photo: Getty Images

What properties of the internal combustion engine are required to make it run?

Early in the development of the internal combustion engine, engineers recognized this problem. The solution seemed simple enough. Because the specific heat of water is high, it absorbs a great deal of heat before its temperature rises; therefore, water could be circulated through the engine to carry away unwanted heat. However, water freezes at temperatures that are common in winter months; ice would not only cool an engine, it would crack it. The temperature of an engine while it is operating can rise above the boiling point of water in summer months; steam cannot cool the engine any better than ice can, and it can also risk damaging the engine by putting too much pressure on it. The engineering challenge, therefore, was to find a substance to use as an engine coolant that would simultaneously lower the freezing point of water and raise its boiling point.

Chemists understood that if a solute were dissolved in a liquid to form a solution, the freezing point of the solution would go down. At the same time, the boiling point will go up. Chemists developed liquids that are commonly called antifreezes. A 50/50 solution of water and antifreeze, composed mainly of a substance called ethylene glycol, retains the heat-absorbing properties of water. This solution has a lower freezing point and higher boiling point than water, making it useful as an engine coolant. As a matter of fact, the information label of one commercial antifreeze-water mixture states that it will stay liquid down to –84°F and keep from boiling up to 276°F.

In addition to formulating antifreeze, people working in applied chemistry and chemical engineering develop formulas for the fuels used to run cars, the fluids used in car systems, the paint used to protect car bodies, the plastics used to make car interiors, the compounds used for tires, and pretty much any other material you can think of that has to do with cars. An understanding of the physical behavior of matter helps people working in chemical engineering to keep people safe while they are on the move.

photo: Getty Images

Antifreeze solution in a car's cooling system will not freeze, even in frigid weather. What happens to water when antifreeze is added to a vehicle?

Physical Behavior of Matter

The physical behavior of a substance can be depicted in a phase diagram like the one shown. Use the following graph and other sources as needed to answer the questions.

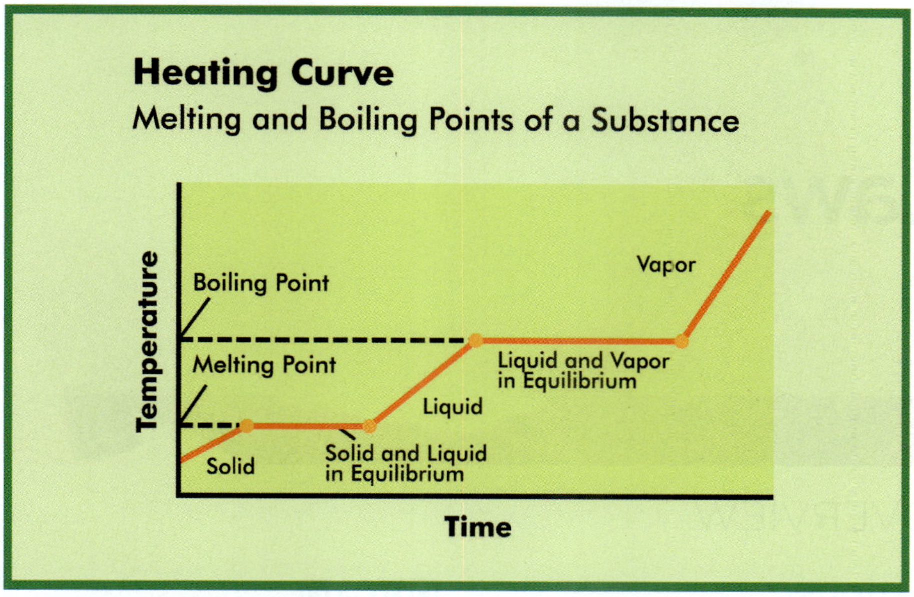

1. Describe what happens to a substance as it changes from a liquid to a gas, based on the heating curve. Include in your answer how temperature is affected during this change.

2. While a substance is melting, what is happening to the energy that is being added to it, since this heat energy is not changing its temperature?

3. Sublimation is a process that occurs when a solid substance changes phase directly into a gas. Describe this process in terms of heat energy, relating what happens to the behavior of matter.

Gas Laws

photo: Getty Images

LESSON OVERVIEW

Lesson Questions

- What is Boyle's law?
- What is Charles's law? How do Charles's law and Gay-Lussac's law describe gas behavior?
- How is the combined gas law used to calculate changes in pressure, temperature, and/or volume for a fixed amount of gas?
- What is the ideal gas law?
- What is Graham's law of effusion?
- What is Dalton's law of partial pressures?

Lesson Objectives

By the end of the lesson, you should be able to:

- Explain Boyle's law.
- Explain Charles's law and Gay-Lussac's law.
- Use the combined gas law to calculate changes in pressure, temperature, and/or volume for a fixed amount of gas.
- Explain the ideal gas law.
- Explain Graham's law of effusion.
- Explain Dalton's law of partial pressures.

Key Vocabulary

Which terms do you already know?

- ☐ **average kinetic energy**
- ☐ **Boyle's law**
- ☐ **Charles's law**
- ☐ **combined gas law**
- ☐ **Dalton's law of partial pressures**
- ☐ **diameter**
- ☐ **directly proportional**
- ☐ **effusion**
- ☐ **elastic collision**
- ☐ **Graham's law of effusion**
- ☐ **ideal gas law**
- ☐ **inelastic collision**
- ☐ **intermolecular force**
- ☐ **inversely proportional**
- ☐ **standard temperature and pressure (STP)**
- ☐ **thermo-dynamics**
- ☐ **total pressure**

ENGAGE

Have you ever wondered why geysers or volcanoes erupt, why weather patterns change, or why basketballs or balloons uniformly increase in size when inflated? Visit the Engage page online to learn more.

Explain Question

What are the physical properties of gases, and how do they change under certain conditions?

EXPLORE

What Is Boyle's Law?

How Gases Behave

The kinetic theory of gases explains that all particles have kinetic energy. When matter is in the gas state, its particles spread evenly throughout the container in which the gas is held so that the gas can be said to take the shape of the container. The gas particles, usually molecules, collide with other particles and with the container itself. During these perfectly elastic collisions, no energy is lost due to friction. Kinetic energy is maintained.

The force of these collisions exerts pressure on the container. Pressure is measured by determining the amount of force exerted on a specific area of a container's surface. Pressure is often measured in millimeters of mercury (mm Hg) or in atmospheres (atm). One atmosphere is the standard amount of pressure exerted by the atmosphere at sea level. The SI unit of pressure is the pascal (Pa), which is equal to one Newton per square meter (N/m^2).

Unit Conversions for Pressure

$$1 \text{ atm} = 1.01 \times 10^5 \text{ Pa}$$
$$1 \text{ mm Hg} = 133 \text{ Pa}$$
$$1 \text{ Pa} = 1 \text{ N/m}^2$$
$$1 \text{ atm} = 760 \text{ mm Hg}$$
$$1 \text{ atm} = 760 \text{ torr}$$

Atoms or molecules are everywhere except in a vacuum. How do molecules behave in a gas as compared with their behavior in a solid or liquid?

The behaviors of a gas are dependent on several factors, of which pressure is just one. The size of the container in which the gas is held, the temperature of the system, and the number of gas particles contained in the system must also be considered. If any one of these variables changes, at least one of the others will change as well. The gas laws describe the relationships between these variables.

Boyle's Law

A bicycle pump is used to fill a tire with air. When you push the handle, a plunger inside the cylinder moves the air into a smaller space. The farther you push the handle, the more resistance you feel.

This phenomenon can be explained by Boyle's law, which states that the volume and pressure of a gas are inversely proportional at a constant temperature. If one variable decreases, the other increases. In other words, when you push the handle of a bicycle pump, the volume of the cylinder in which the air molecules are contained decreases. As a result, the pressure exerted by the gas inside the cylinder increases, and you can feel the force of that pressure against the bicycle pump's handle. This is because the gas molecules are being held in a tighter space. They collide with each other and the sides of the container more frequently, exerting more force on the cylinder.

Boyle's law is summarized by the following equation:

$$P_1V_1 = P_2V_2$$

Boyle's Law: Sample Problem 1

A sample of a gas has a volume of 3.5 liters at a pressure of 760 mm Hg. What will the volume of the gas be if the pressure is reduced to 450 mm Hg?

Solution

The equation relating pressure and volume for a gas is:

$$P_1V_1 = P_2V_2$$

where P_1 is the initial pressure of the gas, V_1 is the initial volume of the gas, P_2 is the final pressure of the gas, and V_2 is the final volume of the gas.

$$P_1V_1 = P_2V_2$$
$$V_2 = \frac{P_1V_1}{P_2}$$

To find the final volume, rearrange the equation to solve for V_2:

$$\begin{aligned} V_2 &= \frac{P_1V_1}{P_2} \\ &= \frac{(760\ \cancel{\text{mmHg}})(3.5\ \text{L})}{(450\ \cancel{\text{mmHg}})} \\ &= 5.9\ \text{L} \end{aligned}$$

Substitute the given values and solve.

This answer is logical; as pressure decreases, volume should increase. The pressure decreased by slightly less than half, so the volume should increase by a factor of slightly less than two.

Boyle's Law: Sample Problem 2

A gas sample has an initial volume of 250 milliliters (mL) at 760 mm Hg. Calculate the pressure required to change the volume to 150 mL.

Solution

The equation relating volume and pressure for an ideal gas is:

$$P_1V_1 = P_2V_2$$

Where P_1 is the initial pressure of the gas, V_1 is the initial volume of the gas, P_2 is the final gas pressure, and V_2 is the final gas volume.

To find the final pressure, rearrange the equation to solve for P_2:

$$P_2 = (760 \text{ mm Hg})(250 \text{ ml}) / 150 \text{ ml}$$

Substitute the given values and solve for P_2:

$$P_2 = 1{,}300 \text{ mm Hg}$$

The answer is logical because of the inverse relationship between gas volume and pressure. As the volume decreased (from 250 mL to 150 mL) the amount of pressure had to increase.

What Is Charles's Law? How Do Charles's Law and Gay-Lussac's Law Describe Gas Behavior?

Charles's Law

An important part of Boyle's law is that the temperature of a system is held constant. We know that the temperature of a system is related to its energy; the kinetic theory of gases explains that the energy in particles is responsible for their movement. What happens to the other properties, then, when thermal energy is added to a system that contains gas particles?

Charles's law discusses the relationship between temperature and volume. In his experiments, Jacques Charles found that temperature and volume of a gas are directly proportional when pressure is held constant. In other words, as the temperature increases, the amount of volume required to keep the gas at the same pressure must increase as well.

As a hot air balloon is being filled, it demonstrates Charles's law. As fuel is burned underneath the balloon, the gas inside the space warms and expands, inflating the balloon into a distinct shape.

photo: Getty Images

An inflating hot air balloon exhibits Charles's law. Why is it necessary to continue burning fuel under the balloon even when it is in the air?

Charles's law is expressed by the equation:

$$\frac{T_1}{V_1} = \frac{T_2}{V_2}$$

Charles's Law: Sample Problem

A sample of a gas is heated from 124 Kelvin to 185 Kelvin. At 185 Kelvin, the gas has a volume of 0.743 liters. What was the initial volume of the gas?

Solution

The equation relating temperature and volume for a gas is:

$$\frac{T_1}{V_1} = \frac{T_2}{V_2}$$

where T_1 is the initial temperature of the gas, V_1 is the initial volume of the gas, T_2 is the final temperature of the gas, and V_2 is the final volume of the gas.

To find the initial volume, rearrange the equation to solve for V_1:

$$\frac{T_1}{V_1} = \frac{T_2}{V_2}$$

$$T_1 = \frac{V_1 T_2}{V_2}$$

$$V_1 = \frac{T_1 V_2}{T_2}$$

Substitute the given values and solve:

$$
\begin{aligned}
V_1 &= \frac{T_1 V_2}{T_2} \\
&= \frac{(124\ \mathrm{K})(0.745\ \mathrm{L})}{(185\ \mathrm{K})} \\
&= 0.498\ \mathrm{L}
\end{aligned}
$$

This answer is logical; as temperature increases, volume should also increase.

Gay-Lussac's Law

What happens if the temperature increases but the volume is not changed? This might happen if an empty water bottle or a beach ball is left in the sun.

Remember that temperature reflects the average kinetic energy of particles in a system. As a system increases in temperature, the gas particles in the system gain kinetic energy. As a result, the particles inside collide with the container with greater force, and pressure increases as well.

Gay-Lussac's law explains that an increase in thermal energy will cause an increase in pressure. Temperature (K) and pressure, then, are directly proportional.

Gay-Lussac's law can be expressed by the equation:

$$\frac{P_1}{T_1} = \frac{P_2}{T_2}$$

When working with equations using this law, temperature should always be expressed in Kelvin (K). The temperature conversions between SI units of Celsius (°C) or Kelvin (K), and the standard units of Fahrenheit (°F), are described following.

Unit Conversions for Temperature

$$
\begin{aligned}
T\ (\mathrm{K}) &= T\ (^\circ\mathrm{C}) + 273.15 \\
T\ (^\circ\mathrm{F}) &= 1.8T\ (^\circ\mathrm{C}) + 32 \\
T\ (^\circ\mathrm{C}) &= 0.56\ [T\ (^\circ\mathrm{F}) - 32)]
\end{aligned}
$$

Gay-Lussac's Law: Sample Problem

A sample of gas in a steel tank at 101 kilopascal and 25.0°C is heated to 55.0°C. What is the air pressure in the heated tank?

Solution

The equation relating pressure and temperature for a gas is:

$$\frac{P_1}{T_1}=\frac{P_2}{T_2}$$

where P_1 is the initial pressure of the gas, T_1 is the initial temperature of the gas, P_2 is the final pressure of the gas, and T_2 is the final temperature of the gas.

Convert the temperatures from degrees Celsius to Kelvin.

$$T_1 = 25°\text{C} + 273 = 298\ \text{K}$$
$$T_2 = 55°\text{C} + 273 = 328\ \text{K}$$

To find the final pressure, rearrange the equation to solve for P_2:

$$P_2 = \frac{P_1 T_1}{T_1}$$

Substitute the given values and solve:

$$P_2 = \frac{(101\ \text{kPa})(328\ \text{K})}{298\ \text{K}} = 111\ \text{kPa}$$

This answer is logical; as the temperature increases, the pressure increases. The temperature increased by about 10 percent, so the pressure should increase by approximately 10 percent also.

How Is the Combined Gas Law Used to Calculate Changes in Pressure, Temperature, and/or Volume for a Fixed Amount of Gas?

Scientific Method and the Gas Laws

The gas laws were formulated after a great deal of careful research by early physicists who applied the scientific method. Precise measurement and accurate record keeping were required to ensure that results were valid and that the experiments could be replicated.

The researchers who conducted these experiments set the stage for later advances in physical chemistry. They provided information that has enabled all of us to better understand our world.

The Combined Gas Law

The first three gas laws all work together. In fact, scientists have combined Boyle's, Charles's and Gay-Lussac's laws in one equation. This is referred to as the combined gas law. Changing any variable in the equation means another variable must also be changed. These changes keep the system in balance.

$$\frac{P_1V_1}{T_1} = \frac{P_2V_2}{T_2}$$

The Combined Gas Law: Sample Problem

A sample of an ideal gas is cooled from 341 Kelvin (K) to 275 K. During the cooling process, the volume of the gas decreases from 0.381 liters (L) to 0.274 L. If the initial pressure of the gas was 0.750 atm, what is the final pressure of the gas?

Solution

The equation relating pressure, volume, and temperature for a gas is:

$$\frac{P_1V_1}{T_1} = \frac{P_2V_2}{T_2}$$

where P_1, V_1, and T_1 are the initial pressure, volume, and temperature of the gas, respectively, and P_2, V_2, and T_2 are the final pressure, volume, and temperature of the gas, respectively.

To find the final pressure, rearrange the equation to solve for P_2:

$$\frac{P_1V_1}{T_1} = \frac{P_2V_2}{T_2}$$

$$P_2 = \frac{P_1V_1T_2}{T_1V_2}$$

Substitute the given values and solve:

$$\begin{aligned} P_2 &= \frac{P_1V_1T_2}{T_1V_2} \\ &= \frac{(0.750\ \text{atm})(0.381\ \text{L})(275\ \text{K})}{(341\ \text{K})(0.274\ \text{L})} \\ &= 0.841\ \text{atm} \end{aligned}$$

What Is the Ideal Gas Law?

Moles

Some properties of a gas depend on the number of particles present. The more particles present at constant pressure, the more volume is needed to contain them. Think about a balloon. The more air you put into the balloon, the more the balloon must expand to hold that air.

Avogadro's law states that equal volumes of gases at the same temperature and pressure contain equal numbers of particles. Scientists use a unit of measurement called a mole to describe quantities of gas particles. One mole of any substance always contains the same number of particles: 6.02×10^{23}.

The number of moles of a gas, usually represented by the variable *n*, is directly proportional to volume when pressure and temperature are constant. Another way to express this relationship is through Avogadro's law using the equation following:

$$V = kn$$

That is, if pressure and temperature are held constant, and you know either the volume or the amount in moles, you can determine the other. For example, at standard temperature and pressure, the value of the constant *k* is 22.4 liters per mole.

The Ideal Gas Law

Many factors affect the physical properties of gases. For example, intermolecular forces may cause gas particles to attract or repel each other and affect the behavior of the gas. Scientists have found, however, that at relatively low pressures the effects of these intermolecular forces are negligible. This is because distance between particles, compared with the diameter of each particle, is very large.

Scientists have used this information to create a model of a theoretical gas that has no intermolecular forces. This model can be applied to all gases and is useful in calculations. It is called an ideal gas. An ideal gas is a gas that meets the following criteria:

- The gas consists of point particles.
- Gas particles move randomly and do not interact unless they collide.
- The gas does not condense into a liquid at low temperatures.

In commonly found conditions, most real gases behave like ideal gases. The ideal gas model only fails at very high pressures or very low temperatures where gas particles are pushed close enough to exert forces upon each other.

To make experimental calculations easier, scientists find it useful to use standard sets of conditions in their experiments. When studying gases, they standardize temperature and pressure. Standard Temperature and Pressure (STP) is a temperature of 273.15 K (that is 0° Celsius) and a pressure of 100 kPa. Another common standard, used because it is close to room temperature, is the Normal Temperature and Pressure (NTP). This is slightly different and is a temperature of 293.15 K (25° Celsius) and 101.325 kPa (1 atmosphere). Of these, STP is the most commonly used. For most of your calculations, this can be approximated to 273 K and 1 atmosphere (1 atm). A mole of an ideal gas has an approximate volume of 22.4 liters at STP.

When using this model, scientists can use the ideal gas law:

$$PV = nRT$$

The ideal gas law relates pressure, volume, the number of moles, and the temperature of a gas. The constant *R* is the proportionality constant. The value used for *R* depends on the units in which the other variables are measured; it is important to look carefully at units before beginning calculations.

$$R = \frac{8.314\ \mathrm{L} \times \mathrm{kPa}}{\mathrm{mol} \times \mathrm{K}}$$

or

$$R = \frac{0.0821\ \mathrm{L} \times \mathrm{atm}}{\mathrm{mol} \times \mathrm{K}}$$

Using the ideal gas law, you can calculate any one of the variables if you are given the other three. The ideal gas law can usually be used to describe the behavior of real gases at room temperature and atmospheric pressure.

The Ideal Gas Law: Sample Problem

A sample of an ideal gas is contained in a rigid, 2.50-L cylinder. It is at a pressure of 1.43 atm. The temperature of the gas is 248 K. How many moles of gas are in the cylinder?

Solution

The ideal gas law equation is:

$$PV = nRT$$

P is pressure, *V* is volume, *n* is number of moles, *R* is the ideal gas constant, and *T* is temperature.

To determine the number of moles of gas present, rearrange the equation to solve for *n*:

$$PV = nRT$$
$$n = \frac{PV}{RT}$$

The ideal gas constant has a value of 0.0821 L × atm/mol × K.

Substitute the given values and solve:

$$\begin{aligned} n &= \frac{PV}{RT} \\ &= \frac{(1.43\ \cancel{\text{atm}})(2.50\ \cancel{\text{L}})}{(0.08206\ \frac{\cancel{\text{L}} \times \cancel{\text{atm}}}{\text{mol} \times \cancel{\text{K}}})(248\ \cancel{\text{K}})} \\ &= 0.176\ \text{mol} \end{aligned}$$

What Is Graham's Law of Effusion?

Graham's Law of Effusion

Many behaviors of gases are dependent on volume, pressure, temperature, and the number of moles of gas present. Some behaviors, however, depend on the specific types of gas molecules present.

Consider a can of soda. What happens when you open the container? Carbon dioxide, a common gas, has been dissolved in a liquid, which is then sealed in a can. The carbon dioxide molecules have kinetic energy. When you pop open the top, some of the carbon dioxide molecules escape from the can. You might hear the gas as it moves out of the small opening in the can and into the air in the room. This movement of gas through a small opening in a container is referred to as effusion.

Gas molecules do not lose their energy when they leave a container. They continue to move and spread apart. The carbon dioxide from a soda can, for example, spreads throughout the room, and mixes with the gas (air) in the room. The movement of the molecules throughout a container is called diffusion.

Another example of diffusion can be observed when spraying cologne or perfume in a room. Once the scented particles leave their container, they enter room. Eventually, these scented particles spread out across the entire room. Once this occurs, the scent can be smelled throughout the room.

Gaseous Effusion

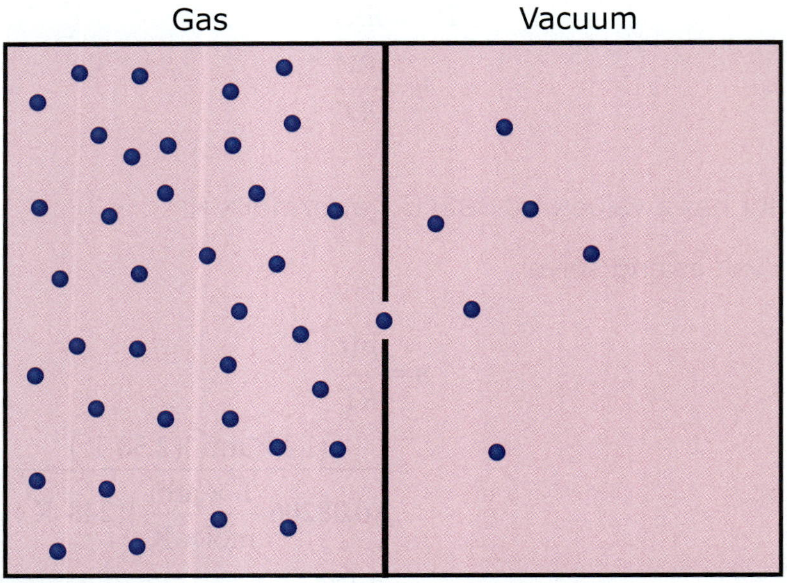

In effusion, gas molecules move from a region of high pressure to a region of low pressure through a narrow opening. What are some real-life examples of effusion?

Some gases diffuse more quickly than carbon dioxide, and some diffuse at a slower rate. The diffusion and effusion of a gas are related to the molar mass of the gas. The molar mass of a gas is the mass of 1 mole of its molecules. It is almost always measured in grams per mole (g/mol).

Molar mass is calculated by adding the atomic mass of the elements present. For example, carbon dioxide gas contains one carbon atom (with atomic mass of 12 g) and two oxygen atoms (each with atomic mass of 16 g). One mole of carbon dioxide has a molar mass of:

$$12 \text{ g/mol} + 2\,(16 \text{ g/mol}) = 44 \text{ g/mol}$$

Graham found that heavier gas molecules move more slowly. The relationship is described by Graham's law of effusion, which states that the rate of effusion of a gas is inversely proportional to the square root of its molar mass. The law also applies to the diffusion of gas molecules.

The rate of effusion or diffusion can be described as the velocity of the molecules, and so is assigned the variable *v*. Molar mass is represented by the variable *M*.

$$\frac{v_1}{v_2} = \sqrt{\frac{M_2}{M_1}}$$

Graham's Law of Effusion: Sample Problem

The molar mass of oxygen is 32 g/mol. The molar mass of carbon dioxide is 44 g/mol. Compare the rates of effusion of these two gases.

Solution

The equation for Graham's law of effusion is:

$$\frac{v_1}{v_2} = \sqrt{\frac{M_2}{M_1}}$$

v_1 = oxygen

v_2 = carbon dioxide

showing the relationship between the four variables. Substitute the given information and calculate the ratio of the two rates of effusion:

$$\frac{v_1}{v_2} = \sqrt{\frac{44\text{g/mol}}{32\text{g/mol}}}$$

$$\frac{v_1}{v_2} = \sqrt{1.375} = 1.7$$

Oxygen diffuses at a rate 1.17 times faster than carbon dioxide does.

What Is Dalton's Law of Partial Pressures?

Dalton's Law of Partial Pressures

Another important gas law was discovered by John Dalton (English chemist, 1766–1844). In many cases, the gas in a container is a mixture of more than one type of molecule or atom. Dalton found that each gas in a mixture exerts the same amount of pressure as it would if separated from the mixture. The pressure of each gas in the mixture is referred to as partial pressure, while the pressure of the system as a whole is the total pressure. Dalton's law of partial pressures is expressed by the equation:

$$P_{total} = P_1 + P_2 + P_3 \ldots$$

Dalton's Law of Partial Pressures: Sample Problem 1

A rigid gas cylinder contains a mixture of nitrogen gas, oxygen gas, and water vapor at a total pressure of 15.30 atm. If the partial pressure of nitrogen in the mixture is 3.87 atm and the partial pressure of oxygen in the mixture is 8.20 atm, what is the partial pressure of water vapor in the mixture?

Solution

The equation relating partial pressures in a mixture to the total pressure of the mixture is:

$$P_T = P_1 + P_2 + \ldots$$

where P_T is the total pressure of the mixture and P_1, P_2..., are the partial pressures of the gases in the mixture.

To find the partial pressure of water vapor, rearrange the equation to solve for the desired quantity:

$$P_T = P_{N_2} + P_{O_2} + P_{H_2O}$$
$$P_{H_2O} = P_T - P_{N_2} - P_{O_2}$$
$$P_{H_2O} = 15.30\text{ atm} - 3.87\text{ atm} - 8.20\text{ atm}$$
$$P_{H_2O} = 3.23\text{ atm}$$

Dalton's Law of Partial Pressures: Sample Problem 2

The Martian atmospheric pressure is now known to consist of many gases. There are three predominant gases: carbon dioxide, nitrogen, and argon. Calculate the total gas pressure given the following partial pressures: carbon dioxide = 4.28 mm Hg, nitrogen = 0.14 mm Hg, argon = 0.07 mm Hg.

Solution

The equation relating partial pressures in a mixture to the total pressure of the mixture is:

$$P_T = P_1 + P_2 + \ldots$$

where P_T is the total pressure of the mixture and P_1, P_2..., are the partial pressures of the gases in the mixture.

To find the total pressure of the gases, use Dalton's law equation to solve for P_T:

Substitute the given values and solve:

$$P_T = 4.28\text{ mm Hg} + 0.14\text{ mm Hg} + 0.07\text{ mm Hg}$$
$$P_T = 4.49\text{ mm Hg}$$

Consider the Explain Question:

What are the physical properties of gases, and how do they change under certain conditions?

Go online to complete the scientific explanation.

Check for Understanding:

Can you provide real-world examples of Boyle's law, Charles's law, and Avogadro's law?

STEM in Action

STEM and Gas Laws

Engineers design processes, tools, and equipment that help us work safely and more efficiently. They must consider the properties of matter in everything they create, along with the factors that can affect those properties. Of course, air surrounds the entire planet and fills even the smallest spaces in our environment. Air pressure is one factor that engineers must nearly always consider. Other gases inside a system might also be an important factor.

The gas laws are used, therefore, during many engineering and research tasks. For example, automobile safety engineers must consider the volume and pressure of the air that fills an airbag when it is inflated. When an airbag is deployed, there is a series of chemical reactions that take place so that the airbag increases and protects the individuals in the car. How are they able to complete this task? Chemical engineers are able to accurately measure the amount of gases given off during a controlled reaction They apply their knowledge to the volume of the airbag. What is the result? When the car is in a collision, sensors within the car are immediately activated to release the chemicals that trigger instantaneous chemical reactions in each airbag. The gases resulting from these reactions are produced in just the right quantity so that the airbag inflates to the correct size and protects the individuals from harm. After a car accident where airbags have been deployed, people will often say they have "chemical burns." This is from the chemicals that caused the airbag to deploy and save their life.

For this activity, you will research free diving. You will research free diving and use the gas laws to apply their practical use as individuals free dive. Use what you have learned and researched to answer the questions.

1. What is free diving?
2. How do you train to complete free diving? You must include at least five explanations.
3. What is considered the safest depth for a free diver? What is the maximum depth that a free diver can travel?
4. What is the pressure at average depth and the deepest depth?
5. Apply Boyle's law to the ability of the diver to complete the dive at that depth. What will happen to the lung volume of the individual? What impact does this have on the person, especially when they push the limits of the free dive?
6. Why do free divers never have to worry that their lungs will rupture if they surface at a fast pace?

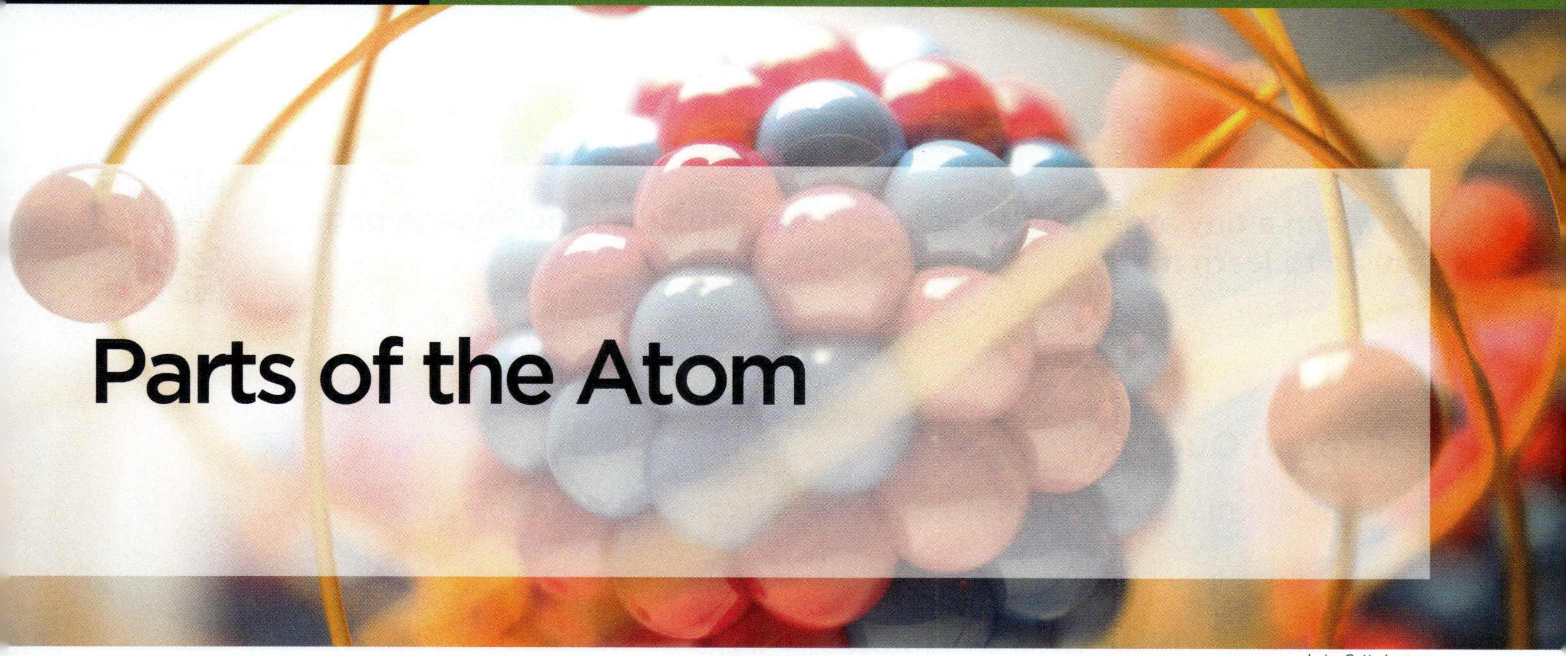

Parts of the Atom

photo: Getty Images

LESSON OVERVIEW

Lesson Questions

- How can the subatomic parts of an atom be used to identify it?
- How can isotopes of a given element be identified?
- How can the quantity of protons, neutrons, and electrons in an atom be determined?
- What is the difference between atomic mass and atomic number?
- How can the average atomic mass for an element be determined given the masses of its isotopes and their relative abundance?

Key Vocabulary

Which terms do you already know?

- ☐ **atom**
- ☐ **atomic mass unit**
- ☐ **atomic number**
- ☐ **average atomic mass**
- ☐ **electron**
- ☐ **electron cloud**
- ☐ **isotope**
- ☐ **mass number**
- ☐ **negative charge**
- ☐ **neutron**
- ☐ **nucleus (atom)**
- ☐ **positive charge**
- ☐ **proton**
- ☐ **subatomic**

Lesson Objectives

By the end of the lesson, you should be able to:

- Determine the quantity of protons, neutrons, and electrons in an atom.
- Distinguish between atomic mass and atomic number.
- Identify isotopes of a given element.
- Given masses of isotopes and relative abundance, determine the average atomic mass for an element.

ENGAGE

How can a tiny atom power an enormous ship? Visit the Engage page online to learn more.

Explain Question

What information about an atom can be determined from its parts?

EXPLORE

How Can the Subatomic Parts of an Atom Be Used to Identify It?

Protons, Neutrons, and Electrons

Scientific thinkers long believed that atoms were the smallest particle that could be identified, as had been proposed in John Dalton's atomic theory in the early 1800s. It was not until the early 1900s that scientists discovered this to be false. We now know that atoms are made up of subatomic particles that have their own distinct properties.

Two kinds of subatomic particles exist in the core, or nucleus, of an atom—protons and neutrons. Protons are positively charged particles. Neutrons do not have a charge.

photo: Getty Images

Subatomic particles are constantly in a state of movement. What subatomic particle circles around the atom's nucleus?

All atoms contain at least one proton. With the exception of the stable hydrogen atom, all atoms also contain at least one neutron. Protons and neutrons have the same mass, which is defined as 1 atomic mass unit (amu).

The space outside of an atom's nucleus is occupied by electrons. An electron is a negatively charged particle. The magnitude of the negative charge on an electron is equal to the amount of positive charge of a proton. However, the mass of an electron is extremely small; it is only 1/1836 the mass of a proton or a neutron. Thus, almost all of the mass of an atom is found in its protons and neutrons.

Electrons orbit the nucleus in a relatively large space surrounding the nucleus. This space is called an electron cloud, and it accounts for most of an atom's size. The configuration of electrons within this space depends on the number of electrons present and their energy.

If an atom of any given element contains the same number of electrons as protons, its overall charge is zero. An atom with an equal number of electrons and protons is said to be in a neutral state.

However, under certain conditions an atom of an element will gain or lose electrons. In this case, the charges in the atom become unbalanced and the particle has a net charge. The charge is negative if the number of electrons is greater than the number of protons. The charge is positive if the number of electrons is less than the number of protons. Such charged particles are called ions.

Identifying Atoms

The atoms of each element differ from one another. These differences account for the various properties that the elements exhibit. The primary detail that identifies the atoms of a specific element is the number of protons in its nucleus.

For example, an atom of the element hydrogen contains exactly one proton in its nucleus. This is true of all hydrogen atoms. An atom of the element helium contains exactly two protons in its nucleus. Again, this is true of all helium atoms.

Thus, the atoms of each known element contain a specific and unique number of protons. Put another way, the atoms of different elements contain different numbers of protons in their nuclei.

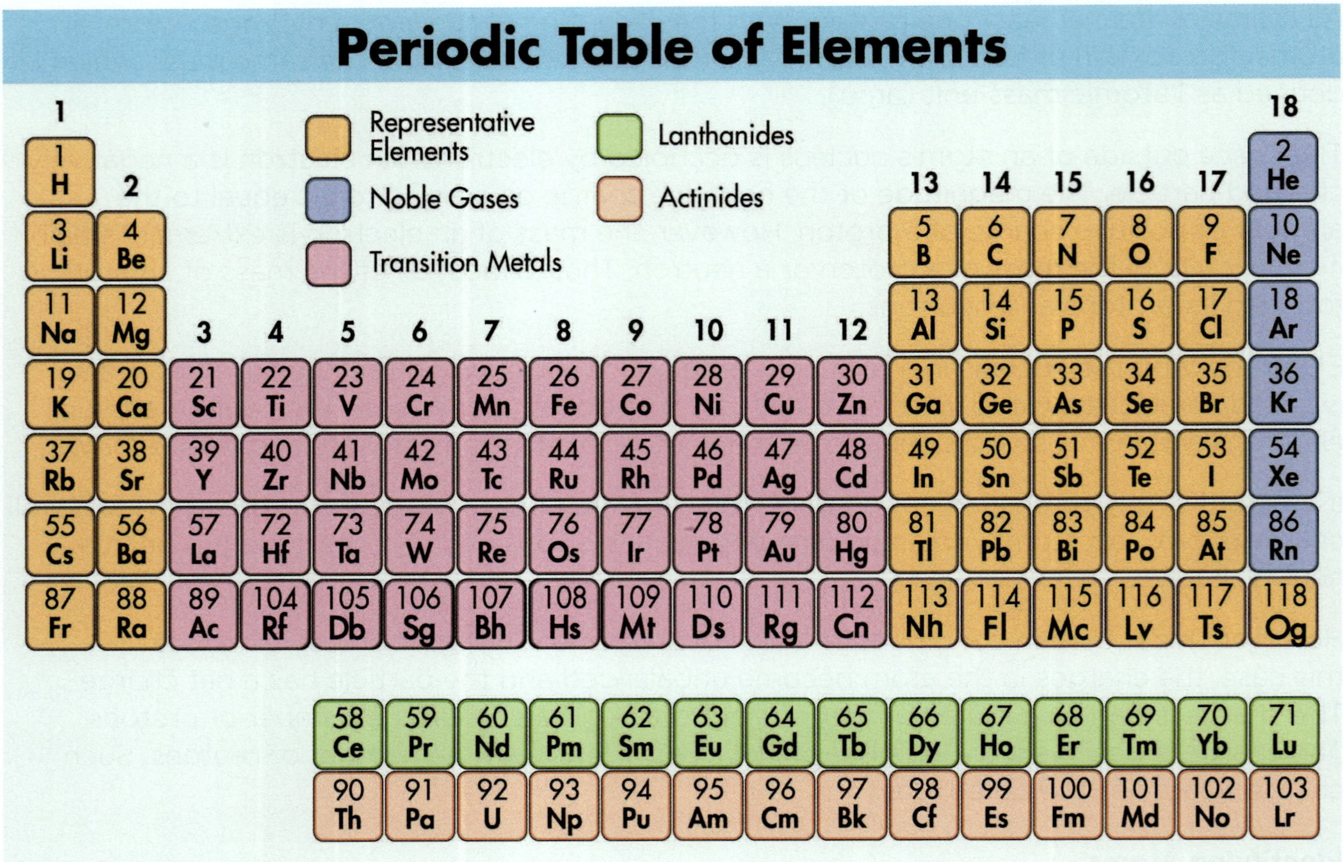

The numeral above each symbol of an element in the periodic table represents its atomic number. Which part of the atom determines the atomic number of an atom?

How Can Isotopes of a Given Element Be Identified?

Isotopes

As it turns out, not all atoms of a given element are completely identical. The most abundant hydrogen isotope, hydrogen-1, is also called protium. It contains one proton and no neutrons. The nucleus of hydrogen-2, also called deuterium or “heavy hydrogen,” consists of one proton and one neutron. The nucleus of hydrogen-3, also called tritium, consists of one proton and two neutrons. Hydrogen-1 and hydrogen-2 are naturally occurring isotopes of hydrogen. Hydrogen-3 is an unstable radioactive isotope that is extremely rare on Earth but is created synthetically.

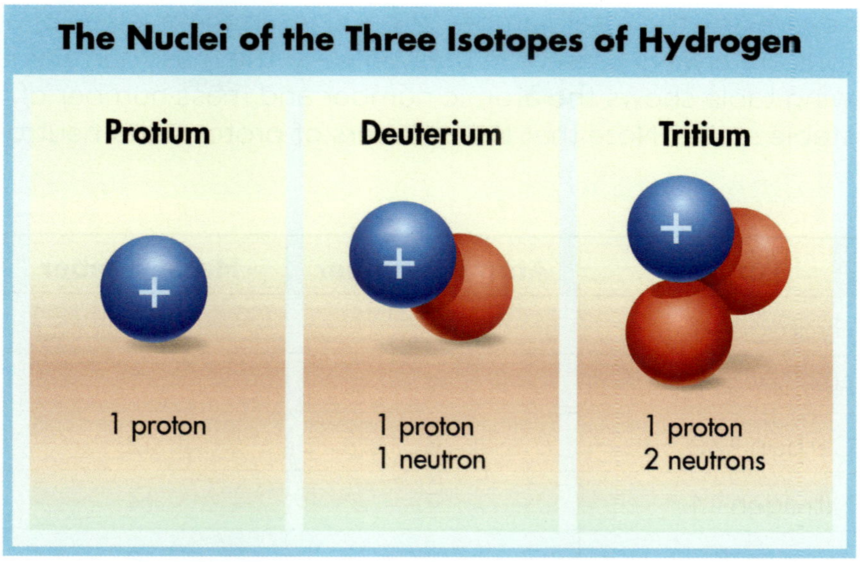

There are three isotopes of hydrogen. The nucleus of hydrogen-1, the most common hydrogen isotope, consists of one proton and no neutrons. What are the relative masses of these three isotopes of hydrogen?

Note that all three isotopes of hydrogen contain a single proton in their nucleus. As discussed earlier, the number of protons present is what distinguishes one element's atoms from another element's atoms.

A change in the number of neutrons in an atom's nucleus does not change the atom's identity. It does, however, change the atom's mass.

How Can the Quantity of Protons, Neutrons, and Electrons in an Atom Be Determined?

Counting Protons, Neutrons, and Electrons

The atomic number of an element equals the number of protons in the nucleus of the element's atoms. Every element has a different atomic number because the elements' atoms contain different numbers of protons.

The number of electrons in a neutral atom equals the number of protons in the atom's nucleus. In other words, in addition to describing the number of protons in an atom, the atomic number also describes the number of electrons in a neutral atom. This only applies, however, if the atom has not lost or gained any electrons.

Recall that each proton and neutron has a mass of about 1 atomic mass unit. Because the electrons in the electron cloud have very little mass, the entire mass of the atom can be considered to equal the mass of all the protons and neutrons in the nucleus. A neutral atom of carbon, for example, has six protons and six neutrons, each with a mass of 1 atomic mass unit. The approximate mass of the stable carbon atom is therefore 12 atomic mass units. This atom has a mass number of 12.

The mass number expresses the total number of particles in the nucleus of an atom. Because protons and neutrons are always present in their entirety, the mass number is always a whole number. The following table shows the atomic number and mass number of various common elements in their stable states. Note that the numbers of protons and neutrons are not always equal.

Isotope	Atomic Number	Mass Number
Hydrogen-1	1	1
Helium-3	2	3
Carbon-12	6	12
Nitrogen-14	7	14
Oxygen-16	8	16
Fluorine-19	9	19
Sodium-23	11	23
Calcium-40	20	40
Iron-56	26	56
Silver-108	47	108
Gold-197	79	197
Lead-207	82	207

The atomic number is equivalent to the number of protons in the atoms of an element. The mass number is equivalent to the number of protons and the number of neutrons in the atoms. Do protons and neutrons have the same mass?

If the mass number for an element is given, the number of neutrons in its atoms can be calculated by subtracting the atomic number from the mass number.

mass number - atomic number = number of neutrons

Gold-197 has a mass number of 197 and an atomic number of 79, which means it has 118 neutrons in the nucleus of an atom.

$$197-79=118$$

Hydrogen-1 has an atomic number of 1 and a mass number of 1, so the number of neutrons in a hydrogen atom is calculated to be 0.

$$1-1=0$$

Hydrogen is the only element with atoms that contains no neutrons in the stable state.

The atomic number and the mass number are used in symbols that represent the isotopes of all the elements using the following form:

$$^{A}_{Z}X$$

Here, *X* represents any element. *Z* represents the atomic number for the isotope, and *A* represents the mass number. For example, the isotope described above, gold-197, would be represented by the following symbol:

$$^{197}_{79}Au$$

The symbols for other isotopes of gold would have the same atomic number but different mass numbers.

What Is the Difference between Atomic Mass and Atomic Number?

Atomic Mass vs. Atomic Number

The atomic number of an element equals the number of the protons in the nucleus of its atoms. The atomic number is the same for every atom of a particular element. For example, the atomic number of hydrogen and all its isotopes is 1.

The mass number will vary for different isotopes of the same element because different numbers of neutrons are present in the atoms. Mass number is usually used to describe a specific isotope that a scientist is working with in a lab. The isotopes of hydrogen have mass numbers of 1 and 2 for hydrogen-1 and hydrogen-2, respectively.

Modern technology has enabled scientists to calculate the exact masses of protons and neutrons. While the atomic mass unit (amu) is still a relative value, it has been defined as 1/12 the mass of a stable carbon atom. The precise measurement of the mass of an atom of a specific element is referred to as its atomic mass. The atomic mass of hydrogen-1 is 1.008 atomic mass units and the atomic mass of hydrogen-2 is 2.014 atomic mass units.

How Can the Average Atomic Mass for an Element Be Determined Given the Masses of Its Isotopes and Their Relative Abundance?

Average Atomic Mass

It is often useful to consider the general properties of an element, even if several isotopes are present. For example, an environmental scientist might want to study the behavior of oxygen in the atmosphere. There are three naturally occurring isotopes of oxygen: oxygen-16, oxygen-17, and oxygen-18. The scientist's work will be concerned with the oxygen that actually exists in the atmosphere rather than with each isotope separately.

Scientists have found that they can use the average atomic mass of a specific element to be accurate in quantitative measurements. The average atomic mass takes into account the masses of all the naturally occurring isotopes of a particular element, weighted according to their abundance. Because it is a calculated value, it is likely to include decimals. These are the values typically listed on the periodic table of the elements. The average atomic mass of oxygen is 15.999. The whole number atomic mass of radioactive elements shown on the periodic table is the mass of the most stable isotope of that element.

The average atomic mass of hydrogen is 1.008. This is very similar to the atomic mass of hydrogen-1 because over 99.9 percent of hydrogen occurs in the stable form.

The weighted averages take into account the abundance of the naturally occurring isotopes of an element. Abundance is usually expressed as a percent of the element of each isotope. To find the atomic mass of an element, follow these steps:

1. Find the percent abundance and atomic mass of each naturally occurring isotope of the element.
2. Multiply the percent abundance by the atomic mass of each isotope.
3. Find the sum of the products.
4. Divide the sum by 100.

The following sample problem illustrates how weighted averages are used to calculate the atomic mass of an element.

Average Atomic Mass: Sample Problem

Determine the average atomic mass of the three naturally occurring isotopes of magnesium given the following data:

Isotope	Abundance (%)	Mass of Isotope (amu)
Magnesium-24	79	23.985
Magnesium-25	10	24.986
Magnesium-26	11	25.983

Solution

Magnesium has three isotopes, the masses of which are shown in the table. The data in the table indicates that in a sample of 100 magnesium atoms, 79 will be magnesium-24, 10 will be magnesium-25, and 11 will be magnesium-26. Thus, the average atomic mass of magnesium can be calculated as follows:

$$\text{Average atomic mass of magnesium} = \frac{(79 \times 23.985\text{ amu}) + (10 \times 24.986\text{ amu}) + (11 \times 25.983)}{100}$$

$$= \frac{2{,}430.5\text{ amu}}{100} = 24.305\text{ amu}$$

Consider the Explain Question:

What information about an atom can be determined from its parts?

Go online to complete the scientific explanation.

Check for Understanding:

How do subatomic particles determine the chemical identity of an element?

STEM in Action

STEM and Parts of the Atom

They form numbers on the faces of digital clocks and with the push of a button they light up watches. They send signals from remote controls and produce images on high-definition television screens. Electrical engineers help design and manufacture them for consumer use. What does each of these examples describe? They refer to a type of semiconductor light source called LED, or light-emitting diodes.

You might think of an LED as an extremely small light bulb. Similar to their relative the incandescent light bulb, LEDs are activated by a flow of electric current. However, the current does not flow through a metal filament as in traditional light bulbs. Instead, in LEDs the flow of an electric current causes the movement of electrons through a diode. A diode is a device that allows electricity to move in only one direction. In general, an LED is made of a material called a semiconductor, which is a type of material that regulates the flow of an electric current. Semiconductors include the metalloid elements, such as silicon and germanium.

Moving by the electric current, electrons will interact with atoms of the semiconductor, forcing them to emit packets of light called photons. The emitted light enables you to see the images on a high-definition television screen and makes the numbers on a digital clock visible. It is also the light you see shining from an LED light bulb.

photo: Getty Images

Light-emitting diodes, or LEDs, are widely used to light homes and our television screens. What makes an LED different from a traditional incandescent light bulb?

An LCD television produces its picture by shining light produced by LEDs through color filters. Relative newcomers to the LED scene are OLEDs, organic light-emitting diodes, which use LEDs to generate both light and color. The term organic refers to the fact that the materials are carbon-based. By using electroluminescence, OLEDs are amazingly thin and can be applied to flexible surfaces. They can be bent and shaped.

An LED differs from other technological sources of light in a number of important ways. For one thing, LEDs do not produce very much heat. Your television screen does not become too warm, nor does your illuminated watch, the digital numbers on your kitchen clock, or an LED light bulb. Another important advantage of LEDs is that when compared with most other forms of lighting, they consume little energy—making them very inexpensive to operate.

Decades ago, a light bulb was simply a light bulb. There were not many options available, and incandescent lighting ruled. Today, there are many options all with the goal to save lots of energy, last for a very long time, and be inexpensive. Besides the well-known and outdated incandescent light bulb, there are two additional types: LED and CFL. The CFL, or compact fluorescent light, is the squiggly shaped light bulb you commonly see sitting next to the other type, called an LED. Each light bulb has different characteristics as shown in the table. Use the following table to answer the questions.

	 LED	 Incandescent	 Compact Fluorescent (CFL)
Life span	50,000 hours	1,200 hours	8,000 hours
Watts used to generate 800 lumens (the amount of light a traditional 60W incandescent bulb emits)	6–8	60	13–15
Initial bulb cost	~$24	~$.49	~$2.50
Annual cost	$32.85 (per year)	$328.59 (per year)	$76.65 (per year)

photo: Getty Images

1. Why do you suppose an LED bulb would be a better light option than an incandescent light bulb?
2. What light bulb works the best at saving energy, in a cost-effective way? Which bulb is the most cost effective over its life span?
3. Someone asks for your opinion as to what light bulb to choose for an outdoor house light fixture. What do you recommend, and why?

Development of the Atomic Theory

photo: Getty Images

LESSON OVERVIEW

Lesson Question

- How did the atomic theory evolve from ancient Greece to the modern day, and which key scientists made which contributions?

Lesson Objective

By the end of the lesson, you should be able to:

- Trace the evolution of the atomic theory from ancient Greece to the modern day, including the contributions of key scientists.

Key Vocabulary

Which terms do you already know?

- [ ] **Aristotle**
- [ ] **Aristotle's four elements**
- [ ] **atomic theory**
- [ ] **cathode-ray tube**
- [ ] **Democritus**
- [ ] **electron cloud**
- [ ] **energy levels**
- [ ] **Ernest Rutherford**
- [ ] **gold foil experiment**
- [ ] **Heisenberg uncertainty principle**
- [ ] **J.J. Thomson**
- [ ] **John Dalton**
- [ ] **law of conservation of mass**
- [ ] **Niels Bohr**
- [ ] **orbital model of the atom**
- [ ] **photoelectric effect**
- [ ] **plum pudding model**
- [ ] **proton**
- [ ] **quantum mechanics**

ENGAGE

How is the process of figuring out a mystery pocket dial similar to figuring out a scientific theory? Visit the Engage page online to learn more.

Explain Question

How did scientists use the work of earlier scientists and the contributions of their peers to develop the atomic theory?

EXPLORE

How Did the Atomic Theory Evolve from Ancient Greece to the Modern Day, and Which Key Scientists Made Which Contributions?

Ancient Atomic Theory

Ancient Greek philosophers were among the first to discuss atoms and matter. This became known in time as atomic theory. Atomic theory states that matter is composed of discrete units called atoms. It was not until the late 1800s, however, that this theory was well researched and became widely accepted.

Thales (c. 620–c. 546 BCE) believed that everything came from water. However, Democritus (460–370 BCE) proposed that all matter in the world consisted of invisible and indivisible corpuscles (or small particle-like objects). He called these corpuscles atoms. Democritus used his preliminary atomic theory to explain the structure of nonliving objects and biological processes. His understanding of atoms was based purely on his own imagination because he lacked the scientific tools to experiment with atoms.

Democritus thought that there was empty space between atoms. He also thought atoms were in constant motion and had various shapes and sizes. He theorized that the properties of the material (such as weight, taste, and smoothness) corresponded with the properties of the atom.

Aristotle (384–322 BCE) disagreed with Democritus' ideas. Instead, Aristotle believed there were four main elements: earth, water, air, and fire. Aristotle's four elements were based on Thales's belief in water as the originator of all other substances. Epicurus (341–270 BCE) reintroduced the ideas of Democritus. He proposed that atoms had mass, moved at a consistent speed, and were able to change direction.

photo: Getty Images

Epicurus developed atomic theory where Democritus left off. How do the ideas of Epicurus compare with ideas we hold today about the atom?

John Dalton's Atomic Theory

John Dalton (1766–1844) was an English chemist and physicist. He is best known for his work on modern atomic theory, which came about during his development of partial pressure gas laws.

Dalton based his atomic theory on observations from his own experiments in gases and meteorology. He also considered evidence from experiments done by other scientists. Here are some examples of the kind of questions and facts Dalton would have considered when composing his theory:

- When water evaporates to form water vapor, it mixes with the air. How can these two gases occupy the same space at the same time? This suggests that gases are made of particles that can be mixed together.
- Different elements, such as silver and gold, have different properties. This suggests that particles of different elements are different from each other.
- Matter cannot be destroyed. Even when a substance is completely burnt, products such as gases and ash result.

- Matter cannot be made out of nothing.
- The same volumes of gases always react with each other. For example:

  2 volumes of hydrogen gas + 1 volume of oxygen gas ⟶ 2 volumes of water.

  This suggests that particles can combine in fixed ratios to form new compounds.

Dalton used these and other experimental observations and facts to compile a list of ideas describing how matter works at the atomic level. Dalton's ideas are sometimes referred to as postulates.

Dalton's Postulates on Atomic Theory

In 1808, John Dalton first theorized that elements contained atoms of a single type, and that these atoms combined to form more complex structures.

Dalton believed that elements combined in the simplest way in whole number ratios. For this to be true, chemical combinations would have to include atoms of different mass. This aspect of Dalton's theory differentiated his work from that of Democritus and other Greek philosopher scientists.

John Dalton's atomic theory had five points:

1. All matter is composed of particles called atoms.
2. Atoms of a given element have unique properties. While atoms of the same element are identical, atoms of different elements differ in size, mass, and properties.
3. Atoms cannot be subdivided, created, or destroyed.
4. Atoms of different elements combine in simple, whole-number ratios to form chemical compounds.
5. During chemical reactions, atoms are combined, separated, or rearranged.

The main principles of Dalton's theory remain critical to the understanding of atomic theory. They also support other important physical laws, such as the law of conservation of mass, the law of definite composition, and the law of multiple proportions. The finding that atoms are indeed made up of smaller components greatly transformed the study of matter. However, scientists have since discovered that Dalton's theory was incomplete, and they have added subatomic particles and a well-defined structure to his model.

Early Developments in Atomic Theory

J. J. Thomson discovered the first subatomic particle a little more than a century ago. He used a cathode-ray tube to determine that atoms contain negatively charged particles, or electrons.

A cathode-ray tube is a glass tube containing wires attached to each end. The tube forms a vacuum from which as much air has been removed as possible. Attaching the wires to a battery causes a beam of electrons to travel across the tube. Scientists in Thomson's day did not understand this beam. Thomson studied it by extending the tube and then directing the beam onto the far end. He coated this end with a substance that glowed when the electrons struck it. He found that he could deflect the beam up or down by passing it through oppositely charged metal plates. He could also deflect it by moving a magnet near it. The direction of the deflection showed that the beam in the cathode-ray tube had a negative charge. Further studies led Thomson to conclude that the beam consisted of negatively charged particles that were originally from atoms.

Thomson used his results to draw conclusions about the structure of an atom. He believed these negatively charged particles were dispersed throughout an atom, embedded in matter that was positively charged. This model was called the "plum pudding" model of the atom because it resembled a common dessert of the time that consisted of bits of fruit spread throughout a pudding.

In 1909, physicist Robert Millikan performed the now-famous oil-drop experiment that revealed the mass and charge of electrons. Millikan suspended tiny oil droplets between metal electrodes and measured the size of the droplets in response to forces acting on the droplet. Because the basic physical properties of the oil drops were known, Millikan was able to calculate the electric charge and mass of oil droplets of varying size. Repetition of this experiment using oil droplets of varying sizes revealed the mass and charge of electrons, which was a major step forward in understanding atomic theory.

Ernest Rutherford's Gold Foil Experiment

Ernest Rutherford was a student of Thomson's who was also intrigued with the idea of subatomic particles. Rutherford contributed a great deal to the understanding of the structure of the atom. In fact, he is considered by some to be the "Father of Modern Physics." He is perhaps best known for his gold foil experiment. This research showed that atoms contained a dense central region. He found that most of the mass of an atom was located in this nucleus, as he labeled it, and that it carried a positive charge.

In his famous gold foil experiment, Rutherford directed a narrow beam of positively charged particles toward a thin piece of gold foil. Surrounding the foil was a specially coated screen that flashed when the particles hit it.

Because of the mass of the charged particles, Rutherford predicted that most particles would pass through the foil. He also predicted that a few would be deflected by the electrons in the atoms of the gold foil, but that the path of these particles would not change much because they had much greater mass than the electrons.

One part of his prediction was correct: most of the particles passed right through the foil or were deflected just a bit. Some of the particles, however, were deflected right back at the particle source! This proved that something larger than an electron was blocking its path through the foil.

Rutherford concluded that an atom's positive charge and its mass are concentrated in a very small area of the atom. He believed the electrons occupied the space around the nucleus, but he did not propose a specific electron structure. His model was called the nuclear model of the atom. However, Rutherford could not explain why the negatively charged electrons were not pulled into the positively charged nucleus.

Rutherford was able to determine the percentage of deflected particles, which allowed him to estimate the size of the nucleus. He determined the radius of the nucleus to be at least 10,000 times smaller than the radius of the entire atom.

Rutherford and his colleagues made many other contributions to modern atomic theory. They studied the effects of X-rays on various materials and explained radioactive decay. They continued to study the atomic nucleus. Their efforts later led to the discovery that the nucleus is composed of smaller particles, filling in some of the missing pieces of the atomic model. Rutherford showed that the nucleus contains positively charged protons. James Chadwick, a colleague of Rutherford's at Cambridge University, was responsible for making observations that later led to the discovery that the nucleus also contains a neutral particle, the neutron.

Refining the Atomic Theory

John Dalton's atomic theory was rapidly being refined. By the early 1900s, physicists understood that atoms contained a nucleus that was surrounded by electrons. But they did not understand why atoms of different elements bonded and reacted differently. Over time, it was shown that there were some interesting connections between the properties of energy and matter. One idea proposed was that all matter and energy are quantized. This means they can be divided only into certain definite amounts.

Niels Bohr, who worked with Rutherford, studied the changes that occurred within an atom when energy was added to it or removed. He found that the amounts of energy absorbed and emitted by hydrogen atoms were related to the energies of their electrons. His theory claimed that electrons orbit the nucleus of an atom. The amount of energy determines the electron's distance from the nucleus. If the atom has not absorbed extra energy, it is said to be in its "ground state." The greater the energy, the farther from the nucleus the electron's orbit will be.

When Bohr studied the light energy given off by an atom of hydrogen, he found that each electron is associated with a specific amount of energy. These are referred to as energy levels. When an electron absorbs a quantum of energy, it becomes excited and moves to a higher energy level, one that is farther from the nucleus. When the electron falls from the excited state back to its ground state, it emits energy in the form of electromagnetic radiation.

Bohr's theory gave rise to the planetary or orbital model of the atom because it stated that electrons orbited the atom's nucleus in circular patterns, similar to those of planets in the solar system. As with the models created by his predecessors, though, Bohr's model was later proven to be incomplete. The specific calculations he used only applied to the hydrogen atom, and it was found that the structure of other atoms was more complicated. His ideas about the relationship between electrons and their energy, however, formed the basis of today's understanding of the atom.

In 1905, Albert Einstein used Bohr's ideas to explain the photoelectric effect. The photoelectric effect occurs when some metals are exposed to light. If the light has a certain frequency, electrons are emitted from the metal's surface. Einstein theorized that the energy of these "escaped" electrons is related to the atoms they come from. This idea led to a Nobel prize for Einstein. It also supported future understanding of the properties of electrons. The photoelectric effect is the basis for many techniques now used to identify materials based on the energies of the electrons they emit.

Modern Quantum Theory

Today's understanding of the atom is built on Bohr's model. It is now believed that electrons move around the nucleus in a "cloud" instead of circular paths. Quantum theory, as it is called, attempts to explain the fact that energy and electrons have properties similar to both particles and waves. The view that electrons exist as both particles and waves was first described by French physicist Louis de Broglie, earning him the Nobel prize in physics in 1929. The mathematical description of the wave properties of electrons was later described by Erwin Schrodinger, whose ideas were instrumental in developing the modern-day "electron cloud" model of the atom.

Quantum theory is also based on the Heisenberg uncertainty principle, which states that the exact location and speed of an electron are impossible to determine at the same time. Scientists can only describe specific regions in the atom called "orbitals" where electrons are most likely to be located. Unlike Bohr's model, quantum theory describes the energy levels for all the elements, not just hydrogen, as well as the probable location of electrons within those energy levels.

Consider the Explain Question:

How did scientists use the work of earlier scientists and the contributions of their peers to develop the atomic theory?

Go online to complete the scientific explanation.

Check for Understanding:

Can you describe two experiments that support the development of the atomic theory?

STEM in Action

STEM and Development of the Atomic Theory

The search for the structure of the atom continues, using new technologies that allow scientists to probe for subatomic particles far smaller than protons, neutrons, and electrons. The 2012 announcement of the discovery of the Higgs boson, a subatomic particle that gives matter mass, was a major breakthrough that provided significant support to the Standard Model of particle physics. This model is a mathematical description of the most basic particles of matter, quarks and leptons, and three fundamental forces, strong, weak, and electromagnetic forces, that cause them to interact. Gravity, the fourth fundamental force, is not a part of the Standard Model.

But how did scientists discover the Higgs boson? It's not like discovering a new species of insect or bird. They couldn't pick it up, or take a picture of it. Technology has advanced enough that scientists can now make images of atoms. However, no technology is able to make images of particles like the Higgs boson, which are orders of magnitude smaller than an atom. Instead, scientists used a giant machine, the Large Hadron Collider (LHC) at the CERN facility near Geneva, Switzerland. The LHC is a supercollider that smashes atomic particles together at an incredibly high speed. Data from the resulting collisions is analyzed to reveal secrets of subatomic particles.

photo: Getty Images

The Large Hadron Collider helps scientists study subatomic particles. How do new technologies contribute to the continual evolution of the atomic model?

Three thousand physicists are engaged in research at CERN. But physicists are just the tip of the iceberg when it comes to people who have been involved in designing, constructing, and maintaining this 17-mile wide circular tunnel, and all the equipment that goes with it. Building a tunnel is a feat of civil engineering that requires expertise both in planning and in operating the equipment used to dig. The superconducting magnets that make the supercollider possible require ultralow temperatures to operate. Workers in the field of cryogenics maintain a system that uses liquefied helium to keep the magnets cool. If there is a helium leak, as there was in 2008, then there is the risk that the magnets could heat up enough to start a fire. Therefore, CERN has its own specially trained firefighters who know how to respond. These specialists also conduct regularly scheduled helium leak drills with everyone who works in the area. Processing data from CERN requires so much computing power that some of it happens off-site. Regardless of the location, both software architects and hardware engineers are needed to provide the physicists with the tools they need to analyze the data that results.

These are just a few of the many jobs that have to be done in order to keep a particle accelerator up and running. Whenever you hear of a scientific discovery, think of all the people whose names may not be in the news but who nonetheless are essential contributors.

Whether you call it a barbecue, a cookout, or just cooking on the grill, the reactions are the same: carbon in charcoal combines with oxygen to release carbon dioxide, heat, and a mixture of compounds called ash. When the fire is burning low, incomplete combustion of the charcoal also produces carbon monoxide. Complete the activity "Applying Dalton's Law to Identify Compounds" on the STEM in Action page online. Then, answer the questions:

1. Demonstrate that the ratio of oxygen in these two compounds is a simple whole number ratio. Explain the answer.
2. Explain how these compounds confirm the law of multiple proportions.

Nuclear Chemistry

photo: Getty Images

LESSON OVERVIEW

Lesson Questions

- How do alpha, beta, and gamma radiation compare?
- What conditions support the stability of a nucleus?
- How do transmutation and chemical change compare?
- How do fission and fusion compare?
- How are chain reactions controlled in nuclear reactors?
- How is the half-life of radioactive materials used?

Key Vocabulary

Which terms do you already know?

- ☐ **alpha particle**
- ☐ **alpha radiation**
- ☐ **beta particle**
- ☐ **beta radiation**
- ☐ **chain reaction**
- ☐ **fission**
- ☐ **fusion**
- ☐ **gamma radiation**
- ☐ **gamma ray**
- ☐ **half-life**
- ☐ **isotope**
- ☐ **neutron**
- ☐ **nuclear fission**
- ☐ **nuclear reaction**
- ☐ **nuclear reactor**
- ☐ **nuclide**
- ☐ **radiation**
- ☐ **radioactive decay**
- ☐ **radioactivity**
- ☐ **transmutation**

Lesson Objectives

By the end of the lesson, you should be able to:

- Compare the three basic types of radiation (alpha, beta, and gamma) in relation to penetrating power and relative energy.
- Explain what conditions support the stability of nuclei.
- Distinguish between transmutation and chemical change.
- Compare fission and fusion.
- Explain chain reactions and describe how they are controlled in nuclear reactors.
- Explain the concept of half-life.

ENGAGE

What would you think if you learned that some of the food you eat had been exposed to nuclear radiation? Is the food still okay to eat? Visit the Engage page online to learn more.

Explain Question

What are the different types of radiation, and how are they produced by isotopes of different elements?

EXPLORE

How Do Alpha, Beta, and Gamma Radiation Compare?

Three Types of Radiation

In 1896, the French scientist Henri Becquerel discovered that certain uranium minerals could create images on photographic plates wrapped in paper by giving off high-energy particles and rays called nuclear radiation. Marie Curie, a scientist working with Becquerel, coined the term radioactivity to identify the process by which certain nuclei give off radiation. To this day, scientists use the becquerel as the SI unit for radiation. A becquerel (Bq) is defined as the activity of a quantity of radioactive material in which one nucleus decays per second.

As scientists studied different materials that give off radiation, three types of radiation soon were identified. Alpha radiation, beta radiation, and gamma radiation were identified, and their properties determined. Alpha radiation is composed of alpha particles. An alpha particle is a helium nucleus. It is composed of two protons and two neutrons. As a result, an alpha particle has an atomic number of 2 and a mass number of 4. Its symbol is written:

$$\alpha \text{ or } {}^{4}_{2}\text{He}$$

Beta radiation is composed of beta particles. A beta particle is a high-energy or high-speed electron, released from the atom's nucleus when a neutron is split. The symbol for a beta particle is written:

$$\beta \text{ or } {}^{0}_{1}e$$

A neutron split can lead to beta radiation that can be either a beta minus (β^{-}) decay in which a neutron is converted to a proton, or a beta plus (β^{+}) decay, in which a proton is converted to a neutron and the process creates a positron. The β^{+} decay also is known as positron emission.

Gamma radiation consists of gamma rays. Gamma rays are electromagnetic radiation, or light waves, that have short wavelengths and high frequencies. The symbol for a gamma particle is written:

$${}^{0}_{0}\gamma$$

Different Penetrating Powers

Each type of radiation has a different penetrating power related to its mass, charge, and energy. Low penetrating power means that the rays are stopped easily, by materials that are not particularly dense. Rays with high degrees of penetrating power, however, are not stopped as easily.

Alpha radiation has relatively low energy and relatively low penetrating power. Alpha particles have a greater mass and charge than either beta particles or gamma rays, so they do not penetrate matter easily. Alpha particles are stopped by paper, clothing, and a few centimeters of air.

Beta radiation has higher energy. Because electrons have very little mass and a charge of only –1, beta particles have about 100 times the penetrating power that alpha particles have. Beta particles are stopped by a thin sheet of aluminum.

Gamma rays have large amounts of energy and have extremely powerful penetrating power. Gamma rays are light waves, so they have no mass or charge. They can penetrate most materials easily. Thick, dense materials, such as lead or concrete, must be used to construct shields to block gamma rays.

What Conditions Support the Stability of a Nucleus?

Stability of a Nucleus

Why are some nuclei stable while other nuclei are radioactive? Within the nucleus of an atom, protons and neutrons are packed very densely. Protons are positively charged and repel one another. The strong nuclear force acts over the very short distances within the atomic nucleus and helps to hold protons and neutrons together. The balance between the repulsive electrostatic force and the attractive strong nuclear force determines the stability of the nucleus.

For nuclei that are composed of up to 20 protons, having about one neutron per proton creates a balance of charges and a stable nucleus. In other words, a neutron–proton ratio of 1 : 1 results in stable nuclei for atoms of the elements from hydrogen to calcium on the periodic table. For heavier nuclei, a slightly larger neutron–proton ratio is needed.

When the atomic number and mass number of an atom are identified, the particle is referred to as a nuclide. (Isotopes are examples of nuclide groups that all contain the same number of protons but have different numbers of neutrons.) A carbon-12 nuclide, for example, has an atomic number of 6 and a mass number of 12. It also has six protons, and a neutron–proton ratio of 1 : 1. A carbon-12 nuclide is not radioactive. However, a carbon-14 nuclide has a neutron–proton ratio of 8 : 6, or about 1.3 : 1. This slight difference in the neutron–proton ratio results in the instability of the carbon-14 nucleus and makes it more radioactive.

How Do Transmutation and Chemical Change Compare?

Transmutation

What happens to an unstable nuclide? It undergoes radioactive decay. The nucleus of the radioactive isotope releases energy and particles and becomes a new nuclide.

The change of one chemical element or isotope into another is known as nuclear transmutation. Through transmutation, an unstable, radioactive nucleus can become stable.

Transmutation Versus Chemical Change

Transmutations are very different than chemical changes. During a chemical change, chemical bonds break and reform; atoms are rearranged into new substances. Changes to chemical bonds affect the electrons in the atoms. Because the identity of an atom depends on the protons in its nucleus, each atom involved in a chemical change keeps its identity. An atom of carbon found among the reactants is still an atom of carbon in the products. A chemical equation is balanced by making sure that an equal number of atoms of each element exist in the products and in the reactants.

Transmutation, however, is a nuclear change. Nuclear changes affect the particles in the nucleus, not just electrons. During transmutation, the number of protons in an atom may change. As a result, an atom that undergoes a transmutation (the parent nuclide) can change into an atom of a different element (the daughter nuclide).

Transmutation can occur in several different ways. These include alpha decay, beta decay, and gamma decay. Alpha decay occurs when a nuclide emits alpha particles. Notice that this is the same configuration found in the nucleus of a helium atom; therefore, alpha particles often are represented by the symbol:

$$^{4}_{2}\text{He}$$

Alpha decay is most common for elements with more than 83 protons, which are rarely stable in nature. Uranium-238 is one such isotope. As described by the following equation, uranium-238 undergoes alpha decay to form thorium-234:

$$^{238}_{92}\text{U} \rightarrow {}^{234}_{90}\text{Th} + {}^{4}_{2}\text{He}^{2+}$$

Beta decay involves the emission of beta particles from a nuclide that has a high neutron–proton ratio. Beta particles include high-speed electrons (β^- or ε^-) or positrons (β^+ or ε^+). The electrons are released when neutrons break up into a proton and an electron. Positrons have the same mass as electrons, but have a positive charge.

During beta decay, fundamental particles are rearranged to change the number of protons in a nucleus. This often involves a process called electron capture, in which an electron from an orbital near the nucleus is pulled in and combined with a proton to form a neutron.

Carbon-14 is an unstable nuclide that breaks down into nitrogen.

$$^{14}_{6}\text{C} \rightarrow {}^{14}_{7}\text{N} + \beta^-$$

During gamma decay, light energy is emitted in photons. Photons have no mass or charge and consist mostly of energy. Gamma decay often occurs along with alpha or beta decay.

Because different elements form, a nuclear equation is not balanced in the same way that a chemical equation is balanced. Instead of having the same number of atoms on each side of the equation, in a balanced nuclear equation, the sum of the atomic numbers on each side must be equal. The sum of the mass numbers on each side of the equation also must be equal. If the starting nucleus and the means of transmutation are known, the product can be predicted and the equation balanced.

Transmutation Versus Chemical Change: Sample Problem

Write a balanced equation for the transmutation that occurs when a carbon-14 nucleus undergoes beta decay.

Solution

Given:

- Starting nucleus: carbon-14
- Means of transmutation: beta decay

Carbon atoms have an atomic number of 6. A carbon-14 nucleus has a mass number of 14. The symbol for a carbon-14 nucleus is

$$^{14}_{6}\mathrm{C}$$

The nucleus undergoes beta decay, so it must give off a beta particle. The symbol for a beta particle is

$$^{0}_{1}e$$

The equation must show the carbon nucleus as a reactant and the beta particle as a product.

$$^{14}_{6}\mathrm{C} \rightarrow {}^{0}_{-1}e + {}^{A}_{Z}X$$

The unknown product is written as X with an unknown mass number A and an unknown atomic number Z. The sum of the mass numbers on the right and left must be the same, so the missing product must have a mass number A of 14.

$$^{14}_{6}\mathrm{C} \rightarrow {}^{0}_{-1}e + {}^{14}$$

The sum of the atomic numbers on the right and left must be the same, so:

$$6 = (-1) + Z$$

The missing product must have an atomic number of 7. The element that has atomic number 7 is nitrogen (N).

The balanced equation for the beta decay of a carbon-14 nucleus is:

$$^{14}_{6}\mathrm{C} \rightarrow {}^{0}_{-1}e + {}^{14}_{7}N$$

How Do Fission and Fusion Compare?

Nuclear Fission Versus Nuclear Fusion

Each day, people on Earth make use of energy released by nuclear changes. The main way in which people use nuclear energy is through sunlight. However, over the last century, scientists have learned how to produce and use nuclear energy.

In 1938, German scientists Otto Hahn and Fritz Strassmann made a surprising discovery while working with uranium atoms. At that time, scientists thought that shooting neutrons at uranium atoms would produce heavier elements. They expected the products to have atomic numbers higher than 92, the atomic number of uranium. Instead, Hahn and Strassmann obtained barium as a product, with an atomic number of only 56. They had discovered nuclear fission.

Nuclear fission is a process in which the nucleus of a massive atom splits into smaller, lighter nuclei. This process often produces neutrons and gamma rays. Today, nuclear fission is used in nuclear reactors to produce energy. For example, a nucleus of uranium-235 that is struck by a neutron might split into barium-144 and krypton-89 and three additional neutrons, according to the equation:

$$^{1}_{0}n + {}^{235}_{92}\mathrm{U} + {}^{144}_{56}\mathrm{Ba} + {}^{89}_{36}\mathrm{Kr} + 3^{1}_{0}n$$

Uranium-235 and plutonium-239 are the most common nuclear reactor fuels. Inside a nuclear reactor, a controlled series of reactions takes place. The neutrons produced by each reaction start more of the same reactions. This process continues, generating large amounts of energy. The heat energy produced by a reactor can be used to generate electricity in a nuclear power plant. The Explore section titled "How Are Chain Reactions Controlled in Nuclear Reactors?" has more information on nuclear reactors.

Uranium, plutonium, and the products of fission are radioactive. Radioactive waste from nuclear fission processes is hazardous and must be stored for long periods of time so it does not contaminate the environment.

Nuclear fission is not the only way to produce nuclear energy. Energy from the sun is released during nuclear fusion. How do these processes compare?

Nuclear fusion is a process in which two or more small atomic nuclei are joined or fused together to form a single heavier nucleus. In the sun, hydrogen nuclei are fused to form a helium nucleus in several steps. Two positrons also are produced, along with energy. The overall reaction is:

$$4^{1}_{1}\mathrm{H} \rightarrow {}^{4}_{2}\mathrm{He} + 2\beta^{+}$$

As a star uses up its hydrogen, it fuses more massive nuclei. Hydrogen, however, is abundant on Earth. It can be formed from the decomposition of water. Also, fusion produces little radioactive waste. Why, then, isn't fusion used as an energy source on Earth? Remember that hydrogen nuclei are positively charged and repel each other. They must be forced to fuse at very high temperatures. Thus, large amounts of energy are needed to cause and sustain the fusion reaction, which makes it impractical with our current level of technology.

Supplies of the uranium typically used in nuclear fission reactors are limited, especially compared with the hydrogen that nuclear fusion would use. Also, while a fission reaction requires less energy to maintain, the process is potentially more dangerous than fusion.

Both nuclear fission and nuclear fusion are used to make nuclear weapons. A nuclear fission bomb is called an atomic bomb. Inside the bomb, energy is produced by uncontrolled chain reactions of uranium or plutonium. A nuclear fusion bomb is called a hydrogen bomb. In this kind of bomb, energy is produced by both fission and fusion chain reactions. The fission reactions produce the energy needed to start the fusion reactions. The fusion reactions take place between isotopes of hydrogen.

The atomic bomb was the first human-made weapon of mass destruction. Its invention changed human history forever. Today, organizations around the world monitor nuclear weapons and their production very carefully. In some regions, such as Latin America and the Caribbean, nuclear weapons are banned.

Conservation of Matter and Energy

Like chemical reactions, nuclear reactions must balance. In chemical reactions, the numbers and types of atoms are the same on both sides of the equation. In nuclear reactions, the total number of protons plus neutrons is the same on both sides of the equation. This is true even when a neutron changes into a proton or vice versa. The total number of neutrons and protons always is conserved. You apply this concept when you balance nuclear equations.

However, careful measurements of the mass of materials before and after a nuclear reaction show that the products have slightly less mass than the starting materials had. What happened to this mass?

During nuclear reactions, some mass is converted directly into energy. The change in mass is the result of a very small change in the mass of protons and neutrons. The "missing" mass, known as the mass defect, can be accounted for by calculating the mass equivalent of the energy released during the change using Einstein's famous equation:

$$E = mc^2$$

In this equation, E is the energy expressed in Joules (J), m is the mass expressed in kilograms, and c is the speed of light in a vacuum (3.0×10^8 meters per second). Because the mass is multiplied by a very large number, even small amounts of matter are converted into large amounts of energy.

How Are Chain Reactions Controlled in Nuclear Reactors?

Chain Reactions

Knocking over a single domino can set off a chain reaction that eventually causes millions of dominos to fall. At least one domino that falls must cause at least one other domino to fall to keep the chain going. In chemistry, a chain reaction is a sequence of reactions in which a product or by-product causes additional reactions to happen. In nuclear fission, neutrons are the key to forming a chain reaction. A neutron collides with a uranium nucleus and causes it to split. Neutrons also are released in the reaction, as shown in the equation:

$$^{1}_{0}n + ^{235}_{92}\text{U} + ^{144}_{56}\text{Ba} + ^{89}_{36}\text{Kr} + 3^{1}_{0}n$$

A chain reaction occurs when at least one of the neutrons produced in the reaction in turn splits another uranium nucleus. If too many neutrons split additional nuclei, an uncontrolled fission reaction would happen. Uncontrolled fission is responsible for the large destructive force of an atomic bomb. To use nuclear fission to generate electricity, the chain reaction must be controlled.

Controlling Nuclear Fission

A nuclear reactor is a device that is used to initiate and control a sustained nuclear chain reaction. The chain reaction requires a minimum amount of fissionable material called the critical mass. This ensures that neutrons released during fission are able to hit other fissionable nuclei to continue the reaction. In a nuclear reactor, fuel rods contain the uranium fuel. The fuel rods are surrounded by a moderator, such as water, that slows down the neutrons. Fast-moving neutrons have too much energy to split the uranium nuclei; they cause side reactions instead. To prevent too many neutrons from causing additional fission reactions, control rods absorb some of the neutrons. Control rods are used in nuclear reactors to control the fission rate of uranium and plutonium. They are composed of chemical elements such as boron, silver, indium, and cadmium that are capable of absorbing many neutrons without themselves fissioning.

Although fission occurs in the core of the reactor, to generate electrical energy requires several additional steps. Water often is used as a coolant to absorb thermal energy generated in the reactor core. This energy is transferred to a separate water supply through the heat exchange. Water becomes steam, which turns a turbine that is connected to a generator.

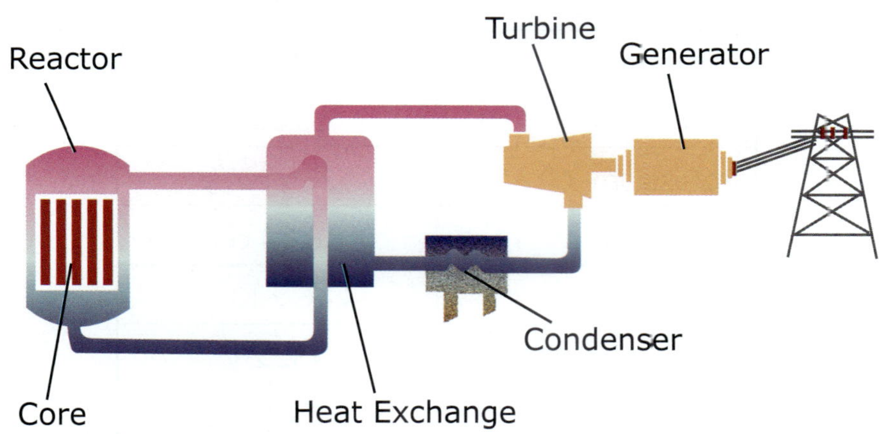

Generating electrical energy from nuclear energy involved converting water to steam. Why isn't the steam released by the nuclear reactor radioactive?

Advantages and Drawbacks

Generating electrical energy using nuclear fission has some benefits compared with using fossil fuels. The combustion of fossil fuels, such as coal, produces air pollutants. Some air pollutants are known as greenhouse gases because they hold heat in Earth's atmosphere and are causing the planet's temperature to increase. Air pollutants include carbon dioxide (a greenhouse gas) and sulfur oxides (which contribute to acid precipitation). A nuclear reactor does not produce these gases. If equal masses of uranium and coal are used, more energy can be generated from the uranium because more energy per gram is released during fission.

However, there are challenges associated with using fission. The collection, preparation, and use of uranium produce radioactive waste. These waste materials must be stored for long periods of time before the levels of radioactivity drop to safer levels. As with fossil fuels, uranium is a limited resource. Another concern is the release of radioactive materials into the environment as a result of nuclear accidents or natural catastrophes. A tsunami that damaged a nuclear reactor in Japan in 2011 resulted in the release of radioactive materials to the environment.

How Is the Half-Life of Radioactive Materials Used?

Half-Life

When deciding whether to use nuclear fission to generate electrical energy, the problem of radioactive waste is a large drawback. Radioactive materials undergo nuclear changes, such as alpha decay and beta decay. The rate at which these nuclear changes happen cannot be changed. They are not affected by temperature or pressure, as many chemical reactions are. Instead, each nuclide decays at its own pace.

Half-life is the time it takes for one-half of a sample of a radioactive isotope to decay. The more unstable a nuclide is, the shorter its half-life. Half-lives range from billions of years to fractions of a second.

Half-Lives of Some Radioactive Nuclides

Nuclide	Half-Life
Hydrogen-3	12.26 y
Flourine-18	1.83 h
Carbon-14	5,730 y
Potassium-40	1.25×10^{9} y
Iodine-131	8.04 d
Uranium-238	4.46×10^{9} y
Americium-241	432.2 y
Fermium-244	3.7×10^{-3} s

This table shows the half-lives of some radioactive nuclides. Why do these different nuclides have different half-lives?

The half-life of a nuclide is helpful in determining the amount of radioactive nuclides that should be present in a sample and the time elapsed. Half-lives of several nuclides are used to help determine when past events happened. Radioactive dating relies on knowledge of half-lives to determine when a volcanic rock formed or when a mammoth died. To answer these questions requires knowing the half-life of the nuclide and the amount of the nuclide that was present originally. One also must understand that during each half-life, one-half of the sample decays.

Half-Life: Sample Problem

Archaeologists analyze a sample of wood from a tomb that originally contained 2.0 grams of carbon-14. If 0.5 grams of carbon-14 remains, how old is the tomb? The half-life of carbon-14 is 5,730 years.

Solution

Given:

- Original mass = 2.0 g
- Remaining mass = 0.5 g
- Half-life of carbon-14 = 5,730 y

Because one-half of the sample decays during each half-life, the sample has undergone two half-lives.

first half-life: $2.0\ g \times \frac{1}{2} = 1.0\ g$ remaining

second half-life: $1.0\ g \times \frac{1}{2} = 0.5\ g$ remaining

Each half-life takes 5,730 years, so 2 half-lives would equal 11,460 years.

$$2\ \cancel{\text{half-lives}} \times \frac{5{,}730\,y}{1\ \cancel{\text{half-life}}} = 11{,}460$$

The tomb is 11,460 years old.

Consider the Explain Question:

What are the different types of radiation, and how are they produced by isotopes of different elements?

Go online to complete the scientific explanation.

Check for Understanding:

Can you model a nuclear chain reaction which occurs in a nuclear reactor?

STEM in Action

STEM and Nuclear Chemistry

A mission to Jupiter, Saturn, and the far reaches of the solar system requires energy. These planets are such a great distance from the sun that arrays of solar panels would be too large and heavy to attach in a practical way to a spacecraft. Instead, nuclear energy is used. NASA engineers are not sending nuclear fission power plants into space aboard these probes, however. The engineers are using radioisotope thermoelectric generators (RTGs), which use the energy released by radioactive decay to generate electrical energy. Nuclear chemists and engineers have been fundamental in the creation and development of radioisotope thermoelectric generators.

RTGs convert the thermal energy generated during the decay of radioactive nuclides into electrical energy. They primarily use plutonium-238, which is a non-weapon-grade isotope of plutonium. They are compact and have no moving parts, so they are extremely reliable and can provide energy for years. However, they are not very efficient.

Engineers are currently developing the advanced Stirling radioisotope generator (ASRG). This is the next step in powering space probes. These generators are much more efficient and can use one-fourth the plutonium to generate the same energy as an RTG.

photo: NASA

This radioisotope thermoelectric generator is being prepared for use in the New Horizons spacecraft that will explore Pluto and Charon. How does it use nuclear energy?

Nuclear Chemistry

A scientist used a Geiger counter to measure the radiation given off by a sample over the course of 10 days. The scientist then graphed the collected data.

Use the graph to answer the following questions.

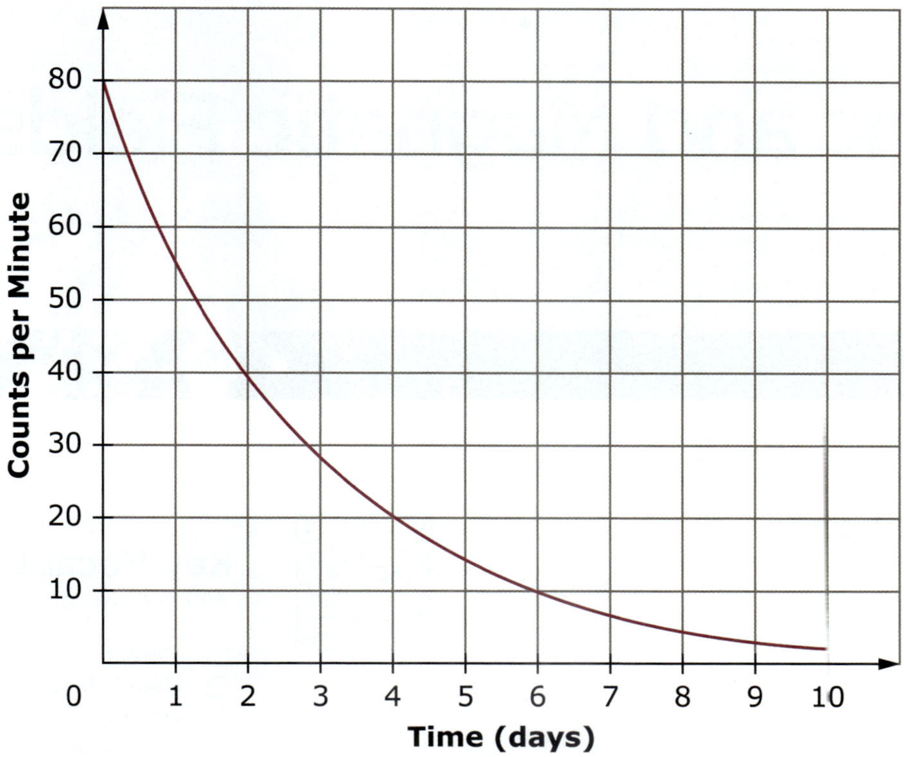

1. What is the half-life for this substance?
2. How many half-lives will it take for the substance to decay to 25 percent of its original amount?
3. What percentage of the substance remains after one day?

Electric and Magnetic Fields

photo: Getty Images

LESSON OVERVIEW

Lesson Questions

- What is an electric field, and how does it interact with charged particles?
- What produces and affects the strength of a magnetic field?
- How do you describe and determine the direction of a magnetic field produced by an electric current?
- What is a vector field, and how does it relate to electric and magnetic fields?
- How do the quantities of electric force, electric field, and electric charge relate to one another?
- What is a static electric field, and how does the arrangement of different electric charges affect it?
- How can you describe the magnetic force acting on a charged particle that is traveling through a magnetic field?
- How do the motions of charges and magnets relate to the production of electric and magnetic fields, and how does this relate to electric motors and generators?

Key Vocabulary

Which terms do you already know?

- ☐ **current**
- ☐ **electric field**
- ☐ **electro-magnetism**
- ☐ **force**
- ☐ **induction**
- ☐ **Lorentz force**
- ☐ **magnetic field**
- ☐ **negative charge**
- ☐ **parallel**
- ☐ **positive charge**
- ☐ **right-hand rule**
- ☐ **vector**
- ☐ **velocity**

Electric and Magnetic Fields

Lesson Objectives

By the end of the lesson, you should be able to:

- Cite evidence to show that charged particles possess electric fields and are affected by the force of other electric fields.
- Cite evidence to show that both electric currents and magnets give rise to magnetic fields that are affected by the force of other magnetic fields.
- Model the direction of a magnetic field produced by an electric current.
- Demonstrate how magnetic and electric fields act as vector fields.
- Construct solutions to problems involving the charge of a particle within an electric field and the force exerted by the field on that particle.
- Cite evidence to show that static electric fields arise from the arrangement of electric charges.
- Construct solutions to problems involving the magnitude and direction (found by using the right-hand rule) of the force of a magnetic field on a moving, charged particle within the field.
- Analyze evidence that moving magnets generate electric fields and moving electric charges produce magnetic fields. Recognize that electric motors and generators are powered by these interactions.

ENGAGE

Have you listened to music in a car, or been to a rock music concert? How did the speakers produce the sounds you heard? Visit the Engage page online to learn more.

Explain Question

What is the cause/effect relationship between electric and magnetic fields?

EXPLORE

What Is an Electric Field, and How Does It Interact with Charged Particles?

Electric Fields and Particles

A charged particle is a particle with a positive or negative electric charge. A proton is a particle with a positive charge, and an electron is a particle with a negative charge. Recall that charged particles exert forces on each other; like charges repel, and opposite charges attract. How can we describe the nature of these forces? What determines the force that two charges will experience when they are moved closer to one another?

The answer to these questions is found in the electric field. An electric field, or static electric field in the case of static electricity, is the region surrounding any charged particle that will exert an electric force on another particle. The strength of this field is related to the force that a charged particle would experience if it were placed in the field. The electric field is strong near a charged particle, and it becomes weaker at greater distances from the particle. The strength of the electric field also relates to the electric field energy. Two similarly charged objects with interacting electric fields have the potential to do work. This interactional field energy is a potential energy function. It is directly related to the amount of charge on each of the objects and indirectly (inversely) related to the distance between the charged objects. Therefore, when the objects change relative position, the field energy also will change. As they move closer together, the potential energy increases because the distance decreases. As they move farther apart, the potential energy decreases because the distance increases. If the charges are opposite, the potential energy decreases as the distance decreases.

This potential energy can be transformed into kinetic energy. In the case of two similarly charged particles, they repel each other. As they move farther apart, some of their potential energy is converted into kinetic energy, and the potential energy decreases.

Comparing Electric Fields and Gravitational Fields

The interaction between electric fields and charged particles is analogous in many ways to the interaction between gravitational fields and mass. Just as every massive object possesses its own gravitational field, every charged particle possesses its own electric field. Also, just as the fields of two massive objects exert gravitational forces on each other, the fields of charged particles exert forces on each other.

Let's elaborate on this analogy with an example. A small piece of debris in Earth's gravitational field experiences a force that pulls it toward Earth. It is important to understand that the debris also possesses its own gravitational field; however, because Earth is so much more massive, the debris' gravitational field is negligible compared with Earth's. As a result, Earth does not move toward the debris with any real acceleration. In contrast, if a very large, planet-sized object in space moves toward Earth, the gravitational fields of the large object and Earth would exert significant forces on each other, giving both bodies significant accelerations and velocities.

The same thing happens with charged particles and electric fields. A small, charged particle in a strong electric field of another charge (or group of charges) would experience an attractive (or repulsive) force, similar to the gravitational force the debris experiences. This force would cause the small particle to move in the field. The stronger charge, however, wouldn't be affected as much. On the other hand, if two particles with charges of equal magnitude move close to one another, their electric fields would interact, causing both particles to experience large forces. This result is similar to the way Earth and a very large space object would exert forces on one another.

The key idea is that all charged particles are surrounded by electric fields that interact and exert forces on other charges. The strengths of these fields increase as the charged particles move closer together.

What Produces and Affects the Strength of a Magnetic Field?

Magnetic Fields

Similar to how an electric field surrounds all charged particles, a magnetic field surrounds all magnets. The magnetic field extends in a closed-loop path pointing from the north pole of a magnet to the south pole of the magnet. Magnetic fields exert forces on charged particles moving through them. If a charged particle moves close to one of the magnet's poles, it will experience a strong magnetic force. If the particle moves farther from the magnetic poles, it will experience a weaker magnetic force. Magnetic fields are typically shown as magnetic field lines. The magnetic field is strongest in regions where the field lines are drawn close together. The interaction between magnetic fields and moving charges is so important that the magnitude of magnetic field strength incorporates the particle's charge and velocity into its unit, the tesla (T).

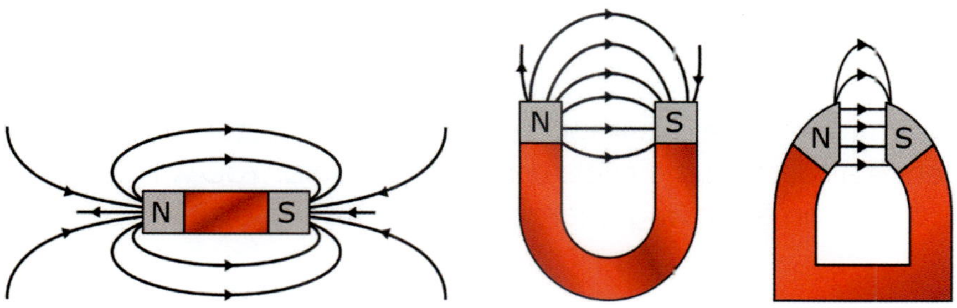

Magnetic fields extend in a closed-loop path from the north pole to the south pole of a magnet. How does a diagram show the strength and direction of the electric field?

Magnetic fields only act on charged particles when the particles are not moving parallel to the field. The force on the particle is greatest when the velocity of the particle is perpendicular to the field.

Particle accelerators use both electric and magnetic fields to control charged particles. The electric fields do work to speed up the charges, greatly increasing their kinetic energies. The magnetic fields control the direction in which the charged particles move. In circular accelerators, very powerful electromagnets make the particles travel in a circle around a large loop.

Comparing Electric Fields and Magnetic Fields

Magnetic fields are not just produced by magnets. Magnetic fields are also produced by moving charged particles, or electric current. In fact, most magnetic fields used in electrical devices are produced by current through loops of wire. The current in the wire forms a temporary magnet, also known as an electromagnet. You will learn more about the properties of electromagnets later in this lesson.

Magnetic fields are similar to electric fields. For example, magnetic fields from different sources interact with each other. Consider two wires with electric current that are moved toward one another. The magnetic field from each wire will exert a force on the other wire.

Magnetic fields are also different from electric fields. Recall that an electric field can surround a single proton or an electron on its own. In contrast, magnetic fields are always found with both a north and a south pole. That is, scientists have never observed a magnetic field surrounding a single north pole or a single south pole. The north and south poles of magnets are always found in pairs.

How Do You Describe and Determine the Direction of a Magnetic Field Produced by an Electric Current?

The Right-Hand Rule

The direction of a magnetic field produced by an electric current is related to the direction of the current and the shape of the current-carrying wire. For a long, straight wire carrying electric current, the magnetic field is directed along a circular path surrounding the wire.

You can determine the direction of the magnetic field around a straight wire using the right-hand rule, also called Fleming's right-hand rule. To use this rule, take your right hand and point your thumb along the direction of the current. Then, curl your fingers around the straight wire. Your fingers curl in the direction of the magnetic field surrounding the wire.

The direction of the magnetic field is important because it determines how, or if, a moving charge will experience a force as it travels through the field. You will learn more about the force of a magnetic field on a moving charge later in this lesson.

Right-Hand Rule

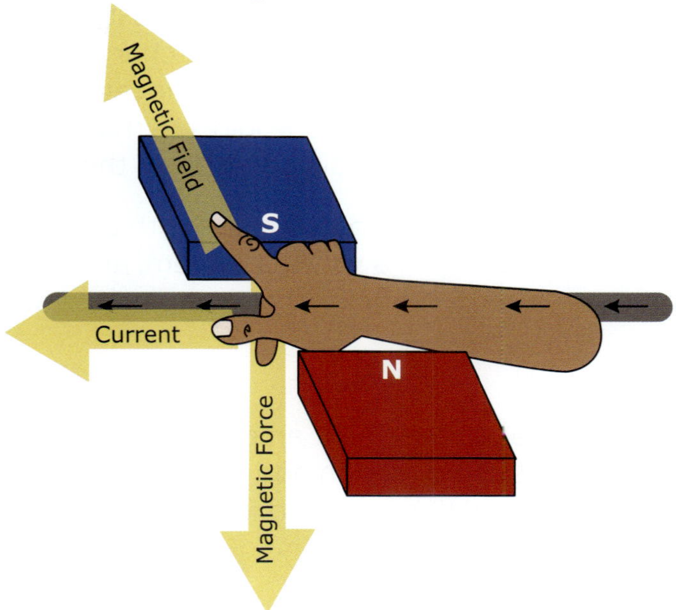

Use the right-hand rule to determine the direction of a magnetic field produced by this electric current. In what direction is the magnetic field?

What Is a Vector Field, and How Does It Relate to Electric and Magnetic Fields?

Representing Electric Fields with Vectors

A vector field represents the strength and direction of a field by using vectors, or arrows that have specific length and direction. When scientists talk about the "strength" and "direction" of an electric field, they are really asking two things: How much force would a test charge experience if it were placed at a particular position in that field, and in which direction would it move? Here, a test charge is defined as a particle with a positive charge that is too small to affect the electric field significantly and too small to produce a noticeable magnetic field when moving. Different vectors in a diagram can represent the strength and direction of the force that a test charge would experience at different locations in the electric field. The electric field vector is denoted by the vector:

$$\vec{E}$$

The practice among scientists is to indicate the direction of an electric field around a charge as the direction a positive charge would move if placed in the field. Consider the vector field representing the electric field of a positive charge. If we were to place a positive test charge in the field near the positive charge, the test charge would experience a very strong repulsive force because like charges repel and an electric field is quite strong near its source. So, we would draw a long arrow pointing away from the positive charge at that location in the field. If we repeated this process at every location around the charge, we would notice the vectors all point outward radially. The vectors also get shorter at greater distances from the charge.

Next, consider the electric field of a negative charge. What would happen if we were to place a positive test charge near the negative charge? The test charge would experience a very strong attractive force pulling it toward the negative charge. So, we would draw a long arrow pointing inward toward the negative charge at that location in the field. If we continue to draw arrows at different points in the field, we would see that the arrows all point inward radially, and the arrows get shorter at greater distances from the negative charge.

Scientists refer to a "uniform electric field" when the strength and direction of the field is assumed to be the same at all locations. In a vector diagram, all the vectors representing a uniform electric field would point in the same direction and would have the same length.

Electric Field Lines

We can also draw electric field lines to represent an electric field. Like vector diagrams, electric field line diagrams indicate the direction and strength of a magnetic field. Instead of the arrows used in a vector representation, the field line representation uses lines.

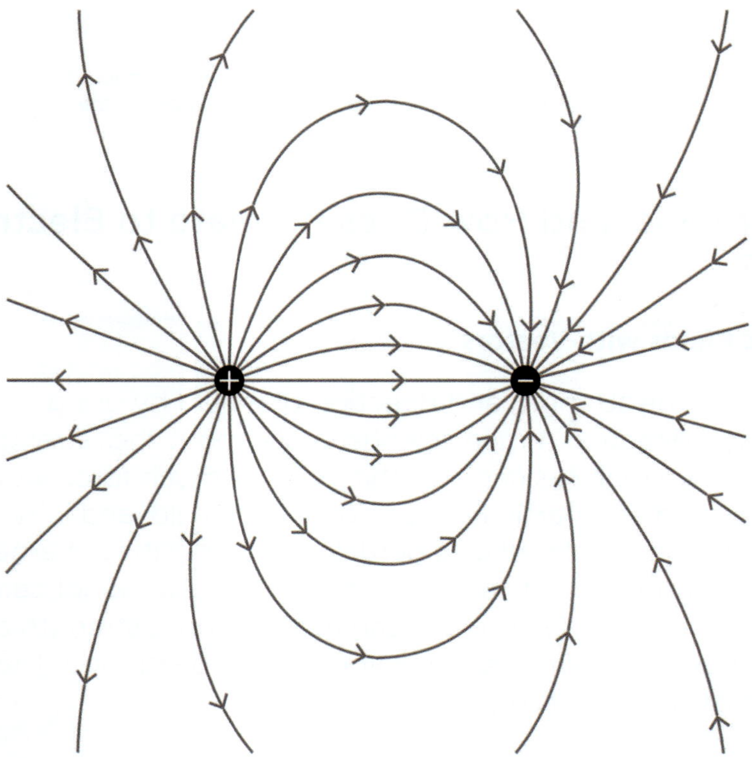

The electric field represented by this diagram consists of two opposite but equal charges: a positive charge on the left and a negative charge on the right. What do you think would happen to the lines in the diagram if the charge on the right was increased?

As with electric fields, the arrows are defined as if a positive charge were passing through the field. So, the arrows on the field lines point away from positive charges and point toward negative charges. The density of lines also corresponds to the magnitude of the force exerted at that point. Note that the field lines are densest near the particles. That is because the force is greatest near the particles. In an electric field line diagram, the number of field lines emanating from a particle is roughly proportional to the magnitude of the charge of that particular particle.

Representing Magnetic Fields with Vectors

We can also draw field diagrams for magnetic fields. The vectors in a magnetic field diagram correspond to the strength and direction of the magnetic field at different locations in the field. What do we mean when we talk about the direction of a magnetic field? We are really asking, "If we were to place a tiny bar magnet, such as a compass needle, in a magnetic field, which direction would the north pole of the needle point?" Think of the tiny bar magnet as a very weak magnet with a magnetic field of its own that is too small to notice. When we discuss the strength of the magnetic field, we are asking, "How much force would the tiny magnet experience in the field?" A vector representing the strength and direction of a magnetic field is denoted by the variable $\vec{B}$.

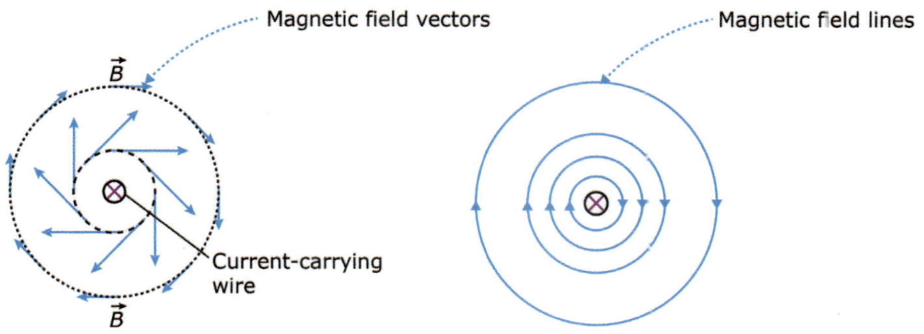

Magnetic field vectors are tangent to circles around the wire. How can you determine the direction of these vectors?

Consider a vector-field diagram for a typical bar magnet. What would happen if we were to place a small compass needle next to the north pole of the magnet? The north pole of the compass needle would point away from the magnet (because like poles repel). Because the magnetic field is very strong near the poles of a magnet, the vector will be long. Similarly, if we were to place the compass needle next to the south pole of the magnet, the north pole of the compass would point in toward the magnet (because opposite poles attract). The magnetic field is also strong at the magnet's south pole, so the arrow of our vector will be long. If we were to draw a vector at every point in the field, we would see that the vectors point outward away from the north pole and loop around toward the south pole. We would also notice that the arrows are shorter toward the middle of the magnet, where the field is not as strong. If we place iron filings around a bar magnet, we would see them align to form the same pattern as our vector field arrows.

The Lorentz Force

To understand the relationship between a magnetic field and the forces it generates, consider a charge moving through a uniform magnetic field. As the charge moves, it will generate its own magnetic field. The two magnetic fields will interact and the charge will follow a path perpendicular to the original uniform magnetic field.

Now, consider a charged particle moving simultaneously through a magnetic field and an electric field. The force experienced by the particle under these conditions is called the Lorentz force. The Lorentz force is caused by the interaction of magnetic and electric fields, and the field generating the Lorentz force is known as an electromagnetic field. In other words, the Lorentz force is the vector sum of the electric forces and magnetic forces acting on a charged particle in an electromagnetic field. For a positive charge, the direction of the electric force is the same as the direction of the electric field. The direction of the magnetic force acts perpendicular to the direction of the external magnetic field because of the interaction between the external magnetic field and the magnetic field generated by the moving charged particle.

How Do the Quantities of Electric Force, Electric Field, and Electric Charge Relate to One Another?

Solving Problems

Let us now discuss how to solve problems involving electric force, electric field, and electric charge. Suppose an electric field, $\vec{E}$, exists in a region. We are interested in the force, $\vec{F}$, that a charged particle with charge q would experience if it were placed into that field. This relationship is given by the following equation:

$$\vec{F}=q\vec{E}$$

Here, the quantity $\vec{F}$ is a vector with units of newtons, q is a scalar quantity with units of coulombs (C), and $\vec{E}$ is a vector with units of newtons per coulomb (N/C).

Recall earlier when we discussed the electric-field vectors. We defined the direction of the electric field as the direction that a positive test charge would move if it were placed in that field. In which direction would a negative charge move? Here are two important rules to remember when approaching problems involving an electric field:

1. The direction of an electric field is the direction in which a positive charge would move. Positive charges experience a force in the same direction as the electric-field lines.
2. If a negative charge were placed into the same field, it would move in the opposite direction. Negative charges experience a force in the opposite direction of the electric-field lines.

Electric and Magnetic Force: Sample Problem: Determining Electric Force

A uniform electric field between two metal plates points downward and has a strength of 6.0×10^4 N/C.

(a) What force would a 2.0×10^{-5} C charge experience if it were placed in that field?

(b) What force would a -5.0×10^{-9} C charge experience if it were placed in that field?

Solution

(a) We know that the electric field has strength 6.0 × 10^4 N/C and we also know that the charge placed in the field is 2.0 × 10^{-5} C. So:

$$\vec{E} = 6.0 \times 10^4 \text{ N/C}$$
$$q = 2.0 \times 10^{-5} \text{ C}$$

We can substitute these values into the following equation to calculate $\vec{F}$:

$$\begin{aligned} \vec{F} &= q\vec{E} \\ &= (2.0 \times 10^{-5} \text{ C}) \times (6.0 \times 10^4 \text{ N/C}) \\ &= 1.2 \text{ N} \end{aligned}$$

So, the magnitude of the force on the charge is 1.2 N. What is the direction of this force? Recall that positive charges move in the same direction as the electric-field lines, so the direction is downward. Thus, the charge experiences a force of 1.2 N, downward.

(b) For part b, we follow the same process. We know that the electric field has a strength of 6.0 × 10^4 N/C, and we also know that the charge placed in the field is –5.0 × 10^{-9} C. So:

$$\vec{E} = 6.0 \times 10^4 \text{ N/C}$$
$$q = -5.0 \times 10^{-9} \text{ C}$$

We can substitute these values into the following equation to calculate $\vec{F}$:

$$\begin{aligned} \vec{F} &= q\vec{E} \\ &= (-5.0 \times 10^{-9} \text{ C}) \times (6.0 \times 10^4 \text{ N/C}) \\ &= -3.0 \times 10^{-4} \text{ N} \end{aligned}$$

This time, we've obtained a negative value for the force on the charge. What does this mean? It means that the force will act in the opposite direction of the electric-field lines, which is what we should expect because the charge is negative. Because the field lines act downward, the force on the charge is 3.0 × 10^{-4} N, upward.

Solving Equations with Known Quantities

We can use the equation relating the electric field, electric force, and electric charge to solve for any of the three quantities. For example, if we know the electric force and charge, we can determine the strength and direction of the electric field. Or, if we know the strength of an electric field and the electric force, we can determine the charge of the particle in the field.

Solving Equations with Known Quantities: Sample Problem: Finding the Charge of a Particle

A charged particle experiences a force of 1.9×10^{-17} N to the left. The strength of the electric field is 120 N/C to the right. What is the charge of the particle?

Solution

We know the following information:

$\vec{F} = 1.9 \times 10^{-17}$ N to the left, or -1.9×10^{-17} N

$\vec{E} = 120$ N/C to the right, or +120 N/C

We can see that the force acts in the opposite direction to the electric field. This must mean that the particle is negatively charged. We'll define the direction of the electric field (to the right) as the positive direction. This means that the direction of the force takes on a negative value. We want to solve for ***q***. We can rewrite the following equation, solving for ***q***:

$$\vec{F} = q\vec{E}$$

$$q = \frac{\vec{F}}{\vec{E}}$$

Now, we can plug in the values that we know to solve the problem:

$$\begin{aligned} q &= \frac{\vec{F}}{\vec{E}} \\ &= \frac{-1.9 \times 10^{-17}\,\text{N}}{120\,\text{N/C}} \\ &= -1.6 \times 10^{-19}\,\text{C} \end{aligned}$$

So, the charge on the particle is -1.6×10^{-19} C. This is a negative charge, as we expected. In fact, this value is equal to the charge of an electron.

What Is a Static Electric Field, and How Does the Arrangement of Different Electric Charges Affect It?

Static Electric Fields

Recall that all electric charges are surrounded by an electric field. When a charge is at rest, the field around it is a static electric field. In contrast, moving charges, or electric currents, do not produce static electric fields.

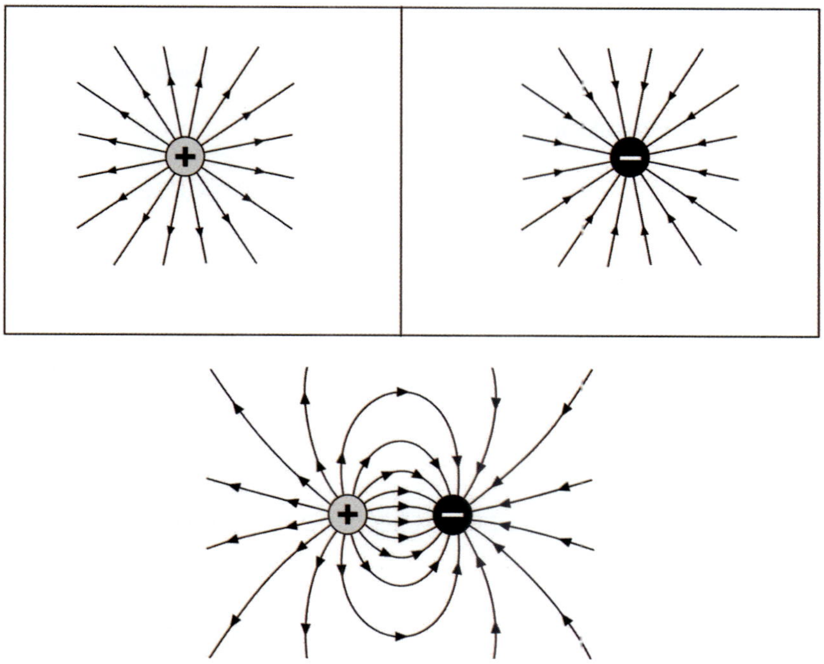

When a positive and negative charge are brought close together, the electric fields produced by the charges interact to form a new field. How do you think the strength of this new field is related to the distance from the two charges?

If many charges are in the same region, the fields surrounding them will combine to produce a net electric field, similar to how multiple forces combine to produce a net force on an object. If many forces push on a box, we add the vectors of the individual forces to determine the net force on the box. Similarly, when there are many charges in a region, we add up the strength of each field to determine the net field in that region.

The arrangement of charged particles, as well as the strength of each charge, will determine how the fields add together. As you can see in the image, when a positive and negative charge of equal strength are moved near one another, the fields combine to form a new field. If you were to move the charges closer or farther apart, the field lines would change. Or, if one of the charges were replaced with a stronger one, the field lines would be different.

How Can You Describe the Magnetic Force Acting on a Charged Particle That Is Traveling through a Magnetic Field?

Exerting a Force on a Particle

Although electric and magnetic fields have similarities, the manner in which they exert force on particles differs significantly. Recall that an electric field will exert an electric force on any charged particle within the field. The magnetic field, on the other hand, exerts forces a bit more selectively.

Consider a charge q moving at velocity v through a magnetic field $\vec{B}$. Assume also that the charge moves at an angle θ to the magnetic field lines. The magnitude of the magnetic force, F, on the charge is described by the following equation:

$$\vec{F} = qv\vec{B}\,sin(\theta)$$

In this equation, $\vec{F}$ has units of Newtons (N), q has units of coulombs (C), v has units of meters/second (m/s), and $\vec{B}$ has units of teslas (T).

Electric and Magnetic Force Equations

Compare this equation for magnetic force with the equation describing the electric field force:

$$\vec{F} = q\vec{E}$$

We can see two immediate differences. First, the magnetic force is related to the velocity of the charge. Second, the force is related to sin(θ). We can conclude three things from this:

1. If the velocity of the charge is zero, then the magnetic force on the charge is zero. Thus, if the charge is at rest relative to the magnetic field, it experiences no force.
2. Recall that sin(0°) = 0 and sin(180°) = 0. So, if the angle between the velocity of the charge and the magnetic field is 0° or 180°, the force on the charge is 0. When is the angle between the velocity of a charge and the magnetic field 0° or 180°? This happens when the charge is moving parallel to the magnetic field lines. Thus, if the charge is moving parallel to the magnetic field lines, it experiences no force.
3. Recall that sin(90°) = 1. This is the maximum value of a sine function. This means that the magnetic force is at a maximum when θ = 90°. Note that θ = 90° when the charge is moving perpendicular to the magnetic field lines. Thus, if the charge is moving perpendicular to the magnetic field lines, it experiences a maximum magnetic force.

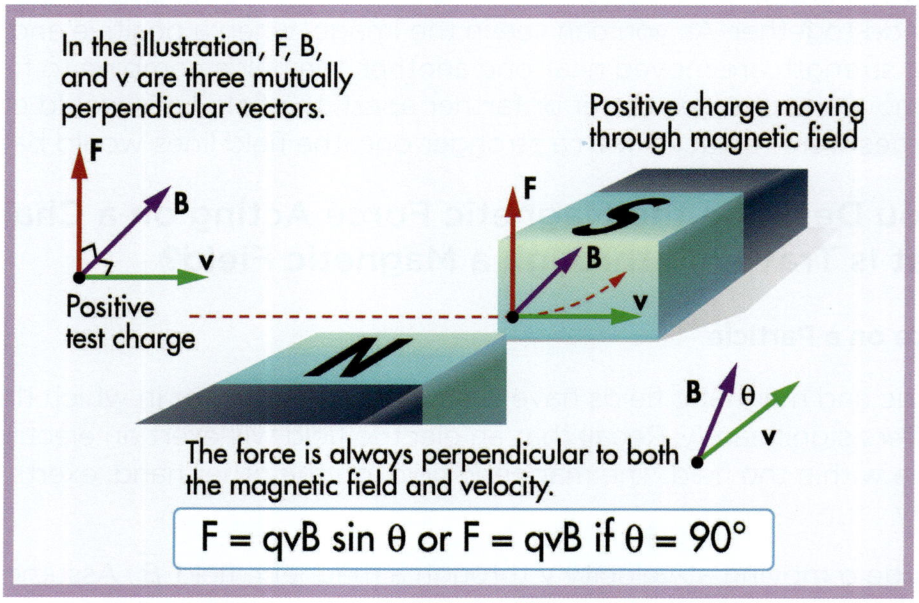

When a test charge moves perpendicular to a magnetic field, it experiences the maximum magnetic force, the magnitude of which is described by the equation F = qvB. In what direction does the force act on both the magnetic field and the velocity vector?

Direction of Magnetic Force

The last difference between magnetic and electric forces is the direction of the magnetic force. A magnetic field applies a force that is perpendicular to the motion of the charge. This force is also perpendicular to the magnetic field lines. You can use the r ght-hand rule to determine the direction of the magnetic force. To do this, point the thumb of your right hand in the direction of the velocity vector. Then, point the index finger of your right hand in the direction of the magnetic field. The middle finger of your right hand, he d so that it points outward from the palm of your hand, will point in the direction of the magnetic force. If the charge is positive, the force points in the direction of your palm or along your middle finger. If the charge is negative, the force acts in the opposite direction. That is, the force is directed (into the palm).

Direction of Magnetic Force: Sample Problem

A charge of 1.6×10^{-19} C enters a uniform magnetic field in a laboratory. The $\vec{B}$ field has a strength of 1.5×10^{-3} T, and it points upward vertically. (This is in the direction of the *z*-axis.) The charge is traveling at a speed of 2.3×10^{5} m/s.

(a) If the charge is traveling to the right (in the direction of the *y*-axis), what is the magnitude and direction of the force?

(b) If the charge is traveling downward vertically (along the negative *z*-axis), what is the magnitude and direction of the force?

Solution

(a) First, list the quantities that are given:

- $q = 1.6 \times 10^{-19}\,\text{C}$
- $\vec{v} = 2.3 \times 10^{5}\,\text{m/s}$ to the right
- $\vec{B} = 1.5 \times 10^{-3}\,\text{T}$ upward

Because the $\vec{B}$ field points upward and the charge is traveling to the right, the value of è = 90°. Now, plug the known values into the equation for the magnitude of the magnetic force.

$$\vec{F} = (1.6\times10^{-19}\,\text{C})\,(2.3\times10^{5}\,\text{m/s})\,(1.5\times10^{-3}\,\text{T})\sin(90°)$$
$$= 5.5\times10^{-17}\ \text{N}$$

Next, we must determine the direction of the force. We can use the right-hand rule to do this. If you point your thumb in the direction of the velocity vector (to the right), and then point your index finger in the direction of the $\vec{B}$ field (upward), your palm or middle finger points toward you, along the direction of the *x*-axis in the following image. Thus, the force on the charge is 5.5×10^{-17} N in the *x*-direction.

(b) First, think about the angle between the $\vec{B}$ field and the velocity vector. If the charge is traveling straight down, then it is traveling in the direction opposite the $\vec{B}$ field. In other words, the charge is traveling at a 180° angle with respect to the $\vec{B}$ field, so θ = 180°. Note that sin(180°) = 0. If we plug this value into the equation for force, we will find that the force on the particle is zero:

$$\vec{F} = (1.6\times10^{-19}\,\text{C})\,(2.3\times10^{5}\,\text{m/s})\,(1.5\times10^{-3}\ \text{T})\sin(180°)$$
$$= 0\ \text{N}$$

Thus, the force on the charge traveling in the downward direction is zero.

How Do the Motions of Charges and Magnets Relate to the Production of Electric and Magnetic Fields, and How Does This Relate to Electric Motors and Generators?

Electric Motors and Generators

We've already discussed how moving electric charges, or electric current, can produce magnetic fields. It turns out that an analogous effect happens for moving magnetic fields. That is, a moving magnetic field will generate an electric current in a wire! These phenomena are the basic principles behind motors and generators.

The key function of a motor is to convert electrical energy into mechanical energy. That is, we want to convert the energy of current through a circuit to energy of motion, such as turning a fan blade. We can create this motion using electric current and magnets. For example, if we bend a wire into a loop and then run electric current through it, the electric current will produce a magnetic field through the center of the loop. If we place a magnet at the center of this loop, the magnetic field from the electric current will cause the magnet to spin. This spinning motion can be used to turn many devices, such as the bristles in a vacuum cleaner or the drums of washing machines.

A generator uses a reverse process. The key function of a generator is to turn mechanical energy into electrical energy. We want to take the energy of motion, such as wind, or position, such as water falling from a height, and convert it to useful electrical energy. Consider a simplified version of a wind-powered generator. The wind turns the blades of a turbine, causing it to spin. These blades are connected to a large magnet that also begins to spin. That magnet can be placed at the center of a loop of wire. As the magnetic field moves, it causes electric charges to flow through the wire. This electric current can then be sent to homes to power lights, computers, and other electrical devices.

AC and the Modern Power Supply

Electric motors and generators use the relationship between magnetism and electricity to convert electricity into motion and vice versa. The relationship between magnetism and electricity is also used in transformers. Transformers change the voltage in a power supply. They are widely used in modern power supply systems; however, transformers need a special type of electric current to work. Modern electrical systems widely use alternating current, AC. Alternating current is different from the direct current, DC, that is produced by batteries. AC is produced using a special generator called an alternator. Transformers only work using AC. Unlike the current from a battery, which flows in only one direction, AC flows backwards and forwards around a circuit. In most electrical systems, this change in direction occurs between 50 and 60 times per second (50-60 cycles per second).

AC current is transmitted along power lines, over long distances, at high voltages and relatively low currents, which helps to conserve energy because it reduces the loss of energy due to resistance. Upon arriving in a neighborhood, the voltage is decreased, or stepped down, by a transformer. In a transformer, two separate coils of insulated wire are wound around an iron core. When an alternating current is passed through one coil, it generates a magnetic field that, in turn, induces an electric current in the second coil. There is no electrical connection between the coils. The output voltage in the second coil depends upon the number of turns in each coil. To step the voltage down the input, or primary coil, must have more turns than the output, or secondary coil. To step the voltage up, the primary coil must have fewer turns than the secondary coil. At the power station, the voltage is stepped up and sent down the power lines. At the neighborhood transformer, the voltage is stepped down again to a voltage that is safer to use. This process does not defy the law of conservation of energy because when the voltage is stepped up, the current goes down, and vice versa.

Consider the Explain Question:

How have electric and magnetic forces shaped the world in which we live?

Go online to complete the scientific explanation.

Check for Understanding:

How does an electric motor work?

STEM in Action

Switching Magnetic Poles

Volcanic rock that forms on the seafloor captures the orientation of Earth's magnetic poles at the time that it formed. How does it do this, and how often do the poles reverse?

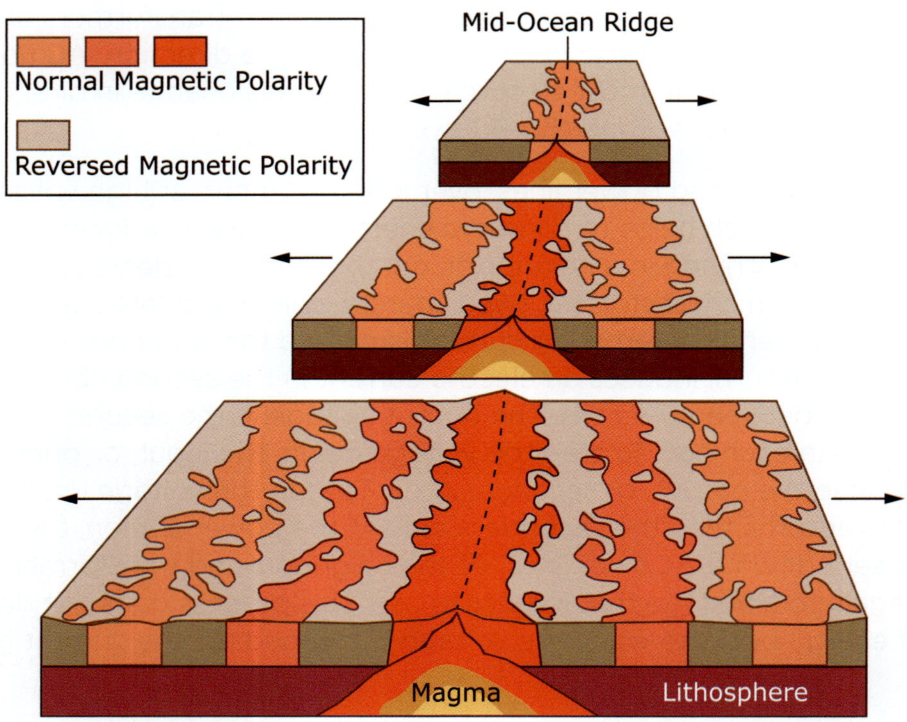

Study the image, which shows the spread of the seafloor along a mid-ocean volcanic ridge. Notice the pattern of colors indicating Earth's magnetic polarity in rocks comprising the oceanic crust. Then, answer the questions:

1. Explain how the history of Earth's magnetic polarity can be preserved in rock.
2. Approximately how many years probably separate the colored bands in the diagram?

Electricity and Magnetism

photo: Getty Images

LESSON OVERVIEW

Lesson Questions

- How did the study of the photoelectric effect lead to a better understanding of the dual nature of light?
- What are some common examples of electric and magnetic forces at work?
- What is Coulomb's law, and how can it be used to describe the relationship between attractive and repulsive forces?
- How are electromagnetic waves generated?
- How do electricity and magnetism relate to the more general phenomenon of electromagnetism?

Key Vocabulary

Which terms do you already know?

- ☐ **Coulomb's law**
- ☐ **distance**
- ☐ **electric field**
- ☐ **electromagnetic wave**
- ☐ **electromagnetism**
- ☐ **electron**
- ☐ **energy**
- ☐ **frequency**
- ☐ **induction**
- ☐ **magnetic field**
- ☐ **photoelectric effect**
- ☐ **photon**
- ☐ **plasma**
- ☐ **transverse waves**
- ☐ **vector**
- ☐ **voltage**

Lesson Objectives

By the end of the lesson, you should be able to:

- Relate the study of the photoelectric effect to advances in understanding the dual nature of light.
- Cite evidence that demonstrates common examples of electric and magnetic forces.
- Relate the acceleration of a charged particle to the generation of electromagnetic waves.
- Distinguish the behavior and interactions of electricity and magnetism as related forces of electromagnetism that can be used in technological devices to transmit and capture information and energy.
- Analyze electrostatic forces between objects using mathematical representations of Coulomb's law to describe and predict the effects of electrostatic forces between objects.

ENGAGE

Have you ever seen the aurora borealis or perhaps the aurora australis? What causes these awe-inspiring phenomena? Visit the Engage page online to learn more.

Explain Question

How have electric and magnetic forces shaped the world in which we live?

EXPLORE

How Did the Study of the Photoelectric Effect Lead to a Better Understanding of the Dual Nature of Light?

The Photoelectric Effect

In the photoelectric effect, a metal emits electrons when light shines on it. The incident light transfers energy to the electrons of the atoms that make up the metal. The energy causes the electrons to overcome their attraction to atomic nuclei, making them "free" electrons. These free electrons can produce a current in a circuit connected to the metal.

This phenomenon, called the photoelectric effect, may sound simple. But the explanation provided by Einstein revolutionized physics in the early 1900s. Named because of the connection between light and electricity, the photoelectric effect could be explained only by thinking about light as distinct energy packets, called photons, not as waves. Why?

Researchers observed that a curious aspect of the photoelectric effect is that only certain frequencies of light cause electrons to be emitted. For example, red light, no matter how intense, does not produce free electrons in a metal. On the other extreme, even the dimmest of blue lights cause electron emission. If light is purely a wave, then increasing the intensity of red light should eventually provide enough energy to free electrons. However, scientists found that even very intense red light did not cause the metal to release electrons. This fact reveals the particle nature of light.

As well as being a wave, light should be thought of as photons, or packets of energy. That would explain why the different frequencies of light affected metal in the experiment differently. This shift in thinking opened the door for quantum theory and launched one of the most persistent scientific conundrums in history: the wave-particle duality of light.

The Compton effect provides more evidence of the particle nature of light. In 1922, the physicist Arthur Compton did experiments in which he crashed gamma-ray and X-ray photons into electrons. After the collisions, the photons had lower energy and the electrons had higher energy, just as you would expect from a collision between two particles.

Dual Nature of Light

Einstein's explanation of the photoelectric effect (that light is made of particles called photons) consistently explains experimental data. On the other hand, a long history of experiments proves that light is a wave. Light can interfere and diffract. These are wavelike behaviors. Yet, the photoelectric effect and other experiments seem to confirm that light is composed of discrete packets with energy determined by the light's frequency. Both interpretations seem accurate. Light behaves like both waves and particles. Scientists have found that the conditions of an experiment determine which characteristic of light will be revealed.

It is not just light that can be a particle or a wave. With the right experimental setup, matter itself can be coaxed into acting like a wave. Electrons and other small-scale particles can act like waves by interfering and diffracting and by showing other wavelike properties. Larger-scale objects, such as a bowling ball, would have such small wavelengths that only their particle nature can be observed. In theory, all matter and all light act like waves and particles.

What Are Some Common Examples of Electric and Magnetic Forces at Work?

Electric Forces

An electrical force is an attraction or repulsion between two charged objects. Every charged particle is surrounded by an electric field. Any other charged particle in that field will experience an electrical force due to the original charged particle. The intensity of the electric field at a point is defined to be the force a positive test charge would experience at the position.

The magnitude of the electrical force that a particle exerts on another particle is directly proportional to the product of their charges and inversely proportional to the square of the distance between them. The direction of the force is along the line joining the particles. As with other forces, the motion and acceleration caused by electrical forces can be described by Newton's laws of motion.

One big difference between electrical forces and gravitational forces is that gravitational forces act only to attract two bodies to one another, whereas electrical force can be attractive or repulsive. Gravity, like electricity, is a force that acts on two bodies.

An example of electrical force in everyday life is when someone's feet rub on the floor before touching a metal doorknob. A spark will jump. The spark can cross through air due to energy from the electrical force.

Electrical forces all trace back to the fundamental electrical force between electrons and protons. In an atom, the electrical force is many orders of magnitude stronger than the gravitational force, and many of the fundamental characteristics of atoms and molecules are determined by electrical forces.

What Is Coulomb's Law, and How Can It Be Used to Describe the Relationship between Attractive and Repulsive Forces?

Coulomb's Law

Charges in electric fields are subject to forces. Coulomb's law describes the electrostatic force between two charged particles.

When charged particles are in motion, the forces become complex. For charges at rest, the formulation of Coulomb's law states: the magnitude of the force of interaction between two point charges is directly proportional to the product of the magnitudes of charges and inversely proportional to the square of the distance between them. In the equation form, Coulomb's law is written as follows:

$$F = \frac{kq_1q_2}{d^2}$$

where F is the electrical force, q_1 and q_2 are the charges, d is the distance between the charges, and k is a constant (approximately 8.99×10^9 N × m^2/C^2) called Coulomb's constant.

The sign of the calculated value of the force can be used to determine whether the force is attractive or repulsive. If the two charges have the same sign, the calculated force will be positive. This means that a positive electrical force is repulsive. If the two charges have opposite signs, the calculated force will be negative. This means that a negative electrical force is attractive.

Electrostatic force, like gravitational force, is inversely proportional to the square of the distance between the two charges. This inverse square "law" is shared with other physics phenomena that include gravitation, electrostatics, electromagnetic radiation, acoustics, and other domains. One reason for the similarity is that because point sources of energy or fields spread out in all directions, the total area covered represents an increasingly larger sphere. Because the surface area of a sphere increases with the square of its radius, the intensity of the source would decrease with the square of the distance.

This means that the electrostatic force between any two charged objects quickly becomes smaller as the distance increases. Suppose the distance between two charges is twice as big. Then, the electrostatic force becomes $1/(2)^2$, or one-fourth, as strong as it was before. Making the distance three times as large makes the force one-ninth as large. The force on charges that are four times as far away as two similar charges is only one-sixteenth as big. On the other hand, if the distance is one-half as large, the force is four times larger. Two charges that are a tenth as far apart as two similar charges will experience a force 100 times as big.

Coulomb's Law: Sample Problem

Two charges experience a repulsive force of 64 Newtons. What will the force be if the distance between the two charges is quadrupled?

Solution

The magnitude of force that a particle exerts on another particle is directly proportional to the product of their charges and inversely proportional to the square of the distance between them:

$$F = \frac{kq_1q_2}{d^2}$$

If q_1 and q_2 are the charges and d is the original distance between them, then:

$$F = \frac{kq_1q_2}{d^2} = 64\ \text{N}$$

If the distance is quadrupled, the new force F_2 can be calculated as follows:

$$\begin{aligned} F_2 &= \frac{kq_1q_2}{(4d)^2} \\ &= \frac{kq_1q_2}{16d^2} \\ &= \left(\frac{1}{16}\right)\frac{kq_1q_2}{d^2} \\ &= \left(\frac{1}{16}\right)F \\ &= \left(\frac{1}{16}\right)(64\ \text{N}) \\ &= 4.0\ \text{N} \end{aligned}$$

This can also be verified by thinking about the fact that the force is inversely proportional to the square of the distance. If r is replaced by $4r$, the new force will be one-sixteenth of the old force. Thus, the new force will be 4.0 N.

Magnetic Forces

Magnetic forces affect magnetic objects. Magnetic fields arise either from permanent magnets or from the flow of electrons in a current. The motion of electrical charge and, therefore, the motion of an electric field give rise to a magnetic field. Any moving charge creates a magnetic field.

Single electrical charges can exist, but no one has ever found an isolated magnetic pole. In a permanent magnet, the motion of electrons that resembles "spin" gives each atom a magnetic field, and the alignment of many of these atoms forms a permanent magnet. There are always two poles to a magnet: a north pole and a south pole. Magnetic field lines always loop around and close in on themselves. Like poles of different magnets repel, while opposite poles attract.

An everyday example of magnetic force is the force that holds a magnet to a refrigerator. During manufacturing, a refrigerator magnet has a strong magnet applied that shapes its magnetic orientation and strength. A compass, which works based on Earth's magnetic field lines, is another example of a common use of magnetic forces.

photo: Paul Fuqua

The magnetic dial in a compass indicates Earth's magnetic north. How could the compass be used to show the direction of Earth's magnetic field?

Just as the moving electric field around a moving charge produces a magnetic field, the movement of electrical charge in a conductor produces a magnetic field. An electromagnet is an example of this kind of magnetic field. Current passes through continuous loops of wire in a coil, each loop contributing its magnetic field. Just as a permanent magnet is made stronger by the alignment of many small atomic magnetic fields, so an electromagnet is made stronger by more wire loops that carry the electric current.

This is not the only connection between electric and magnetic fields. An electric field is present with one or more electrical charges, but an electric field can be produced by a magnetic field that is in motion. This can be observed if a coil of wire is in a circuit that has no source of electric current. If a constant magnetic field is present in the coil, nothing will happen. But if the magnetic field changes its strength or moves with respect to the coil, an electric current will appear in the coil.

How Are Electromagnetic Waves Generated?

Acceleration of Charged Particles

Charged particles create electric fields. When a charged particle moves, that electric field changes. A changing electric field induces a magnetic field. If a charged particle accelerates, then the induced magnetic field also changes and a changing magnetic field induces an electric field. The acceleration does not have to be an increase in the magnitude of the velocity. It could also mean a change in direction, as would be the case if a charge moves in a circle.

While electric and magnetic fields have different characteristics, they are coupled together. The mutual interaction between electric and magnetic fields is known as electromagnetism.

Particle accelerators take advantage of the relationship between charged particles and electromagnetic fields. They tune electromagnetic fields to create tight beams of charged particles that are accelerated to speeds close to the speed of light. These charged particles collide with a target or other particles. The resultant collisions can be studied to reveal the structure and makeup of subatomic particles.

How Electromagnetic Waves Propagate

Electricity and magnetism are related through electromagnetism. Electromagnetic waves are generated when electric and magnetic fields fluctuate simultaneously.

Several principles combine to help explain electromagnetic propagation. Faraday's law of induction explains how a changing magnetic field induces an electric field. This law of induction explains how electric generators work. These devices use moving magnets to induce electrical current.

This law of induction is not enough to explain how electromagnetic waves propagate. Maxwell's equations, a set of four mathematical relationships, describe the underpinnings of electric and magnetic fields. A summary of these four points is as follows:

- Electrical charges give rise to electric fields.
- Magnets always come in pairs with opposite poles.
- Electric current induces a magnetic field.
- A changing magnetic field can induce a current in a wire.

These concepts explain how electric and magnetic fields interact to create electromagnetic waves that propagate as light.

How Do Electricity and Magnetism Relate to the More General Phenomenon of Electromagnetism?

Electricity

Electrical charge is a fundamental property of particles. Examples of charged particles include electrons, which have a negative charge, and protons, which have a positive charge. Charged particles give rise to an electric field whether they are in motion or not.

The strength of an electric field at any point depends on the magnitude of the charge as well as the distance from the charge. Any charge in an electric field will experience a force. Another way to think about this is that any two charged particles experience an electrical force. In this way, gravitational forces, in which two masses experience a resultant gravitational force, can be a helpful analogy.

When charged particles move, a current is created. Electrical current is precisely the flow of charged particles, so electricity is related to moving charges.

A changing magnetic field can make charged particles move. This means that shifting magnets around near a metal wire can induce an electric current in the wire. The reason the wire needs to be a metal is that metals have a molecular structure that freely releases electrons and allows them to flow. Materials that do not release electrons will not sustain a current.

Magnetism

Some materials are naturally magnetic, such as lodestone. Other materials can become magnetic if placed in an external magnetic field. Still other materials, such as wires, can have a magnetic field induced around them. Electromagnets, which work as an electric current passes through coils of wire, use this principle.

When charged particles move, their moving electric fields create magnetic fields. That is why a current in a wire is surrounded by a magnetic field. The strength and structure of the magnetic field depend on the strength and direction of the current. Like electric fields, magnetic fields can exert forces on charged particles.

One important difference between electricity and magnetism is that a single charged particle can create an electric field. The charge can be positive or negative. Magnetic poles always come in pairs: a north pole and a south pole.

Earth is a giant magnet, having north and south magnetic poles. These are near, but not actually at the geographic North Pole and South Pole. Research has shown that the planet's magnetic field has flipped several times in its history.

Electromagnetism

Concepts of electromagnetism describe the combined effect of electric and magnetic fields as they induce and respond to forces. Electromagnetic waves are transverse energy waves that propagate in a direction perpendicular to both the magnetic and electric fields.

An electromagnet is a coil of wire wrapped around a metal (usually iron) core. When current flows through the wire, the core becomes a magnet. More turns of the wire create a stronger magnet. Generators, transformers, and many other modern devices depend on electromagnets.

THE ELECTROMAGNETIC SPECTRUM

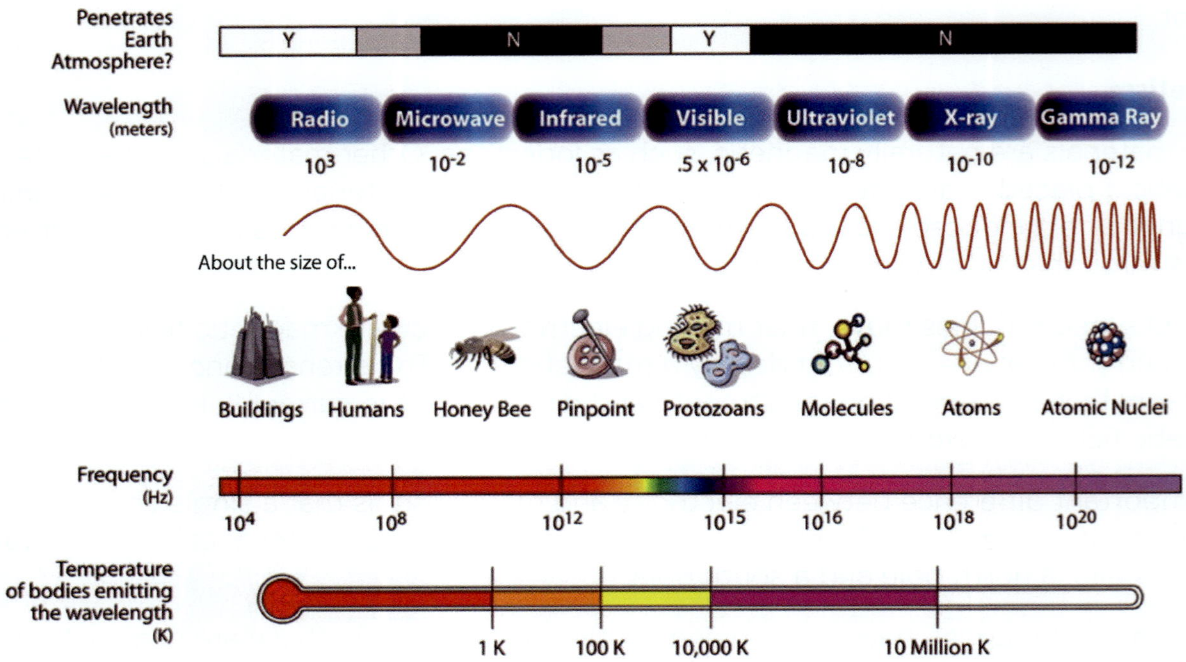

photo: NASA

Electromagnetic waves are detectable by various devices, and some by the human eye. How do scientists use the electromagnetic spectrum for research?

Consider the Explain Question:

How have electric and magnetic forces shaped the world in which we live?

Go online to complete the scientific explanation.

Check for Understanding:

Can you provide evidence that electricity and magnetism are really only one force: electromagnetism?

STEM in Action

STEM and Electricity and Magnetism

Without an understanding of electromagnetism, scientists would not be able to study a critical piece of the puzzle of nature, and we would not know nearly as much about our solar system, galaxy, and beyond.

Light is an electromagnetic wave. Visible light makes up only a small part of the much larger electromagnetic spectrum that ranges from radio waves to gamma rays. Every part of the electromagnetic spectrum plays a role in our understanding of the origin and structure of the universe. Think about the awe-inspiring pictures that you have seen of outer space. These pictures are possible because scientists know how to interpret and manipulate data hidden inside electromagnetic waves.

People who study the universe outside of Earth are called astronomers. Because astronomers study distant stars, planets, and galaxies, they are unable to collect data and specimens directly. Instead, they must rely on information contained in the electromagnetic radiation that comes to Earth from far reaches of outer space. Astronomers use telescopes to collect their data. These telescopes are much more sophisticated than the telescope that you might use to look at the moon from your backyard. Professional telescopes are equipped with digital cameras and computers that allow astronomers to analyze the electromagnetic radiation and use the results to map unknown regions of outer space.

The table shows the frequency of several types of electromagnetic waves. Use this information to help you answer these questions.

Wave Type	Frequency (Hz)
Radio wave	4×10^{8}
Microwave	2×10^{11}
Visible light wave	5×10^{14}
Gamma wave	7×10^{19}

1. Which type of wave has the longest wavelength? Explain your answer.
2. Which type of wave has the most energy? Explain your answer.
3. Which type of wave travels the fastest in a vacuum? Explain your answer.

Electric Circuits

photo: Getty Images

LESSON OVERVIEW

Lesson Questions

- What are the common elements that make up series and parallel circuits, and how are circuits diagrammed?
- What are series and parallel circuits, and how can you design and build them?
- What is the relationship between current, voltage, and resistance?
- What is the relationship between electrical power, current, and voltage?
- How can you determine the relationship between power and resistance in a circuit?
- How do current, potential difference, resistance, and power relate to one another?
- Why do resistors in DC circuits increase in temperature?
- How can the voltage or current in a simple DC circuit be described and predicted?
- What are the characteristics of transistors that make them useful in electric circuits?

Key Vocabulary

Which terms do you already know?

- ☐ **capacitor**
- ☐ **conductor**
- ☐ **current**
- ☐ **electron**
- ☐ **Ohm's law**
- ☐ **parallel circuit**
- ☐ **potential difference**
- ☐ **power**
- ☐ **resistance**
- ☐ **series circuit**
- ☐ **transistor**
- ☐ **velocity**
- ☐ **voltage**

Lesson Objectives

By the end of the lesson, you should be able to:

- Categorize the functions of common elements that make up series and parallel circuits and interpret circuit diagrams.
- Apply Ohm's law to describe the relationship between current, voltage, and resistance.
- Explain the relationship between electrical power, current, and voltage.
- Apply Ohm's law to solve problems involving power in resistive circuit elements.
- For both series and parallel circuits, calculate the current, the potential difference, the resistance, and the power of different circuit elements.
- Explain that resistors in DC circuits increase in temperature because they dissipate energy.
- Predict the voltage or current in a simple DC circuit that includes wires, resistors, capacitors, and batteries.
- Design and build both series and parallel circuits.
- Explain how the characteristics of transistors can be applied to design electric circuits.

ENGAGE

How does a crew know when an airplane needs deicing? How can they test whether it is safe to fly? Visit the Engage page online to learn more.

Explain Question

What happens when an electric charge moves through a circuit?

EXPLORE

What Are the Common Elements That Make Up Series and Parallel Circuits, and How Are Circuits Diagrammed?

Components of a Circuit

Circuits have three basic components: (1) a power source such as a battery, also known as a voltage source; (2) a load, which is a component such as a resistor or motor that transforms it into different forms; and (3) a conducting pathway, usually wires, that allows current to flow and connect the power source to the rest of the circuit in a closed loop. For example, a battery might provide voltage, copper or aluminum wires might serve as conductors, and an electric motor or light bulb might serve as the load.

Circuits can be wired in series, in parallel, or in a combination. A series circuit provides one, and only one, pathway for the flow of current. A circuit wired in parallel provides multiple pathways for the flow of current. A key requirement for a working circuit is that the conducting pathway is closed, meaning that there is at least one closed loop through which current can flow. However, a circuit can have many different configurations and still meet that requirement. Series and parallel circuits share the same basic components. Where they differ is in how those components are physically arranged.

To make it easier to describe circuits, a standard symbolic language is used to depict specific components and the relationships among them. Each type of circuit element has its own symbol. For example, lines represent the conducting pathways that carry current.

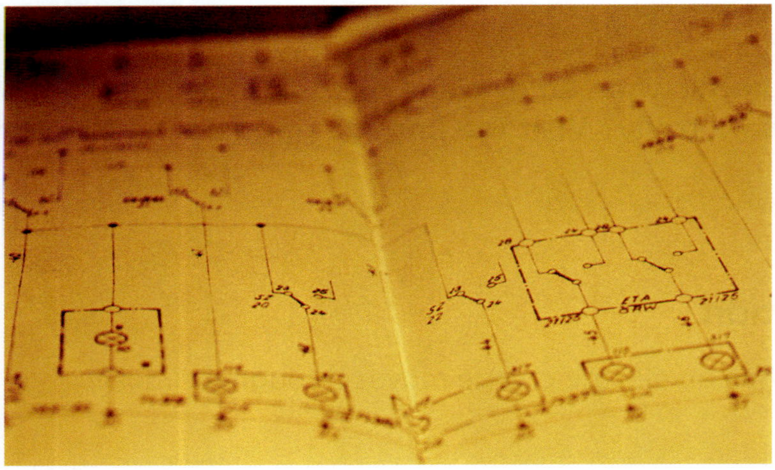

photo: Getty Images

What are the properties of circuits?

What Are Series and Parallel Circuits, and How Can You Design and Build Them?

Designing and Building Circuits

All circuits need a voltage source. This could include many devices, with a battery being the most obvious. Circuits also require a conducting material, such as a wire. Beyond those basic requirements, circuits can have many configurations, and circuit design can be driven by the intended function.

A circuit that needs to merely control a simple on/off function can have a switch that opens and closes the connection with nothing else required. A more complex circuit, such as one controlling an electrical system in a house, would need to include resistors, capacitors, and other elements configured in parallel to branch out to many destinations along the pathway.

One important difference between a series and a parallel circuit is that for a series circuit to work, all the elements must work—meaning that if one breaks, the circuit cannot function. For a parallel circuit, one branch can be cut or broken and the circuit can still operate. A good analogy might be bridges across a ravine. In one of the crossings, there are three bridges hinged one after the other. This is like a set of resistors in series. If one of the bridges weakens and cannot be used, then no one can cross the ravine along that line of bridges.

In another part of the ravine, three bridges are lined up next to each other, with each bridge spanning the ravine. In this case, which is analogous to resistors in parallel, even if one weakens and cannot be used, there are still options available to get across the ravine.

What Is the Relationship between Current, Voltage, and Resistance?

Ohm's Law

Voltage is a measure of the electrical potential difference across or between two points. A voltage drop is the difference in electrical potential. If a circuit had no voltage drop or no electrical potential difference, there would be no flow of current. This axiom establishes the correlation between voltage and current. Similarly, because a resistor reduces the amount of energy available to move electrons around the circuit, it also makes sense that voltage and resistance are related.

Ohm's law describes the relationship among voltage, current, and resistance. Ohm's law states, "Voltage is equal to the product of current and resistance, or $V = IR$." Ohm's law is a powerful mathematical tool for solving problems related to simple circuits.

Current, I, is measured in amperes (A). Voltage, V, is measured in volts (V). Resistance, R, is measured in ohms (Ω).

There are many ways to interpret and think about Ohm's law. For a given voltage, current and resistance are inversely proportional. This means that if one increases, the other decreases to keep the product constant. On the other hand, for a given current, voltage and resistance are directly proportional, and for a given resistance, voltage and current are directly proportional. When two things are directly proportional to each other, they increase or decrease together.

Ohm's Law: Sample Problem

A CD player has a current of 0.1 amperes flowing through it. If the source voltage is 4 volts, what is the resistance?

Solution

Starting from the equation $V = IR$, use algebra to solve for R:

$$\begin{aligned} R &= \frac{V}{I} \\ &= \frac{(4\text{ V})}{(0.1\text{ A})} \\ &= 40\ \Omega \end{aligned}$$

What Is the Relationship between Electrical Power, Current, and Voltage?

Power, Current, and Voltage

Power is a measure of the amount of energy delivered or used over a period of time. Power is a rate. It is used to measure how quickly or how slowly energy is generated, used, dissipated, or transformed.

In circuits, power is the product of the current and the voltage. In mathematical terms, $P = IV$. This relationship is known as Joule's law. The units for power are watts (W), which are equivalent to amp-volts (1 W = 1 AV).

To use a common analogy, imagine a hose pumping out water that turns a waterwheel. The waterwheel's power could be increased in two ways. First, increased water pressure in the hose could produce more power. In another case, a higher flow rate from the hose could increase the power. In terms of an electric circuit, this would be analogous to changing the voltage or the current, respectively.

How Can You Determine the Relationship between Power and Resistance in a Circuit?

Relationships among Components in a Circuit

Voltage is the potential difference that drives a circuit. Power is the rate at which energy is transferred within a circuit. Power is not the same as current, and it is not the same as voltage. A circuit with high voltage and low current can have the same power as a circuit with low voltage and high current. Neither the voltage nor the current alone indicate a circuit's power.

As an example, consider a circuit with a very high voltage drop across two terminals but with no conducting wires. In this case, voltage would be very high, but there would be no current. The resultant power would be zero.

Ohm's law ($V = IR$) offers a simple way to directly link power with resistance. Using simple algebra, the equation for electrical power within a circuit becomes $P = I^2R$.

Relationships among Components in a Circuit: Sample Problem

What is the power of a shaver that draws 6.0 amperes of current from a 120-volt outlet?

Solution

$$\begin{aligned} P &= IV \\ &= (6.0\text{ A})(120\text{ V}) \\ &= 720\text{ W} \end{aligned}$$

How Do Current, Potential Difference, Resistance, and Power Relate to One Another?

Comparing Equations for Series and Parallel Circuits

Circuits can be configured with resistors in series or in parallel. A series connection is one in which an individual charge would pass through each resistor along its path around the circuit. In a parallel configuration, a single charge would pass through only one of the resistors—not all three. Parallel circuits create branches or path options. The physical setup of a circuit has a major impact on the total effective resistance, the total current, and the current across each resistor.

Adding resistors in series adds to the overall resistance because all the charges traversing the circuit must pass through all the resistors. This would be like adding checkpoints along a one-lane highway.

Adding resistors in parallel, however, does not add to the overall resistance. In fact, parallel resistors reduce overall resistance. The reason is that because charges pass through only one of the parallel resistors instead of all of them, the flow is enhanced by having more "forks" in the path.

This difference between resistors in series and resistors in parallel can be described through mathematical relationships. For resistors in series, the total resistance R is given by the equation:

$$R = R_1 + R_2$$

Adding more resistors in series increases the total resistance. For resistors in parallel, total resistance R satisfies the equation:

$$\frac{1}{R} = \frac{1}{R_1} + \frac{1}{R_2}$$

Combining Series and Parallel Circuits

Sometimes, circuits are a combination of series and parallel. A segment of the circuit might have resistors in series, while another segment has resistors arranged in parallel with the segments themselves being connected either in series or in parallel. Solving problems involving resistors in circuits requires careful and detail-oriented sequencing.

Besides differences in total resistance, series and parallel configurations differ in other ways. For example, when a circuit is connected in series, the current is constant across the whole circuit. When a circuit is connected in parallel, the voltage is constant.

The quantities described can be used to figure power spent in each of the resistors. In a series circuit, the power spent in each resistor is the square of the current times its resistance:

$$P = RI^2$$

In a parallel circuit, the power spent in each resistor is the square of the voltage over its resistance:

$$P = \frac{V^2}{R}$$

Comparing Equations for Series and Parallel Circuits: Sample Problem

For each circuit

(i) find the total resistance,

(ii) find the current at each resistor,

(iii) find the voltage at each resistor, and

(iv) find the power consumed by each resistor and by the entire circuit.

(a)

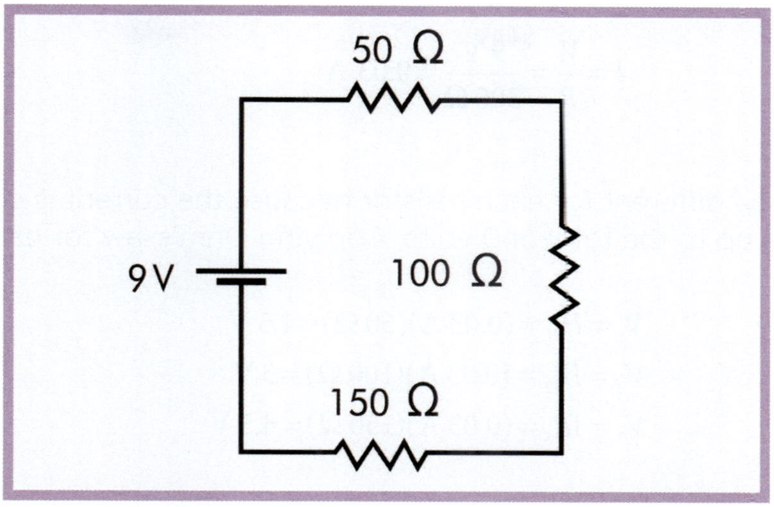

This simple circuit consists of a 9-volt battery and three resistors in series. What is the power consumed by each resistor and by the entire circuit?

(b)

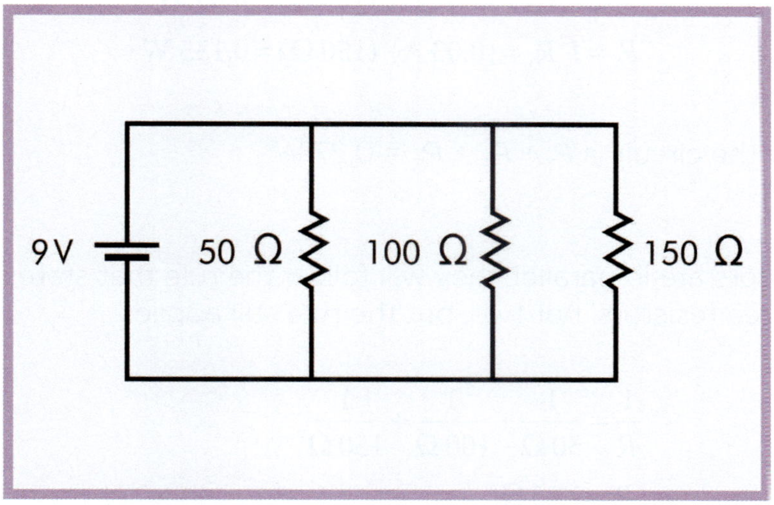

This parallel circuit consists of a 9-volt battery and three resistors in parallel. What is the power consumed by each resistor and by the entire circuit?

Solution

(a)

(i) The total resistance of resistors in series is the sum of the component resistors, so:

$$R = R_1 + R_2 + R_3 = 50\,\Omega + 100\,\Omega + 150\,\Omega = 300\,\Omega$$

(ii) For a circuit configured with resistors in series, the current is constant everywhere. Use $V = IR$ (Ohm's law) to find the circuit's current. V = 9V and R, from part (i) is 300 Ω.

Thus:

$$I = \frac{V}{R} = \frac{9\text{ V}}{300\,\Omega} = 0.03\text{ A}$$

(iii) The voltage will be different for each resistor because the current is constant. The voltages will add up to the total of 9 volts. Applying Ohm's law for each resistor:

$$V_1 = IR_1 = (0.03\text{ A})(50\,\Omega) = 1.5\text{ V}$$
$$V_2 = IR_2 = (0.03\text{ A})(100\,\Omega) = 3\text{ V}$$
$$V_3 = IR_3 = (0.03\text{ A})(150\,\Omega) = 4.5\text{ V}$$

(iv) Power consumed by each resistor is:

$$P_1 = I^2R_1 = (0.03\text{ A})^2(50\,\Omega) = 0.045\text{ W}$$
$$P_2 = I^2R_2 = (0.03\text{ A})^2(100\,\Omega) = 0.09\text{ W}$$
$$P_3 = I^2R_3 = (0.03\text{ A})^2(150\,\Omega) = 0.135\text{ W}$$

Power consumed by the circuit is $P_1 + P_2 + P_3 = 0.27$ W.

(b)

(i) Because the resistors are in parallel, they will follow the rule that states $\frac{1}{R} = \frac{1}{R_1} + \frac{1}{R_2}$. In this case, there are three resistors, not two, but the rule still applies:

$$\frac{1}{R} = \frac{1}{50\,\Omega} + \frac{1}{100\,\Omega} + \frac{1}{150\,\Omega}$$
$$R = 27.3\,\Omega$$

(ii) The current will be different at each resistor, but the voltage will be constant. Rearrange Ohm's law to $I = \frac{V}{R}$ and apply it for each resistor:

$$I_1 = \frac{V}{R_1} = \frac{9\text{ V}}{50\,\Omega} = 0.2\text{ A}$$
$$I_2 = \frac{V}{R_2} = \frac{9\text{ V}}{100\,\Omega} = 0.1\text{ A}$$
$$I_3 = \frac{V}{R_3} = \frac{9\text{ V}}{150\,\Omega} = 0.06\text{ A}$$

(iii) Because the resistors are arranged in parallel, the voltage will be the same across the circuit, 9 volts.

(iv) Power consumed by each resistor is:

$$P_1 = \frac{V^2}{R_1} = \frac{(9\ \text{V})^2}{50\ \Omega} = 1.62\ \text{W}$$

$$P_2 = \frac{V^2}{R_2} = \frac{(9\ \text{V})^2}{100\ \Omega} = 0.81\ \text{W}$$

$$P_3 = \frac{V^2}{R_3} = \frac{(9\ \text{V})^2}{150\ \Omega} = 0.54\ \text{W}$$

Power consumed by the circuit is $P_1 + P_2 + P_3 = 2.97$ W.

Why Do Resistors in DC Circuits Increase in Temperature?

Why Resistors Heat Up

As electrons pass through a resistor, they collide with matter in the resistor. These collisions impede the flow of current. Resistors transform the electrical energy into thermal energy.

Some everyday appliances are actually resistors. For example, ovens, toaster ovens, room heaters, hair dryers, curling irons, and electric teapots all depend on the heating effects of resistors. Circuits that have low resistance heat up less and are therefore more heat efficient. However, as car brakes are designed to dissipate mechanical energy in the form of thermal energy, the circuits of such appliances are designed to create useful thermal energy due to resistance.

How Can the Voltage or Current in a Simple DC Circuit Be Described and Predicted?

Voltage and Current in DC Circuits

Predicting the voltage and current in a circuit and in its components can seem like a confusing and complex task. However, circuits and their elements can be described by some simple rules and equations.

A first basic step is to identify the voltage source for the circuit and resistances at each resistor. Usually, there is a battery attached to a circuit, and its voltage will be the circuit's only source of voltage. Sometimes, though, the voltage is the missing variable, in which case other variables need to be identified and used to solve for the voltage.

Sometimes, the resistances are labeled or obvious, but sometimes finding resistances requires some tricky math. Nonetheless, the rules always apply. For resistors in series, the total resistance is the sum of the component resistors. For resistors in parallel, the reciprocal of the total resistance is the sum of the reciprocals of the component resistances. Ohm's law ($V = IR$) and Joule's law ($P = IV = I^2R$) can be used to solve a wide variety of problems related to circuits.

Voltage and Current in DC Circuits: Sample Problem

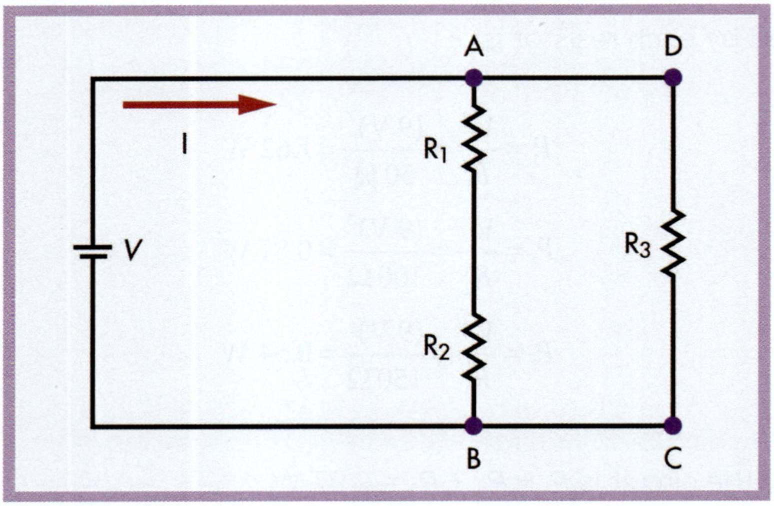

A DC circuit can have resistors in parallel and in series. How are the current and power for the circuit calculated?

Find the current and the power for the pictured circuit. Use the values: R_1 = 20 Ω, R_2 = 200 Ω, and R_3 = 2 Ω. The power source supplies 12 volts.

Solution

It is imperative to find the total resistance. This circuit is a combination of series and parallel resistors. The branch with R_1 and R_2 has a total resistance of:

$$R_1 + R_2 = 220\,\Omega$$

In conjunction with R_3, though, the configuration is in parallel, not in series. Therefore, the equation used is the sum of the reciprocals of the resistances rather than the sum of the resistances. This means:

$$\frac{1}{R} = \frac{1}{220\,\Omega} + \frac{1}{2\,\Omega}$$

$$\text{Thus, } R = 1.98\,\Omega$$

The current at the battery is:

$$I = \frac{V}{R} = \frac{12\text{ V}}{1.98\,\Omega} = 6.06\text{ A}$$

To find the power, apply the equation:

$$P = IV$$

$$P = IV = (6.06\text{ A})(12\text{ V}) = 72.7\text{ W}$$

Kirchhoff's Junction Rule

What happens when a circuit branches? The point at which the branch occurs is called a node or junction. Because energy is always conserved, the current coming into a node junction must equal the current leaving a junction. Electrons are not destroyed at any junction in a circuit. Every electron that enters the junction must leave it.

Kirchhoff's Current (Junction) Rule: Sample Problem

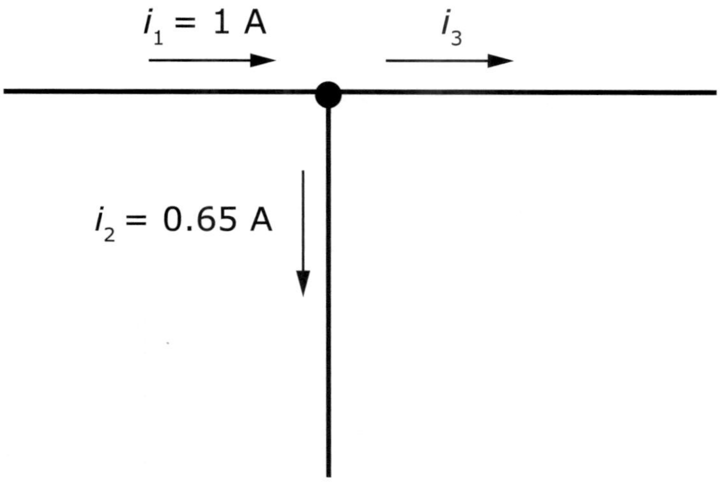

This junction in a circuit consists of three wires carrying a current. The first wire brings 1 ampere to the junction, the second wire takes 0.65 amperes of current from the junction, and the third wire takes an unknown amount of current from the junction. Use Kirchhoff's junction rule to find the current in the third wire.

Solution

Kirchhoff's junction rule says that the algebraic sum of currents at a junction is zero. This can be represented by the following equation:

$$-i_1 + i_2 + i_3 = 0$$

where i_1 is the current into the node and i_2 and i_3 are currents leaving the junction. The quantity we want to find is i_3. To make the math simpler, we rearrange this equation to state that the current leaving the junction is equivalent to the current entering the junction.

$$i_2 + i_3 = i_1$$

We know that i_2 = 0.65 A, and i_1 = 1 A, so we can substitute these values to obtain:

$$0.65\ \text{A} + i_3 = 1\ \text{A}$$
$$i_3 = 1\ \text{A} - 0.65\ \text{A} = 0.35\ \text{A}$$

Kirchhoff's Voltage (Loop) Rule

One of the fundamental rules in physics is that energy is always conserved. This rule also applies to the electrons moving through a circuit, and it can be used to help analyze the voltage in a circuit. Specifically, around any closed loop in a circuit, the sum of the potential difference across all elements in the loop must be zero.

A circuit loop is any area in the circuit where the current starts and ends at the same place. For example, the circuit in the following sample problem consists of two smaller loops (through points A and B and resistors 1 and 2, and through points A, B, C, and D and resistors 1, 2, and 3) and one larger loop (through points A, B, C, and D and resistor 3).

Kirchhoff's Voltage (Loop) Rule: Sample Problem

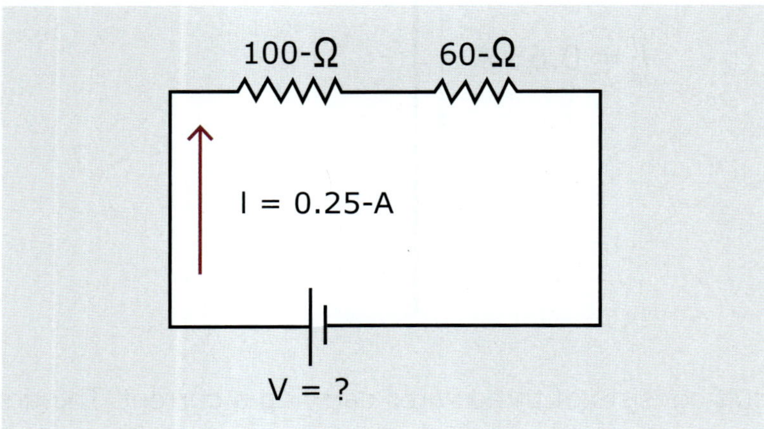

This circuit consists of a battery supplying an unknown voltage, a 100-Ω resistor, and a 60-Ω resistor. A 0.25-ampere current flows through the circuit. Use Kirchhoff's loop rule to find the voltage supplied by the battery.

Solution

Kirchhoff's loop rule says that the sum of the potential differences across all three elements in this circuit equals zero. This can be represented by the following equation:

$$V_b - V_{100} - V_{60} = 0$$

Where V_b is the voltage across the battery, V_{100} is the voltage across the 100-Ω resistor, and V_{60} is the voltage across the 60-Ω resistor. V_b is the quantity we want to find, and the two resistor voltages are given by Ohm's law:

$$V_b - (0.25\ \text{A})(100\ \Omega) - (0.25\ \text{A})(60\ \Omega) = 0$$
$$V_b = (0.25\ \text{A})(100\ \Omega) + (0.25\ \text{A})(60\ \Omega)$$
$$V_b = 25\ \text{V} + 15\ \text{V}$$
$$V_b = 40\ \text{V}$$

Capacitors

Another circuit element that should be considered when analyzing circuits is a capacitor. A capacitor stores energy between two charged plates, separated by some distance and a nonconducting substance called a dielectric. Capacitors can build up charge and then dissipate charge back into the circuit, so they greatly affect a circuit's performance and are at the core of most electronic devices. The capacitance of a capacitor, C, is given by Q/V, where V is the voltage between the plates and Q and $-Q$ are their charges. The SI units for capacitance are farads (F). One farad is equal to one second per ohm. In a circuit that has capacitors and resistors, the voltage does not remain constant. Capacitors affect the voltage and current of a circuit as they either charge or discharge their stored potential energy. For example, for a circuit that has only one capacitor and one resistor, the voltage across the resistor as a function of time, $V(t)$ is given by the equation:

$$V(t) = V_0 e^{\frac{-t}{RC}}$$

where R is the resistance, C is the capacitance, and V_0 is the initial voltage across the capacitor. The current across the resistor also varies, but the formula $V = IR$ still holds. Using the formula for voltage, a formula for current at time t is:

$$I(t) = I_0 e^{\frac{-t}{RC}}$$

There are similar formulas for the voltage and current across the capacitor in the sample circuit. Typically, the calculations become quite cumbersome in more complicated circuits and will not be approached at the high school level. Just like resistors, the result of including multiple capacitors in a circuit depends on whether they are connected in series or in parallel. The difference between the total capacitance of these two types of circuits can be described using mathematical relationships. For capacitors in series, the total capacitance C is given by the equation:

$$C = \frac{1}{\frac{1}{C_1} + \frac{1}{C_2} + \ldots \frac{1}{C_n}}$$

For capacitors in parallel, the total capacitance *C* is given by the equation:

$$C = C_1 + C_2 + \ldots C_n$$

Adding more capacitors in a series circuit decreases the total capacitance, whereas adding them to a parallel circuit increases the total capacitance.

What Are the Characteristics of Transistors That Make Them Useful in Electric Circuits?

Transistors

Transistors are circuit elements that amplify or switch signals. They do so using semiconductors, which are specifically configured materials along the range between conductors and insulators. Modern technology is based on transistors and semiconductors, and, without them, no computer or cell phone would be able to function.

Transistors include a base, a collector, and an emitter. When included in a circuit, transistors allow a small amount of current to the base to relay into a large amount of current between the collector and the emitter. This means that a transistor alters circuit characteristics based on inputs and specific material characteristics, leaving designers with a lot of control to adjust and finesse how the circuit functions. This helps explain why transistors are the fundamental building blocks of modern electronics.

Consider the Explain Question:

What happens when an electric charge moves through a circuit?

Go online to complete the scientific explanation.

Check for Understanding:

You have a 12-V battery, a resistor R_1 = 12 Ω, and another component that requires a current of 2.0 A and a voltage drop of 6.0 V. What value of resistor would you connect in parallel with R_1, with both resistors in series with the other component, to meet the requirements?

STEM in Action

STEM and Electric Circuits

The physics behind electric circuits includes conservation of energy, energy change, and other principles. Electric circuits also reflect more complex topics such as semiconductors, transistor design, and efficiency.

Our wired world depends on the physics of electric circuits. Computers and computing lie at the heart of modern technologies. They depend on moving electrons from one place to another. Computers are complex, but underneath their power lies the basic concepts related to circuits.

Computer engineers have the challenging task to combine electrical engineering and computer science. Their job is to develop computer hardware and software. One of the challenges computer engineers face is the fact that resistors heat up. That is why microchip design continuously advances toward greater energy efficiency, smaller-scale volume, and innovative materials.

A 12-volt battery is connected in parallel with resistor $R_1 = 12\ \Omega$ and resistor $R_2 = 8.0\ \Omega$.

1. On a separate sheet of paper, draw the circuit diagram and label it with each component.
2. Find the equivalent resistance.
3. Find the total current in the circuit.
4. Find the current through each resistor.
5. Find the power in the circuit.

Conductors and Insulators

photo: Getty Images

LESSON OVERVIEW

Lesson Questions

- What are differences and similarities between insulators, conductors, semiconductors, and superconductors?
- What materials make good conductors, semiconductors, or insulators?
- How does magnetic induction create an electric current in a conductor?
- What are the features of the electric charge on an insulator?

Lesson Objectives

By the end of the lesson, you should be able to:

- Compare insulators, conductors, semiconductors, and superconductors.
- Classify materials as conductors, semiconductors, or insulators.
- Explain how magnetic induction creates electric current in conductors.
- Provide evidence that electric charge is generally static on insulators.

Key Vocabulary

Which terms do you already know?

- ☐ **conductor**
- ☐ **current**
- ☐ **electrical conductivity**
- ☐ **electro-magnetism**
- ☐ **electron**
- ☐ **induction**
- ☐ **insulator**
- ☐ **magnetic field**
- ☐ **metal**
- ☐ **metals**
- ☐ **resistance**
- ☐ **semiconductor**
- ☐ **static electricity**
- ☐ **superconductor**
- ☐ **transistor**

ENGAGE

Have you ever wondered why birds on a telephone wire do not get electrocuted? What would happen to a person who touched a wire with one hand while his feet were still on the pole? Visit the Engage page online to learn more.

Explain Question

How does the structure of a material explain its behavior as a conductor or an insulator?

EXPLORE

What Are Differences and Similarities between Insulators, Conductors, Semiconductors, and Superconductors?

Conductors and Insulators

Different materials can be classified according to how easily an electric current can pass through them. An electric current is a flow of electrons through a material. Different materials have different forces between the particles inside that affect how easily electrons can move about inside the material. In some materials, such as metals, negatively charged electrons can move about easily. In others, such as most plastics, electrons cannot move easily through the material.

The electrical conductivity of a substance expresses how well the material conducts current. The higher the electrical conductivity, the more easily a current can flow through the substance. Copper has a very high electrical conductivity, while diamond has a very low electrical conductivity.

Conductors are materials that conduct electricity well. Metals are good conductors. For this reason, the electric current in circuits usually runs along a metal wire. Electrons can move easily through a conducting material.

Insulators are materials that do not conduct electricity or that conduct electricity very poorly. They are the opposite of conductors. Insulators can be usec for safety reasons to prevent the flow of a current. Usually, metal wires in circuits are coated with a plastic insulating jacket. Electrons cannot move easily through an insulator.

Semiconductors

Semiconductors are materials with an electrical conductivity in between conductors and insulators. Computers, televisions, cell phones—in fact, nearly all modern electrical devices—use semiconductors. Semiconductors are used in the chips that a computer uses to operate. They are used in circuit components called diodes and transistors. These devices are used as tiny switches in circuits that control the flow of current. The microprocessor chips used in computers hold thousands of these tiny diodes and transistors.

Superconductors

Some materials have a very high level of conductivity. In theory, if their temperature is low enough, they have an infinite conductivity because they have no resistance to the flow of electrons. These materials are called superconductors, and they are materials that conduct electricity with no resistance.

Superconductors are therefore very efficient for transmitting electric current. One difficulty with superconductors is that all currently known superconducting materials work only at very low temperatures. In fact, the temperature at which most materials can act as a superconductor is only –138°C, or equivalently 135 K. However, recently, there have been advances in room temperature superconductors. This class of superconductor operates at temperatures of 35°C or 308.15 K.

Mercury is an example of a superconductor, which becomes a superconductor at the extremely low temperature of –269°C, or equivalently 4.2 K.

What Materials Make Good Conductors, Semiconductors, or Insulators?

Materials and Conductivity

Materials that conduct electricity effectively are conductors. Examples of conductors include metals. Electrical devices use conductors to transmit current through the device. Common electric wires usually are made from copper or aluminum, both of which are metals. Copper and aluminum have particularly high electrical conductivities.

Materials that do not conduct electricity are insulators. Examples of insulators include glass, paper, and plastic. Power cords for electric devices such as computers or cell phone chargers are made from a rubber or plastic insulating coating around a metal conducting wire. Rubber and plastic have particularly low electrical conductivities.

Semiconductors are materials with a moderate level of electrical conductivity. Examples of semiconductors include silicon and germanium. In fact, pure silicon and pure germanium are better insulators than conductors. Semiconductors are made by combining the pure silicon or germanium with small amounts of certain other materials, including arsenic or boron. The additional material allows electrons to move slightly more freely through the silicon or germanium. Silicon chips used in computers are made from this sort of impure silicon.

Material	Conductivity	Examples
Insulator	Low or none at all	Glass, paper, plastic
Conductor	High	Metals
Semiconductor	Intermediate	Impure silicon, germanium

This table summarizes the properties of insulators, conductors, and semiconductors. What feature of a material determines whether it is a good, intermediate, or poor conductor of electricity?

How Does Magnetic Induction Create an Electric Current in a Conductor?

Magnetic Induction

Motors and electric generators are devices that power many things, including cars, electric fans, and automatic doors. Nearly all the electrical power generated by commercial energy companies used in homes is thanks to electric generators. Electric generators use magnetic induction and properties of conductors.

Magnetic induction occurs when a changing magnetic field produces a current in a conductor. The laws of electromagnetism state that a changing magnetic field causes charged particles to feel a force. Placing a conductor in a changing magnetic field causes the electrons in the conductor to feel a force. Electrons that feel a force from a magnetic field will flow through a conductor. The flow of electrons creates an electric current. Thus, a changing magnetic field induces a current.

Faraday's Law

Faraday's law is the physical law that describes how current is induced by a changing magnetic field. Faraday's law states that the electromotive force, or emf, induced by a changing magnetic field is related to how fast the magnetic flux changes with time, or, more specifically:

$$\varepsilon = \frac{-N(\phi_2 - \phi_1)}{t_2 - t_1}$$

In this equation, ε is the emf, N is the number of turns of the wire, and t_1 and t_2 are the initial and final times, respectively. The phi symbols, ϕ_1 and ϕ_2, are the initial and final values, respectively, for the magnetic flux.

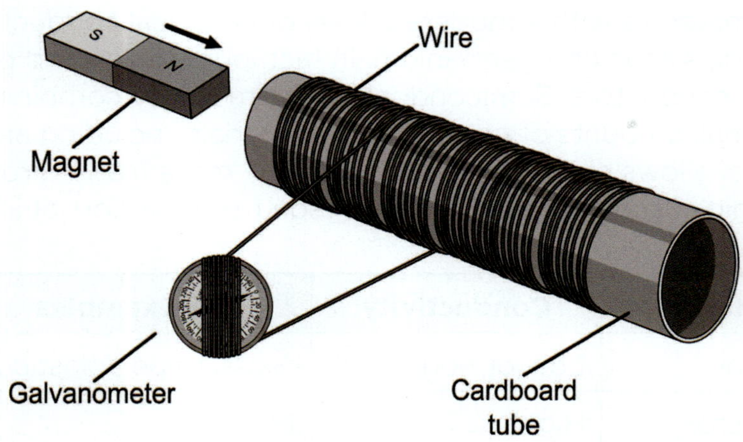

Faraday's model is easy to re-create. If you do not have a galvanometer, how else might you determine that an electrical current was being produced?

Magnetic flux is defined as the strength of the magnetic field multiplied by the surface area of the wire loop. It is important to remember that this definition applies only for uniform magnetic fields that are perpendicular to the area of the wire loop. The direction of the induced current is dictated by the "right-hand rule." This means if you point the thumb of your right hand in the direction of the current, the direction in which your fingers curl is the direction of the magnetic field. If the emf from the equation is negative, the current flows in the opposite direction.

Faraday's Law: Sample Problem

The following diagram shows a wire, looped three times into a circle with a radius of 10 centimeters, and placed in a magnetic field (B) of 0.1 tesla. The tesla (symbol T) is a unit of measure of the strength of a magnetic field.

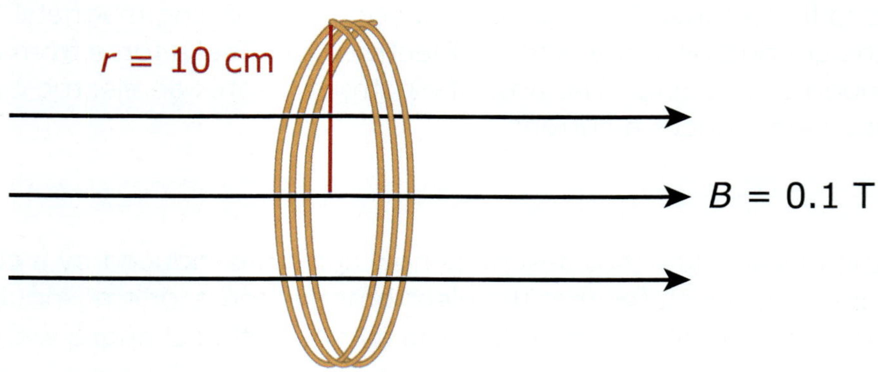

A 0.1 tesla magnetic field passes through a three-turn wire loop with a radius of 10 centimeters. How does the number of turns of this type of wire loop affect the emf generated in the loop?

What would be the emf induced in this wire if the magnetic field were increased from 0.1 tesla to 1 tesla over a period of 30 seconds?

To solve this problem, use the equation for Faraday's law. In this setup, *N* is 3 and the initial flux is 0.1 tesla times the surface area of the circle:

$$\varphi_1 = (0.1\ \text{T})\,[\pi(0.1\ \text{m})^2] = 3.14\times10^{-3}\ \text{T}\times\text{m}^2$$

The surface area of the wire is not changing, but the strength of the magnetic field is. The final magnetic field is 10 times that of the initial magnetic field, meaning the final flux will be 0.031 T × m^2. The initial time can be set to zero, meaning the final time is 30 seconds. Putting all of this into the equation for emf results in the following calculation:

$$\varepsilon = -3\left(\frac{0.031\,\text{T}\times\text{m}^2 - 3.14\times10^{-3}\,\text{T}\times\text{m}^2}{30\ \text{s}}\right) = -\left(\frac{0.028\ \text{T}\times\text{m}^2}{10\ \text{s}}\right) = -0.003\ \text{V}$$

The units for emf are volts (V), so the induced emf in this wire is -0.003 volts.

What Are the Features of the Electric Charge on an Insulator?

Insulators and Static Charge

Static electricity is the accumulation of electric charge in a substance or object.

Electric charges can move freely in a conductor, so conductors are good carriers of electric current. Electric charges in insulators cannot move freely, which is why insulators are very poor conductors of electricity.

Insulators can store an electric charge even though the charge cannot move freely. Electric charge can be built up on insulators in the form of static electricity because a charge on an insulator does not move, or does not move very much. It often is easier to store a charge on an insulator than on a conductor because the charge cannot easily transfer from the insulator onto another object. The charge remains static on the insulator.

Consider the Explain Question:

How does the structure of a material explain its behavior as a conductor or an insulator?

Go online to complete the scientific explanation.

Check for Understanding:

A 12-V battery is connected in parallel with resistor R_1 = 12 Ω and resistor R_2 = 4.0 Ω. Can you calculate the power of each resistor in the circuit?

STEM in Action

STEM and Conductors and Insulators

Have you ever had a job to do, but did not have the correct tool to complete it? A tool—or any product—should be constructed of the right material to do its job well, and these materials must be able to perform as needed. Materials engineers require a firm understanding of physics to develop, produce, and test materials that are used in many products, including superconducting magnets. Research into superconductivity continues to turn up surprising features and new technological applications.

The work of a materials engineer involves both fundamental science and applied science, and focuses on superconducting materials and their formation, structure, and magnetic and electrical properties. Materials engineers develop devices that use superconducting magnets because these magnets can create an extremely strong magnetic field. The medical technology called magnetic resonance imaging (MRI) uses superconducting magnets that have been designed by materials engineers. The strong magnetic field helps create images of organs and other soft tissues inside the body. Medical doctors can then interpret these images to help diagnose and manage health issues.

Superconducting magnets can also be used in maglev technology to levitate heavy objects such as passenger trains. Maglev is named for and is based on magnetic levitation. Objects such as the JR-Maglev in Yamanashi, Japan, are suspended and accelerated using magnets rather than traditional methods. The magnets are used to keep the trains levitated above and to move the trains along and above the tracks. Because there is no physical contact between the trains and the tracks, there is no friction between them, so the amount of wear and tear on the trains and tracks is greatly reduced. Theoretically, the contactless trains are able to travel much faster than traditional trains.

The speed record for maglev trains is 603 kilometers per hour, attained by the CJR's MLX01 superconducting maglev in 2015 in Japan. This speed is 28 kilometers per hour beyond the speed record of conventional trains set by the French TGV (*Train à Grande Vitesse*, or high-speed train) in 2007.

Complete the activity "Find the Solution" on the STEM in Action page online. In this activity, you will test several liquids of your choice and classify them according to their conductivity. Continue testing until you have three liquids that conduct and three that do not. Then, answer the questions and complete the table. Start by either making a conductivity tester from wire, a battery, and a bulb, or obtaining a preassembled tester from your teacher.

1. What liquids did you find were conductors? Name at least three examples.
2. What liquids did not conduct, or were poor conductors? Name at least three examples.
3. Provide a data table that summarizes the results for the liquids you tested.
4. What statement can you make about the conductivity of ammonia solution? What statement can you make about the conductivity of vinegar? Mix some ammonia solution and some vinegar together, and check the conductivity of the resulting solution. Explain the results.

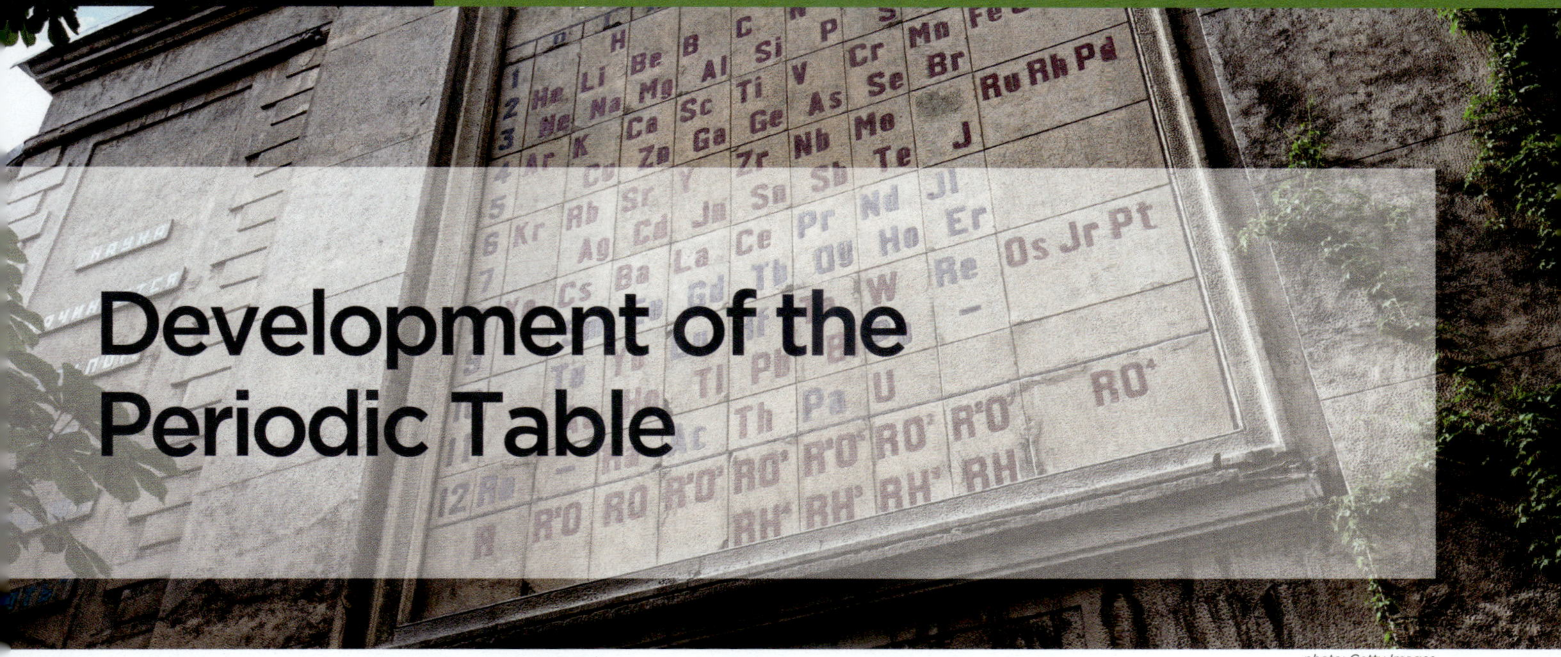

Development of the Periodic Table

photo: Getty Images

LESSON OVERVIEW

Lesson Questions

- How did early scientists attempt to organize the elements?
- How and why did Mendeleev organize the elements the way he did?

Lesson Objectives

By the end of the lesson, you should be able to:

- Summarize early attempts at organizing the elements.
- Explain Mendeleev's basis for his organization of the elements.

Key Vocabulary

Which terms do you already know?

- [ ] **atom**
- [ ] **element**
- [ ] **families**
- [ ] **groups**
- [ ] **inert gas**
- [ ] **ionic bond**
- [ ] **Mendeleev**
- [ ] **Pauli exclusion principle**
- [ ] **period**
- [ ] **periodic table of the elements**
- [ ] **periodicity**
- [ ] **valence electron**

ENGAGE

Have you ever noticed how organized some things around you tend to be? How is organizing items in a grocery store similar to organizing the elements of which all matter is composed? Visit the Engage page online to learn more.

Explain Question

Explain early attempts at organizing the elements, what Mendeleev's basis was for his organization of the elements, and how early versions of the periodic table compare with the modern version.

EXPLORE

How Did Early Scientists Attempt to Organize the Elements?

The Predictable Properties of Elements

Early scientists and philosophers in ancient Greece proposed that all matter was composed of only four elements: earth, water, air, and fire. As scientists observed matter, however, they began to suspect this was not true. Over time, new theories emerged.

Early scientists recognized that some elements, such as metals, had similar properties and, therefore, similar uses. But they also noticed that not all metals are the same. Alkali metals (such as sodium) are very reactive, while transition metals (such as gold) are very unreactive.

In 1817, Johann Wolfgang Dobereiner proposed the law of triads. He arranged elements in groups of three based on similarities in their appearance, other physical properties, and chemical reactivity. This was the first proposal of grouping elements together based on recognized patterns. For example, Dobereiner grouped chlorine, bromine, and iodine together in a triad. Each of these elements reacts with an alkali metal such as sodium to form a salt. He observed that the atomic mass of bromine is about halfway between the masses of chlorine and iodine. Similarly, the density of bromine is about halfway between the densities of chlorine and iodine.

Many other scientists contributed to the development of the periodic table. In 1828, Jöns Jacob Berzelius proposed symbols for the elements based on their Latin names. These are the same names and symbols that are used today.

Development of the Periodic Table

As more elements were discovered, scientists attempted to organize the elements according to their observable physical and chemical properties. They assessed physical properties such as melting point, boiling point, atomic mass, density, and color. Early scientists also assessed chemical properties. For example, they studied whether an element would react with other elements such as oxygen, sodium, or chlorine. They looked at which elements reacted with water and which could be burned to form new products. Early scientists examined how strongly an element reacted and what types of products were formed from each reaction. These observations led to the identification of patterns, which meant the properties of elements could be predicted. Scientists began to try to organize the elements, often using tables or charts.

Around 1863, John Newlands proposed an early version of the periodic table that arranged the known elements in order of increasing atomic mass along rows. Each row contained seven elements, but Newlands believed that elements would be discovered to fill an eighth spot in each row. He proposed the "law of octaves" in which elements, like notes on a musical scale, are arranged in sets of eight. This arrangement in periods (horizontal rows) was the first of its kind.

	Element	Atomic Mass	Properties
Group I	Cl Br I	35 80 127	Exist as diatomic molecules and form simple salt compounds
Group II	S Se Te	32 79 128	Form salt compounds
Group III	Ca Sr Ba	40 88 137	Reactive metals
Group IV	Li Na K	7 23 39	Very reactive metals

Dobereiner arranged the elements in groups of three based on their similarities. What properties did he consider when making these arrangements?

How and Why Did Mendeleev Organize the Elements the Way He Did?

Predicting the Properties of Undiscovered Elements

As early as the 1700s, French scientist Antoine Lavoisier developed a periodic table list that separated metals from nonmetals. His table contained only the 33 elements known at the time. Just like the periodic table that is used today, early versions incorporated scientists' observations about the periodic nature of elements. Periodicity is the pattern of chemical and/or physical similarities of elements that occur at regular intervals in the periodic table.

A Russian scientist and teacher named Dmitri Mendeleev was planning a lesson about the elements when he developed his version of the periodic table in 1869. He attempted to arrange the 63 known elements in order of increasing atomic mass. Also, using an element's density, Mendeleev could calculate its atomic volume. He noted the periodicity of the volume as he organized the elements.

Mendeleev noticed other repeating patterns: Some elements had distinct similarities and differences in their physical and chemical characteristics. For example, lithium, sodium, and potassium are all metals that react strongly with water and with oxygen. Each one reacts with chlorine to produce a salt. Similarly, chlorine, bromine, and iodine are all reactive nonmetals. Each one reacts with sodium to produce a salt. Based on the physical and chemical properties of these elements, Mendeleev placed the elements in columns (now called groups or families).

			Ti = 50	Zr = 90	? = 180
			V = 51	Nb = 94	Ta = 182
			Cr = 52	Mo = 96	W = 186
			Mn = 55	Rh = 104,4	Pt = 197,4
			Fe = 56	Ru = 104,4	Ir = 198
			Ni = Co = 59	Pd = 106,6	Os = 199
H = 1			Cu = 63,4	Ag = 108	Hg = 200
	Be = 9.4	Mg = 24	Zn = 65,2	Cd = 112	
	B = 11	Al = 27,4	? = 68	Ur = 116	Au = 197?
	C = 12	Si = 28	? = 70	Sn = 118	
	N = 14	P = 31	As = 75	Sb = 122	Bi = 210?
	O = 16	S = 32	Se = 79,4	Te = 128?	
	F = 19	Cl = 35,5	Br = 80	J = 127	
Li = 7	Na = 23	K = 39	Rb = 85,4	Cs = 133	Tl = 204
		Ca = 40	Sr = 87,6	Ba = 137	Pb = 207
		? = 45	Ce = 92		
		?Er = 56	La = 94		
		?Yt = 60	Di = 95		
		?In = 75,6	Th = 118?		

Mendeleev organized the elements in order of increasing atomic mass. How many elements were on Mendeleev's periodic table?

In the same year, Lothar Meyer published a table similar to the one that Mendeleev had proposed. There was one significant difference between the works of the two men. When Mendeleev proposed his table of 63 elements, he left gaps where he believed new elements would fit. He knew that known elements did not fit into these gaps based on the predicted chemical and physical properties of those elements.

Mendeleev took a risk by presenting his table with empty spaces, and he received a great deal of criticism for doing so. Not long after Mendeleev's periodic table was published, however, new elements were discovered that fit into the gaps. Mendeleev's table was such a success because he could use it to predict the chemical and physical properties of missing elements. For example, before germanium was discovered, Mendeleev used his table to predict correctly its atomic mass, density, reactivity with oxygen, and reactivity with chlorine. Because of this, he is often called the "father" of the periodic table.

The Modern Periodic Table

In 1894, William Ramsay discovered a new group of elements called the inactive, or inert, gases. Today, these elements are called the noble gases. Newlands had predicted these elements would be discovered. Based on their properties, the noble gases also fit into the periodic table proposed by Mendeleev.

In 1900, Rutherford studied how certain elements naturally decay into other elements. He later developed a nuclear model of the atom. His research helped confirm that the atomic structure of elements contributes to their periodicity, but it also revealed a problem. The properties of some elements do not vary periodically when the elements are ordered by atomic mass.

Mendeleev knew of this problem when he published his periodic table. Based on properties, he could order most elements by atomic mass, but a few he could not. Around 1913, Henry Moseley, a student of Rutherford's, solved this problem. Moseley studied the emission spectra of elements. Emission spectra are the wavelengths of light emitted by atoms that have been energized. Each element emits a different set of wavelengths. Studying the spectra, Moseley could determine the number of protons in an atom of each element. The number of protons is the atomic number. This solved Mendeleev's problem. It showed that the elements should be arranged by atomic number on the periodic table, not by atomic mass. This was an important step for the periodic table. It also allowed chemists to refine the periodic law:

When elements are arranged in order of increasing atomic number, their physical and chemical properties show a periodic pattern.

In 1916, Gilbert Newton Lewis developed the valence bond theory. Valence electrons are those in the outermost energy level of an atom. They determine many properties of an element. This important finding supported the organization of the periodic table, with elements in each column sharing similar properties. Lewis found that, in most cases, the number of valence electrons in the atoms of an element corresponds with the column in which the element appears.

Modern researchers have continued to refine the periodic table. For example, in 1940, Glenn Seaborg synthesized nine elements that, although unstable, exhibited properties as predicted by their placement in the periodic table. These elements had properties like those of actinide (atomic number 89), and they filled places in period seven of the table.

The so-called lanthanides (elements 58–71) and actinides (elements 88–103) have properties different from other period six and seven elements. They are usually displayed below the main part of the periodic table to keep the size of the table manageable.

Today, scientists still debate some of the details regarding recently discovered elements. In fact, recently the periodic table has been adjusted so that lanthanide (element 57) and actinide (elements 89) are displayed below the table with the others in their respective series. However, many versions still place these elements up in the main body of the table.

Consider the Explain Question:

Explain early attempts at organizing the elements, what Mendeleev's basis was for his organization of the elements, and how early versions of the periodic table compare with the modern version.

Go online to complete the scientific explanation.

Check for Understanding:

Why was Mendeleev's organization of the elements different from those of other scientists before Mendeleev?

STEM in Action

STEM and the Development of the Periodic Table

In 2010, a significant change to the modern periodic table was made. The atomic masses of 10 elements were updated. The previous version of the periodic table included atomic masses based on the relative abundance of isotopes for each specific atom. Historically, atomic mass (which the IUPAC calls the atomic weight) was considered a fixed value. Scientists discovered, however, that atomic masses might actually vary slightly in nature.

Now, the periodic table includes a range of atomic masses for the 10 elements identified, including hydrogen, boron, carbon, silicon, and chlorine. The ranges are very small, but the implications are huge.

Nuclear scientists worked to design experiments, collect data, and analyze it to support this update. Their findings were then submitted to the International Union of Pure and Applied Chemistry's (IUPAC) Commission on Isotopic Abundances and Atomic Weights, the official keeper of the periodic table. The IUPAC Commission then reviewed all the supporting evidence and consulted with the scientific community on the proposed changes. After careful consideration, the next evolution of the periodic table of the elements was approved.

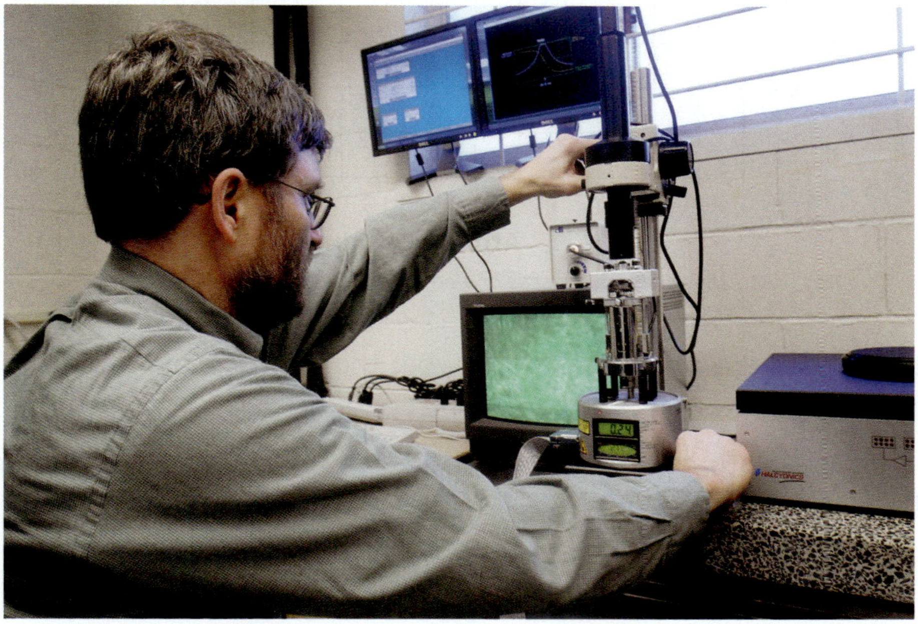

photo: Getty Images

This microscope is able to use the repulsion force between atomic forces and a probe to image individual atoms. What type of information could this reveal about elements?

Predicting Atomic Masses of New Elements

In December 2015, IUPAC announced another significant change in the periodic table. The discoveries of four new chemical elements were verified, and the elements were added to period 7 of the periodic table. The atomic numbers of the new elements are 113, 115, 117, and 118. One factor missing from the periodic table blocks for each of the elements, however, was their atomic masses.

The following graph shows the atomic masses of elements near the four that were added.

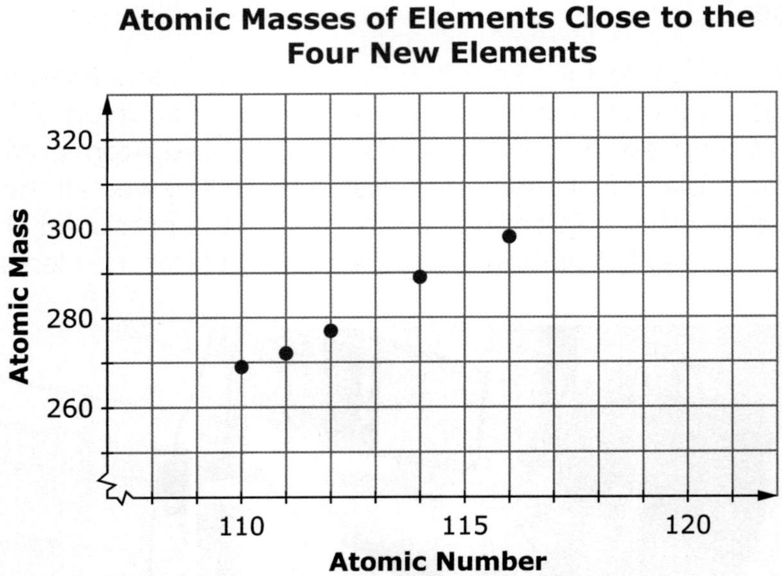

This graph shows atomic masses of elements close to the four elements.

1. Use the data in the graph to predict what the approximate atomic masses of each of the new elements will be.
2. Based on these atomic masses, determine the approximate number of protons and neutrons atoms of the elements will have.

Structure of the Periodic Table

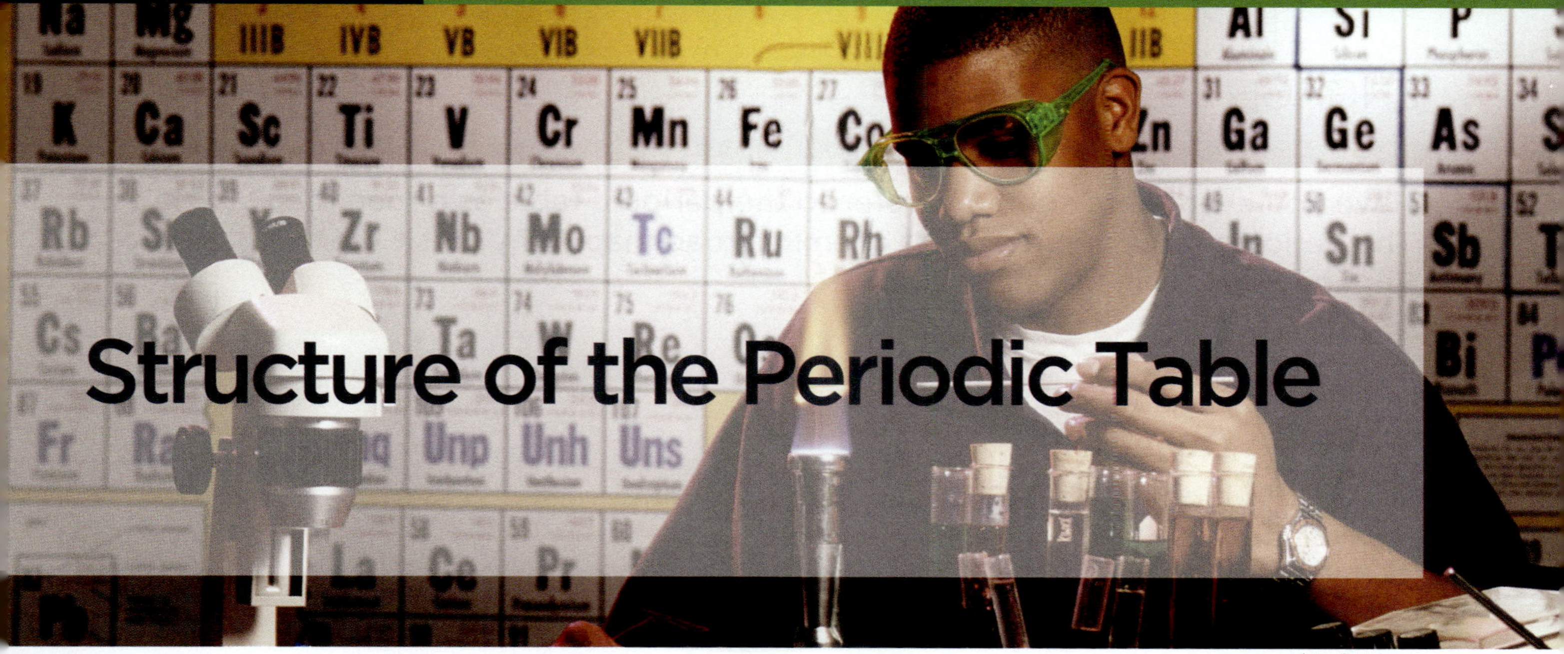

photo: Getty Images

LESSON OVERVIEW

Lesson Questions

- How do scientists distinguish between metals, metalloids, and nonmetals?
- How are the columns of the periodic table organized, and what do the elements in each group have in common?

Lesson Objectives

By the end of the lesson, you should be able to:

- Distinguish the structures and properties of metals, metalloids, and nonmetals.
- Interpret the columns of the periodic table as groups or families that contain elements with similar properties.
- Relate the location of an element in the periodic table to its general properties for: the alkali metals, the alkaline earth metals, the halogens, the transition metals, and the noble gases.
- Show how the result of a chemical reaction is related to atomic structure, trends in the periodic table, and patterns of chemical properties.
- Cite evidence to show how the structure of substances relates to the strength of electrical forces between particles.
- Show how the functioning of designed materials depends on molecular-level structure.

Key Vocabulary

Which terms do you already know?

- ☐ actinide series
- ☐ alkali metals
- ☐ alkaline earth metals
- ☐ atomic radius
- ☐ conductor
- ☐ ductility
- ☐ electron configuration
- ☐ electro-negativity
- ☐ families
- ☐ groups
- ☐ halogens
- ☐ ionic bond
- ☐ ionization energy
- ☐ lanthanide series
- ☐ malleability
- ☐ melting point
- ☐ metalloids
- ☐ metals
- ☐ noble gas
- ☐ nonmetals
- ☐ Pauli exclusion principle
- ☐ period
- ☐ periodic table of the elements
- ☐ periodic trend
- ☐ transition metals

ENGAGE

Have you ever heard someone refer to the periodic table as a road map? The periodic table is the perfect chemical road map. Why? Visit the Engage page online to learn more.

Explain Question

How can general properties influence element location on a periodic table?

EXPLORE

How Do Scientists Distinguish between Metals, Metalloids, and Nonmetals?

The Periodic Table

The periodic table includes information about all of the elements that have been discovered. Elements with similar chemical and physical properties are placed together within groups or families (vertical columns). There are also patterns in the way elements appear within periods (horizontal rows). The period (or row) number indicates the number of electron energy levels in an atom of that element.

photo: Getty Images

The properties of elements played a large role in the development of the periodic table. What elements are you familiar with in your everyday life?

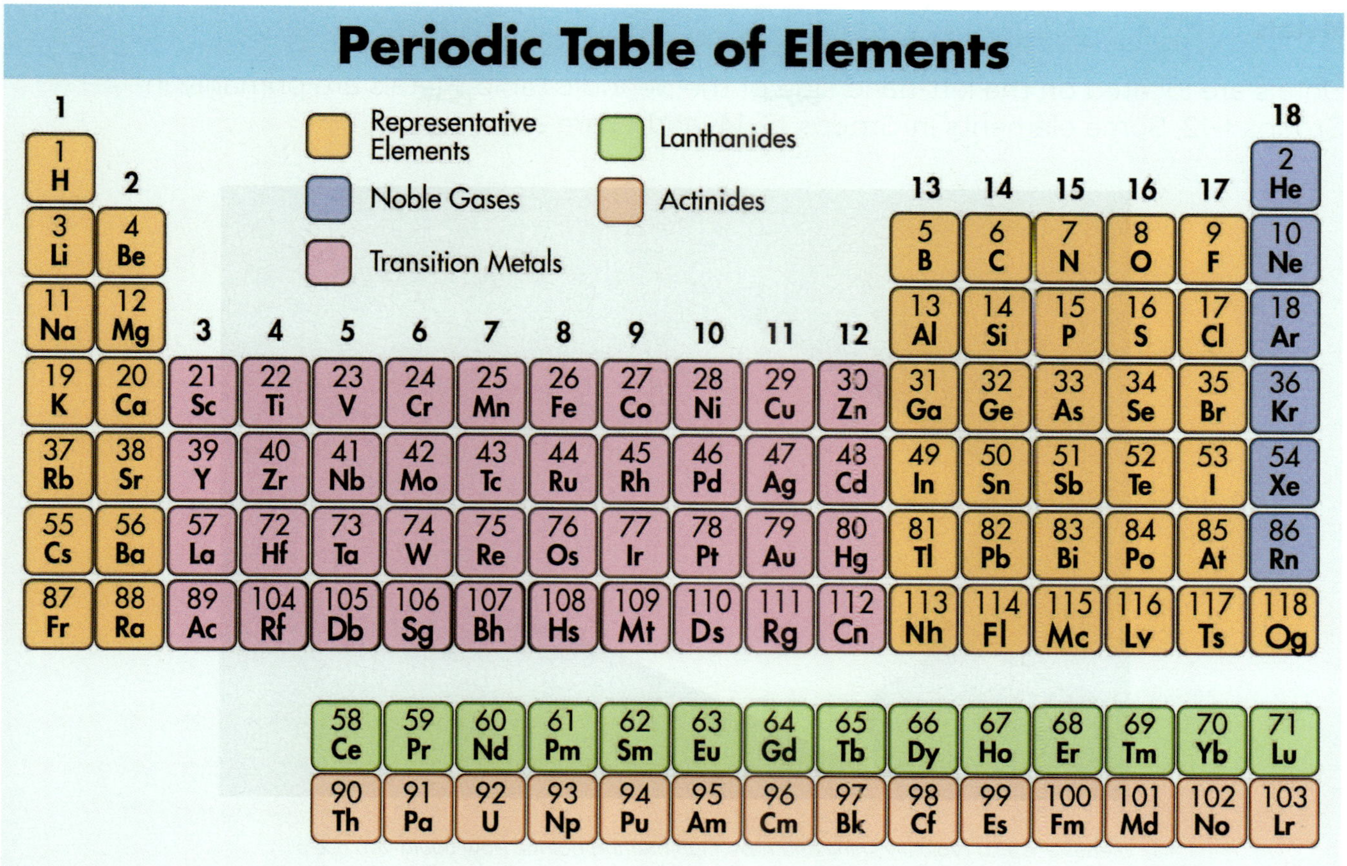

The periodic table is divided into groups of elements with similar electron configurations and properties. What patterns in the periodic table would allow you to predict the chemical properties of specific elements?

There are many versions of the periodic table that include slightly different information. The designations for the groups, for example, are presented in three main styles. The European and U.S. conventions use "A" and "B" labels for different parts of the periodic table. In the U.S. model, the "A" groups include the "main group elements" in columns 1, 2, and 13–18. These are often called the "representative elements." The "B" groups are the transition metals. The International Union of Pure and Applied Chemistry (IUPAC), a worldwide organization working to standardize chemistry conventions, simply numbers the groups 1–18.

In all versions of the table, elements are placed in the periodic table according to increasing atomic number, which is the number of protons in an atom's nucleus. The configuration of the electrons in an atom can also be predicted using the periodic table. These details about atomic structure affect the chemical and physical properties of an element.

Elements that appear in certain places on the periodic table exhibit similar properties. For example, elements can be classified as metals, nonmetals, or metalloids depending on their properties.

Metals

Metals are located on the left-hand side of the periodic table. Metals are primarily in Groups 1–12. Some elements in Groups 13, 14, and 15 are also metals.

photo: Paul Fuqua

Metals are also typically shiny and have high melting points. How could you use a metal's position on the periodic table to predict these properties compared with other elements?

Metals exhibit similar properties. They have high melting points. Melting point is the temperature at which a solid substance transforms into a liquid. Metals are usually solid at room temperature because of this property. Mercury is a highly toxic metal. It is one of only two elements that exist in the liquid state at room temperature.

Valence electrons are an important factor in determining the properties of an element. Valence electrons are those that atoms tend to gain, lose, or share during a chemical reaction. The group number corresponds with the number of valence electrons in the atoms of many elements. For representative elements, they are electrons in the outer energy level. Elements in Group 1 contain one valence electron. Elements in Group 2 contain two valence electrons. For elements in Group 13–18, the number of valence electrons is 10 less than the group number. An element in Group 15, for example, has five valence electrons. The elements in Groups 3–12 have varying numbers of valence electrons.

Valence electrons produce many of the familiar properties of metals. Most metals are quite reactive because they easily gain, lose, or share valence electrons. Metals are typically conductive, malleable, and ductile, and they have shiny surfaces. A metal is a good conductor of heat and electricity. In solid metals, the valence electrons are shared equally among positive metal ions. These electrons flow freely between the nuclei of the metal atoms. They form a "sea of electrons" and move easily through the outer shells of metallic atoms, which explains why metals are good conductors. However, the attraction between the positive metal ions and the sea of electrons is strong. This explains why most metals have high melting points.

Malleability, ductility, and luster also are physical properties exhibited by metals. Malleability is the ability of an element to be molded into another shape without breaking or cracking. Ductility is the ability of an element to be pulled into a long, thin wire. The sea of electrons can flow with metal ions when they move. As a result, the metal ions slide past each other without breaking the bonds that hold them together. Luster refers to the shine reflected from an element's surface. Physical and chemical properties are important to consider when choosing a metal for a specific project.

Nonmetals

The nonmetal elements are located on the upper-right side of the periodic table. Atoms of most nonmetals tend to gain valence electrons during chemical reactions. The noble gases are found in the far-right column. They are stable and tend not to react with other substances.

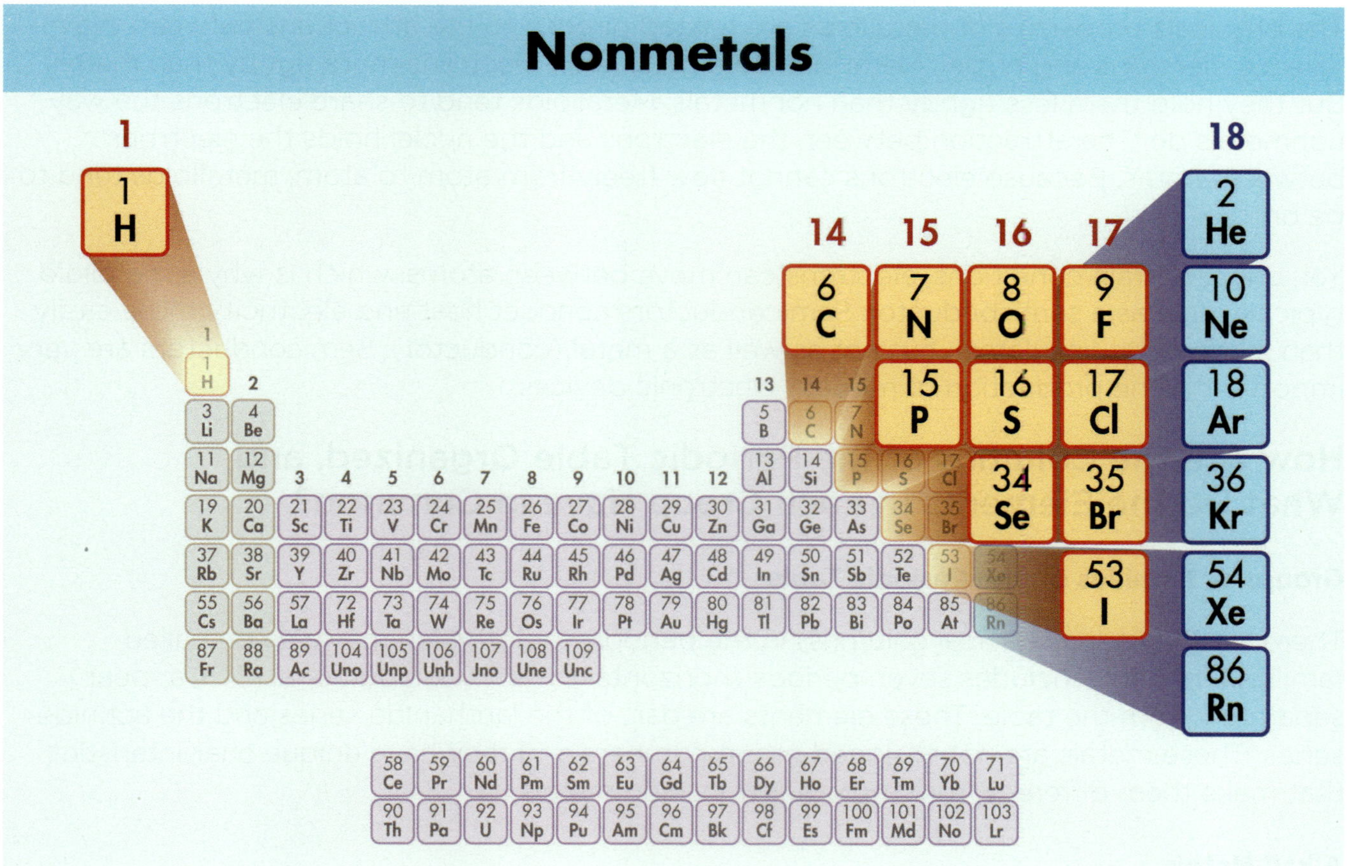

Nonmetals are stable and not very reactive. How do these features relate to the number of their valence electrons?

Nonmetals tend to have low melting points. This means that nonmetals are more likely to exist as gases at room temperature. When solid, nonmetals often exhibit properties such as brittleness, poor conductivity, and dull surfaces. The term "brittleness" is used to describe how easily something crumbles or breaks. Glass is brittle and will often break if dropped. Steel is not brittle and will not break if dropped.

The properties of nonmetals are also explained by attractions between the valence electrons and the nuclei of the atoms. Unlike metals, nonmetals hold their valence electrons tightly. Nonmetals share electrons between a few atoms to make molecules. However, the molecules do not share electrons with each other. They are held together by weaker forces, which is why nonmetals often have low melting and boiling points. Nonmetals are brittle because the electrons do not flow from atom to atom. Bonds break when the atoms move.

Metalloids

There are only seven elements that are considered metalloids. The metalloids appear in Groups 13–16. They form a jagged line between the metals and nonmetals.

The metalloids exhibit properties of both metals and nonmetals. Metalloids are solid at room temperature and have luster, like metals. However, metalloids are usually brittle and behave like nonmetals.

The physicals properties of metalloids are related directly to the attractions between the valence electrons and nuclei. Metalloids hold on to their electrons more tightly than metals. But they hold them less tightly than nonmetals. Metalloids tend to share electrons the way nonmetals do. The attraction between the electrons and the nuclei holds the electrons between atoms. Because electrons cannot flow freely from atom to atom, metalloids tend to be brittle.

Yet, under certain conditions, electrons can move between atoms, which is why a metalloid typically acts as a semiconductor. Semiconductors conduct heat and electricity more easily than a nonmetal (insulator), but not as well as a metal (conductor). Semiconductors are very important in the production of modern electronic devices.

How Are the Columns of the Periodic Table Organized, and What Do the Elements in Each Group Have in Common?

Groups or Families on the Periodic Table

There are 18 groups (vertical columns) in the periodic table, which may also be called families. The table includes seven periods (horizontal rows). Two additional rows appear separately from the table. These elements are part of the lanthanide series and the actinide series. These metals are not assigned group numbers and they have unique characteristics that make them different from the other transition metals.

Alkali Metals

The alkali metals make up Group 1 of the periodic table.

The atoms of these elements each contain a single valence electron, which makes them quite reactive. They tend to lose this electron during chemical reactions. Alkali metals easily react with Group 17 elements, such as chlorine or bromine. The element sodium (Na), for example, is an alkali metal. It combines with chlorine (Cl) to form sodium chloride (NaCl). Each sodium atom gives its valence electron to a chlorine atom during the reaction.

Alkali metals also quickly combine with oxygen when exposed to air. They react vigorously when placed in contact with water and form strong bases such as sodium hydroxide (NaOH). In fact, the term "alkali," for which this group of elements was named, means "basic."

Alkaline Earth Metals

Alkaline earth metals are found in Group 2 of the periodic table. The atoms of these elements contain two valence electrons, which makes them fairly reactive. These elements tend to lose electrons during chemical reactions.

The behavior of alkaline earth metals is fairly predictable, though the properties of beryllium and magnesium vary a bit compared with the other metals in the group. These metals also tend to form weak bases when placed in water, and the salts that contain alkaline earth metals tend to display a distinct color when placed in a flame.

When exposed to oxygen, the alkaline earth metals tend to form oxides with high melting points. (An oxide is a compound made of oxygen and another element.) They often exist in very hot environments such as volcanoes.

The Halogens

The halogen group, Group 17, includes five nonmetallic elements. This group is located on the right-hand side of the periodic table. Because the atoms of these elements contain seven valence electrons, they are very reactive. The atoms of these elements usually need to gain one valence electron to make them stable. Because elements from Group 1 contain one valence electron, halogens easily bond with these metals.

The word "halogen" means "salt-former." This refers to the fact that when halogens react with metals, salts are produced. Sodium chloride (NaCl) is a common example of this, as shown by its following chemical equation:

$$\mathrm{Na^{+} + Cl^{-} \rightarrow NaCl}$$

The halogen group includes elements that exist in different forms at standard temperature and pressure. Astatine and iodine most commonly occur in solid form, while bromine is usually a liquid, and chlorine and fluorine are usually gases. This is the only group that includes elements commonly found in each of three states of matter.

The Noble Gases

The six noble gases in Group 18 are in the far-right column of the periodic table. These elements are generally unreactive because they each contain a stable set of valence electrons. The valence shell of a helium atom holds two electrons. The valence shells of the other noble gases hold eight.

Because the noble gases tend not to bond with other elements, they typically appear as monatomic elements (as a single atom). As a result, noble gases have low boiling points and usually exist in the gas state. However, the noble gases make up only a small percentage of Earth's atmosphere.

photo: Pixabay

Hydrogen was first used in dirigibles and balloons, before helium was used. What properties of helium make it a suitable alternative to hydrogen?

The Transition Metals

There are many elements in Groups 3 through 12, found in the middle block of the periodic table. These are referred to as the transition metals. Like other metals, the transition elements are ductile, malleable, and conductive. Some transition metals (for example, iron, cobalt, and nickel) are able to produce a magnetic field.

The atomic structure of the transition metals allows not only electrons from the valence shell but also some from the next inner shell to participate in bonding. This makes it more difficult to predict an element's properties.

The Lanthanide Series

Two rows of elements appear in a separate section below the main body of the periodic table. This is done only to keep the table from being too wide. These elements are sometimes referred to as the inner transition metals because the atomic numbers for these elements fall within the transition metals.

The lanthanide series is part of Period 6. It contains 15 elements with atomic numbers from that of lanthanum (La, 57) to lutetium (Lu, 71). These metals have properties similar to those of other metals.

One of the lanthanide metals, Promethium (Pm), was synthesized in a laboratory. The rest commonly occur in nature, usually mixed together. This series is therefore sometimes called the "rare earth metals." The name likely came about because the properties of these metals are so similar, and it is difficult to isolate the inner transition metals in their pure elemental forms.

The Actinide Series

The actinide series is part of Period 7 on the periodic table. It contains 15 elements with atomic numbers from that of actinium (Ac, 89) to lawrencium (Lr, 103). Only three of these metal elements occur naturally. The others must be synthesized in a laboratory, and most are quite unstable. All of the actinide series metals are radioactive.

Uranium and thorium are widely used in nuclear reactors because of their radioactive properties. Americium-241 is a common radioactive isotope that is used in smoke detectors and in medical diagnostic equipment.

Consider the Explain Question:

How can general properties influence element location on a periodic table?

Go online to complete the scientific explanation.

Check for Understanding:

How are groups or families organized on the periodic table of the elements?

STEM in Action

STEM and Structure of the Periodic Table

Can we take what we know about an element, based on its location on a periodic table, and use it for our technological benefit? This is what manufacturers think about as they design many of the items we use in products such as electronics. Let's take semiconductors for example.

Semiconductors are chemical compounds used in electronics. These compounds are special because they conduct electricity only when needed. One type of semiconductor, silicon, is very good at gaining or losing up to four electrons. In fact, most modern technology is highly dependent on purified silicon, including the production of computer chips, computer circuits, and electronics. One property that contributes to the use of silicon for a variety of purposes is its abundance in Earth. Silicon is extremely common, and compounds containing a high proportion of silicon (such as sand) make up more than 90 percent of the planet's crust. Silicon is similar to boron (another metalloid) and carbon (a nonmetal) in its chemical and physical properties. Now that you are an expert about all things silicon, where is this element located on the periodic table?

How would life be without the semiconductor? There is a large positive impact of semiconductors on our everyday living. Computer engineers use computer science and electronic engineering to develop a wide range of computer systems or technological devices. Their research work on semiconductors has paved the way for a lot of the modern technology we see today. Years ago, portable radios were the hot item. Now, music through a portable headphone is the way to go. Back then, there was only a desktop computer available for use. Now, there are tablets, smartphones, and smart watches that function similarly to the desktop computer. As leaders in pioneering the future of computer hardware, such as semiconductors, computer engineers make sure we stay on the cusp of modern technology.

1. On a separate sheet of paper, draw the atomic structure of sodium, making sure to arrange the electrons properly. Based on your model drawing, how many electrons and electron shells surround the nucleus of the sodium atom, and what can you say about the location of sodium on the periodic table?

2. Why do you suppose there is a difference in the relationship between sodium and lithium when compared with sodium and magnesium? Use the periodic table as a reference if needed.

Periodic Trends

photo: Getty Images

LESSON OVERVIEW

Lesson Questions

- What is the relationship between the observed chemical reactivity of an element and its position on the periodic table?
- How can the observed periodic trends in reactivity of the elements be explained?
- What other periodic trends besides reactivity occur, and how are they explained?

Lesson Objectives

By the end of the lesson, you should be able to:

- Relate an element's chemical reactivity to its position on the periodic table.
- Explain the basis for the observed periodic trends in reactivity of the elements.
- Describe and explain the basis for trends in atomic radius, ionic size, metallic character, ionization energy, and electronegativity of the elements across groups and periods within the periodic table.

Key Vocabulary

Which terms do you already know?

- ☐ atomic radius
- ☐ chemical reactivity
- ☐ electron
- ☐ electron configuration
- ☐ electro-negativity
- ☐ energy levels
- ☐ families
- ☐ groups
- ☐ ion
- ☐ ionic bond
- ☐ ionic radius
- ☐ ionic size
- ☐ ionization energy
- ☐ metallic character
- ☐ nucleus (atom)
- ☐ orbital model of the atom
- ☐ Pauli exclusion principle
- ☐ periodic table of the elements
- ☐ periodic trend
- ☐ shielding effect
- ☐ shielding effect
- ☐ valence electron

ENGAGE

Have you ever noticed that some of your family members share the same characteristics? How is this similar to what scientists have observed about atoms in families? Visit the Engage page online to learn more.

Explain Question

How do trends in the periodic table help predict the properties of an element?

EXPLORE

What Is the Relationship between the Observed Chemical Reactivity of an Element and Its Position on the Periodic Table?

Reactivity of Metals: The Alkali Metals

The periodic table that we use today was developed after a great deal of experimentation and observation. A fundamental observation was that elements in the same group had similar physical and chemical properties.

A good example is found in Group 1. The elements in Group 1 are known as the alkali metals and include lithium, sodium, potassium, rubidium, cesium, and francium. These metals have similar physical properties such as low melting points and low densities. These properties make the alkali metals very soft and easy to cut with a knife. The alkali metals are also highly reactive when placed in water.

Not all alkali metals have the same chemical reactivity, however. Sodium reacts more vigorously with water than lithium. Potassium is even more reactive than sodium. Going down the group, the alkali metals become more and more reactive when placed in water. The most reactive metal is francium (Fr), in the lower-left corner of the periodic table.

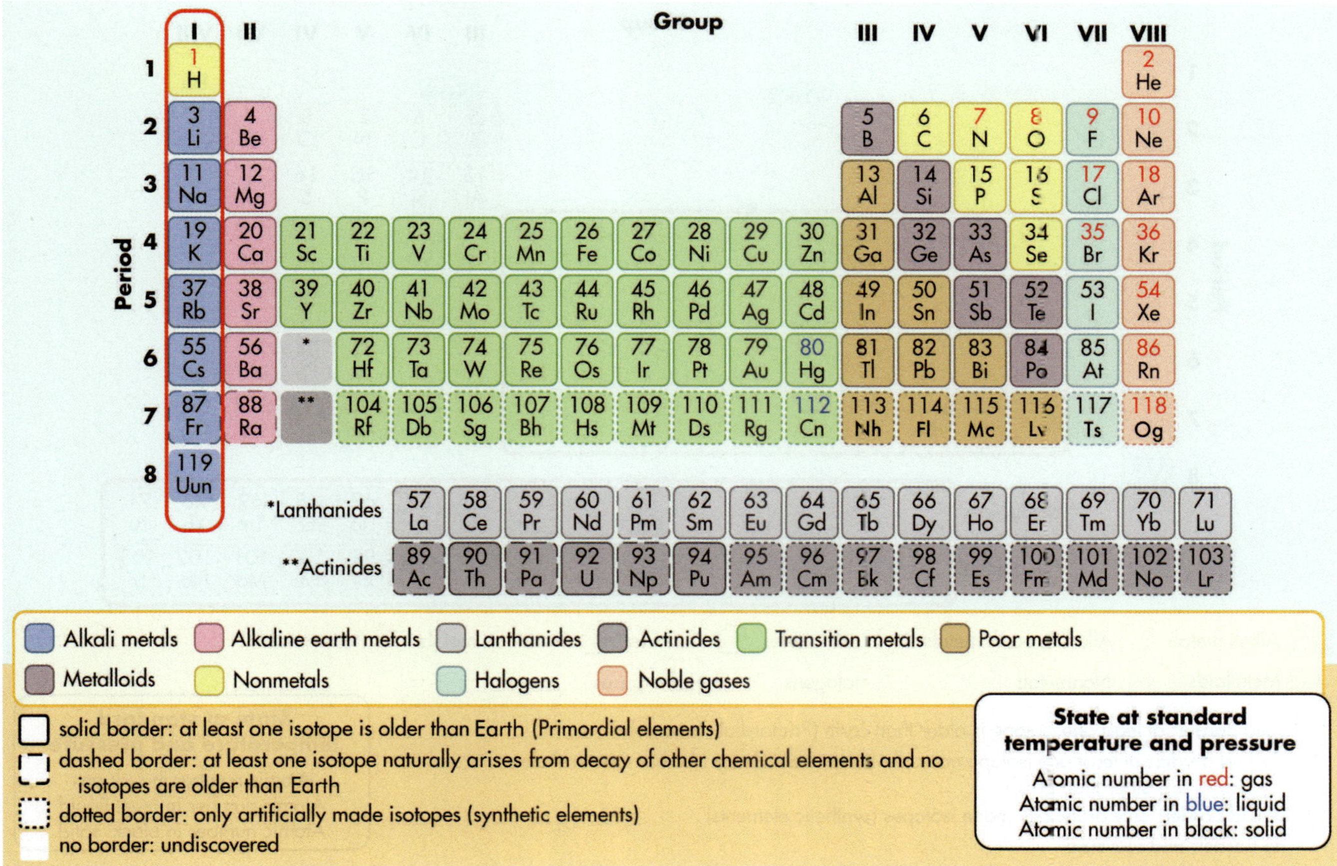

The alkali metals (shown circled in red) are the most reactive of all metals. Why do alkali metals have such high reactivity?

Reactivity of Metals: The Alkaline Earth Metals

Group 2 is known as the alkaline earth metals. These elements are nearly as reactive as the alkali metals. However, compared with the alkali metals, these elements are harder and have higher melting points.

We see a similar pattern of reactivity as we saw with the alkali metals as we go down within Group 2. The farther down within the group, the more reactive the alkaline earth metal.

Reactivity of Metals: The Transition Metals

Moving further to the right on the periodic table, we encounter the transition metals, which include the lanthanide and actinide metals at the very bottom.

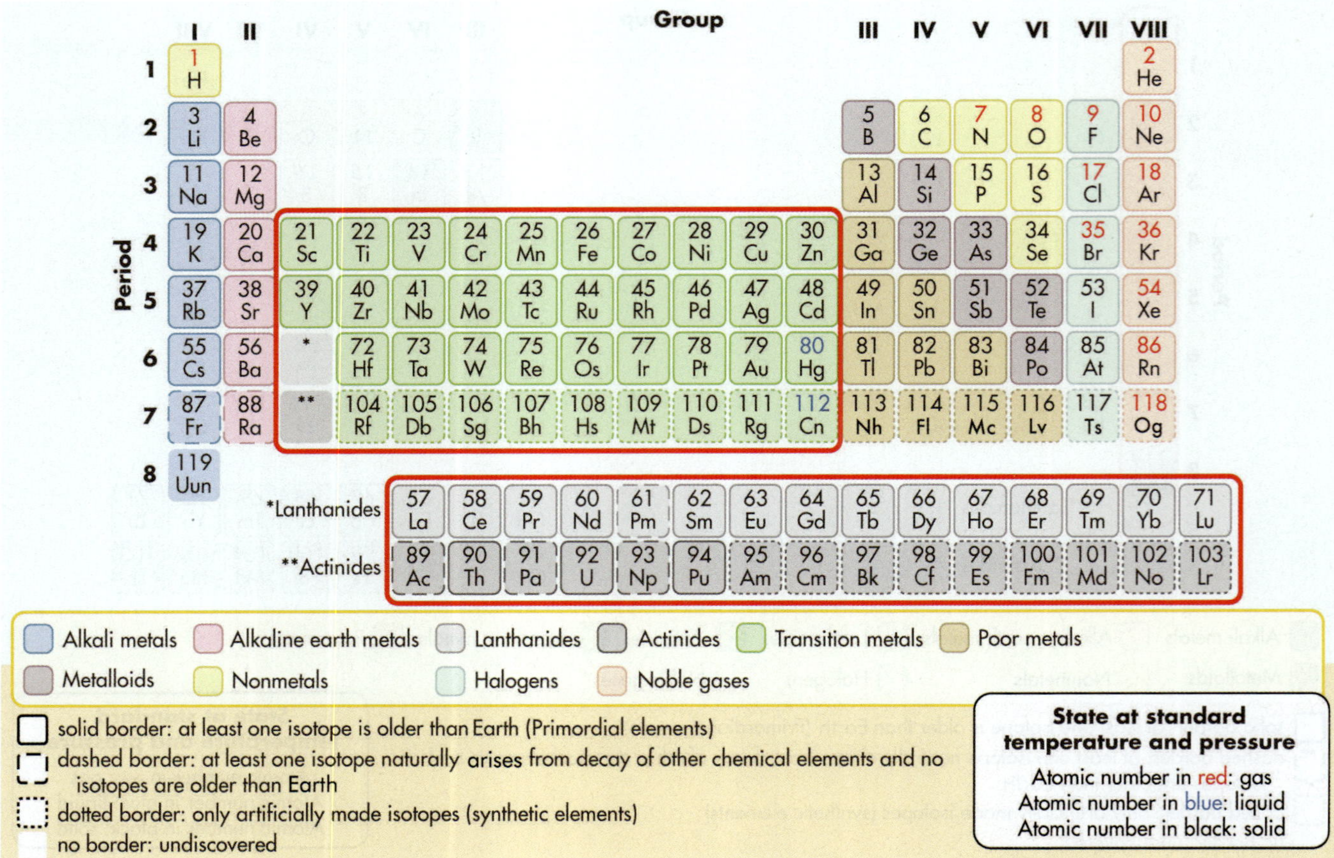

The transition metals, including the lanthanides and actinides, are much less reactive than the alkaline earth metals and the alkali metals. What other properties do they possess?

Transition metals tend to be very hard and have high densities. They are also much less reactive than either the alkaline earth metals or alkali metals. This low reactivity makes them suitable for making products that must resist change over long time periods.

photo: Getty Images

Gold metal is not very reactive. What is another characteristic of gold that makes it so desirable for jewelry?

Reactivity of Metals: Periodic Trends

In summary, the reactivity of metals tends to decrease from left to right across the periodic table. The reactivity of metals also tends to increase as you move down within a group. The most reactive metal is francium (Fr) in the lower-left corner of the periodic table.

Summary of Metal Reactivity with Water, Hydrochloric Acid, and Air

Metal	Reaction with Water	Reaction with Dilute Hydrochloric Acid	Reaction with Air
Potassium	Reacts violently with cold water; hydrogen ignites	Very violent dangerous reaction producing hydrogen gas that ignites	Rapidly forms a white oxide; burns violently with a lilac flame
Sodium	Reacts violently with cold water; hydrogen may ignite	Very violent dangerous reaction producing hydrogen gas that ignites	Rapidly forms a white oxide; burns violently with a yellow flame
Calcium	Reacts slowly producing bubbles of hydrogen gas	Reacts vigorously producing hydrogen gas	Slowly forms a white oxide; burns with a red flame
Magnesium	Reacts slowly producing bubbles of hydrogen gas; burns in steam	Reacts fairly vigorously producing hydrogen gas	Burns rapidly with a bright white flame
Aluminum	No reaction*	Reacts slowly*	No reaction*
Zinc	Reacts slowly with steam	Reacts fairly vigorously producing hydrogen gas	Zinc wool burns with a blue green flame; forms an oxide very slowly in air
Iron	Reacts slowly with steam	Reacts slowly	Forms a reddish oxide but iron wool and iron filings burn quickly with sparks; forms an oxide in the presence of air and water
Lead	No reaction	No reaction	Slowly forms an oxide film; oxidizes readily on heating
Copper	No reaction	No reaction	Slowly forms an oxide film; oxidizes slowly on heating
Silver	No reaction	No reaction	No reaction

**Aluminum is protected by a thin film of aluminum oxide. If the film is removed, it behaves in a manner intermediate between magnesium and zinc.*

Reactivity of Nonmetals: Groups 14 and 15

How do the nonmetals vary in reactivity? Do they follow the same periodic trend as the metals? Again, we can turn to observations of the elements to find answers to these questions.

Carbon is representative of Group 14. This element is not highly reactive, although in some forms it can easily form bonds with other elements. Two familiar forms of carbon are graphite, used to make pencils, and diamond, a crystalline form of carbon. Neither of these carbon forms is very reactive.

Nitrogen is in Group 15. In its elemental form, nitrogen atoms bond to one another to form diatomic nitrogen, N_2. This gas is found in abundance in Earth's atmosphere and tends to be nearly inert, or nonreactive.

photo: Getty Images

Graphite is one of the forms of carbon. What other forms does carbon take?

Reactivity of Nonmetals: Groups 16 and 17

As we move to the right in the periodic table, we find that reactivity of the nonmetals increases. Whereas carbon in Group 14 and nitrogen in Group 15 are relatively nonreactive, oxygen in Group 16 is more reactive.

Moving still further to the right are the halogens in Group 17. These elements are the most reactive nonmetals.

Reactivity of Nonmetals: Noble Gases

The least reactive of all the elements are the noble gases that make up Group 18. These elements exist as gases and are colorless, odorless, and inert.

Reactivity of Nonmetals

If we exclude the noble gases, which are the most inert of all the elements, the reactivity of nonmetals tends to decrease from right to left across the periodic table. The reactivity of nonmetals also tends to increase as you move up within a group. These trends are opposite to those observed for the metals. The element fluorine (F) at the top of Group 17 is the most reactive of all the elements.

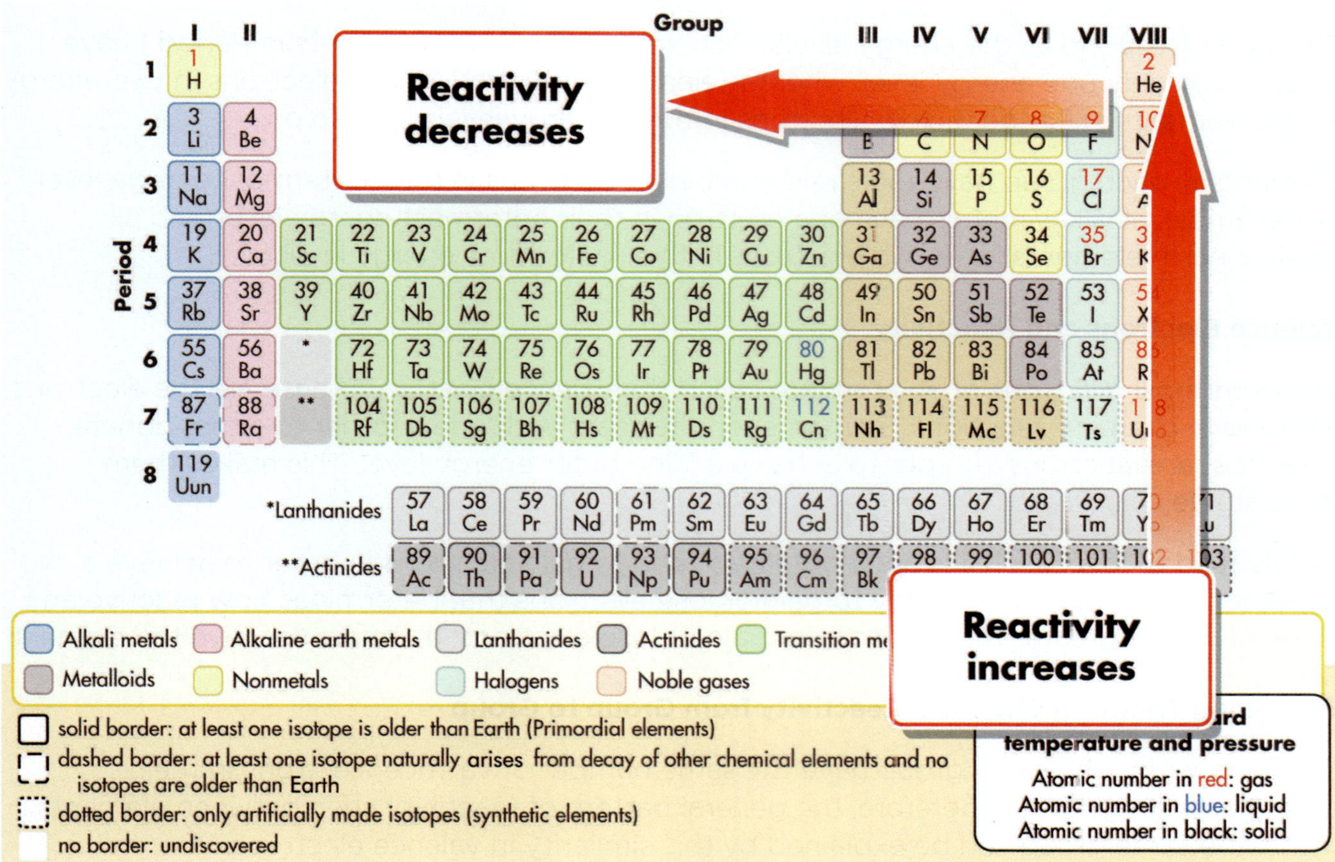

Excluding the noble gases, the reactivity of nonmetals tends to increase moving up a group and decrease from right to left across a period. Why is the periodic table organized in this manner?

Periodic Trends in Reactivity

In summary, we can say that metals and nonmetals both show a periodic trend for reactivity, although these trends are opposing. Metal reactivity increases down within a group and decreases from left to right across a period. Reactivity for nonmetals other than noble gases varies in the opposite way. It increases as you move up within a group and decreases from right to left across a period.

How Can the Observed Periodic Trends in Reactivity of the Elements Be Explained?

Electron Configurations and the Periodic Table

The periodic table was originally developed as a way to organize elements according to their properties. As chemists learned more about atomic structure, they began to understand that the arrangement of the periodic table also reflected the arrangement of electrons in atoms.

Periods correspond to the energy levels occupied by electrons. Elements in Period 1 have electrons in only one energy level, whereas elements in Period 2 have electrons in two energy levels, elements in Period 3 have electrons in three energy levels, and so on.

Elements in each group have the same number of electrons in their outermost energy level. For example, all alkali metals have one electron in their outermost energy level, and all alkaline earth elements have two electrons in their outermost energy level.

Valence Electrons and Reactivity

The outermost electrons of an atom are called the valence electrons. These are the electrons involved in chemical reactions. Elements tend to react with one another to either donate, take, or share electrons in order to achieve a filled outer energy level. This makes them more stable.

Metals tend to lose electrons and nonmetals tend to gain electrons in order to achieve a full outer level. It is this tendency to gain or lose electrons that determines how reactive an element is. The greater the tendency to gain or lose electrons, the more reactive the element.

Explaining Trends in Chemical Reactivity from Group to Group

Elements within the same group have the same number of valence electrons in their outermost energy level. Therefore, the general pattern of reactivity seen between elements within the same group can be explained by this similarity in valence electrons.

For example, all alkali metals have one electron in their outermost energy level. We can explain their highly reactive nature from this feature. Because they have only one s orbital electron to lose to achieve a filled outer energy level, these elements lose this electron readily. This explains their high reactivity.

In the case of transition metals, the tendency to react is much less. Most of these elements have filled s and p orbitals. Only their d orbitals are partially filled. It is energetically less favorable for atoms with partially filled d orbitals to react. As a result, these metals are comparatively stable.

With respect to nonmetals, the most reactive elements are the halogens, which gain one electron to form a filled outer energy level. Nonmetals in Group 15 must gain three electrons, and those in Group 16 must gain two electrons, which is less likely to happen. That makes these atoms less reactive.

The pattern that emerges from these observations is that, for many atoms, the closer the atom is to being able to achieve eight valence electrons in its s and p orbitals, the more reactive it is.

Of course, at the very far extreme are the noble gases. These elements have filled s and p orbitals. Their inert character results from this feature because they have a tendency to neither lose nor gain electrons. As a result, noble gases are very nonreactive.

Explaining Trends in Chemical Reactivity from Period to Period

We saw that reactivity increases as we go down in a group of metals and decreases as we go down in a group of nonmetals. Why is that?

The answer has to do with the ease of gain or loss of valence electrons by an element. We can take the alkali metal group as an example. As you move down this group, the number of energy levels increases. As these energy levels add layers to the structure of the atom, the valence electrons are located farther away from the positively charged nucleus. More electrons are present in the inner energy levels. Those inner electrons shield the outer electrons from the positive nuclear charge. With so much shielding, the valence electrons have less pull from the nucleus and can move away with greater ease. Because the alkali metals lose electrons to achieve stability, greater shielding results in greater reactivity.

If we examine the halogens as an example of the nonmetals, we observe something different. As we move down the group, the number of energy levels increases just as with the metals. This again increases the shielding of the outermost energy levels from the nuclear charge. The nonmetals, however, gain electrons to achieve stability instead of losing them, and the increase in shielding decreases reactivity in the lower periods of a group because the pull of the nucleus is limited.

What Other Periodic Trends Besides Reactivity Occur, and How Are They Explained?

Atomic Radius: Observed Periodic Trends

One measure of atomic size is the distance from the center of the nucleus to the outermost electrons. This distance is defined as the atomic radius. The size of the radius is hard to describe because the electron cloud does not have a well-defined surface. A common way to describe it is half the distance between the nuclei of neighboring atoms. These might be identical atoms that are bonded together or atoms next to each other in a crystal. The atomic radius does not vary much for transition elements but it does for main group (representative) elements. The chart "Atomic Radii in Picometers" shows the changes.

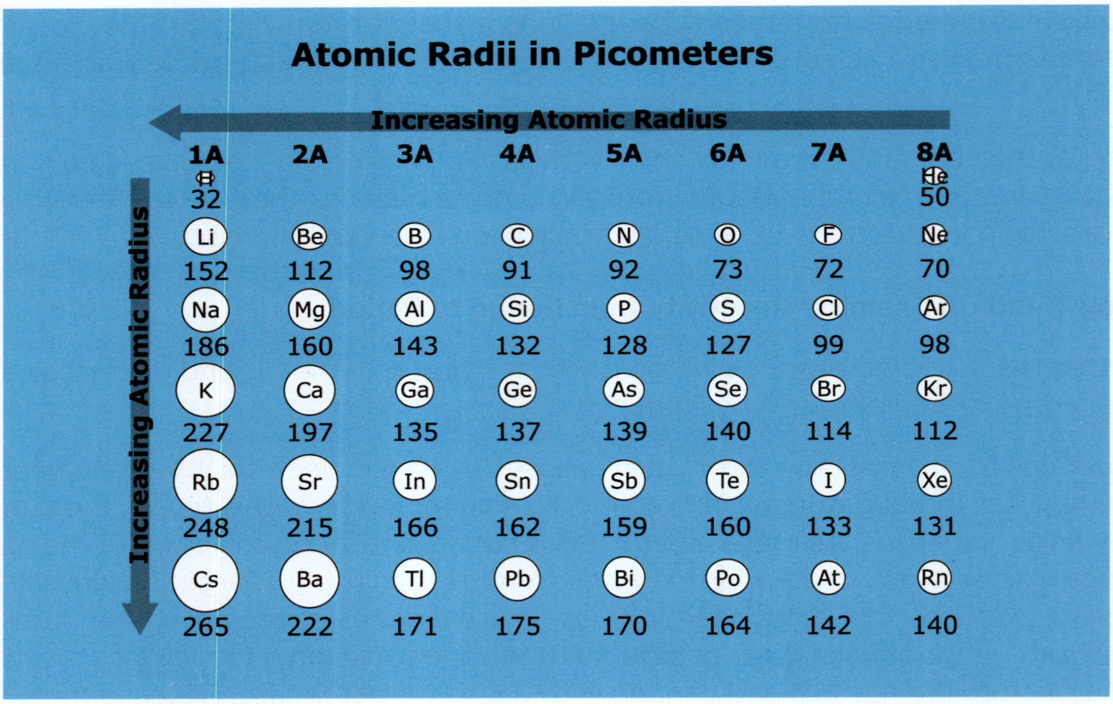

For main group elements, atomic radius increases as you move down within a group but decreases as you move from left to right across a period. What does this pattern reveal about the properties of the elements?

For main group elements, the atomic radius increases as you go down a group. This increase results because a new energy level is added as you move from one period down to the next. The outermost electrons occupy energy levels that are farther and farther from the nucleus with each shift to a new period.

Look at the "Atomic Radii" diagram again to see what happens to the atomic radius of main group element atoms as you move from left to right within the same period. With each move to the right, one more proton and one more electron are added. You might think that the addition of these particles would result in an atom of larger size. However, the diagram shows that for most main group elements, atomic radius decreases from left to right across a period. Something else besides a physical increase in numbers of subatomic particles affects the radius. What might explain this observation?

Atomic Size: The Effect of Nuclear Charge

The decrease in atomic size across a period can be explained by the effect of increasing nuclear charge. As you move from left to right within a period, the numbers of protons and electrons increase. This increases the amount of positive charge in the nucleus. Because the electrons are being added to one energy level, they occupy the same region. As a result, they are held more tightly by the increasing positive charge of the nucleus.

Factors Underlying Periodic Trends

The factors that explain the observed trends in chemical reactivity and atomic radius apply to other properties of the elements. Along with chemical reactivity and atomic radius, the important factors are:

- total number of electrons in an atom
- the size of the nuclear charge
- the distance between the nucleus and the outermost valence electrons
- amount of shielding between the positively charged nucleus and negatively charged electrons in the outermost energy level

All other periodic trends can be explained using these four factors.

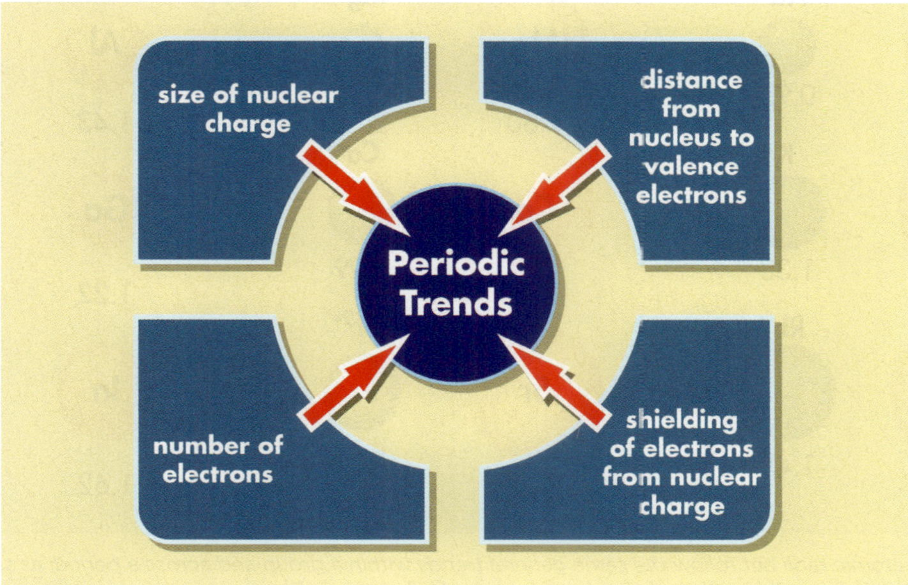

Four main factors contribute to the trends exhibited in the periodic table. How do these factors impact the organization of elements?

Periodic Trends in Ionic Radius

An ion is a charged particle that results when an atom loses or gains electrons. The charge results from an imbalance in the numbers of protons and electrons present.

Ionic radius is the distance from the center of the nucleus to the outermost electrons of an ion. The ionic radius, like the atomic radius, is hard to describe because the electron cloud does not have a well-defined surface.

Metal atoms become ions by losing electrons. In these ions, the net charge is positive because protons outnumber electrons. These ions are smaller than their neutral atoms because there is more pull by the nucleus on the electrons in the outer energy level.

Nonmetal atoms become ions by gaining electrons. The net charge is negative because electrons outnumber protons. These ions are larger than their neutral atoms because the pull of the nucleus on the electrons is weaker. Among metals and among nonmetals, ionic radii tend to decrease as you move left to right across a period. The nonmetal ions, however, tend to be larger than the metal ions.

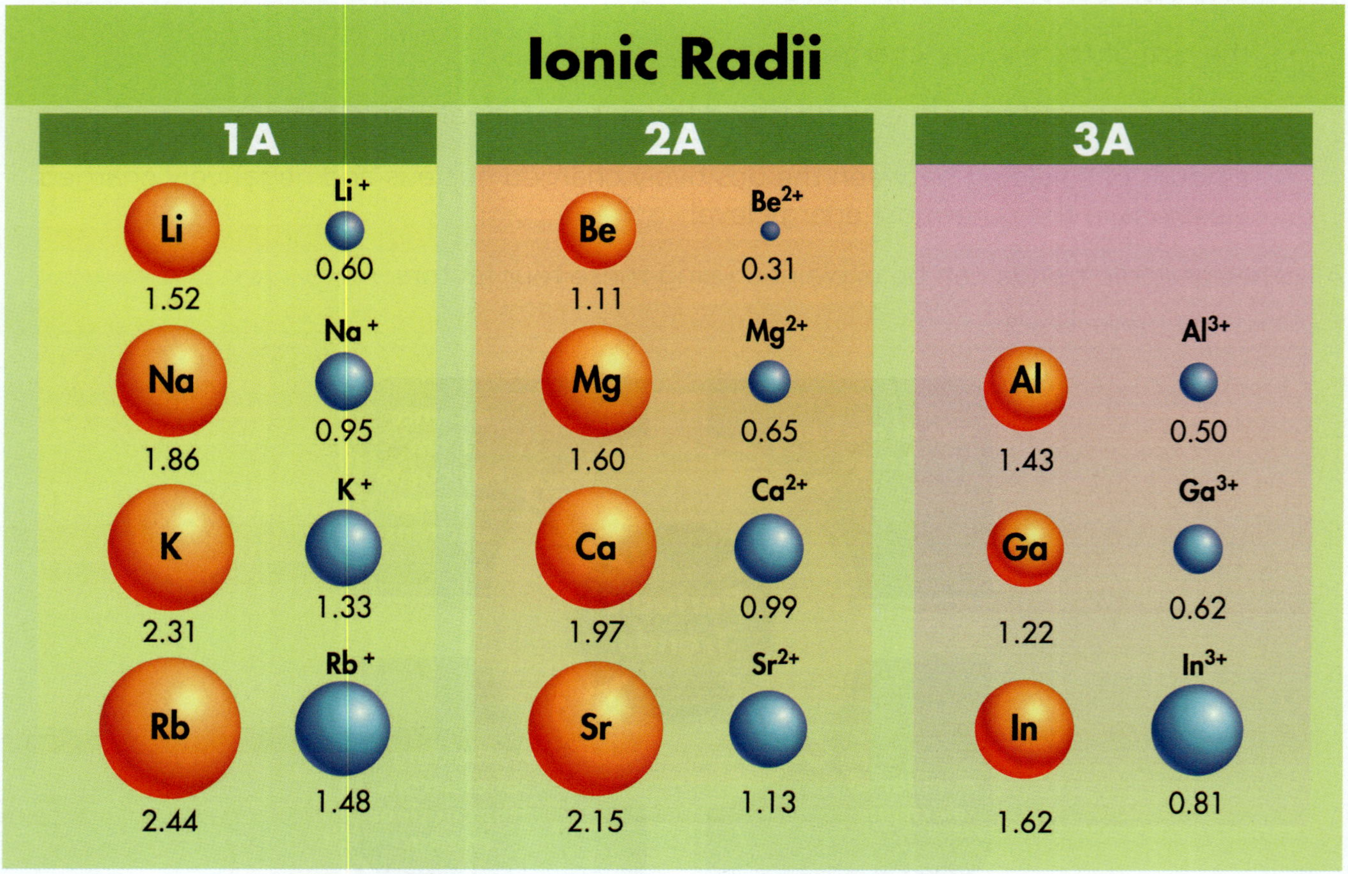

Ionic radii differ from atomic radii but follow the same general trends within a group and across a period, as shown by this chart for metals. What are the ionic radii for other elemental categories likely to reveal?

Periodic Trends in Metallic Character

Metallic character is the measure of how easily an atom loses an electron.

The diagram summarizes the periodic trends in this property. Metallic character decreases across a period, and increases down a group. The four factors noted earlier can explain these trends.

The addition of new energy levels as you move down a group has two effects. The distance between the nucleus and the outermost electrons increases, and greater numbers of energy levels occupy this distance. The shielding effect of the inner electrons and the increasing distance reduce the effective nuclear charge experienced by the outermost electrons. As a result, less energy is needed to remove one outer electron, and so metallic character increases.

Within a period, however, only the nuclear charge factor is at work. The number of protons increases as you move from left to right, but no new energy levels are introduced. Therefore, as you move from left to right, the outermost electrons experience a greater and greater positive charge from the nucleus, and it becomes more difficult for an electron to be removed. Metallic character decreases as a result.

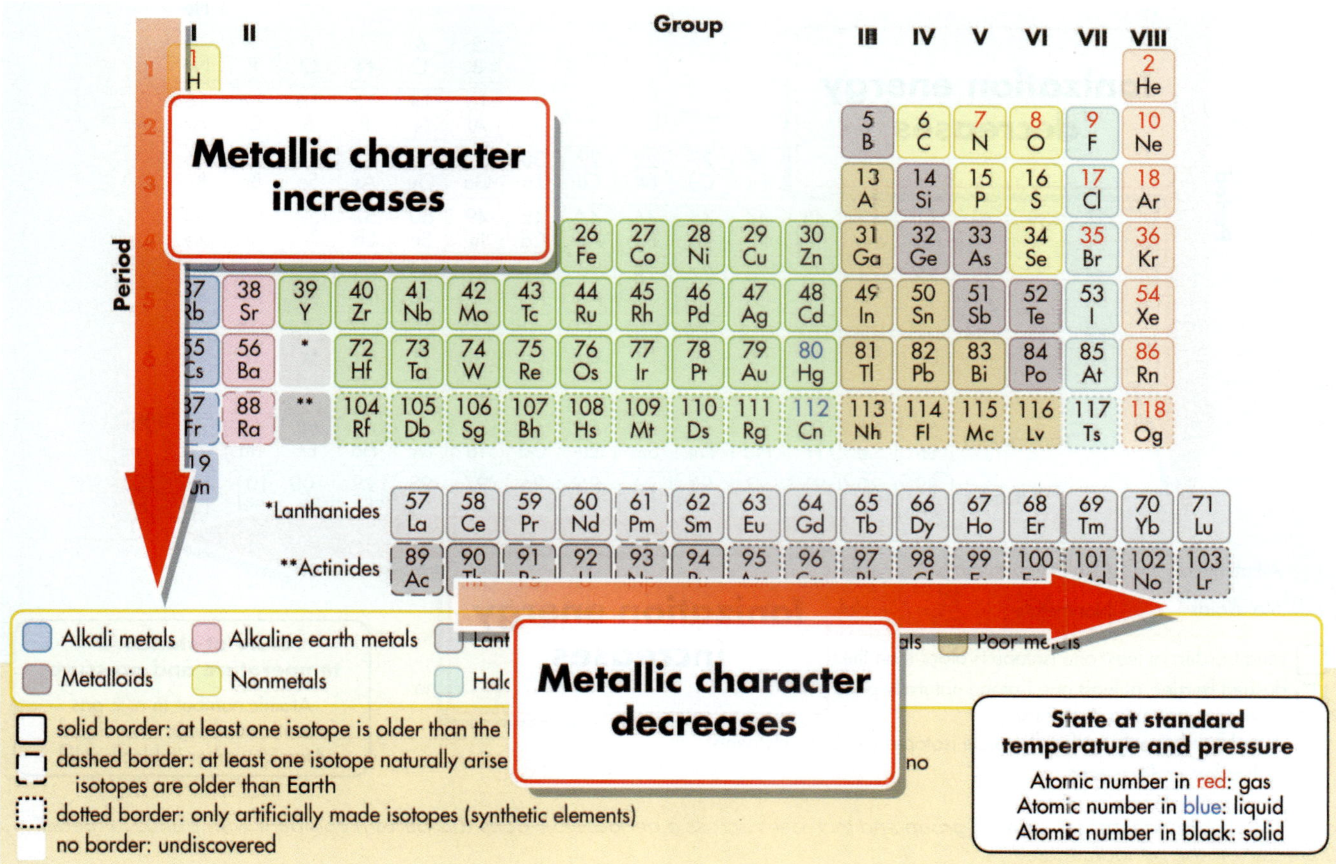

Metallic character increases down a group and decreases across a period. How do variations in the energy levels of the elements explain these changes?

Periodic Trends in Ionization Energy

Ionization energy is the energy needed to remove one electron from the outermost energy level of a gaseous neutral atom to form an ion having +1 charge.

$$Na(g) \rightarrow Na^{+}(g) + e^{-}$$

The energy required to remove the outermost electron from the gaseous +1 ion is called the second ionization energy. It is larger than the first ionization energy and the third ionization energy is higher than the second. This is because fewer electrons remain to shield the nucleus from the electrons and its pull is stronger on the remaining electrons. Thus, each successive electron removed from the atom feels an increasingly stronger effective nuclear charge.

As shown in the Ionization Energy Chart, ionization energy decreases as you go down within a group and increases as you go from left to right across a period. Ionization energy behaves opposite to metallic character.

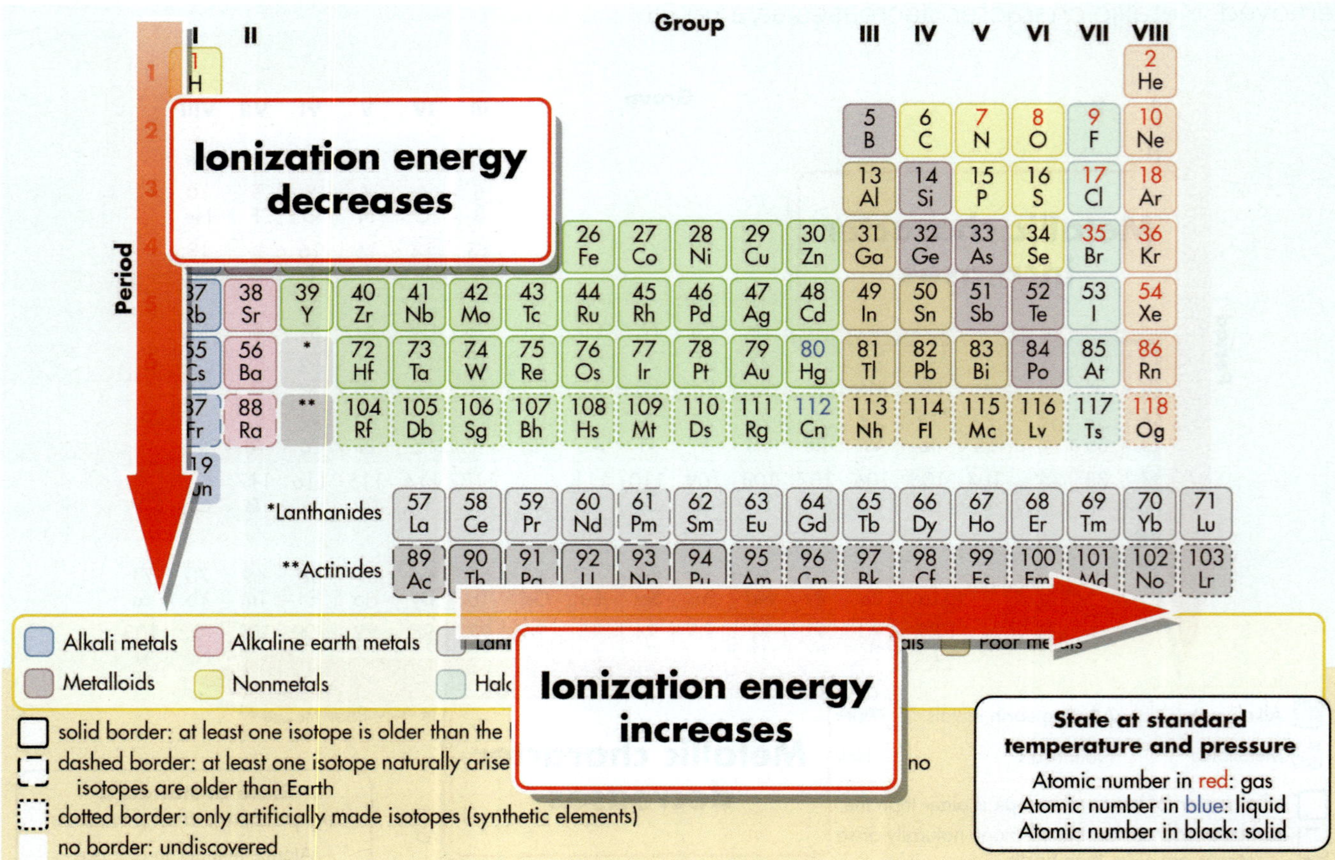

Ionization energy decreases down a group and increases across a period. How does this pattern compare with trends in metallic character on the periodic table?

The addition of new energy levels as you move down a group has two effects. The distance between the nucleus and the outermost electrons increases. Greater numbers of energy levels occupy this distance between them as well. The shielding effect of the inner electrons and the increasing distance reduce the effective nuclear charge experienced by the outermost electrons. Less energy is required to remove an outer electron as a result. This increases metallic character, but decreases ionization energy.

Within a period, only the nuclear charge factor affects metallic character. The number of protons increases as you move from left to right, but no new energy levels are introduced. Therefore, as you move from left to right, the outermost electrons experience a greater positive charge from the nucleus. More energy is required to overcome this pull. Therefore, metallic character decreases, but ionization energy increases.

Periodic Trends in Electronegativity

Electronegativity is defined as the tendency of an atom to attract electrons.

The Electronegativity Chart shows the periodic trends observed for this property. Electronegativity decreases as you go down within a group and increases from left to right across a period. Once again, the factors discussed earlier become important when explaining this trend.

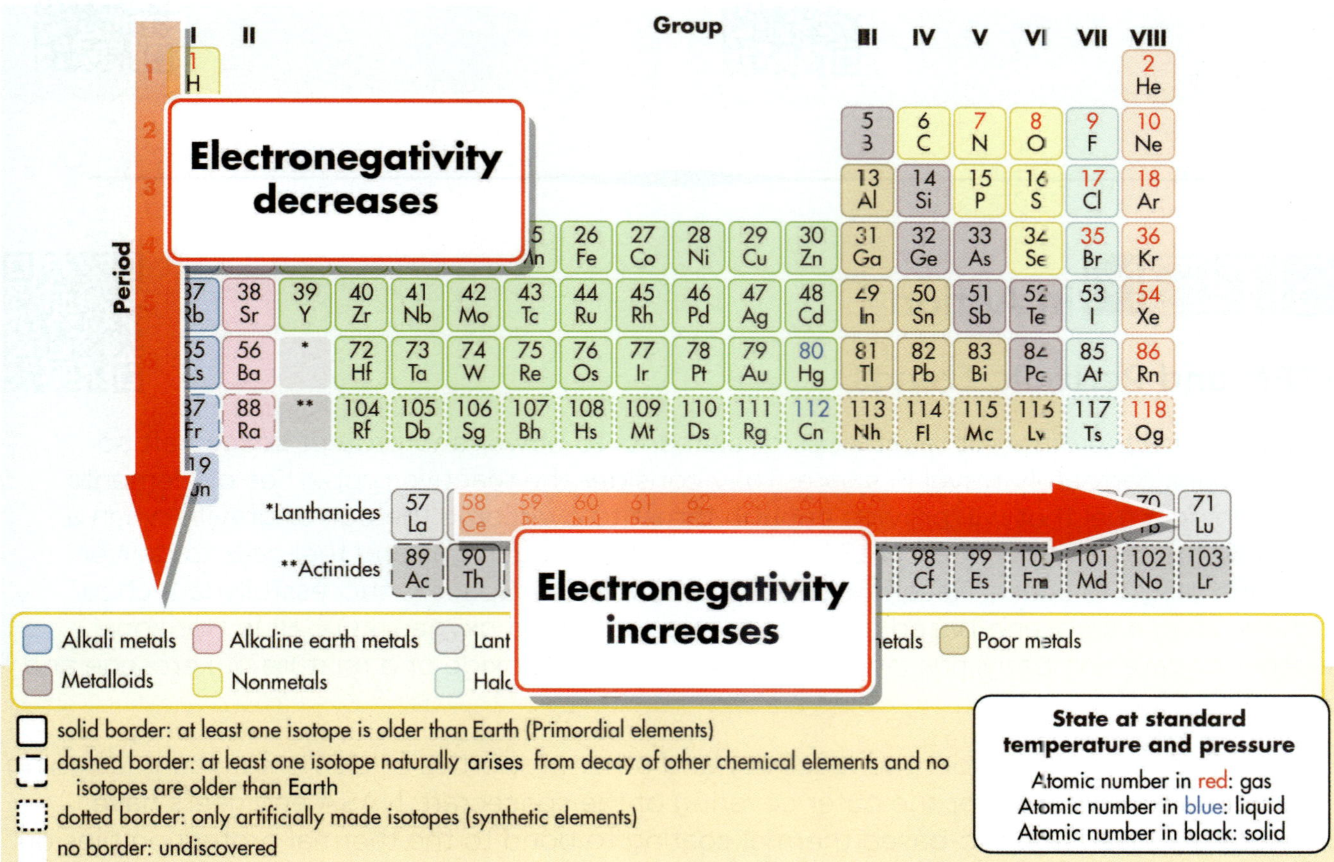

Electronegativity decreases down a group and increases across a period. How does shielding play a role in electronegativity?

As you move down a group, the distance between the nucleus and the outermost electrons increases. Also, more energy levels filled with electrons shield the outermost electrons from the nucleus. Both factors decrease the effective nuclear charge experienced by the outermost electrons. This decreases the positive pull of the nucleus on electrons, which decreases an element's electronegativity.

Moving from left to right within a period, however, the atoms gain protons and electrons, but no new energy levels form. Therefore, as you move from left to right, the outermost electrons are added to the same energy level. The added protons in the nucleus cause a greater positive charge, which pulls on the electrons. This increasing positive pull makes the elements more electronegative.

Consider the Explain Question:

How do trends in the periodic table help predict the properties of an element?

Go online to complete the scientific explanation.

Check for Understanding:

How can you use the periodic table to explain the chemical properties of elements?

STEM in Action

STEM and Periodic Trends

Engineers use the different reactivities of elements to power and protect spacecraft so that humans can safely travel in space. They consider the reactive properties of elements in designing a fuel that will provide enough thrust to overcome the pull of gravity when a rocket blasts off. For decades, NASA has relied on hydrogen gas and fuel cells to deliver crew and cargo to space. Recently, one private company, SpaceX, successfully launched rockets using a kerosene-based fuel that reacts with liquid oxygen, LOX. Blue Horizon, another private company, has launched rockets using peroxide or a mixture of kerosene and peroxide as fuel.

Engineers must also consider the reactivity and other properties of elements when designing the materials to be used for the outer covering of the spacecraft. NASA engineers have developed a silver, metallic-based thermal coating to bond to the thermal protection tiles on the outside of its Orion rocket. This will reduce heat loss as well as limit high temperatures inside the rocket.

photo: Getty Images

Engineers design the materials used for a variety of products such as spacecrafts. How do engineers incorporate periodic trends into their development process of spacecraft materials?

A titanium skeleton with a carbon fiber skin will reduce the mass of the heat shield's structure. Periodic trends such as melting point and density are some of the important factors that engineers use in designing effective and safe space vehicles.

The rare earth metal lanthanum is used in energy-efficient batteries for hybrid vehicles and electric-powered vehicles. Each hybrid car requires about 10 kilograms of lanthanum in its battery. Using the periodic table, answer the questions.

1 H																	2 He
3 Li	4 Be											5 B	6 C	7 N	8 O	9 F	10 Ne
11 Na	12 Mg											13 Al	14 Si	15 P	16 S	17 Cl	18 Ar
19 K	20 Ca	21 Sc	22 Ti	23 V	24 Cr	25 Mn	26 Fe	27 Co	28 Ni	29 Cu	30 Zn	31 Ga	32 Ge	33 As	34 Se	35 Br	36 Kr
37 Rb	38 Sr	39 Y	40 Zr	41 Nb	42 Mo	43 Tc	44 Ru	45 Rh	46 Pd	47 Ag	48 Cd	49 In	50 Sn	51 Sb	52 Te	53 I	54 Xe
55 Cs	56 Ba	57–71	72 Hf	73 Ta	74 W	75 Re	76 Os	77 Ir	78 Pt	79 Au	80 Hg	81 Tl	82 Pb	83 Bi	84 Po	85 At	86 Rn
87 Fr	88 Ra	89–103	104 Rf	105 Db	106 Sg	107 Bh	108 Hs	109 Mt	110 Ds	111 Rg	112 Cn	113 Nh	114 Fl	115 Mc	116 Lv	117 Ts	118 Og

57 La	58 Ce	59 Pr	60 Nd	61 Pm	62 Sm	63 Eu	64 Gd	65 Tb	66 Dy	67 Ho	68 Er	69 Tm	70 Yb	71 Lu
89 Ac	90 Th	91 Pa	92 U	93 Np	94 Pu	95 Am	96 Cm	97 Bk	98 Cf	99 Es	100 Fm	101 Md	102 No	103 Lr

1. Write the atomic structure of an atom of lanthanum with an atomic mass of 139 atomic mass units. Calculate the correct number of protons, neutrons, and electrons. Explain the structure of the electrons within their shells.
2. Why is this rare earth metal used to store hydrogen fuel in electric-powered vehicles?
3. How can the periodic table be used to predict the reactive nature of lanthanum?

Ion Formation

photo: Getty Images

LESSON OVERVIEW

Lesson Questions

- How do monatomic ions form?
- How do cations and anions differ?
- How does the formation of ions relate to the octet rule?

Key Vocabulary

Which terms do you already know?

- ☐ **anion**
- ☐ **cation**
- ☐ **ion**
- ☐ **ionic bond**
- ☐ **monatomic ion**
- ☐ **valence electron**

Lesson Objectives

By the end of the lesson, you should be able to:

- Describe how monatomic ions form.
- Distinguish between cations and anions.
- Explain ion formation and its relationship to the octet rule.

ENGAGE

What are electrolytes, and why do you need them to perform? Visit the Engage page online to learn more.

Explain Question

How do ions form, and how does the octet rule determine if an atom becomes a cation or an anion?

EXPLORE

How Do Monatomic Ions Form?

Monatomic Ions

Atoms are the smallest units of an element. A neutral atom does not possess a net charge because it contains an equal number of positive protons and negative electrons. For example, an atom of sodium (Na) contains 11 protons in its nucleus and 11 electrons outside its nucleus. The charges on a proton and an electron are equal in magnitude. So, the negative charges of the 11 electrons outside the nucleus balance the positive charges of the 11 protons inside the nucleus. This balance makes the atom neutral.

A neutral atom acquires a charge if its positive and negative charges become unbalanced. This happens when an atom gains or loses one or more electrons. The atom then becomes a charged particle called an ion. For example, when an atom of sodium loses one electron, it still contains 11 protons, but only 10 electrons. Because the sodium atom now has one more positive charge than negative charges, it has a net charge of +1. The symbol for this ion is written as Na^{1+}, or Na^{+} in shorthand notation.

Many ions consist of a single charged atom. For example, an ion of sodium is one charged sodium atom. An ion that consists of only one charged atom is called a monatomic ion.

photo: Getty Images

Sodium metal reacts violently with water. How does this reaction produce sodium ions?

The following chart shows some examples of monatomic ions. Each ion's charge is either positive (+) or negative (–) and has a value between 1 and 4. A value of 1 is represented by a superscript plus or minus sign. Values greater than 1 are shown by placing a number in front of the + or – sign. For example, the calcium ion has a charge of +2, written as Ca^{2+}.

Common Monatomic Ions	
sodium, Na^{+}	chromium (III), Cr^{3+}
hydrogen, H^{+}	lead (IV), pb^{4+}
potassium, K^{+}	bromide, Br^{-}
silver, Ag^{+}	chloride, Cl^{-}
barium, Ba^{2+}	flouride, F^{-}
calcium, Ca^{2+}	hydride, H^{-}
copper (II), Cu^{2+}	iodide, I^{-}
iron (II), Fe^{2+}	oxide, O^{2-}
aluminum, Al^{3+}	sulfide, S^{2-}

The chart shows some common monatomic ions. Those in the left column are positively charged metallic ions. Those in the right column are negatively charged nonmetallic ions. Can you observe any patterns in the way ions are charged?

In contrast to monatomic ions, polyatomic ions form when a group of atoms are joined together and have an overall positive or negative charge. For example, the hydroxide ion, OH^-, contains an atom of oxygen and an atom of hydrogen. There is one more electron than proton in the group, so this polyatomic ion has a charge of –1. Polyatomic ions will be discussed in more detail in another concept.

Transfer of Electrons

Monatomic ions form when single atoms gain or lose electrons. In general, this happens when one or more electrons are transferred from an atom of one element to an atom of another element. The atom that loses one or more electrons becomes positively charged. The atom that gains electrons becomes negatively charged.

For example, common table salt is a solid compound made up of one atom of sodium (Na) and one atom of chlorine (Cl). The formula NaCl represents this compound. However, to form this compound, each sodium atom transfers one electron to each chlorine atom, so a more accurate representation of table salt would be Na^+Cl^-.

How Do Cations and Anions Differ?

Cations and Anions

An early model of the atom was developed in 1913 by the Danish scientist Niels Bohr. This model shows the atom as a central nucleus containing protons and neutrons, with the electrons in circular electron shells orbiting at specific distances from the nucleus, like planets orbiting a star. Each electron shell has a different energy level. Those shells closest to the nucleus have a lower energy than those farther from the nucleus. The outermost shell is called the valence shell and electrons present in this shell are called valence electrons.

Positively charged ions are called cations. A cation forms when an atom loses one or more valence electrons. Cations usually form from the atoms of metals, such as sodium, potassium, calcium, copper, and iron.

Negatively charged ions are called anions. An anion forms when an atom gains one or more electrons in its valence shell. Anions usually form from the atoms of nonmetals, such as fluorine, chlorine, oxygen, and sulfur.

A useful mnemonic to recall the difference is that A Negative ion is an ANion.

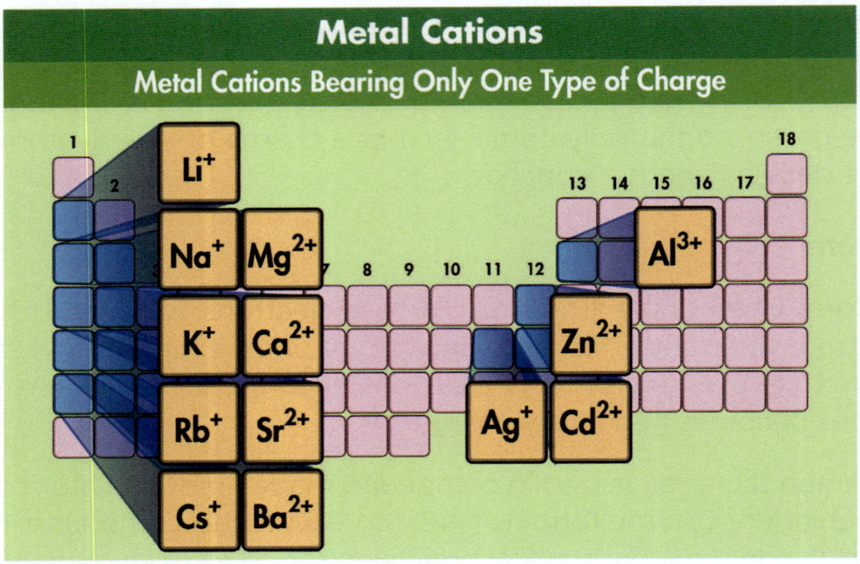

Cations, such as the ions following, are usually formed from the atoms of metals. What type of charge do cations have?

Predicting the Charge on a Cation or Anion

The charge on a cation or anion can be predicted from the number of valence electrons in an atom. The number of valence electrons can be determined by noting the group in which the atom's element appears in the periodic table. Groups toward the left of the periodic table tend to lose electrons and therefore form cations. With the exception of Group 18, groups toward the right of the periodic table tend to gain electrons and therefore form anions.

For example, the atoms of Group 1 elements (Li, Na, K, Rb, Cs, and Fr) have only one valence electron. These atoms readily lose their valence electron. Loss of an electron from an atom of any of these elements will thus produce a cation with a charge of +1. Group 2 elements (Be, Mg, Ca, Sr, Ba, and Ra) have two valence electrons that their atoms tend to lose. In these elements, losing the two valence electrons from their atoms produces cations with a charge of +2.

The charge on an anion is predicted the same way. For example, the atoms of Group 17 elements (F, Cl, Br, I, and At) have seven valence electrons. These atoms tend to gain a single electron. The gain results in the ions of these elements having a charge of –1. The atoms of Group 16 elements (O, S, Se, Te, and Po) have six valence electrons. These atoms tend to gain two electrons, so the ions of these elements have a charge of –2.

How Does the Formation of Ions Relate to the Octet Rule?

The Octet Rule

The elements in Group 18 of the periodic table (He, Ne, Ar, Kr, Xe, and Rn) are the noble gases. Due to their stable outer electron shells, noble gases do not readily react with other elements.

The outer valence shell of each of the noble gas atoms holds a full complement of electrons. Excluding helium, each of the noble gas atoms has an outer valence shell consisting of two *s* orbitals and three *p* orbitals. Each of these four orbitals contains two electrons. Because their valence shells are complete, the atoms are stable. Chemical reactions tend to occur so that the atoms of non-noble elements have a stable outer shell of eight electrons as well. Because a set of eight is an octet, this principle is called the octet rule.

For example, an atom of sodium holds one electron in its outermost valence shell. The next inner shell holds eight electrons. When a sodium atom loses the electron in its outermost shell, a complete valence shell remains. Because the sodium atom has lost a negative charge, its ion has a net charge of +1 and is written as Na^{+}. Metals such as sodium typically lose electrons to form a complete valence shell. Therefore, metals usually form positively charged ions.

How does the octet rule affect elements at the other end of the periodic table? This is where nonmetals reside. An atom of chlorine, for example, holds seven electrons in its valence shell. When it gains an electron, its valence shell will have the stable configuration of eight electrons. The octet rule is again followed.

Nonmetals, such as chlorine, typically gain electrons to form a complete valence shell. Therefore, nonmetals form negatively charged ions.

Exceptions to the Octet Rule

There are many exceptions to the octet rule. Transition metals, for example, may not follow the octet rule because they form bonds that involve *d* or *f* orbitals. A complete outer shell for these elements would therefore include more than eight electrons.

Another exception to the octet rule involves the element helium (He). Helium is a very small atom. It has only two protons in its nucleus, so helium needs only two electrons to make a neutral atom. These two electrons make up helium's valence shell and give helium a complete, stable electron configuration. To achieve this configuration (and become ions) the atoms of metallic elements close to helium in the periodic table lose electrons so that their valence shells contain two electrons.

For example, lithium (Li) has a total of three electrons. Two of these lie in an inner shell. One lies in lithium's valence shell. So, to achieve a complete, stable outer shell of two electrons like that of helium, the lithium atom loses its single outermost electron. The lithium ion then has a charge of +1 (Li^{+}).

An atom of beryllium (Be) has a total of four electrons. Two are in the inner shell, and two are in the outer shell. To achieve a stable valence shell, the beryllium atom can lose its two outermost electrons. The beryllium ion then has a charge of 2 (Be^{2+}).

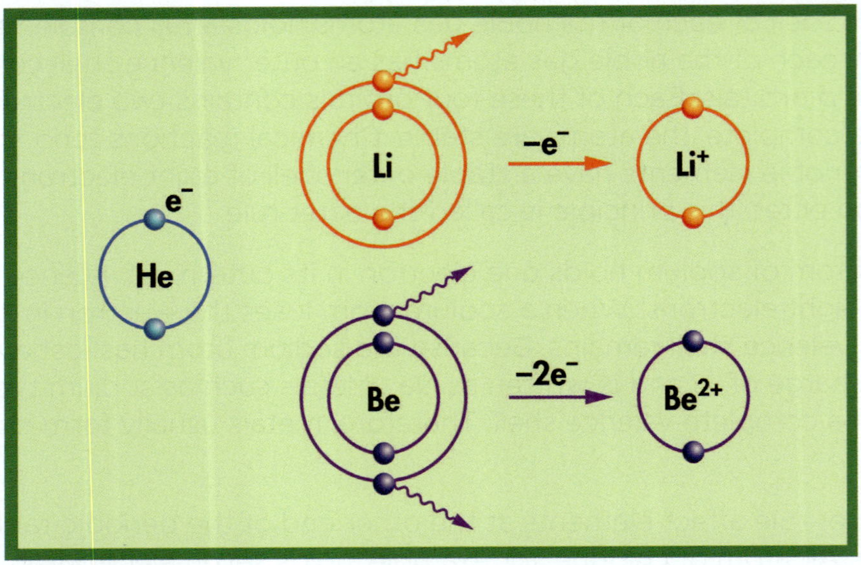

Helium holds only two electrons (e–) in its valence shell. To achieve stability, the atoms of metals close to helium in the periodic table lose electrons, resulting in a valence shell identical to that of helium. Beryllium (Be) loses two electrons to become a positive ion with a charge of +2. How does lithium change to achieve stability?

Consider the Explain Question:

How do ions form, and how does the octet rule determine if an atom becomes a cation or an anion?

Go online to complete the scientific explanation.

Check for Understanding:

How is a neutral atom similar to or different from ions of the same element?

STEM in Action

STEM and Ion Formation

To conserve petroleum and reduce air pollution, scientists and engineers have designed cars that run partly or entirely on electricity. Cars that run only on electricity have no starter under the hood, nor do they have spark plugs or any of the other parts of a conventional internal combustion engine that runs on gasoline. Instead, an electric car is powered by a motor that converts electrical energy into mechanical energy. Electrical cars cost less to operate per mile than gasoline-powered cars.

The electricity that powers the car is produced by a rechargeable lithium-ion battery pack, which is composed of a number of individual lithium-ion batteries connected together. In each battery, lithium (Li) is converted into lithium ions (Li^+) with the release of electrons (e^-). The chemical equation for this reaction is:

$$Li \rightarrow Li^+ + e^-$$

This reaction produces a flow of charged particles, or an electric current, that runs the motor. In turn, the motor makes the car move.

When the battery is depleted, it can be recharged using a public charging station, or it can be recharged at home with electricity from the grid. Solar-powered charging stations are being built that are virtually emission free.

Electric vehicles represent a growing field that requires workers with knowledge not only of electricity, but also chemistry, mechanics, industrial design, materials, and computer software and hardware. Workers who design both the vehicles and the stations required to recharge them generally have an advanced degree in a field of engineering. Technicians and skilled tradespeople typically require a two- to four-year degree, and are valued partners in design, prototype testing, and modification of the vehicles.

photo: Getty Images

Endurance racing for horses includes long distance events of 50 miles or more over natural terrain—no racetracks. Horses racing long distances generate a lot of sweat. What do ions have to do with horse sweat?

Electrolytes are ions that mammalian bodies require in order to regulate nerve and muscle function. Horses that are working hard, as in endurance racing, lose high levels of electrolytes through their sweat, which can then lead to muscle cramping in the horse. A research veterinarian is collecting data on the amounts of specific electrolytes lost from a group of competing horses on a race day, as well as data about the frequency of muscle cramps in the horses throughout the race.

Use a graphing calculator to create a bar graph of the data in the Electrolyte Loss data table, using multiple bars for each horse. Then, answer the questions that follow.

	Electrolyte Loss			
Horse	**Sweat Loss (Liters)**	**Na^+ (Grams)**	**Cl^- (Grams)**	**K^+ (Grams)**
1	30	60	105	14
2	25	100	120	50
3	40	113	100	85
4	50	32	76	10

1. Which horse would be most likely to have problems with muscle cramps during the race? Explain your answer.
2. Which horse is least likely to have muscle cramps? Explain your answer.

Ionic Bonding

photo: Getty Images

LESSON OVERVIEW

Lesson Questions

- How do ionic bonds form in binary compounds?
- How is electronegativity used to determine which atoms form ionic bonds?

Lesson Objectives

By the end of the lesson, you should be able to:

- Illustrate how ionic bonds are formed in binary compounds.
- Use electronegativity to determine which atoms form ionic bonds.

Key Vocabulary

Which terms do you already know?

- ☐ **anion**
- ☐ **binary compound**
- ☐ **cation**
- ☐ **covalent bonding**
- ☐ **crystal lattice**
- ☐ **double bond**
- ☐ **electro-negativity**
- ☐ **evaporation**
- ☐ **formula unit**
- ☐ **ion**
- ☐ **ionic bond**
- ☐ **molecular bonding**
- ☐ **polyatomic ion**
- ☐ **resonance structure**
- ☐ **single bond**
- ☐ **triple bond**
- ☐ **valence electron**
- ☐ **valence shell**

ENGAGE

You are probably very familiar with salt in two different forms: the salt you shake onto your popcorn, and the salt in the ocean. But, have you ever thought about whether those two types of salt are the same? Visit the Engage page online to learn more.

Explain Question

How do atoms form ions that then combine to form crystals?

EXPLORE

How Do Ionic Bonds Form in Binary Compounds?

Binary Ionic Compounds

An ionic compound is formed when two electrically charged atoms, or ions, are held together by their opposite charges. This bond is called an ionic bond.

Valence electrons are those that inhabit the outermost electron shell of an atom. An ionic bond is formed when one or more valence electrons move from a metal atom to a nonmetal atom. In the process of ionic bond formation, the metal atom becomes positively charged because it contains more positively charged protons in its nucleus than negatively charged electrons outside of its nucleus. An atom that has acquired a positive charge is called a cation.

In contrast, the nonmetal atom becomes negatively charged because it contains fewer positively charged protons in its nucleus than negatively charged electrons outside of its nucleus. An atom that has acquired a negative charge is called an anion. When a cation and an anion are chemically bonded through the attraction of their opposite electrical charges, an ionic compound is formed.

The simplest ionic compounds consist of the atoms of only two elements. Such compounds are called binary ionic compounds. The term *binary* refers to the quantity two.

Sodium chloride (NaCl), which is table salt, is a common binary ionic compound. The following chemical equation shows how this ionic compound forms:

$$2\,\mathrm{Na} + \mathrm{Cl_2} \rightarrow 2\,\mathrm{Na^+} + 2\,\mathrm{Cl^-}$$

An electron from each of two sodium atoms is transferred to each of two chlorine atoms.

Magnesium chloride ($MgCl_2$) is another example of a binary ionic compound. In this case, a single magnesium atom transfers two electrons, one to each of two chlorine atoms.

$$Mg + Cl_2 \rightarrow Mg^{2+} + 2\,Cl^-$$

Notice that in both equations, the positive and negative charges are balanced. In other words, overall, no electrons are gained or lost.

photo: Getty Images

How are salts formed from ions?

Ionic Compounds with Polyatomic Ions

Not all ionic compounds are binary compounds. Some ionic compounds are formed by using both ionic bonding and covalent bonding. In order for this to occur, polyatomic ions are used. Polyatomic ions are more complex ions that form when two or more atoms are covalently bonded together. The resulting ion contains a charge because the total number of electrons from the bonded set of atoms does not produce a stable structure. In short, the resulting ion no longer has an equal number of protons and electrons. Even though more than two types of elements are being used to form the ionic bond, the process of ionic bonding, the transfer of valence electrons, is still applied. Chemists use parentheses to indicate that the atoms in the polyatomic ion are bonded together, as seen in the following example.

An example of an ionic compound using a positively charged polyatomic ion is ammonium chloride: NH_4Cl. In this case, two polyatomic ions of ammonium are needed to transfer their available valence electron to the two atoms of chlorine. When this occurs, each atom of chlorine gains an electron and becomes the anion, chloride.

$$2(NH_4)^+ + Cl_2 \rightarrow 2(NH_4)Cl$$

Representing Ions and Ionic Compounds as Lewis Structures

Lewis structure diagrams can be used to represent ions and ionic compounds. These diagrams are a useful tool for visualizing the outer electron configuration of ions. If an atom gains electrons, its ion will have a negative charge. For example, if a chlorine atom gains one electron to become a chloride ion, it will have a charge of −1. If an atom loses electrons, its ion will have a positive charge. For example, when a calcium atom forms an iron, it loses two electrons and therefore has a charge of +2. When drawing a Lewis structure of an iron, the structure is usually placed inside a bracket, and the resulting charge on the ion is placed outside the bracket. Because all the valence electrons are transferred to the cation, the outer shell electrons are only shown on negative ions.

When a chemical reaction occurs between two elements that form a compound with an ionic bond, one element, usually a metal, donates one or more electrons to another (a nonmetal). The metal loses its outer valence electrons. This provides it with a full outer shell. The other element in the compound uses these electrons to fill its outer shell.

Because ionic compounds are held together by electrostatic forces, there is no sharing of electrons. Lewis structures of ionic compounds therefore keep the ions separate; each is usually depicted as being contained within its own set of brackets.

Lewis structures can be used to depict electron configuration in more complex polyatomic ions. In this case, all the atoms that make up the ion are placed within a bracket and the overall charge on the ion is placed outside the bracket.

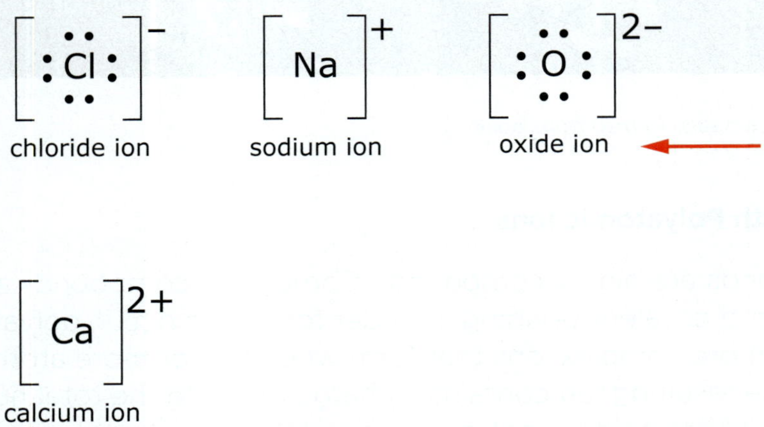

These are the Lewis structures of some common ions. Which structures are anions?

How Is Electronegativity Used to Determine Which Atoms Form Ionic Bonds?

Electronegativity

Which elements are likely to form ionic bonds? The answer depends on the electronegativity of each element. Electronegativity is a measure of the attraction of an atom for electrons in a chemical bond.

If the difference in electronegativity between two atoms is 1.7 or greater, the atoms will form an ionic bond. For example, fluorine (which has an electronegativity of 4.0) will form an ionic bond with nickel (which has an electronegativity of 1.8). The difference between the two values is 2.2. On the other hand, phosphorous (which has an electronegativity of 2.1), will not form an ionic bond with nickel because the difference between these two values is only 0.3.

In general, elements with the lowest electronegativity are found among the metals in Groups 1 and 2 of the periodic table. Elements with the highest degree of electronegativity are found among the nonmetals, especially those in Groups 16 and 17. As a matter of fact, the element with the highest electronegativity (4.0) is fluorine, which is located at the top of Group 17.

Crystal Lattices

Compounds held together by ionic bonds form three-dimensional geometric solid shapes called crystals.

The atoms of an ionic compound are linked in a particular structural pattern. This structural pattern of atoms is called a crystal lattice. In sodium chloride, atoms of sodium and chlorine form a lattice that produces a cubic crystal. A variety of crystal shapes are produced by other ionic substances. The shape of the crystals is determined partly by the ratio and sizes of the positive and negative ions in the compound.

Ionic Compounds

Structure of Solid NaCl

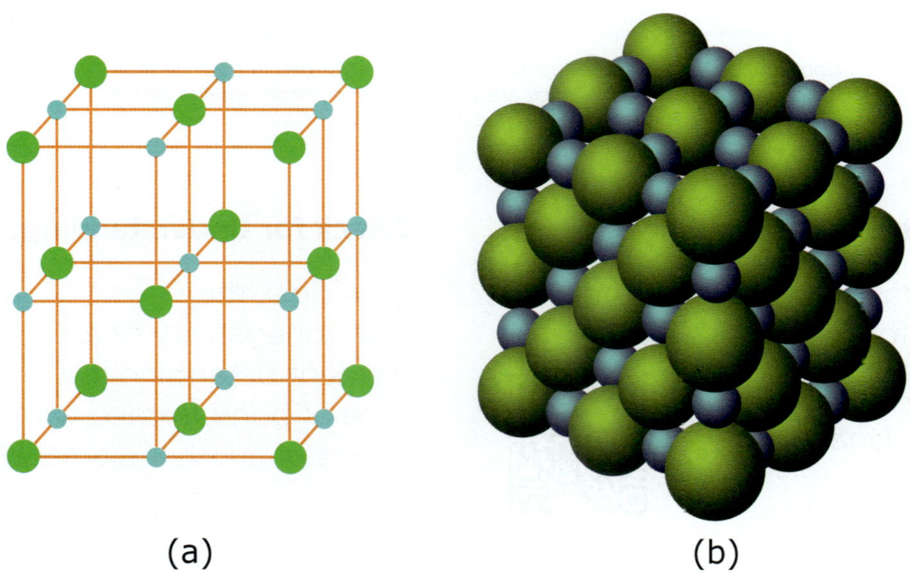

In solid form, ionic compounds such as common salt (NaCl) form arrangements of ions called crystals. Why are there no sodium chloride molecules?

Because multiple ionically bonded atoms are attracted to each other in a crystal lattice, they do not form individual molecules. The smallest unit of an ionic compound is called a formula unit. This is the smallest whole number ratio that exists between the cations and anions in an ionic compound.

Crystals tend to be brittle and have high melting and boiling points. For example, the melting and boiling points of sodium chloride (NaCl) are 801°C and 1,413°C. Those of lithium fluoride (LiF) are 870°C and 1,676°C. In general, the stronger the ionic bonds between atoms in a compound, the higher their melting and boiling points. Based on this principle, scientists have determined that the ionic bonds holding lithium and fluoride ions together are stronger than the ionic bonds holding sodium and chloride ions together.

Another characteristic of ionic compounds is their ability to conduct electric current when melted. The cations flow to one electrode and the anions to another. This produces a flow of electricity. Ionic compounds also conduct electricity if they are dissolved in water because the ions are free to move in the solution.

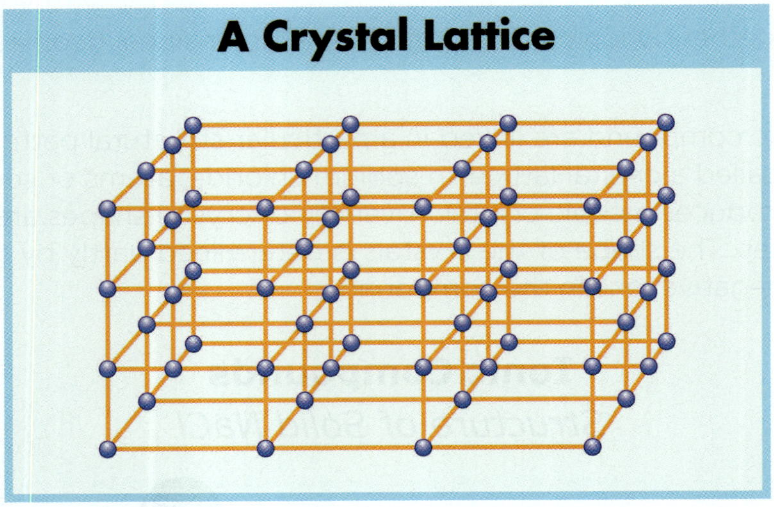

Compounds held together by ionic bonds are arranged in a lattice that produces a cubic crystal. What are the properties of compounds with a crystal lattice structure?

Consider the Explain Question:

How do atoms form ions that then combine to form crystals?

Go online to complete the scientific explanation.

Check for Understanding:

How does the electronegativity of an element determine the type of bond formed with atoms of another element?

STEM in Action

STEM and Ionic Bonds

Crystallography is the study of the arrangement of particles and bonding of atoms in crystalline solids. Compounds formed by ionic bonds form crystal lattices in which the arrangement of the ions is repeated in a regular pattern. The ionic compounds do not contain individual molecules.

Crystals have characteristic geometrical shapes that are related to the arrangements of ions. In early studies, crystallographers learned about the shapes of these compounds by measuring the angles between the faces of crystals. They used the optic properties of crystals, measuring the diffraction angles and frequencies of light that passed through them. Modern X-ray diffraction crystallography, however, changed how scientists study crystals.

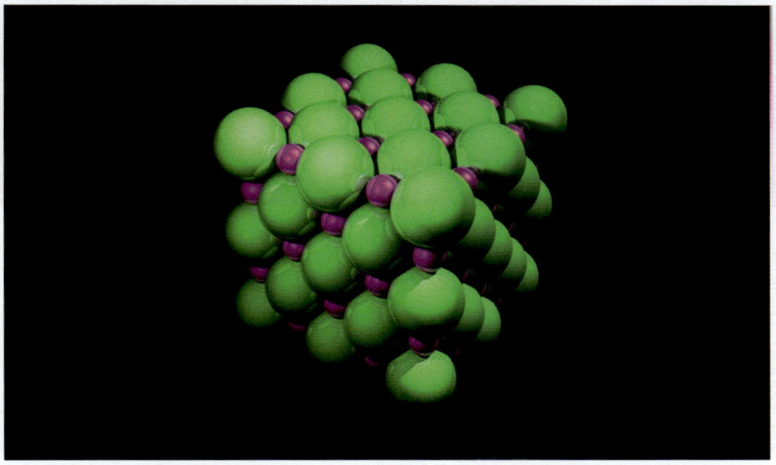

photo: Getty Images

The shape of a sodium chloride crystal is related to its crystal lattice structure. How could you determine which spheres are the sodium atoms?

When an X-ray beam is directed at a crystal, a diffraction pattern is produced. The diffraction pattern is related to the arrangement of the particles in the crystal. Using the diffraction pattern to predict the arrangement of a compound's ions can be tricky. The task has been compared with building a piano based only on the sounds it makes while it is falling down a flight of stairs. It is a complex procedure involving mathematical skills. This once tedious task, however, has been made much easier by advances in computer technology.

Although ionic crystals seem like the obvious subject of study in this field, the technique can even be applied to studies of large, complex organic molecules. In fact, the double helix structure of DNA was deduced, in part, from crystallographic data. Today, crystallographers work in many fields, including chemistry, physics, geology, biology, materials science, medicine, forensics, and metallurgy.

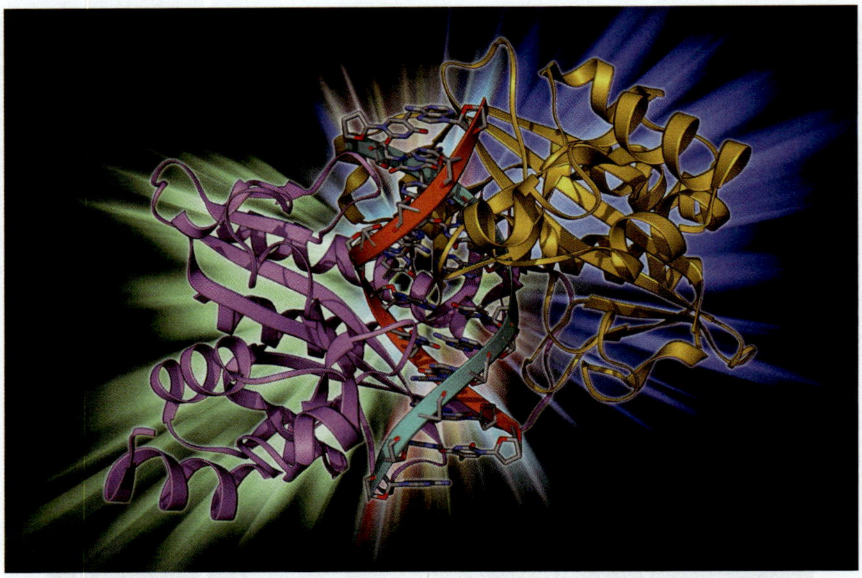

photo: Getty Images

This model of an enzyme (pink and yellow) is based on data from X-ray crystallography. Do you recognize the blue and red structure?

A person choosing a career as a crystallographer can expect to work in a laboratory setting. The work involves sophisticated technical instruments and computers. Chemistry, physics, and biology are important high school courses for a person interested in a career in crystallography. Mathematics, including calculus, is also necessary. Undergraduate college preparation includes chemistry, physics, mathematics, and advanced computer science and programming. Laboratory experience will be very valuable. Many crystallographers acquire an advanced degree, especially those who plan to work as research scientists. Some employment opportunities include research, scientific journal editing, materials engineering, or sales and marketing.

Graph 1 shows the availability of ionic nutrients in soil with changing soil pH. Graph 2 shows the growth of various plant species in soil with different pH levels. Examine the graphs, then answer the questions.

Graph 1

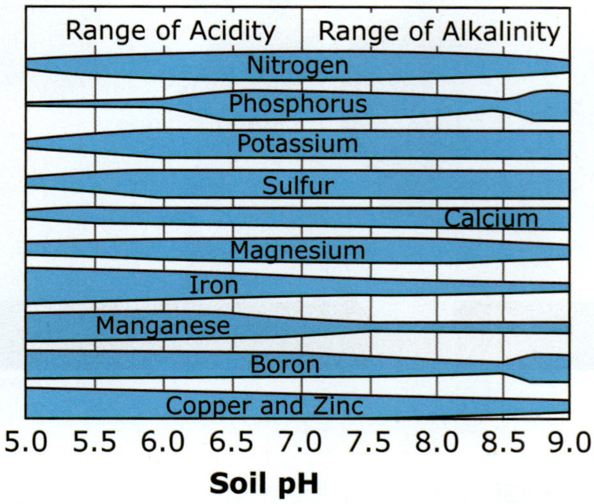

Graph 2

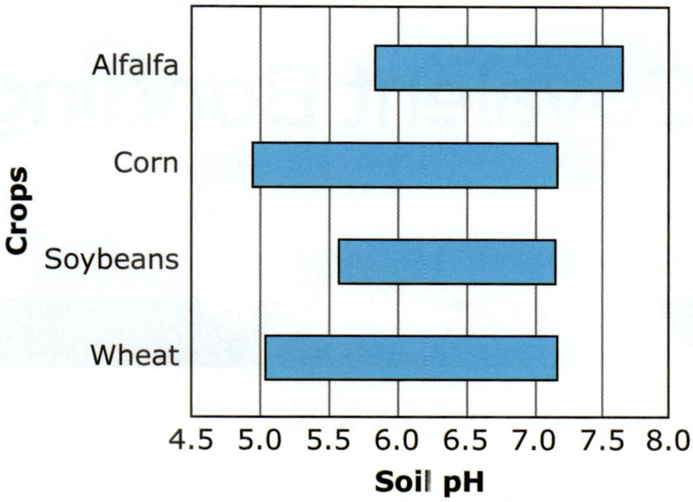

1. Explain why the optimum pH for soil is 5.5 to 7.5.
2. Which fertilizer would provide needed ions for soybeans in acidic soil: magnesium sulfate or copper sulfate? Explain your answer.
3. Why would ionic compound fertilizers need to be applied with water?

Covalent Bonding

photo: Getty Images

LESSON OVERVIEW

Lesson Question

- How does the formation of covalent bonds relate to the octet rule?

Lesson Objective

By the end of the lesson, you should be able to:

- Explain covalent bond formation and its relationship to the octet rule.

Key Vocabulary

Which terms do you already know?

- ☐ carbon compound
- ☐ coordinate covalent bond
- ☐ covalent bond
- ☐ electronegativity
- ☐ hybrid bonding
- ☐ Lewis (electron) dot structure
- ☐ molecule
- ☐ octet rule
- ☐ resonance structure
- ☐ single bond
- ☐ sp orbital
- ☐ sp2 orbital
- ☐ sp3 orbital
- ☐ triple bond
- ☐ valence electron

ENGAGE

How can you make straight hair curly, or curly hair straight? And what does chemistry have to do with it? Visit the Engage page online to learn more.

Explain Question

How does the Lewis structure of an oxygen molecule, O_2, demonstrate the octet rule?

EXPLORE

How Does the Formation of Covalent Bonds Relate to the Octet Rule?

Covalent Bonds and Electronegativity Difference

When two atoms form a chemical bond, the difference in their electronegativities can be used to classify the type of bond that forms. Electronegativity measures the ability of atoms to attract electrons. An electronegativity difference of 1.7 or greater indicates a transfer of one or more electrons and the formation of an ionic bond. Ionic bonds often form when metals bond with nonmetals.

An electronegativity difference below 1.7 means that neither atom attracts electrons strongly enough to allow the transfer of an electron. Instead, the atoms share electrons and form a covalent bond. Covalent bonds form between two nonmetal atoms because the atoms have generally similar electronegativities. The covalent bonds can be classified as nonpolar or polar bonds based on the equal or unequal sharing of the electrons.

Nonpolar covalent bonds form when the difference in electronegativity between bonding atoms is between 0.0 and 0.4. The electrons are shared equally between the atoms.

Polar covalent bonds form when the electronegativity difference is between 0.4 and 1.7. This results in unequal sharing of electrons. One atom with a higher electronegativity draws the bonding electrons toward itself, pulling those electrons away from the atom with the lower electronegativity value. The atom with the higher electronegativity becomes partially negatively charged and the atom with the lower electronegativity becomes partially positively charged. This makes the bond polar. Examples of atoms with a polar bond are water (H_2O) and ammonia (NH_3).

The Octet Rule and Lewis Structures

The elements of Group 18 are the noble gases. They are nonreactive nonmetals. These elements exist naturally as individual atoms. What makes these atoms unlikely to form chemical bonds?

Think about the number of valence electrons that are in the outermost energy level of an atom. Each noble gas atom has eight valence electrons, except for the helium atom, which has only two valence electrons.

A convenient model to show the valence electrons in an atom is a Lewis structure. In this model, an atom's nucleus and inner electrons are represented by its chemical symbol. A dot is used for each valence electron. For example, the Lewis structure for a neon atom begins with the chemical symbol, Ne. This symbol is then surrounded by two dots above, below, to the left, and to the right. The dots represent Neon's eight valence electrons.

Neon is a noble gas and therefore has a full valence shell of eight electrons. What does this mean for neon's ability to create covalent bonds?

Atoms that have eight valence electrons (and helium with two) do not tend to react with other atoms. Many other atoms, however, tend to form chemical bonds that put eight electrons in their outermost energy levels. The octet rule describes the tendency of atoms to gain, lose, or share electrons until they have eight valence electrons.

Single, Double, and Triple Bonds

The Lewis structure of a fluorine atom shows that each fluor ne atom has seven valence electrons.

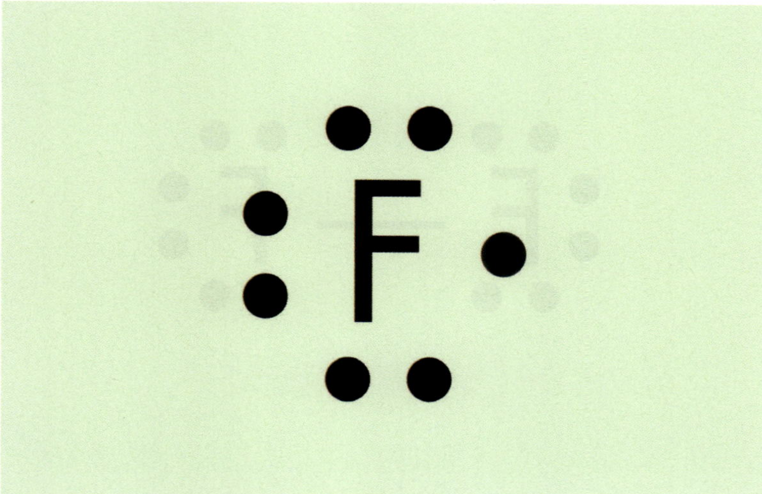

Neon has an almost full valence shell of seven electrons. What does this mean for fluorine's reactivity?

A fluorine atom needs only one more electron to complete its octet. If a fluorine atom shares one of its electrons with another fluorine atom, the valence shells of both atoms become filled. The two shared electrons form a single bond between the atoms.

Fluorine gas consists of two fluorine atoms with a bond allowing them to share a single pair of electrons. How does this bond compare with unbounded atoms?

A single covalent bond is composed of two shared electrons. In the Lewis structure for the fluorine molecule formed by this bond, a line can be used to represent the shared electron pair.

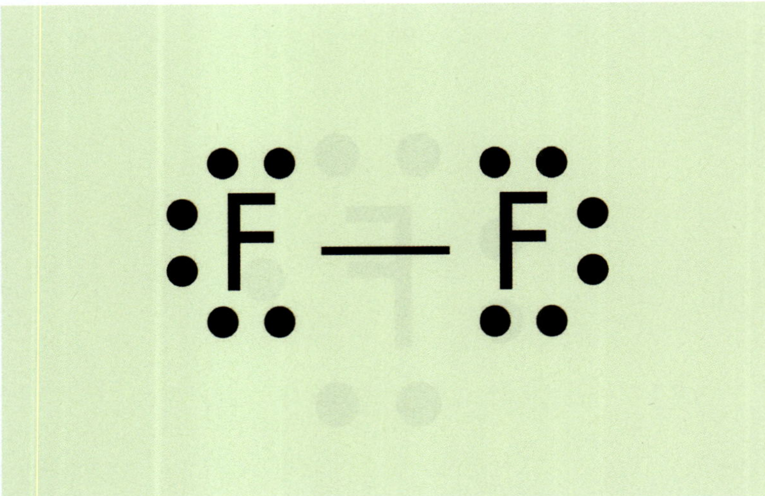

A line can be used to represent the shared electron pair. The bonding between which other elements could be represented this way?

The Lewis structure for oxygen shows that each atom has six valence electrons. An oxygen atom needs two more electrons to complete its octet.

Oxygen has an almost full valence shell of six electrons. What does this arrangement mean for oxygen's reactivity?

A single bond with another oxygen atom would not fill the octet of either atom. Each atom would only have seven electrons in its valence shell. Instead, oxygen atoms form a double bond. They share two pairs of electrons between them. In this way, each oxygen atom completes its octet.

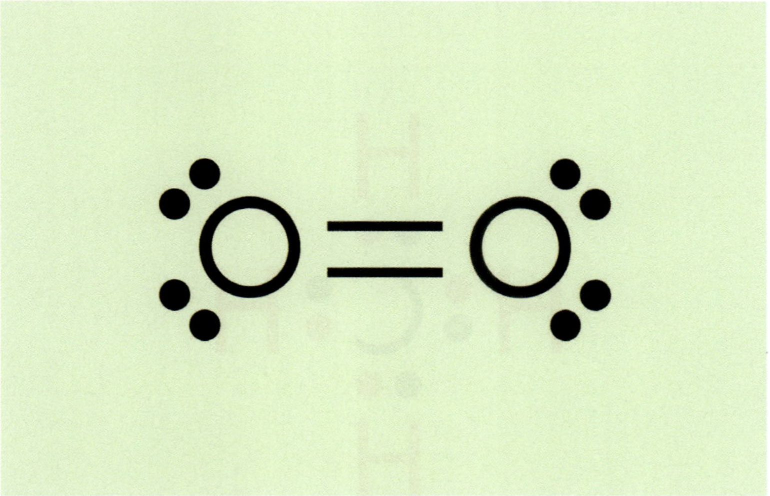

Oxygen gas consists of two oxygen atoms with a double bond connecting them. How does a double bond compare with a single bond?

A triple bond consists of three pairs of electrons shared by two atoms. What kinds of atoms can form a triple bond? An atom must have room for at least three electrons in its valence shell if it is to form a triple covalent bond. A nitrogen atom has five valence electrons, so it needs three more electrons to complete its octet.

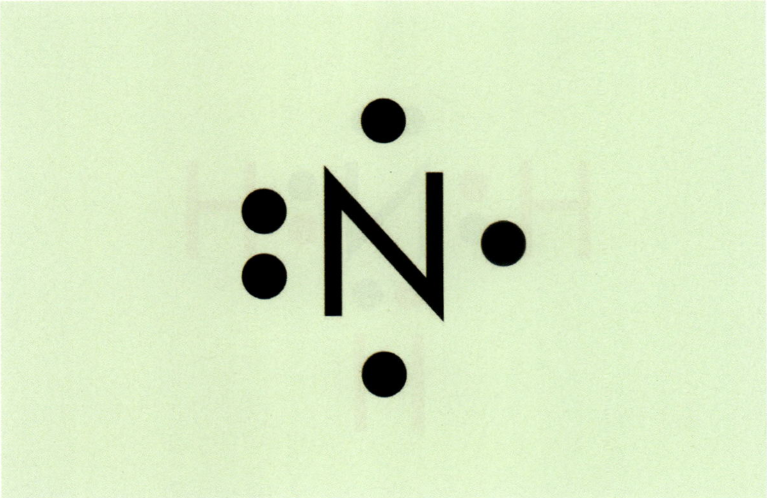

Nitrogen's Lewis structure shows that it only has five electrons in its valence shell. How does this arrangement affect its reactivity?

Two nitrogen atoms can form a nitrogen molecule by sharing three pairs of electrons in a triple covalent bond. The Lewis structure for the nitrogen molecule shows the triple bond as three lines between the nitrogen symbols.

Coordinate Covalent Bonds

Covalent bonds form when an atom shares valence electrons with another atom. For example, in a molecule of methane, CH_4, a carbon atom forms a single bond with each of four hydrogen atoms. Each single bond is made up of one electron from the carbon atom and one electron from a hydrogen atom.

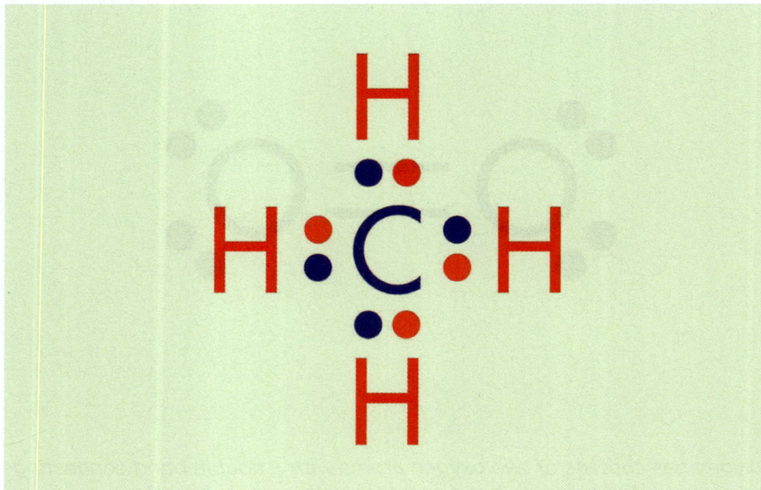

Methane consists of a carbon atom forming covalent bonds with four hydrogen atoms. How many electrons are originally in carbon's valence shell?

Likewise, each single bond in ammonia, NH_3, is composed of one electron from the nitrogen atom and one electron from a hydrogen atom.

Ammonia consists of a nitrogen atom forming covalent bonds with three hydrogen atoms. How many electrons are originally in nitrogen's valence shell?

The nitrogen atom in NH_3 has a complete octet. Can the nitrogen atom form a fourth bond to make the polyatomic ion ammonium (NH^{4+})? Yes. The nitrogen atom in ammonia has a nonbonding pair of electrons. The nitrogen atom uses this pair of its own electrons to form a single bond with a hydrogen ion, which has no electrons. A covalent bond in which both electrons come from one atom is a coordinate covalent bond.

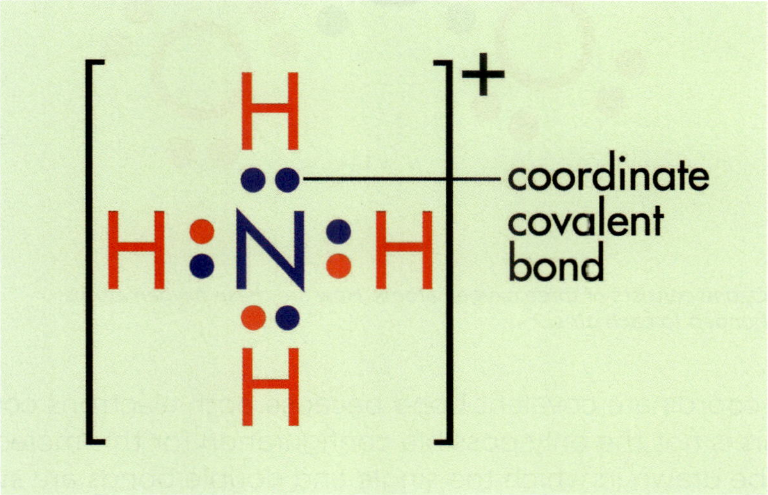

Coordinate covalent bonds consist of bonds in which both electrons come from one atom. How does polyatomic ion ammonium compare with ammonia?

Resonance Structures

Lewis structures often show the correct bonding in a molecule. But as with any model, Lewis structures have their limitations. In some molecules that have double or triple bonds, the locations of the bonds may vary. We say that these molecules exhibit resonance. They cannot be represented by only one Lewis structure. Resonance structures represent all of the possible positions of electrons in a molecule's chemical bonds.

For example, ozone, O_3, is an important molecule for life on Earth. Ozone helps to shield Earth's surface from harmful ultraviolet light. The Lewis structure of an ozone molecule has a central oxygen atom joined to one oxygen atom by a single bond and to another oxygen atom by a double bond.

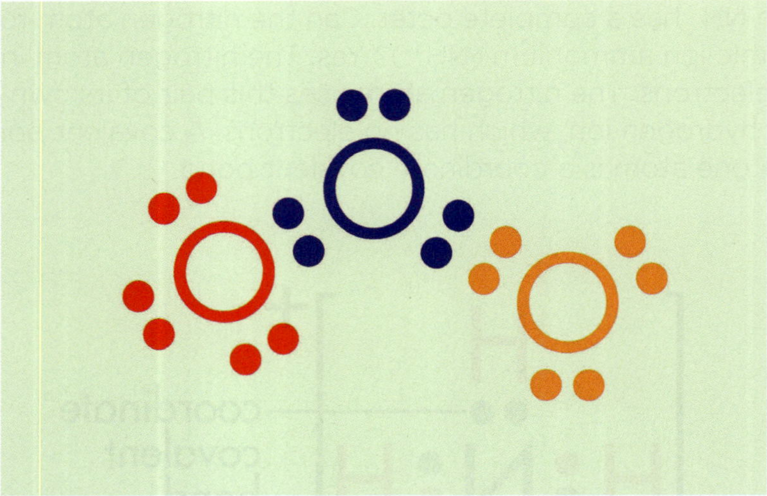

Ozone consists of three oxygen atoms. How are these oxygen atoms bonded to each other?

The single bond is a coordinate covalent bond because both electrons come from the central oxygen atom. But this is not the only possible configuration for this molecule. A second Lewis structure can be drawn in which the single and double bonds are switched.

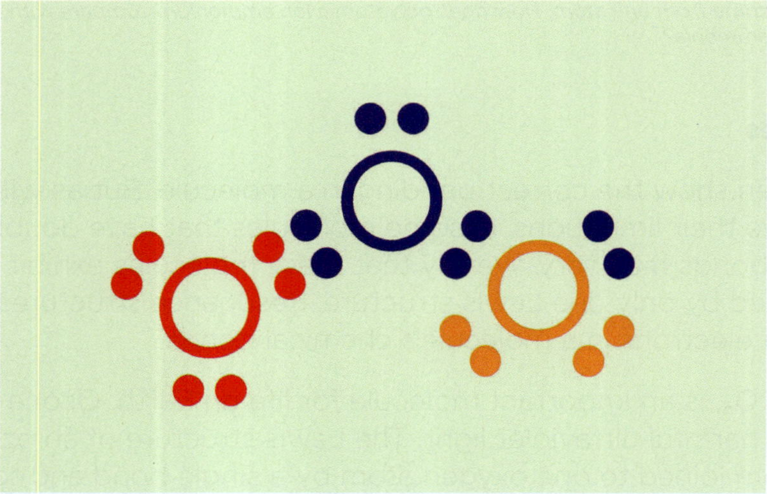

Ozone consists of three oxygen atoms, one pair bonded by a single bond and the other pair bonded by a double bond. How does this ozone structure compare with the prior arrangement?

Thus, ozone has two resonance structures. Studies of ozone show that both bonds are identical. Neither bond is a single bond or a double bond. Instead, an ozone molecule is a mixture of the two resonance structures. The bonds have characteristics that fall somewhere between those of single bonds and those of double bonds.

Covalent Bonding and Properties of Substances

Many covalently bonded compounds, such as methane and carbon dioxide, are gases at room temperature and standard pressure. Others, such as water, are liquids at room temperature and standard pressure. When atoms bond covalently, they can form molecules of simple compounds that have low melting points and low boiling points.

However, some atoms that bond covalently do not form individual molecules. Instead, they form a large network of bonded atoms. These substances are referred to as network solids or network compounds. For example, silicon dioxide or silica, SiO_2, does not exist as individual molecules. Each silicon bonds with four oxygen atoms, and each oxygen atom bonds with two silicon atoms. The result is a network compound that is very hard and has a high melting point, which is why glass is made from silica.

Hybridization in Carbon Atoms

Each carbon atom can form up to four bonds with other atoms. The Lewis structure of carbon shows us how this works. Carbon has four single electrons that can be used to form chemical bonds.

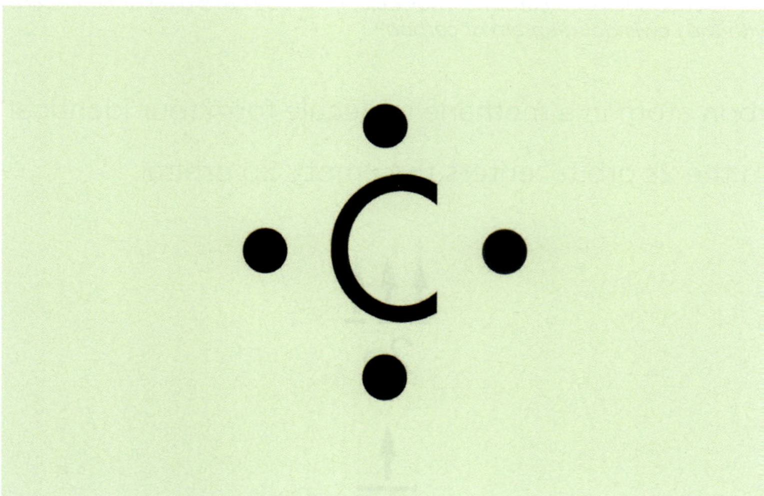

The carbon atom has four electrons in its valence shell. How does this affect carbon's ability to form bonds?

A carbon atom can complete its octet in several ways:

- four single bonds
- two double bonds
- a double bond and two single bonds
- a triple bond and one single bond

However, the orbital diagram for a carbon atom tells a different story. The valence electrons are not arranged as four single electrons. Two electrons are paired in the 2s orbital. The other two electrons are unpaired in two 2p orbitals.

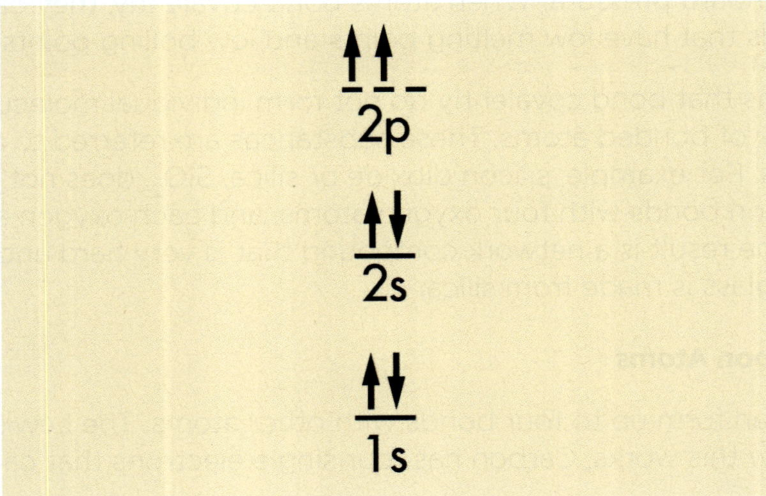

The orbital diagram conveys information about the number of electrons in the valence shell of an atom or molecule. How does this diagram compare with the Lewis dot diagram of carbon?

So, how does the carbon atom in a methane molecule form four identical single bonds?

First, an electron from the 2s orbital enters the empty 2p orbital.

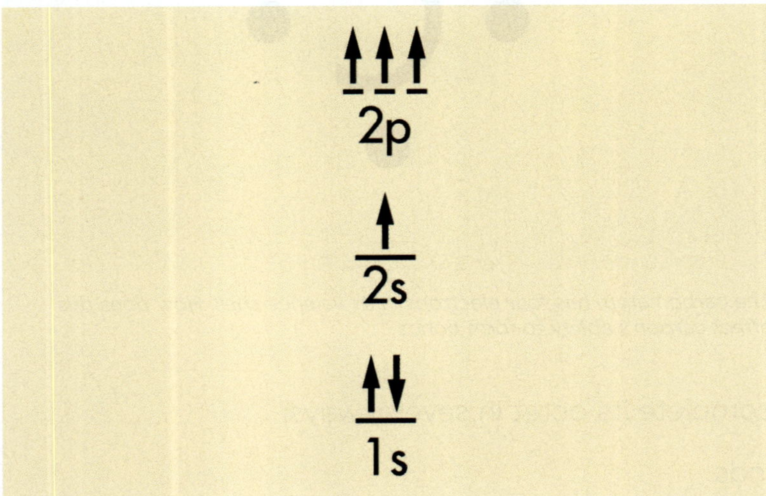

The orbital diagram conveys information about the number of electrons in the valence shell of an atom or molecule. How does this compare with the orbital diagram of carbon?

This gives the carbon atom four unpaired electrons. But these electrons are in different types of orbitals, so they cannot form four identical bonds. However, the orbitals are combined in a process called hybridization. The result is four identical orbitals.

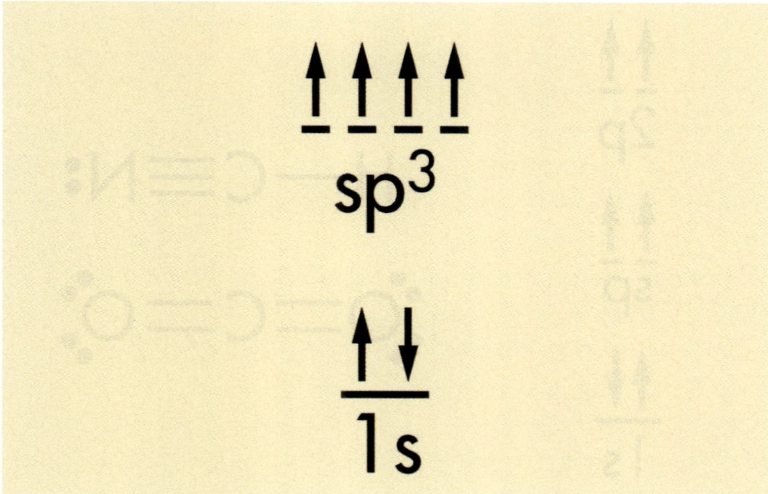

The orbitals here are combined in a process called hybridization. How does this relate to the orbital diagram of methane?

The orbitals are equal in energy. Because they formed from one s and three p orbitals on the same energy level, these four orbitals are called sp^3 hybrid orbitals. They a low the carbon atom to form four identical single bonds.

To form a double bond and two single bonds, a carbon atom needs to kee𝐩 one electron in a p orbital. The s and two of the p orbitals hybridize. So, the carbon atom in methanal, H_2CO, uses sp^2 hybrid orbitals.

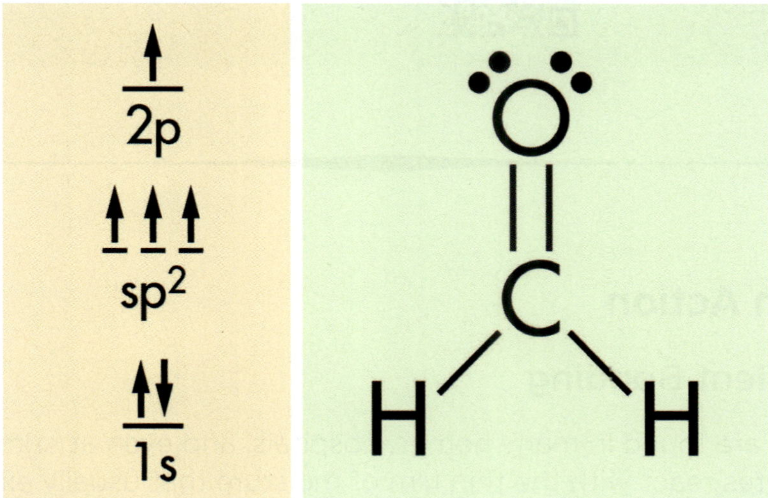

Methanal forms a double bond and two single bonds. What needs to happen to form these bond types?

To form a triple bond or two double bonds, a carbon atom needs to keep two electrons in p orbitals. The s and one p orbital hybridize. In this case, the carbon atoms in hydrogen cyanide, HCN, and in carbon dioxide, CO_2, use sp hybrid orbitals.

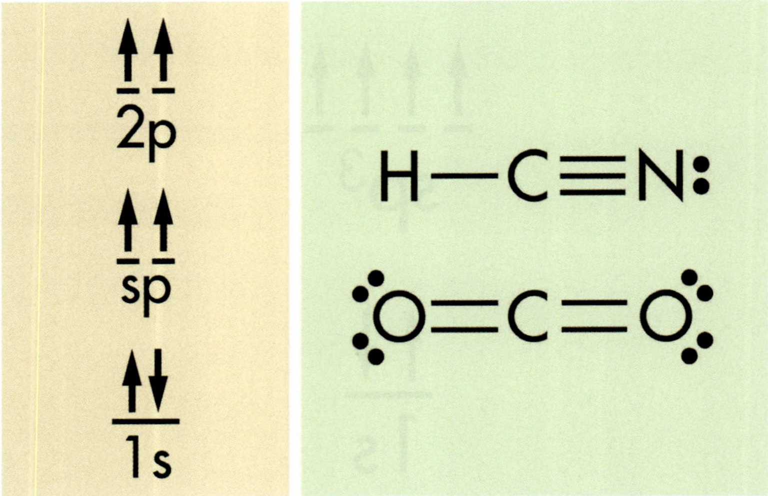

Hydrogen cyanide, HCN, and carbon dioxide use sp hybrid orbitals. How would you describe the properties of these bonds?

Consider the Explain Question:

How does the Lewis structure of an oxygen molecule, O_2, demonstrate the octet rule?

Go online to complete the scientific explanation.

Check for Understanding:

Can you model what happens to electrons during the formation of covalent bonds?

STEM in Action

STEM and Covalent Bonding

Cyanoacrylate glues are found in many homes, hospitals, and even at crime scenes. Once applied, cyanoacrylates react with the thin film of moisture that usually exists on a surface. Double bonds between carbon atoms in the glue are broken during the reaction. Then, one molecule forms a new single bond to another molecule, which forms a new single bond to another molecule, and so on. Long chains, or polymers, of cyanoacrylate molecules join the glued pieces together.

The ability of cyanoacrylate glues to bond to skin led to their use in medicine. These compounds are used to close wounds without stitches. They have become very important in battlefield medicine.

Forensic scientists use chemistry in a number of ways to help solve crimes. A forensic scientist might use cyanoacrylate fuming to develop latent fingerprints and make them visible. A latent fingerprint comes from materials such as sweat and oils that are left behind when a finger touches a surface. Fresh fingerprints can adequately be developed by fingerprint powders because these powders stick to sweat. However, a forensic scientist knows that cyanoacrylate adheres to skin components that do not evaporate, such as oils, proteins, and fatty acids. This glue enables the scientist to lift fingerprints that are old or are on uneven, multicolored, or otherwise difficult surfaces. Covalent bonding between atoms helps cyanoacrylate glues to be truly super.

Fire scene investigations often rely on forensic chemists to determine if certain compounds called accelerants were used by arsonists to accelerate the development of a fire. Commonly used accelerants are covalently bonded compounds such as gasoline or kerosene. When a forensic chemist takes samples at a crime scene, he does not know the nature of the sample until it is analyzed. If chemical tests reveal evidence of accelerants, investigators use this information to narrow the suspects by looking at who had access to the accelerant found at the scene. Sometimes, forensic scientists are asked to testify in court trials and explain their findings.

Research the use of cyanoacrylate glues in medicine. Be sure to look at early trials of the products and the scientific trials that were used during development.

Watch the “Super Glue Prank” video on the STEM in Action page online. Then, answer the questions:

1. How have cyanoacrylate glues been used historically? How does historical use differ from current use?
2. Compare and contrast the use of glue versus the use of sutures.
3. Conduct a survey of local hospitals or emergency rooms to determine how often cyanoacrylate glue is used locally. Use a table or graph to display your results.

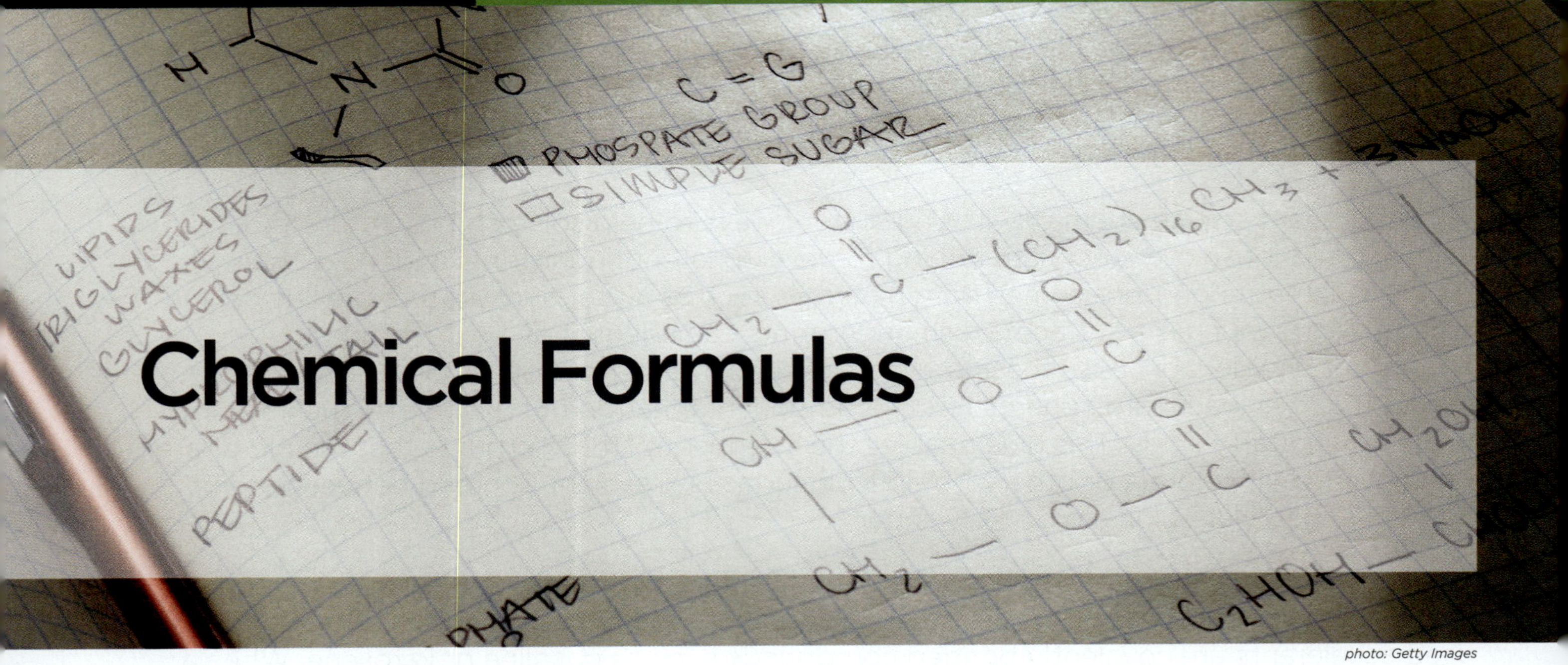

Chemical Formulas

photo: Getty Images

LESSON OVERVIEW

Lesson Questions

- How are the formulas for ionic compounds written?
- How are the formulas for molecular compounds written?
- How are empirical and molecular formulas distinguished?

Lesson Objectives

By the end of the lesson, you should be able to:

- Compose and write formulas for ionic compounds.
- Compose and write formulas for molecular compounds.
- Distinguish between empirical and molecular formulas.

Key Vocabulary

Which terms do you already know?

- ☐ **chemical formula**
- ☐ **chemical symbol**
- ☐ **crystal lattice**
- ☐ **electronegativity**
- ☐ **empirical formula**
- ☐ **formula unit**
- ☐ **ion**
- ☐ **molecular formula**
- ☐ **molecule**
- ☐ **nonmetals**
- ☐ **polyatomic ion**
- ☐ **prefix**
- ☐ **subscript**
- ☐ **transition metals**
- ☐ **valence electron**

ENGAGE

How did scientists identify the formula of penicillin? What steps did they take to derive this formula? Visit the Engage page online to learn more.

Explain Question

How are the formulas for ionic and covalent compounds determined?

EXPLORE

How Are the Formulas for Ionic Compounds Written?

Chemical Symbols and Subscripts

A chemical formula tells what elements make up a compound and how many ions or atoms of each element are present.

An ionic compound consists of cations and anions held together by the attraction between their opposite charges. The chemical formula of an ionic compound consists of two pieces of information. The first is a chemical symbol. A chemical symbol uses one to three letters of the alphabet to represent an element. Each element has its own symbol, as approved by the International Union of Pure and Applied Chemistry (IUPAC).

In an ionic compound, the symbol for the positively charged ion always appears on the left side of a chemical formula. The symbol for the negatively charged ion always appears on the right. The negative and positive ions in neutral ionic compounds are always electrically balanced. For example, consider the formula for calcium fluoride. This ionic compound contains the elements calcium and fluorine. An ion of fluorine is called fluoride. In its chemical formula, the symbol for calcium (Ca) will appear on the left because the calcium ion is positively charged. The symbol for fluorine (F) will appear on the right because the fluoride ion is negatively charged.

The second piece of information included in a chemical formula is the subscript, which follows each ion's chemical symbol. The subscript indicates the number of ions of each element present in each formula unit of the ionic compound. Because ionic compounds are composed of ions, they do not form discrete molecules. The ions in a sample of the compound are ordered in a repeating pattern, forming a crystal lattice. The smallest ratio of ions in any sample of the compound is the formula unit. For example, the compound calcium fluoride consists of one calcium ion bonded to two fluorine ions. Therefore, the formula for calcium fluoride is written as:

$$CaF_2$$

Note that the numeral "1" is not written as a subscript after the symbol Ca. When an ionic compound contains only one ion of a certain element, no subscript follows that ion's chemical symbol.

Consider another example. The ionic formula for common table salt, or sodium chloride, is NaCl. Because no subscripts follow the symbols for sodium (Na) or chlorine (Cl), the formula shows that only one ion of sodium is bonded to only one ion of chlorine.

Determining Subscripts

How are the subscripts in the formula of an ionic compound determined? Remember, all ionic compounds are electrically balanced. The total positive charge of the cations in a formula unit is equal to the total negative charge of the anions in the formula unit. Consider the ionic compound calcium chloride.

The calcium ion has a charge of 2+ (Ca^{2+}). The chloride ion has a charge of negative 1, or 1– (Cl^-). For the charges in calcium chloride to balance, the compound must contain two chloride ions for every calcium ion. This means that a balanced chemical formula for calcium fluoride could be written as Ca_2Cl_4. However, when writing chemical formulas for ionic compounds, the simplified ratio of ions is used. The simplest way to describe the ratio of calcium to chloride ions in calcium chloride is 1 : 2. Thus, the chemical formula is written:

$$CaCl_2$$

This ratio is due to calcium having two valence electrons. Chlorine can only receive one electron per atom; therefore, it takes two chlorine atoms to bond with one Ca atom.

photo: Getty Images

Based on the formula shown here, how many hydrogen atoms are present in a molecule of water?

Subscripts for Transition Metal Cations

The process is a little trickier for cations formed by transition metals such as copper (Cu), iron (Fe), cobalt (Co), chromium (Cr), and mercury (Hg). Many transition metals can have multiple oxidation numbers and therefore can have more than one correct formula. Thus, most transition metals can form more than one ion. For example, copper can form compounds with either copper(I) ions (Cu^+) or copper(II) ions (Cu^{2+}).

To write a formula with a transition metal, the charge of the cation is needed. The name of the compound will contain a Roman numeral in parentheses to indicate the charge, immediately following the transition metal's name.

Using the correct charge, the same steps are used to write a formula that will have a ratio of cations to anions that results in a neutral compound. In the case of copper(II) chloride, the copper(II) ion has a charge of 2+ (Cu^{2+}), and the chloride ion has a charge of 1– (Cl^-). For the charges to balance, the compound must contain one copper(II) ion for every two chloride ions. This 1 : 2 ratio is used to write the formula:

$$CuCl_2$$

Polyatomic Ions

Many ions consist of single atoms that have lost or gained electrons. Ions may also form, however, when a group of two or more bonded atoms loses or gains an electron and thus carries a charge. Such groups of atoms are called polyatomic ions.

The atoms in a polyatomic ion are covalently bonded. Their bonds consist of pairs of shared valence electrons. However, the total number of electrons in a polyatomic ion differs from the total number of its protons.

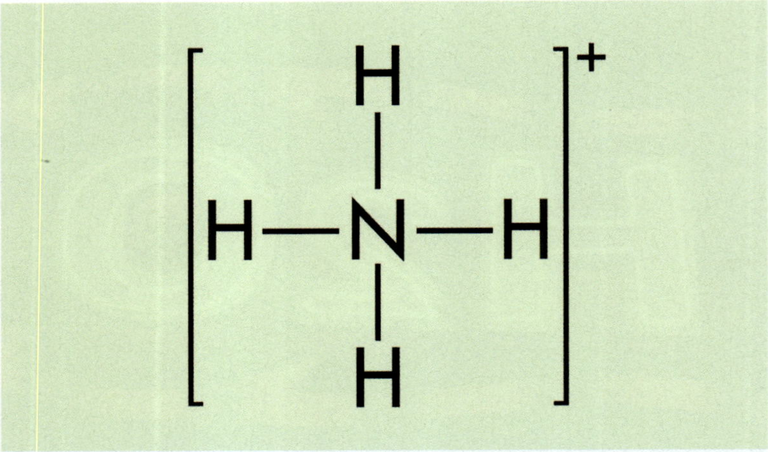

Polyatomic ions are formed by the sharing of electron pairs between two nonmetal atoms. What do the dashes represent in the diagram of the ammonium ion?

This is true because the polyatomic ion has gained or lost at least one valence electron. The electrons may have been transferred to or from another atom or group of atoms. The result is that the polyatomic ion possesses an overall positive or negative charge.

How Are the Formulas for Molecular Compounds Written?

Chemical Symbols and Subscripts

The ions in an ionic compound form a cluster of charged atoms or groups rather than a discrete molecule. Although attracted to one another, the ions are somewhat independent. This is not the case, however, in a molecular compound. Molecular compounds are also called covalent compounds.

In covalent compounds, valence electrons are not gained or lost by the compound's atoms. Instead, a covalent bond forms from a shared pair of electrons. This is because the electronegativity difference between the nonmetals in the bonds is not great enough to cause a transfer of electrons and therefore results in sharing. This is the reason that these groups of atoms form discrete molecules rather than ordered clusters.

Like the formulas for ionic compounds, the formulas for covalent compounds consist of chemical symbols and subscripts. A molecular formula describes the components of a molecule, including the number of atoms of each element that are present in one molecule of the compound.

A molecular formula may describe a molecule of a single element. For example, H_2 is the molecular formula for hydrogen gas. This means that in the gaseous state, hydrogen forms a molecule consisting of two covalently bonded atoms of hydrogen. Similarly, oxygen (with a molecular formula of O_2), fluorine (F_2), and chlorine (Cl_2) do the same. Note that these differ from the chemical symbols of these elements because they contain subscripts. Elements that form diatomic molecules are called diatomic elements.

A molecular formula may also describe a molecule that contains atoms of two or more elements. For example, the molecular formula of the compound hydrogen chloride is HCl.

The molecular formula of the gas methane is CH_4. The molecular formula for a molecule of the sugar glucose is $C_6H_{12}O_6$. In each case, the molecular formula describes the element (the chemical symbols) and quantity of the atoms (the subscripts) that make up a discrete molecule.

Naming Molecular Compounds

Molecular formulas are a convenient way to describe the elements in a compound and the ratios of the ions or atoms present. The chemical name of a molecular compound also often reveals the elements it contains. In some cases, the name may also describe the amounts of each element present.

For example, the name ***hydrogen chloride*** (HCl) reveals that the molecule of this substance contains atoms of the elements hydrogen and chlorine. The name ***hydrogen bromide*** (HBr) reveals that a molecule of this substance contains atoms of the elements hydrogen and bromine.

What, however, do the names ***carbon dioxide*** and ***carbon tetrachloride*** reveal about the components of these molecules? The answer to this question lies in an analysis of the words and word parts in each name.

In the case of carbon dioxide, the names of two elements are mentioned. These are carbon and oxygen. It is clear that carbon dioxide molecules are made up of atoms of carbon and oxygen atoms.

Additionally, the name reveals something more, a quantitative fact. The prefix *di-* means "two." The name ***carbon dioxide*** reveals that a molecule of this compound contains two oxygen atoms. This is shown in the chemical formula for carbon dioxide, CO_2.

What about carbon tetrachloride? What does its name reveal about the components of its molecule? A clue to the answer lies in the prefix ***tetra–***, which means "four." Therefore, a molecule of carbon tetrachloride contains four chlorine atoms. The molecule's chemical formula is CCl_4. Molecular compounds containing three or more elements have complex rules.

Common Prefixes Used in Chemistry

Prefix	Meaning	Example	Formula
mono-,mon-	one	carbon monoxide	CO
di-	two	carbon dioxide	CO_2
tri-	three	phosphorous trichloride	PCl_3
tetra-	four	carbon tetrachloride	CCl_4
penta-	five	dinitrogen pentoxide	N_2O_5

Common Greek prefixes are often used to describe the quantity of atoms of each element in a molecular compound. What formula would you predict for the compound silicon tetrafluoride?

How Are Empirical and Molecular Formulas Distinguished?

Empirical Formulas

A molecular formula describes the structure of a molecular compound. It describes the elements and the number of atoms of each present in one molecule of the compound. Ionic compounds, in contrast, are often composed of ordered clusters of the charged ions. Because these clusters can vary in size, the chemical formulas for ionic compounds are written as empirical formulas.

The empirical formula describes the simplest whole-number ratio of the atoms of each element in a compound. When trying to determine the chemical formula of a newly discovered molecule, scientists can often determine experimentally the empirical formula and the molecular weight. They then use that information to deduce the molecular formula.

For example, the molecular formula of the compound hydrogen peroxide is H_2O_2. The ratio of hydrogen to oxygen in the molecular formula is 2 : 2. The empirical formula of the same compound is HO, with a simplified ratio of 1 : 1. Similarly, the molecular formula of benzene is C_6H_6 and its empirical formula is CH.

Consider the Explain Question:

How are the formulas for ionic and covalent compounds determined?

Go online to complete the scientific explanation.

Check for Understanding:

Why can an empirical formula, alone, not be used to derive the actual molecular formula for a given compound?

STEM in Action

STEM and Chemical Formulas

By the 1930s, rubber had become a vital resource. Because many military machines and civilian vehicles required rubber in order to function, a shortage of rubber would have caused serious consequences. As World War II approached, some nations far from sources of natural rubber in Southeast Asia realized this could impact their military success.

photo: Getty Images

Natural rubber in the form of white sap can be collected and used to make latex. Can you think of a product made from latex rubber?

Part of the answer was found in butadiene, a substance with a molecular formula of C_4H_6. Discovered by scientists in 1863, the structure of butadiene was deciphered in 1886. Butadiene was a curious substance, but lacked practical applications. This all changed when, in 1910, a Russian chemist by the name of Sergei Lebedev found a way to link together many molecules of butadiene to form a chemical chain called a polymer.

The butadiene polymer has an intriguing property. It behaves in many ways like natural rubber. Unfortunately, the polymer is much softer than natural rubber and has a consistency that is nearly liquid.

As the years passed, organic chemists discovered that they could add molecules of other compounds to the butadiene polymer. One of the compounds they added was called styrene, whose molecular formula is C_8H_8.

Polymers made from combinations of butadiene and styrene molecules produce substances with properties very similar to those of natural rubber. This synthetic rubber became a substitute for natural rubber.

The invention of synthetic rubber contributed to the success of the United States in World War II. Today, this form of synthetic rubber is a key ingredient in car and truck tires and other rubber products used throughout the world.

It is estimated that more than seven million tons of synthetic rubber is produced each year. Synthetic rubber is used to make a wide range of products, such as automobile tires and flooring for a building. The following table provides a list of different polymers used in the production of synthetic rubber materials.

Use the information in the table to answer the questions.

Selection of Polymers	
Butadiene	C_4H_6
Styrene	C_8H_8
Styrene-Butadiene Copolymer (SBR)	$C_4H_6 + C_8H_6 \rightarrow (C_{12}H_{14})_n$

1. You have been tasked with deriving the empirical formula of butadiene. What steps would you take to find the formula? What is the empirical formula?
2. Why is the chemical formula for styrene molecular rather than ionic?
3. A polymerization process is used to create a synthetic rubber material known as styrene butadiene copolymer. Explain the difference between a chemical formula and a chemical equation, using this copolymer as an example.

Chemical Reactions and Equations

photo: Getty Images

LESSON OVERVIEW

Lesson Questions

- What are five types of chemical reactions?
- How can you predict the products of each of the five types of chemical reactions?
- How can you represent chemical reactions using chemical equations?

Lesson Objectives

By the end of the lesson, you should be able to:

- Distinguish between five types of chemical reactions.
- Predict the products of a chemical reaction.
- Write balanced equations that represent chemical reactions.
- Write the net ionic equation and identify spectator ions for a chemical reaction.

Key Vocabulary

Which terms do you already know?

- ☐ anion
- ☐ balanced chemical equation
- ☐ cation
- ☐ coefficient
- ☐ combustion reaction
- ☐ decomposition reaction
- ☐ double-displacement reaction
- ☐ double-replacement reaction
- ☐ ion
- ☐ law of conservation of matter
- ☐ net ionic equation
- ☐ oxidation-reduction reaction
- ☐ phase of matter
- ☐ product
- ☐ reactant
- ☐ reactivity series
- ☐ redox reaction
- ☐ single-displacement reaction
- ☐ single-replacement reactions
- ☐ solubility
- ☐ synthesis reaction

ENGAGE

Why are fireworks different colors? What makes them explode? Visit the Engage page online to learn more.

Explain Question

What are the five main types of chemical reactions, and how can chemical equations help you predict the products of each of the five types of reactions?

EXPLORE

What Are Five Types of Chemical Reactions?

Classifying Reactions into Five Types

Even for relatively mundane purposes, we might want to predict the reactions that occur when two substances are put together. We know that it is safe to combine bleach and detergent in the washing machine, for instance. But is it safe to mix bleach with an ammonia-based cleaning product? (No, it is not.) We know that adding baking soda to batter will help give a cake the shape and texture we want. Rather than try to memorize all the chemical reactions that are possible, it makes sense to find patterns that help us predict reactions.

Chemists classify reactions into categories based on the behaviors of the chemicals involved. By understanding the different types of reaction, scientists can predict the outcome of almost any combination of chemical substances.

Synthesis Reactions

A synthesis reaction occurs when two or more reactants combine to form a single product. This type of reaction is represented by the following general equation:

$$A + X \rightarrow AX$$

A and X can be elements or compounds. AX is a compound.

To demonstrate a synthesis reaction, you can use a balloon filled with a mixture of hydrogen gas and oxygen gas. When a flame comes in contact with the balloon, the balloon bursts. However, the bang is much louder than that of just a popped balloon. The hydrogen and oxygen in the balloon react to form water. In this reaction, hydrogen and oxygen are the reactants, or the starting materials. Water is the product. This chemical reaction is represented by the following equation:

$$2H_2(g) + O_2(g) \rightarrow 2H_2O(g)$$

Decomposition Reactions

In a decomposition reaction, a single compound undergoes a reaction that produces two or more simpler substances. The products might be elements, less complex compounds, or a combination of elements and compounds. This type of reaction is represented by the following general equation:

$$AX = A + X$$

A and X can be elements or compounds. AX is a compound.

To release hydrogen for use as a fuel, chemists use electrical energy to break down water into its constituents, hydrogen and oxygen. This reaction, the reverse of the synthesis reaction demonstrated with the balloon, is represented by the following equation:

$$2H_2O(g) \rightarrow 2H_2(g) + O_2(g)$$

Most decomposition reactions only take place when energy, such as heat or electricity, is added. The process of using an electric current to initiate and drive a reaction is called electrolysis.

Combustion

During a combustion reaction, an element or compound rapidly combines with oxygen to form an oxide. The reaction releases heat and light.

The flame of a laboratory burner represents a combustion reaction. Methane, CH_4, supplied to the burner reacts with oxygen, and carbon dioxide and water are produced. Each of these products is an oxide. Combustion reactions are commonly used in heating, cooking, and transportation. This reaction is represented by the following equation:

$$CH_4(g) + 2O_2(g) \rightarrow CO_2(g) + 2H_2O(g)$$

photo: Paul Fuqua

A fire, like the one you might make in a fireplace, is the result of a certain type of chemical reaction. How does the combustion of a log occur?

Some synthesis reactions can also be categorized as combustion reactions. The combustion reaction between hydrogen and oxygen is also a synthesis reaction. Hydrogen combines with oxygen to form an oxide (water) and the reaction releases thermal energy.

Single-Displacement Reactions

The final two types of reactions are displacement reactions. They are also called replacement reactions because atoms or ions in the reactants are replaced with other elements.

A single-displacement reaction is a type of reaction that occurs when an atom or ion in a compound is replaced by a similar atom or ion. Single-displacement reactions can be represented by the following general equations:

$$A + BX \rightarrow AX + B$$

or

$$Y + BX \rightarrow BY + X$$

A, B, X, and Y are elements. AX, BX, and BY are compounds.

For metals, a single-displacement reaction will occur if the element being replaced is lower in the reactivity series than the element that will take its place. When copper is placed in an aqueous solution of silver nitrate, two visible changes occur. The copper metal becomes covered with crystals of silver metal, and the solution becomes blue. This reaction is an example of a single-displacement reaction. In this case, the more active metal, copper, is replacing the less active metal, silver in solution. This reaction is represented by the following equation:

$$Cu(s) + AgNO_3(aq) \rightarrow CuNO_3(aq) + Ag(s)$$

By conducting tests using combinations of elements and compounds, chemists have developed a list to predict when single-displacement reactions will happen A reactivity series, sometimes referred to as an activity series, lists elements from the most reactive to the least reactive.

Reactivity Series of Metals	
K	Most reactive
Ca	↓
Na	
Mg	
Al	
Zn	
Fe	
Ni	
Sn	
Pb	
Cu	
Ag	
Au	Least reactive

A list of metals in order of reactivity is used to predict the products of chemical reactions. Why are some metals more reactive than others?

In the example, the silver cations in the solution are reduced. They gain electrons and become silver atoms. The blue color of the solution is caused by the presence of copper(II) cations. Some copper atoms in the metal are oxidized. They give up electrons and become copper(II) ions. Silver (Ag) is located below copper (Cu) in the reactivity series. In this single-displacement reaction, the copper atoms replace the silver ions in silver nitrate.

photo: Getty Images

These two flasks contain copper and silver nitrate. Which flask was assembled first?

The most active metals, such as those in Group 1, react vigorously with water to replace hydrogen in water. Less active metals, such as zinc, require the water to be in the form of steam for this reaction to take place. These reaction types are represented by the following equations:

$$2K(s) + 2H_2O(l) \rightarrow 2KOH(aq) + H_2(g)$$

$$Zn(s) + H_2O(g) \rightarrow ZnO(s) + H_2(g)$$

In a different type of single-displacement reaction, one halogen replaces another halogen in a compound. Fluorine is the most active halogen. It will replace all other halogens in their compounds. Each halogen is less active than any halogen shown on the periodic table. This reaction type is represented by the following equations:

$$Cl_2(g) + 2KBr(aq) \rightarrow 2\ KCl(aq) + Br_2(l)$$

$$Br_2(l) + 2KCl(aq) \rightarrow \text{no reaction}$$

Double-Displacement Reactions

In a double-displacement reaction, the ions of two compounds exchange places and form a new compound. This means that the cations and anions trade places. Double-displacement reactions can be represented by the following general equation:

$$AX + BY = AY + BX$$

A, B, X, and Y in the reactants represent ions. AY and BX are ionic or molecular compounds.

For example, a sodium chloride solution is composed of sodium ions and chloride ions dissolved in water. A silver nitrate solution is composed of silver ions and nitrate ions dissolved in water. When the two solutions mix, the ions collide with each other. Sodium ions and nitrate ions remain dissolved because sodium nitrate is soluble. However, when silver and chloride ions collide, they bond together and form a precipitate. This reaction is represented by the following equation:

$$NaCl(s) + AgNO_3(aq) \rightarrow AgCl(s) + NO_3^{-1}(aq) + Br^{+1}(aq)$$

Solubility describes how readily a substance dissolves in water. Solubility rules help scientists determine whether an ionic compound will be soluble or insoluble. The formation of a precipitate when two solutions are mixed is often a sign that a double-displacement reaction has occurred. The formation of the compound, water, or a gas are also signs that a double-displacement reaction has occurred.

Reactions between acids and bases are a type of double-displacement reaction. When acids and bases react, the products are usually water and a salt. For example, when hydrochloric acid reacts with sodium hydroxide, the products are the salt sodium chloride and water.

$$HCl_{(aq)} + NaOH_{(aq)} \rightarrow NaCl_{(aq)} + H_2O_{(aq)}$$

Solubility Rules	
Compounds of These Ions Tend to Be Soluble	**Exceptions**
Alkali metal	
Ammonium	
Nitrate	
Acetate	
Chloride, Bromide	Pb^{2+}, Hg_2^{2+}, Ag^+
Iodide	Hg^{2+}, Pb^{2+}, Hg_2^{2+}, Ag^+
Sulfate	Ca^{2+}, Sr^{2+}, Ba^{2+}, Pb^{2+}, and Hg_2^{2+}
Compounds of These Ions Tend to Be Insoluble	**Exceptions**
Carbonate	Alkali metal and ammonium
Chromate	Alkali metal and ammonium
Phosphate	Alkali metal and ammonium

A list of metals in order of reactivity is used to predict the products of chemical reactions. Why are some metals more reactive than others?

Oxidation-Reduction Reactions

Oxidation-reduction reactions, also known as redox reactions, are reactions in which one element loses electrons, or is oxidized, while another gains electrons, or is reduced.

This is a different way of classifying reactions, so many reactions can be classified as both a redox reaction and another type of reaction. For example, single-displacement reactions and combustion reactions are also redox reactions. Many synthesis and decomposition reactions are also classified as redox reactions.

Chemical Bonding and Chemical Reactions

What happens during a chemical reaction? To answer this question, we must observe the action taking place at the atomic level. At this level, we would see that the chemical bonds linking atoms together in a reactant molecule are broken, and that the atoms reorganize in new arrangements that allow new bonds to form between new pairs of atoms. The result is a new substance—the product.

Chemical reactions involve the breaking of one set of bonds and the forming of a new set of bonds. The number and type of atoms remains the same throughout this process, so that only the bonding patterns change. This is an important feature of all chemical reactions—they obey the law of conservation of matter. This law, sometimes referred to as the law of conservation of mass, states that matter cannot be gained or lost in a chemical reaction.

The Forces behind Chemical Bonding

What is involved as chemical bonds in reactants are broken and then new chemical bonds are formed in products? First, the reactant molecules must collide to be able to exchange atoms. Think about the reaction between hydrogen and oxygen:

$$2H_2 + O_2 \rightarrow 2H_2O$$

Even though the molecules bump into each other, a mixture of hydrogen and oxygen gases is stable under ordinary conditions. But when a flame is brought into a mixture of H_2 and O_2, BOOM! The reaction that ensues is explosive. We can understand why the flame causes this reaction when we consider the different forces involved.

photo: Getty Images

In 1937, forces holding together the stable hydrogen atoms filling the airship Hindenburg were disturbed by an ignition event. What did these hydrogen atoms bond with in the explosion?

H_2 and O_2 molecules are normally stable because the molecules have less energy than the separate atoms. Within each molecule, there is a balance of electromagnet c forces operating on the two atoms bonded in each molecule. The two atoms are pushed apart as the positive charges of the atoms' nuclei repel one another. The two atoms are also pulled together as the negatively charged electrons of one atom attract the positive nucleus of the other atom.

In order for a stable atom to break apart, molecules must collide with energy great enough to break the chemical bonds. In the case of the hydrogen-oxygen reaction, a flame provides this energy. The heated molecules have more kinetic energy and collide with each other with more force.

Once the bonds between atoms are broken, the atoms can recombine to form new stable molecules with less energy than the reactant molecules or the individual atoms. In the case of H_2 and O_2, the hydrogen atoms may form bonds with oxygen instead of with each other.

The Energy Associated with Chemical Bonding

As you saw with the reaction between H_2 and O_2 to form H_2O, molecules need to collide with enough energy for atoms to rearrange and form new bonds. When new bonds form, atoms tend to form bonds with the lowest possible potential energy. Energy is released. Because the chemical energy in molecules tends to be at a minimum, we rarely observe individual atoms in nature; instead, we observe atoms joined together by bonds. The stability in the compounds we observe is due to the lower energy of the bonded atoms.

Every reactant and every product has a certain amount of energy associated with its bonds. This energy is called bond energy. The bond energy of the O-H bond is higher than that of an H-H bond, but lower than that of an O-O bond. You can find the total bond energy of a molecule by adding up all of the bond energies for that molecule.

Because energy is never created or destroyed, the sum of all the energy present before a chemical reaction has to be equal to the sum of all the energy present after a chemical reaction. We can write an equation to show this:

Sum of bond energies of reactants + energy input

=

Sum of bond energies of products + energy output

The minimum amount of energy needed for a reaction to begin is called the activation energy. This input of energy occurs in both exothermic and endothermic reactions.

In the case of the reaction between H_2 and O_2, the sum of the bond energies of the reactants was large compared with the sum of the bond energies of the product. A small amount of energy was added in the ignition flame (energy input), and a large amount of energy was released as the products formed (energy output). This reaction is said to be exothermic because energy is released as light and heat.

An endothermic reaction is one in which the sum of all the bond energies of the products is greater than that of the reactants.

How Can You Predict the Products of Each of the Five Types of Chemical Reactions?

Predicting the Products of a Reaction

During a synthesis reaction, two reactants form a single product. When the reactants are elements, the product is usually a binary compound composed of those elements. For example, the reaction of a metal with a nonmetal can produce a binary ionic compound. The formula of the product can be predicted from the ions that are likely to form. A bright white light is given off when magnesium reacts with oxygen. The product of this dramatic reaction between a metal and a nonmetal is the ionic compound magnesium oxide, MgO. When aluminum reacts with chlorine, the ionic compound aluminum chloride, $AlCl_3$, is the product.

Many decomposition reactions are the reverse of synthesis reactions. Therefore, in most cases, we can predict that a binary compound will decompose into its elements. Thus, sodium chloride can be decomposed into the elements sodium and chlorine. Water decomposes to form the elements hydrogen and oxygen.

Some compounds that are composed of polyatomic ions decompose in predictable ways. For example, metal carbonates break down to produce carbon dioxide and an oxide. Calcium carbonate in limestone decomposes upon heating. The products are carbon dioxide gas and calcium oxide.

$$CaCO_3(s) \rightarrow CaO(s) + CO_2(g)$$

Many acids decompose into nonmetal oxides and water. The breakdown of carbonic acid, H_2CO_3, in a carbonated drink produces bubbles of carbon dioxide gas. The other product of this decomposition is water, an oxide of hydrogen.

$$H_2CO_3 \rightarrow CO_2(g) + H_2O(g)$$

If the hydroxide ion is present in a compound, the products of the decomposition of the compound are water and a metal oxide. So, the decomposition of sodium hydroxide produces water and sodium oxide.

Because combustion reactions always involve a reaction with oxygen, the products are always oxides. Combustion reactions that are also synthesis reactions will produce a binary compound. For example, if a hydrocarbon such as methane is combined with excess oxygen, the products are carbon dioxide and water.

In a single-displacement reaction, an element and a compound react. The products are a different element and a new compound. To predict exactly what the products will be, remember that the replacement occurs between elements that have similar properties. Thus, the single-displacement reaction between copper and silver nitrate produces silver and copper(II) nitrate. In this case, metallic copper replaces metallic silver.

What will happen if silver metal is placed in a copper(II) nitrate solution? Nothing! The silver atoms cannot replace the copper(II) ions. This agrees with the information in the reactivity series: a single-displacement reaction will occur if the element being replaced is lower in the series than the element that would take its place. Copper, being above silver in the series (more reactive than silver), cannot be replaced by silver. However, if copper metal is placed in a silver nitrate solution, the reactants produce silver and copper nitrate, caused by a single-displacement reaction where copper replaces silver to produce copper nitrate.

In a double-displacement reaction, the reactants are two compounds, and the products are two different compounds. The products can be predicted by switching the ions in the compounds. Therefore, when silver nitrate and sodium chloride react, the products are sodium nitrate and silver chloride. This can be seen as either switching the cations or switching the anions.

Not all pairs of compounds will undergo a double-displacement reaction. For example, what products form when solutions of sodium chloride and potassium nitrate are combined? It might seem likely that the products would be potassium chloride and sodium nitrate. But when sodium chloride and potassium nitrate are actually combined, there is no reaction. Predictions about a double-displacement reaction must be checked against the solubility rules. If neither product is insoluble, then a reaction will not occur. No precipitate will form, and the ions will all remain in the solution.

Predicting the Products of a Reaction: Sample Problem

What substance forms when aluminum reacts with sulfur?

Solution

The reactants, aluminum and sulfur, are two elements. The type of reaction that occurs is a synthesis reaction. The synthesis reaction between a metal and a nonmetal produces a single, binary ionic compound. The product is aluminum sulfide, Al_2S_3.

How Can You Represent Chemical Reactions Using Chemical Equations?

Writing and Balancing Chemical Equations

Scientists use chemical symbols to represent the elements. They use chemical formulas to represent compounds. In a similar way, they use chemical equations to represent chemical reactions.

One way to represent a chemical reaction is by writing a word equation. A word equation separates the names of the reactants from the names of the products using an arrow. The arrow can be read as "yields" or "produces." If the reaction involves more than one reactant or product, a plus sign is placed between the corresponding reactants or products. The word equation that represents the synthesis of water from hydrogen and oxygen is:

$$\text{hydrogen} + \text{oxygen} \rightarrow \text{water}$$

A formula equation is written by replacing each name with the chemical symbol or formula that represents that substance. Hydrogen and oxygen are both elements that exist at room temperature as molecules rather than single atoms. The formulas H_2 (hydrogen gas) and O_2 (oxygen gas) are therefore used instead of just H and O in the equation.

The formula equation is shorter than the descriptive word sentence. It also provides a visual description of the chemicals involved.

$$H_2 + O_2 \rightarrow H_2O$$

The formula equation is shorter than the descriptive word sentence. It also provides a visual description of the chemicals involved.

A key principle in writing formula equations is to ensure they are balanced. The following formula equation is still incomplete. Notice that there is one less oxygen atom on the product side of the equation than on the reactant side. We know that the oxygen atom did not simply disappear, so it must be accounted for in the equation.

During the synthesis of water, the chemical bonds in the hydrogen and oxygen molecules break. The atoms form new bonds as they reorganize to form water molecules. Each atom from the reactants becomes part of the product. To make sure that all of the atoms in the reactants and the products are accounted for, the formula equation must be balanced. A balanced reaction shows the same number of atoms of each element in both the reactants and the products. To balance a chemical equation, coefficients are placed before chemical formulas.

For the synthesis of water, a coefficient is needed before the water molecule. This balances the oxygen atoms:

$$H_2 + O_2 \rightarrow 2H_2O$$

Notice that the subscript for oxygen in H_2O is not changed. Subscripts should never be changed to balance an equation. Changing a subscript would change a chemical formula, and the equation would no longer represent the same chemical reaction.

Now, we know two water molecules have formed for each molecule of reactant, but the equation is still not balanced. Due to the coefficient, there are now four hydrogen atoms on the product side of the arrow, but only two on the reactant side. Two more hydrogen atoms are needed in the reactants. A coefficient placed in front of H_2 balances the hydrogen atoms:

$$2H_2 + O_2 \rightarrow 2H_2O$$

Writing and Balancing Chemical Equations: Sample Problem

Write a balanced chemical equation for the formation of potassium chloride and oxygen from the decomposition of potassium chlorate.

Solution

The reactant for this decomposition reaction is potassium chlorate. The products are potassium chloride and oxygen. The word equation for the reaction is:

$$\text{potassium chlorate} \rightarrow \text{potassium chloride} + \text{oxygen}$$

The formula equation for the reaction is:

$$KClO_3 \rightarrow KCl + O_2$$

There is one atom of potassium on each side of the equation. There is also one atom of chlorine on each side of the equation. However, there are three oxygen atoms in the reactants and only two in the products.

The least common multiple of two and three is six. Balance the oxygen atoms with coefficients so that six atoms of oxygen are on each side of the equation:

$$2\,KClO_3 \rightarrow KCl + 3\,O_2$$

Now the oxygen atoms are balanced. However, now the potassium and chlorine atoms are not. To resolve this, place a coefficient in front of KCl:

$$2\,KClO_3 \rightarrow 2\,KCl + 3\,O_2$$

Conservation in Chemical Reactions

The equations for chemical reactions always balance because they follow the law of conservation of matter. During a chemical reaction, matter is neither created nor destroyed. Each atom from the reactants is used in the products. Each atom has mass. A balanced equation, therefore, also shows that the mass of the reactants equals the mass of the products. However, the number of molecules is not always conserved during a reaction. This is evident from the equations for synthesis and decomposition reactions.

Writing Net Ionic Equations

A balanced equation provides information about a chemical reaction. One important piece of information is the matter state of each substance involved. By convention, the symbols (s), (l), (g), and (aq) are used to distinguish these states.

Displacement reactions often involve ionic compounds in water, or aqueous solution. Solid ionic compounds, however, may not be reactive at all. When dry, crystals of table salt, for example, do not react with other compounds. When an ionic compound is dissolved in water, however, its ions can move throughout the solution and interact with other ions.

If sodium sulfate and barium nitrate are both dissolved in water and then combined, the reaction is described by the equation:

$$Na_2SO_4(aq) + Ba(NO_3)_2\,(aq) \rightarrow 2\,NaNO_3(aq) + BaSO_4(s)$$

One of the products is in the solid state because it has formed as a precipitate. The sodium nitrate remains dissolved in water. The solubility rules identify barium sulfate as insoluble.

It seems as though two compounds should be formed by this reaction. Remember, however, that when ionic compounds are dissolved their ions move freely through the water. The reaction of sodium sulfate and barium nitrate could also be described with the equation:

$$2\,Na^+(aq) + SO_4^{2-}(aq) + Ba^{2+}(aq) + 2\,NO_3^-\,(aq) \rightarrow 2\,NA^+(aq) + 2\,NO_3^-(aq) + BaSO_4(s)$$

The sodium ions and nitrate ions have not changed at all. Each ion is dissolved in water before and after the reaction. Ions that are unchanged during a displacement reaction are known as spectator ions. These ions "watch" the action, but do not participate. Because they are not involved in the reaction, the spectator ions can be removed from the chemical equation. The resulting net ionic equation includes only the substances that take part in a reaction.

$$SO_4^{2-}(aq) + Ba^{2+}(aq) \rightarrow BaSO_4(s)$$

The two ions in solution react to form a precipitate.

Writing Net Ionic Equations: Sample Problem

Write the net ionic equation for the reaction that occurs when a piece of zinc metal is placed into an aqueous solution of copper(II) sulfate.

Solution

The reactants include an element and an ionic compound in solution. This reaction is a single-displacement reaction. Zinc is above copper in the reactivity series, so a reaction will occur. The zinc replaces copper to form zinc sulfate:

$$Zn(s) + CuSO_4(aq) \rightarrow ZnSO_4(aq) + Cu(s)$$

The total ionic equation is:

$$Zn(s) + Cu^{2+}(aq) + SO_4^{2-}(aq) \rightarrow Zn^{2+}(aq) + SO_4^{2-}(aq) + Cu(s)$$

The sulfate ion is unchanged during the reaction. Remove this spectator ion to find the net ionic equation:

$$Zn(s) + Cu^{2+}(aq) \rightarrow Zn^{2+}(aq) + Cu(s)$$

Consider the Explain Question:

What are the five main types of chemical reactions, and how can chemical equations help you predict the products of each of the five types of reactions?

Go online to complete the scientific explanation

Check for Understanding:

How does a balanced chemical equation show that the law of conservation of mass is obeyed?

STEM in Action

STEM and Chemical Reactions and Equations

Starting in the late 1700s and early 1800s, a large portion of the manufacturing processes in the United States became dependent on machines. This change, called the Industrial Revolution, had enormous social benefits, but over time, it became apparent that there were also environmental drawbacks. A troubling environmental effect of the Industrial Revolution is acid precipitation, also known as acid rain. Acid precipitation is any form of precipitation that has an increased level of acidity. Certain chemicals released by factories and automobiles, as well as natural sources, can undergo synthesis reactions with rainwater to produce acids. The combination of acids and rainwater is acid precipitation.

Excess carbon dioxide released into the atmosphere, for example, dissolves in and reacts with rainwater. The resulting synthesis reaction forms carbonic acid, which causes precipitation to be naturally acidic:

$$H_2O(i) + CO_2(g) \rightarrow H_2CO_3(aq)$$

Other compounds released into the atmosphere, such as nitrogen oxides and sulfur dioxide, can also undergo synthesis reactions with rainwater and result in precipitation that is many times more acidic than normal.

photo: Paul Fuqua

Trees at high altitudes are at increased risk of damage from acid precipitation. What causes the precipitation to be acidic?

Atmospheric scientists monitor the levels of certain compounds that have been released in the air that result in acidic precipitation. Sulfur dioxide, for example, is a product of the combustion reaction of sulfur-containing fuels, such as coal or farm diesel fuel. The sulfur dioxide enters the atmosphere, where it reacts with oxygen in the following synthesis reaction.

$$2SO_2(g) + O_2(g) \rightarrow 2SO_3(g)$$

The sulfur trioxide then reacts with water to form sulfuric acid.

$$H_2O(i) + SO_3(g) \rightarrow H_2SO_4(aq)$$

Acid precipitation damages structures such as bridges and buildings. It also harms plants and animals. It can make lakes too acidic to support aquatic plants and animals. Acid precipitation leaches chemicals from soil and rocks. These chemicals, which may include lead and mercury, are carried to streams and lakes, where they can harm organisms.

Atmospheric Sulfur Dioxide

This graph shows the level of atmospheric sulfur dioxide in different areas of the world since 1850. Use the graph and other sources to help you to answer the questions.

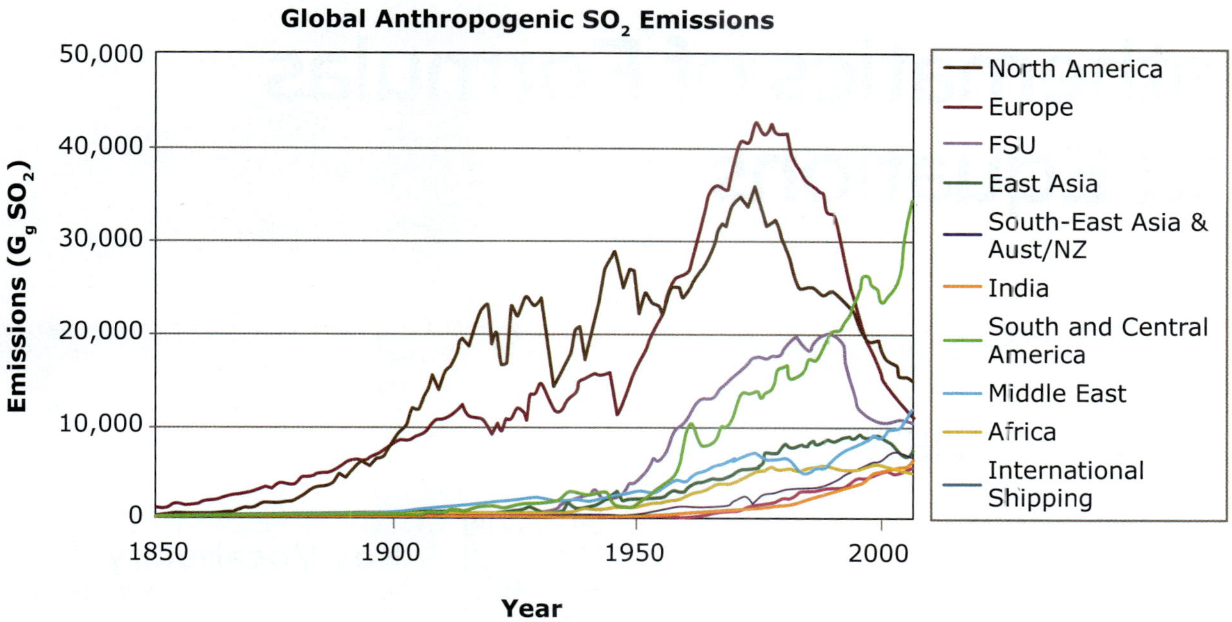

1. Describe the trends in atmospheric sulfur dioxide levels since 1850.
2. What effect do you think changes in sulfur dioxide levels likely had on levels of acid precipitation during this time period? Why?
3. Suggest a possible reason or reasons for the change in sulfur dioxide levels around 1970–1980. Why do you think the changes did not occur worldwide?

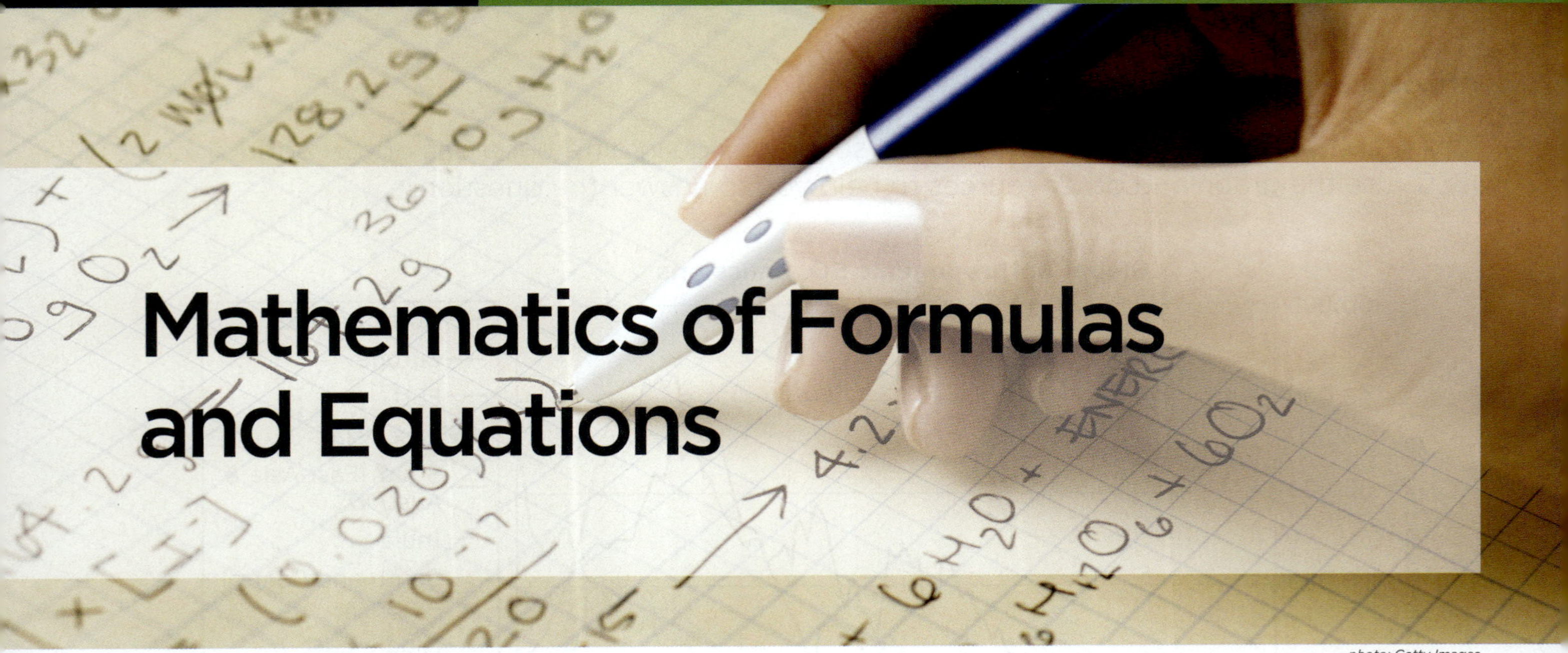

Mathematics of Formulas and Equations

photo: Getty Images

LESSON OVERVIEW

Lesson Questions

- How is molar mass calculated and why is it useful?
- How are percent composition of a compound, empirical formula of a compound, and molecular formula of a compound calculated?
- How are the principles of stoichiometry used to calculate quantities of reactants or products in a chemical reaction?

Lesson Objectives

By the end of the lesson, you should be able to:

- Determine molar mass and formula mass.
- Determine the percent composition of a compound.
- Apply stoichiometric principles.

Key Vocabulary

Which terms do you already know?

- ☐ **atom**
- ☐ **average atomic mass**
- ☐ **Avogadro's number**
- ☐ **balanced chemical equation**
- ☐ **chemical reaction**
- ☐ **compound**
- ☐ **crystal lattice**
- ☐ **element**
- ☐ **empirical formula**
- ☐ **formula mass**
- ☐ **formula unit**
- ☐ **limiting reagent**
- ☐ **mass**
- ☐ **molar mass**
- ☐ **mole**
- ☐ **molecular formula**
- ☐ **molecular mass**
- ☐ **molecule**
- ☐ **percent composition**
- ☐ **percent yield**
- ☐ **periodic table of the elements**
- ☐ **product**
- ☐ **reactant**
- ☐ **standard temperature and pressure (STP)**
- ☐ **stoichiometry**

ENGAGE

Look at all of the things around you right now. Perhaps you see chairs, carpeting, and light fixtures. What role do you think chemical formulas and equations played in the design of these materials? Visit the Engage page online to learn more.

Explain Question

How can you use average atomic masses from the periodic table to find the formula mass and the molar mass of ammonia, NH_3, and the percent composition of each element in ammonia?

EXPLORE

How Is Molar Mass Calculated and Why Is It Useful?

The Mole

In chemistry, it is rare that anyone works with single atoms or molecules. More often, chemists work with amounts of matter that are easily observed. Such matter is measured in gram or kilogram quantities. These quantities contain billions of atoms or molecules.

Recall that a mole is defined as 6.022×10^{23} atoms or molecules. A mole is abbreviated "mol" in calculations. This number is called Avogadro's number in honor of Amedeo Avogadro, the Italian scientist who showed that 1 mole of any gas has the same number of particles as 1 mole of any other gas.

Because 1 mole contains so many particles, it offers a convenient unit for expressing the gram and kilogram quantities of substances that chemists tend to work with. Instead of trying to talk about millions of atoms or molecules, they can talk about moles of atoms or molecules instead.

Molar Mass of an Element

Since Avogadro's time, much evidence has been gathered to support the mole concept. Chemists now use this idea often and have extended it to include all substances. One mole of any pure substance has the same number of particles as 1 mole of any other pure substance.

However, even though 1-mole quantities of two substances contain the same number of particles, they do not have the same mass. Mass can vary depending on the identities of the elements present in each substance.

Suppose a chemist has 1 mole of helium gas and 1 mole of neon gas. An atom of neon is heavier than an atom of helium. This means that a mole of neon will be heavier than a mole of helium.

The mass of 1 mole of an element is easily determined from the average atomic mass of that element. Locate the average atomic mass of the element listed on the periodic table. The value shown is equal to the mass in grams of 1 mole of that element.

Using this approach, the masses of one atom of helium, one atom of neon, and 1 mole of each of these elements can be summarized as shown.

Comparing Mass Table		
	Mass of One Atom	**Mass of One Mole**
Helium	4.002 amu	4.002 g
Neon	20.180 amu	20.180 g

This table shows the relationship between the amu and gram weight of atoms. What is the molar mass of hydrogen, H_2, if its atomic mass is 1.008 amu?

The molar mass of an element is defined as the mass of 1 mole of that element. The units for molar mass are grams per mole, so neon's molar mass is written as 20.180 g/mol. The molar mass of each element can be found on the periodic table.

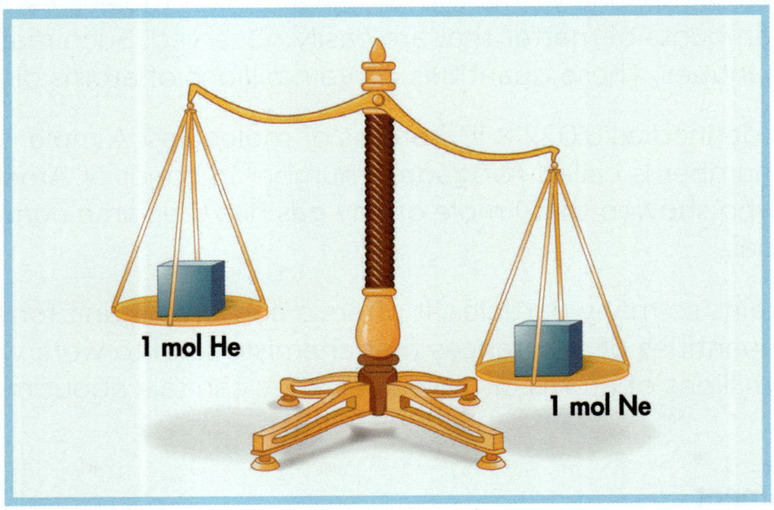

One mole of helium atoms contains the same number of atoms as 1 mole of neon atoms, but has less mass. Why isn't the mass the same?

Molar Mass of a Compound

Recall that compounds are composed of two or more elements whose atoms are bonded together. The molar mass of a compound can be found using the molar masses of its elements.

Consider the compound methane for example. Methane's chemical formula is CH_4. Its molar mass is determined by adding together the molar masses of each carbon and hydrogen atom present in one molecule.

One mole of CH_4 contains 1 mole of carbon atoms and 4 moles of hydrogen atoms. Its molar mass is the sum of the molar masses of these elements:

$$\text{1 mole of carbon: } 12.011 \text{ g/mol}$$

$$\text{4 moles of hydrogen } 4(1.008) \text{ g/mol} = 4.032 \text{ g/mol}$$

$$\text{Total mass} = 12.011 \text{ g/mol} + 4.032 \text{ g/mol} = 16.043 \text{ g/mol}$$

The compound methane has a molar mass of 16.043 grams per mole.

Using Molar Mass in Conversions

It is not easy to measure moles in the lab. It is, however, easy to measure mass in the lab. Electronic balances are standard pieces of equipment in most chemistry labs.

Often, chemists perform calculations to determine the number of moles of a substance needed for a particular reaction. They then do another calculation to convert moles to grams. Knowing the number of grams allows them a means for accurately measuring out the needed quantity of the substance.

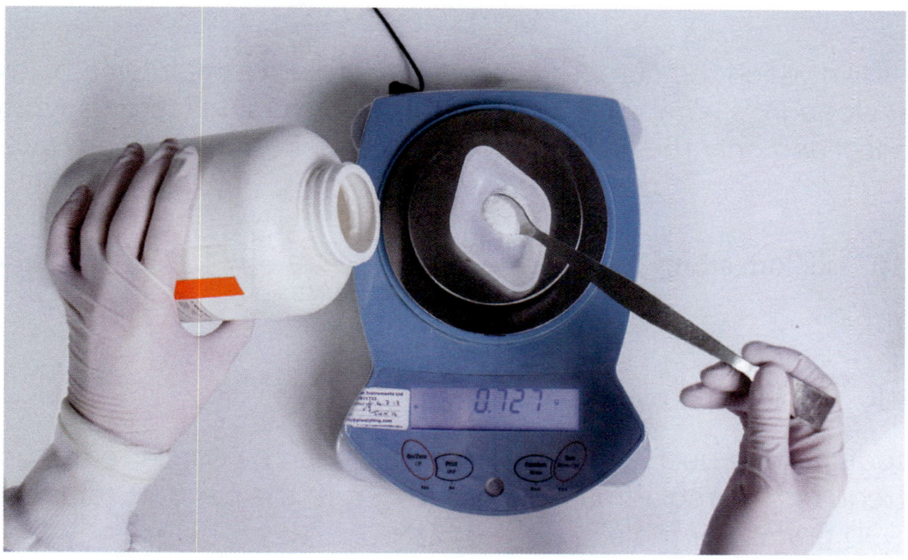

photo: Getty Images

This digital scale is used to measure the mass of a substance. What are the advantages of using digital scales in the laboratory?

To do this type of calculation, chemists use the molar mass as a conversion factor. In general, the following scheme summarizes the types of conversions that are possible:

$$\text{atoms or molecules} \underset{\substack{\text{use} \\ \text{Avogadro's} \\ \text{number}}}{\longleftrightarrow} \text{moles} \underset{\substack{\text{use} \\ \text{molar} \\ \text{mass}}}{\longleftrightarrow} \text{grams}$$

Numbers of atoms or molecules can be converted to moles using Avogadro's number. Moles can be converted to grams using molar mass.

These conversions can also be carried out in reverse. Grams can be converted to moles using molar mass. Moles can be converted to numbers of atoms or molecules using Avogadro's number.

Molar Mass in Conversions: Sample Problem

Calculate the number of moles and the mass in grams of 2.57×10^{18} atoms of sodium.

Solution

The number of atoms of sodium is given. The first step will involve a conversion of this given value to the next logical value. In the following scheme, Avogadro's number can be used to convert numbers of atoms to moles:

$$\text{atoms or molecules} \xleftrightarrow[\text{Avogadro's number}]{\text{use}} \text{moles} \xleftrightarrow[\text{molar mass}]{\text{use}} \text{grams}$$

To convert numbers of atoms to moles, use the conversion factor based on Avogadro's number:

$$1\ \text{mole} = 6.022 \times 10^{23}\ \text{atoms}$$

Set up an equation in which the number of atoms of sodium is multiplied by a conversion factor based on Avogadro's number. Be sure that the conversion factor is written so that like units (atoms) cancel and only the units of moles will be left in the final answer:

$$(2.57 \times 10^{18}\ \cancel{\text{sodium atoms}})\left(\frac{1\ \text{mol sodium}}{6.022 \times 10^{23}\ \cancel{\text{sodium atoms}}}\right) = 4.27 \times 10^{-6}\ \text{mol sodium}$$

There are 4.27×10^{-6} moles of sodium present.

Once moles of sodium have been calculated, this value can be converted to grams of sodium. This conversion requires the molar mass of sodium. From the periodic table, the molar mass of sodium is 22.990 grams per mole.

Set up an equation in which the moles of sodium are multiplied by a conversion factor based on sodium's molar mass. Be sure that the conversion factor is written so that like units (moles) cancel and only the units of grams will be left in the final answer:

$$(4.27 \times 10^{-6}\ \cancel{\text{mol sodium}})\left(\frac{22.990\ \text{g sodium}}{1\ \cancel{\text{mol sodium}}}\right) = 9.81 \times 10^{-5}\ \text{g sodium}$$

There are 9.81×10^{-5} grams of sodium present.

How Are Percent Composition of a Compound, Empirical Formula of a Compound, and Molecular Formula of a Compound Calculated?

Stoichiometry

The term stoichiometry is used to describe numerical relationships in chemistry. Stoichiometric relationships based on the mole are always whole number ratios. There are two types of stoichiometry:

- Composition stoichiometry is related to numerical relationships between elements in chemical compounds. For example, a water molecule contains a 2 : 1 molar ratio of hydrogen atoms to oxygen atoms.
- Reaction stoichiometry is related to molar ratios of reactants and products in chemical reactions. For example, in the reaction of hydrogen with oxygen to form water, there is a 1 : 1 molar ratio of H_2 reactant to H_2O product.

Combustion of Hydrogen

Three Ways of Representing

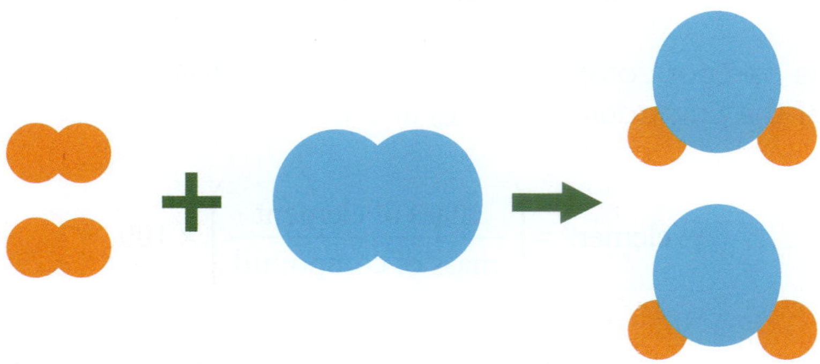

Both composition stoichiometry and reaction stoichiometry are present in the reaction of hydrogen with oxygen to produce water. What example of each type is present?

Examples of Composition Stoichiometry

Composition stoichiometry is related to the numerical relationships between elements in chemical compounds. The different ways that compounds can be described include the following:

- percent water of a hydrate
- percent composition
- empirical formula
- molecular formula

The following first example listed is restricted to compounds that form hydrates, while the other three apply to all compounds. Each of these will be discussed in the order shown.

Calculating the Percent Water in a Hydrate

Hydrates are compounds that bind water molecules in specific ratios when they crystallize. An example is copper(II) sulfate pentahydrate.

$CuSO_4{\cdot}5H_2O$ is bright blue in color. Within the crystal lattice of this compound are five water molecules for every one copper (II) sulfate formula unit. When the compound is heated, the water molecules can be evaporated leaving a white powder, $CuSO_4$.

The percent water in this hydrate is found by the following equation:

$$\%\text{ water} = \left(\frac{\text{grams } H_2O}{\text{grams } CuSO_4 \cdot 5H_2O}\right) \times 100$$

Calculating Percent Composition

Percent composition is defined as the percent by mass of each element in the compound. These values can be determined if the molecular formula is known. Molar masses are used to do these calculations.

For each element, the percent composition is given by the following equation. It is useful to assume that 1 mole of the compound is being analyzed.

$$\%\text{ element} = \left(\frac{\text{mass of element}}{\text{mass of compound}}\right) \times 100$$

The percent by mass of every element in the compound is calculated in this way. When these calculations are complete, check them by adding them together. The sum should add up to 100 percent. This is the theoretical percent by mass of the compound. If the compound is separated into its element experimentally, the mass may be exactly 100 percent due to experimental error.

Calculating Percent Composition: Sample Problem

Calculate the percent composition of water, H_2O.

Solution

First, calculate the molar mass of water. One water molecule contains two hydrogen atoms and one oxygen atom:

$$\begin{aligned}&\text{1 mole of oxygen: } 15.999 \text{ g/mol}\\&\text{2 moles of hydrogen: } 2(1.008) \text{ g/mol} = 2.016 \text{ g/mol}\\&\text{total mass} = 15.999 \text{ g/mol} + 2.016 \text{ g/mol} = 18.015 \text{ g/mol}\end{aligned}$$

Next, assume that 1 mole of water is being analyzed. Calculate the mass of 1 mole of water using the molar mass of water:

$$\text{mass of } H_2O = 1\ \cancel{\text{mol } H_2O} \times \left(\frac{18.015\text{ g } H_2O}{1\ \cancel{\text{mol } H_2O}}\right) = 18.015\text{ g } H_2O$$

Calculate the mass of each element:

$$\text{mass of } H = 2\ \cancel{\text{mol } H} \times \left(\frac{1.008\text{ g } H}{1\ \cancel{\text{mol } H}}\right) = 2.016\text{ g } H$$

$$\text{mass of } O = 1\ \cancel{\text{mol } O} \times \left(\frac{15.999\text{ g } O}{1\ \cancel{\text{mol } O}}\right) = 15.999\text{ g } O$$

Finally, calculate the percent mass of each element by dividing the mass of the element by the mass of the compound. Multiply by 100 to obtain a percentage:

$$\%\, H = \left(\frac{2.016\text{ g } H}{18.015\text{ g } H_2O}\right) \times 100 = 11.19\%$$

$$\%\, O = \left(\frac{15.999\text{ g } H}{18.015\text{ g } H_2O}\right) \times 100 = 88.81\%$$

Check the results by adding the percentages to see if they add to give 100 percent:

$$\begin{array}{r} 11.19\% \\ +\,88.81\% \\ \hline 100.00\% \end{array}$$

The answers appear to be reasonable.

Molecular Formulas and Empirical Formulas

Recall that the molecular formula gives the number of atoms of each element found in one molecule of the compound. If these numbers are divisible by a common factor, they can be simplified to give the simplest ratio of elements in the compound. The simplest molar ratio of elements in a compound is the compound's empirical formula.

Nonmetallic elements often form several different compounds when they bond with another nonmetallic element. For example, sulfur and oxygen can bond to form sulfur monoxide, SO, or disulfur dioxide, S_2O_2. The empirical formula for both compounds is the same, SO, but their molecular formulas are different.

A Crystal Lattice

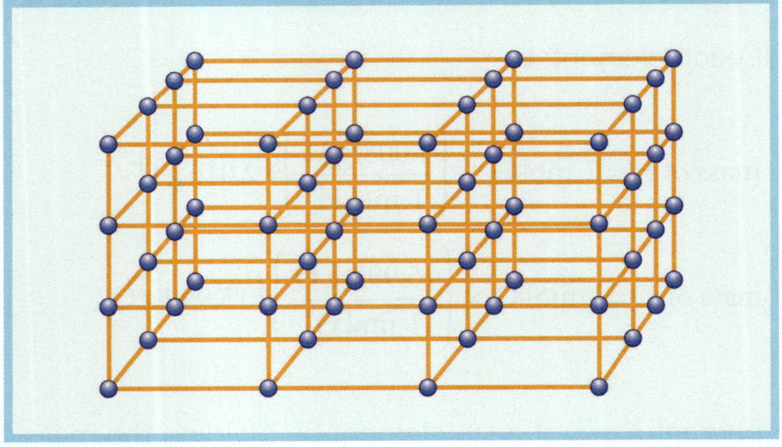

An ionic compound exists as a crystal lattice. How does this differ from a molecule?

Some compounds do not have molecular formulas and can only be described by empirical formulas. Ionic compounds are composed of repeating units of ions that form a crystal rather than individual molecules. Because these compounds do not form discrete molecules, they cannot be said to have molecular formulas. Instead, they are described only by empirical formulas that give the simplest ratios of the elements in the compound.

Empirical and Molecular Formulas		
Compound	**Molecular Formula**	**Empirical Formula**
Benzene	C_6H_6	CH
Acetylene	C_2H_2	CH
Hydrogen peroxide	H_2O_2	HO

This table shows molecular and empirical formulas for three compounds. Why are the empirical formulas of benzene and acetylene the same and yet their molecular formulas are different?

Calculating Empirical Formula from Percent Composition

When chemists synthesize or discover a new compound, they often are able to more easily determine its percent composition than its chemical formula. However, they can calculate the empirical formula of the compound once they know its percent composition. To do the calculations, it is helpful to assume that a 100-gram sample is being used. That way, the percentage values can be assumed to represent grams of each element. Then, follow these steps:

- Use the grams and the molar mass of each element to calculate the moles of each element.
- Compare the moles of all the elements to obtain molar ratios.
- Calculate the simplest ratios to find the empirical formula.

Calculating Molecular Formula from Empirical Formula

The empirical formula gives the simplest molar ratio of elements in a compound. The molecular formula gives the exact amounts of each element in one molecule of the compound. Therefore, the empirical and molecular formulas are related by some whole number factor.

$$x(\text{empirical formula}) = \text{molecular formula}$$

The molecular formula can be determined if the empirical formula and the molar mass of the compound are given. Calculate the formula mass by adding together the molar masses of the elements in the empirical formula. Then, divide the formula mass by the molar mass. They should be related by a factor that is a whole number.

$$\frac{\text{formula mass}}{\text{molar mass}} = x$$

This whole number can be used to multiply the molar ratios in the empirical formula to give the molecular formula.

How Are the Principles of Stoichiometry Used to Calculate Quantities of Reactants or Products in a Chemical Reaction?

Reaction Stoichiometry

Recall that there are two types of stoichiometry: composition stoichiometry and reaction stoichiometry. Composition stoichiometry deals with ratios of elements in chemical compounds. Reaction stoichiometry deals with ratios of all of the compounds involved in a chemical reaction. In the reaction of hydrogen and oxygen to form water, the reaction stoichiometry is given by the balanced equation:

$$2H_2 + O_2 \rightarrow 2H_2O$$

The reaction stoichiometries for this reaction can be summarized as follows:

- There is a 1 : 2 molar ratio of O_2 to H_2.
- There is a 1 : 1 molar ratio of H_2 to H_2O.
- There is a 1 : 2 molar ratio of O_2 to H_2O.

Calculating Reactant or Product Quantities Using Reaction Stoichiometry

Chemists use reaction stoichiometries to do calculations involving chemical reactions. For example, the mass of product expected from a reaction can be calculated using the mass of reactant supplied to run the reaction.

Use the following steps to carry out calculations involving reaction stoichiometries:

- Begin with the balanced chemical equation for the reaction.
- Determine the molar ratio of reactant to product from the balanced equation.
- Convert the mass of reactant to moles.
- Use the molar ratio of reactant to product as a conversion factor. Multiply the molar ratio by the moles of reactant to find the moles of product.
- Convert the moles of product to mass.

Calculating Reactant or Product Quantities Using Reaction Stoichiometry: Sample Problem

Calculate the mass of aluminum oxide, Al_2O_3, formed when 2.00 grams Al is reacted with oxygen from the air. The balanced chemical equation for this reaction is:

$$4\text{Al} + 3\text{O}_2 \rightarrow 2\text{Al}_2\text{O}_3$$

Solution

First, determine the molar ratio of Al to Al_2O_3 from the balanced equation:

$$4 \text{ mol Al} \rightarrow 2 \text{ mol Al}_2\text{O}_3$$

Next, convert the mass of reactant to moles:

$$2.00 \cancel{\text{g Al}} \times \left(\frac{1 \text{ mol Al}}{26.982 \cancel{\text{g Al}}} \right) = 7.41 \times 10^{-2} \text{ mol Al}$$

Use the molar ratio of reactant to product as a conversion factor. Multiply the molar ratio by the moles of reactant to find the moles of product.

$$7.41\times10^{-2}\ \cancel{\text{mol Al}}\times\left(\frac{4\ \text{mol Al}_2\text{O}_3}{2\ \cancel{\text{mol Al}}}\right)=1.48\times10^{-1}\ \text{mol Al}_2\text{O}_3$$

Finally, convert the moles of product to mass.

$$1.48\times10^{-1}\ \cancel{\text{mol Al}_2\text{O}_3}\times\left(\frac{101.961\ g\ \text{Al}_2\text{O}_3}{1\ \cancel{\text{mol Al}_2\text{O}_3}}\right)=15.1\ \text{g Al}_2\text{O}_3$$

Thus, 2.00 grams Al will produce 15.1 g Al_2O_3.

Stoichiometry of Gases: Volume-Volume Calculations: Sample Problem

Calculate the volume of nitrogen monoxide that will be produced in a reaction using 3,500 liters of oxygen gas. The balanced chemical equation for this reaction is:

$$4\text{NH}_{3(g)}+5\text{O}_{2(g)}\rightarrow4\text{NO}_{(g)}+6\text{H}_2\text{O}_{(l)}$$

Solution

First, the volume ratio is the same as the mole ratio, if the two gases are at the same temperature and pressure:

$$\text{Mole ratio of }\frac{\text{NO}}{\text{O}_2}=\frac{4}{5}$$

$$\text{Volume ratio }\frac{\text{NO}}{\text{O}_2}=\frac{4}{5}$$

Next, use the volume ratio of the reactant to product as a conversion factor. Multiply the volume of the reactant by the mole ratio of the product over the reactant to find the volume of product:

$$3{,}500\ \text{L O}_2\times\left(\frac{4\text{L NO}}{5\text{L O}_2}\right)=2800\ \text{L NO}$$

Stoichiometry of Gases: Mass-Mass: Sample Problem

If stoichiometric calculations of gases involve mass, convert the mass of the gas to moles using the molar mass.

How many grams of carbon monoxide gas are produced from the reaction of carbon and oxygen gas, if 12.0 g grams of oxygen are consumed in the reaction? The balanced chemical equation for the reaction is:

$$2C_{(g)} + O_{2(g)} \rightarrow 2CO_{(g)}$$

Solution

First, convert the mass of oxygen to moles of oxygen using the molar mass.

$$12.0\ \text{g O}_2 \times \left(\frac{1\ \text{mol O}_2}{32.00\ \text{g O}_2}\right)$$

Second, use the mole ratio of the reactant to product as a conversion factor. Multiply the mole ratio by the moles of reactant to find the moles of product.

$$12.0\ \text{g O}_2 \times \left(\frac{1\ \text{mol O}_2}{32.00\ \text{g O}_2}\right) \times \left(\frac{2\ \text{mol CO}}{1\ \text{mol O}_2}\right)$$

Next, multiply the moles of carbon monoxide by the molar mass of carbon dioxide to calculate the mass of carbon monoxide produced.

$$12.0\ \text{g O}_2 \times \left(\frac{1\ \text{mol O}_2}{32.00\ \text{g O}_2}\right) \times \left(\frac{2\ \text{mol CO}}{1\ \text{mol O}_2}\right) \times \left(\frac{28.01\ \text{CO}}{1\ \text{mol CO}}\right) = 21.0\ \text{g CO}$$

Limiting Reagent

When a chemical reaction involves two or more reactants, these reactants are not usually present in the exact molar ratio shown by the balanced equation. Usually one of the reactants is present in smaller amounts than the other(s). In this case, this reactant is said to be the limiting reactant or limiting reagent. It limits the amount of product that can form.

If the quantities of all reactants present at the start of the reaction are known, it is possible to use these values to calculate the quantity of product that could be produced from each. The reactant found to produce the least amount of product is the limiting reagent.

The maximum amount of product that will be formed during a chemical reaction can be found by using the limiting reagent. This amount of product is the theoretical yield of the reaction. The actual yield may be smaller than this. This may occur if the reaction does not proceed to completion.

Limiting Reagent: Sample Problem

Calculate the mass of CaO that is possible if 4.8 grams Ca reacts with 10.0 grams O_2.

Solution

Step 1: Write a balanced equation for the reaction:

$$2\text{Ca} + \text{O}_2 \rightarrow 2\text{CaO}$$

Steps 2 & 3: Convert mass to moles and determine how much CaO is possible with each reactant:

$$4.8\text{ g Ca} \times \frac{1\text{ mol Ca}}{40.1\text{ g Ca}} \times \frac{2.00\text{ mol CaO}}{2.00\text{ mol Ca}} \times \frac{56.1\text{ g CaO}}{1.00\text{ mol CaO}} = 6.7\text{ g CaO}$$

$$10.0\text{ g O}_2 \times \frac{1.00\text{ mol O}_2}{32.0\text{ g O}_2} \times \frac{2.00\text{ mol CaO}}{2.00\text{ mol O}_2} \times \frac{56.1\text{ g CaO}}{1.00\text{ mol CaO}} = 17.5\text{ g CaO}$$

Step 4: The reactant that produces the smaller amount of product is the limiting reagent.

Ca produces less CaO than does O_2 (6.7 g MgO vs. 17.5 g MgO). Therefore, Ca is the limiting reagent.

Step 5: The reactant that produces a larger amount of product is the excess reagent.

O_2 produces more CaO than Ca; therefore, O_2 is the excess reagent in this reaction.

Calculating Percent Yield

The maximum amount of product formed using the limiting reagent is the theoretical yield of the reaction. The actual yield may be smaller than this. Reasons for smaller than expected yield of product include:

- the reaction does not proceed to completion
- some product is lost during the steps following reaction in which the product is isolated
- some product decomposes or undergoes some other reaction that converts it to a different compound

The percent yield gives an indication of the success of isolating a product from a chemical reaction. Percent yield is calculated by dividing the actual yield by the theoretical yield and multiplying by 100.

$$\text{percent yield} = \left(\frac{\text{actual yield of product in grams}}{\text{theoretical yield of product in grams}}\right) \times 100$$

Consider the Explain Question:

How can you use average atomic masses from the periodic table to find the formula mass and the molar mass of ammonia, NH_3, and the percent composition of each element in ammonia?

Go online to complete the scientific explanation.

Check for Understanding:

How can molar mass be used to determine the number of molecules in a sample compound, if its mass and molecular formula is known?

STEM in Action

STEM and Mathematics of Formulas and Equations

People have come to depend on synthetic polymers such as polyethylene, polystyrene, and polyacrylate. A polymer is a compound consisting of large molecular chains of smaller molecules, called monomers, bonded together. Natural polymers include proteins in your body, but synthetic polymers can be found everywhere in modern civilization.

For example, have you ever noticed the recycling symbol 1 in a triangle printed on a plastic container? That symbol means the container is produced from a polymer called polyethylene terephthalate (PET). Disposable water bottles and clear plastic containers used for strawberries or blueberries are often made of PET. The plastic recycling symbol 2 means the container is produced from high-density polyethylene. This type of plastic is used for grocery bags as well as milk and juice bottles. Polystyrene, designated by the recycling symbol 6, is used to make packing peanuts and egg cartons. Other recycling numbers refer to other types of polymers. Polymers have numerous familiar uses. Polyacrylate is a polymer used in paints and adhesives. Electrical insulation, toys, home siding, and upholstery are all made of polymers.

Synthetic polymers are developed in chemical laboratories. Much research is carried out to discover new polymers and modify existing ones. Because they are synthesized using chemical reactions, polymers can be described using stoichiometry. Polymer chemists use stoichiometric principles to plan and carry out reactions that they hope will produce new polymers.

photo: Getty Images

This polymer chemist is performing an analysis to determine the composition of a polymer. How is stoichiometry useful in this process?

A Mass of Molar Masses

A polymer chemist conducts an analysis of various polymers. The following table shows the chemical formulas for each repeating unit of the polymers that were tested.

1. Complete the table by calculating the molar mass of each repeating unit. Round each answer to the tenths place. (Record your answers on a separate sheet of paper.)

Polymer	Repeating Unit Formula	Molar Mass of Repeating Unit (g/mol)
A	C_8H_8	
B	C_2H_4	
C	C_3H_6	
D	C_2H_3Cl	
E	$C_2H_2Cl_2$	
F	C_3H_3N	
G	$C_4H_6O_2$	

2. Complete the table by calculating the percent composition of oxygen and hydrogen in each compound. Round each answer to the tenths place. Based on your answers, which compound has the lowest composition of carbon, and which has the lowest composition of hydrogen?

Reaction Rate

photo: Getty Images

LESSON OVERVIEW

Lesson Questions

- What factors influence reaction rate, and what is their impact?
- How do rate laws describe chemical reactions?

Lesson Objectives

By the end of the lesson, you should be able to:

- Identify factors that influence reaction rate and explain their impact.
- Apply rate law equations to predict reaction rates.
- Determine the overall order of a chemical reaction.

Key Vocabulary

Which terms do you already know?

- ☐ **activation energy**
- ☐ **catalyst**
- ☐ **chemical reaction**
- ☐ **collision theory**
- ☐ **enzyme**
- ☐ **inhibitor**
- ☐ **order of a chemical reaction**
- ☐ **oxidation**
- ☐ **reactant**
- ☐ **reaction rate**
- ☐ **surface area**

ENGAGE

How does the temperature at which food is stored affect it? Why does storing food in a refrigerator keep it from spoiling and decomposing? Visit the Engage page online to learn more.

Explain Question

What factors influence a reaction rate, and what is their impact on reaction rate?

EXPLORE

What Factors Influence Reaction Rate, and What Is Their Impact?

Factors That Influence Reaction Rate

The physical and chemical properties of elements determine how atoms interact when they come into contact with one another. Some elements, such as fluorine, are quite reactive, and they form compounds with many other elements. Others, such as the noble gases, do not interact with other elements at all.

Similarly, the specific substances that interact in a chemical reaction may interact quickly, slowly, or not at all. The chemical properties of the reactants play a key role in determining the rate of a chemical reaction. For example, a reaction between potassium and water will progress at a faster rate than a reaction between iron and water.

Another important factor in reaction rate is the amount of each reactant present. For example, in order for rust to form, iron and water must both be present, or oxidation will not occur. If oxygen is not present in a system, the combustion reaction that burns gasoline cannot occur. Often, the amount of a reactant present is described using concentration, particularly when working with solutions.

The effect of concentration on reaction rate is partly explained by collision theory. Remember that, according to kinetic molecular theory, atoms and ions are constantly in motion. Even in a solid, the particles have energy and regularly collide with one another.

During a chemical reaction, chemical bonds break and atoms are rearranged into new compounds. In order for this to happen, the atoms involved must collide with one another with a certain amount of energy. Usually, the more collisions that occur between atoms of the reactants, the faster the reaction proceeds. Logic tells us that the more atoms there are present (the higher the concentration), the more collisions will occur. The size of the reactant's particles is another factor. Differently sized particles have different surface areas. For example, in a cube of sugar, there are six large, exposed sides that might react with another substance. The sugar that lies in the center of the cube is kept out of the reaction, at least for a little while. If the cube is broken apart, however, the surface area of each smaller crystal is exposed. Though each particle is smaller, there are many more of them and the exposed surface area increases. The atoms that make up the smaller crystals of sugar are free to collide with other atoms. This allows the entire reaction to proceed more rapidly.

The conditions of the environment in which a reaction takes place strongly affect the rate of the reaction. These include temperature and pressure. For example, when the temperature of a system is increased, the kinetic energy of all the particles in the system increases as well. The number of collisions between atoms will also increase, causing the reaction to proceed more rapidly. When there is more energy in a system, some of it is absorbed by the molecules present, which makes it easier for molecular bonds to be broken. The atoms can then be rearranged to form new substances more quickly.

If at least one of the reactants is a gas, the pressure of the system may also affect the reaction rate. Remember that when held under high pressure, gas particles cannot move as freely. If a gas is held in a system by pressure, its particles cannot escape. They remain in the system and participate in the reaction. There are more of them present, so the reaction proceeds at a faster rate than it would if some of the gas particles spread out into the space surrounding the system.

There are times when extra substances present in the system might affect the rate of a reaction. Often, this involves the activation energy, or the amount of energy required for a reaction to proceed. Catalysts are substances that speed up reactions, but they remain unchanged when the reaction is complete. A catalyst will reduce the amount of activation energy needed for a reaction to occur. Metals and special proteins, called enzymes, often act as catalysts. Nickel, for example, acts as a catalyst in the reaction used to make margarine. Cars are equipped with catalytic converters, which use metals to remove dangerous substances from engine exhaust. Food digestion, respiration, and most other biological processes rely on catalysts. The use of an inhibitor, on the other hand, will decrease reaction rate. An inhibitor acts in the opposite way that a catalyst does, increasing the energy required for the reaction to proceed. Some inhibitors, in fact, work by limiting the action of the catalysts. Many enzymes can act as inhibitors when they are present during a chemical reaction.

How Do Rate Laws Describe Chemical Reactions?

Rate Laws

Whenever specific chemical compounds come into contact with one another, they will react in a predictable way. Experiments can be completed and repeated to determine the behavior of reactants in a system, including the rate at which the substances react.

As a reaction occurs, the reactants go through a chemical change. Molecular bonds are broken, and atoms are rearranged. The concentration of the remaining reactants can be measured. Monitoring the changes in concentrations provides information that can be used to calculate the reaction rate.

A rate law is a mathematical equation that describes the reaction rate of a specific reaction. The rate law relates the molar concentrations of the reactants to k, the rate constant, which remains the same for a reaction as long as temperature does not change.

A basic chemical reaction can be described using a balanced chemical equation:

$$\mathrm{aA + bB \rightarrow C}$$

The rate law equation uses information about the reaction:

$$r = k[\mathrm{A}]^{\mathrm{x}}[\mathrm{B}]^{\mathrm{y}}$$

Here, the variable *k* is the rate constant. The values in square brackets represent the concentration levels of the reactants A and B. Notice that the coefficients *a* and *b* from the initial reaction have no effect on the rate calculation and can be ignored. The variables *x* and *y* describe the order of the chemical reactions. These values must be determined experimentally and will be discussed in the next section.

Reaction rates for all reactants are determined experimentally. It is not possible to look at a chemical equation and know which reactant will drive the rate of reaction.

photo: Getty Images

The oxidation of iron is a slow reaction that causes corrosion to the metal as iron oxide forms. What can be added as a catalyst to the metal that would prevent or slow the formation of rust?

Rate Laws: Sample Problem

A solution of 0.1 mol acetone (C_3H_6O) reacts with a solution of 0.1 mol bromide (Br_2). Assume that the value of *x* is 0, the value of *y* is 1, and *k* is 1.64 × 10^{-4} /s. Find the reaction rate.

Solution

Substitute the given values into the rate law equation:

$$r = k[\mathrm{A}]^{x}[\mathrm{B}]^{y}$$
$$r = 1.64 \times 10^{-4}\,/\mathrm{s} \times (0.1\ \mathrm{mol/L})^{0x}\ (0.1\ \mathrm{mol/L})^{1}$$
$$r = 1.64 \times 10^{-4}\,/\mathrm{s} \times 1 \times (0.1\ \mathrm{mol/L})$$
$$r = 1.64 \times 10^{-5}\ \mathrm{mol/L{\cdot}s}$$

The reaction will use 1.64 × 10^{-5} moles per liter of reactants per second.

The Order of Reactions

The order of reaction can be determined with respect to each of the reactants as well as for the overall reaction. Recall that the concentration of a reactant can affect the rate of a reaction. The order of reaction describes how changing the concentration of a reactant would affect the reaction rate.

The order is described in the rate law equation using the variables *x* and *y*. When finding the order of reaction with respect to each reactant, however, only one concentration is used in the rate law equation.

$$r = k[\mathrm{A}]^{x}$$

A zero-order reaction has only one reactant. In addition, the reaction rate is independent of the concentration of the reactant. Even if the concentration of the reactant is doubled, the rate of the reaction does not change. This is usually seen in decomposition reactions, such as the decay of a nitrous oxide gas molecule into nitrogen gas and oxygen gas. Because any number raised to the power of 0 is equal to 1, in a zero-order reaction $r = k$.

$$r = k[\mathrm{A}]^{x}$$
$$r = k[\mathrm{A}]^{0}$$
$$r = k \times 1$$
$$r = k$$

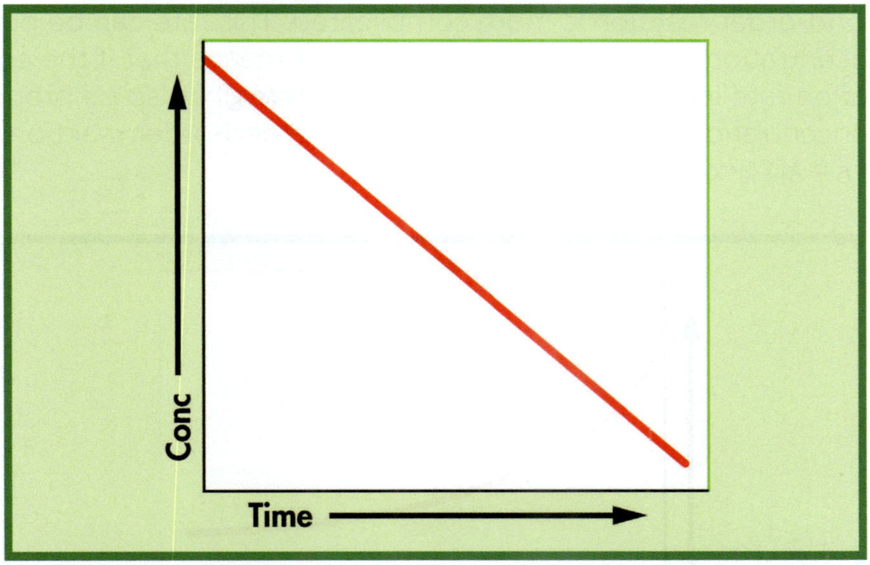

This is the concentration over time of a zero-order reaction. How many reactants are involved in these reactions?

A first-order rate of reaction is directly proportional to the concentration of the reactant. In other words, as the concentration of the reactant increases, the rate of the reaction increases. Because any number raised to the power of 1 is equal to that number, for a first-order reaction, $r = k[\mathrm{A}]$.

$$r = k[\mathrm{A}]^x$$
$$r = k[\mathrm{A}]^1$$
$$r = k[\mathrm{A}]$$

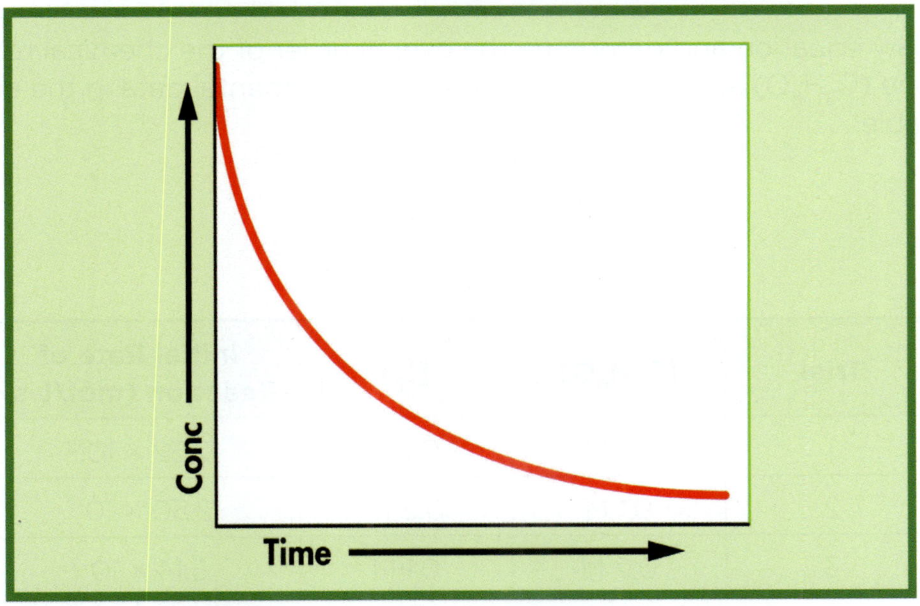

This is the concentration over time for a first-order reaction. How does the concentration of the reactant affect the reaction rate?

The rate of a second-order reaction is more complicated. This rate can be proportional to the square of the concentration of one of the reactants. This means that if the concentration is doubled, the reaction rate increases by a factor of four. It might also be proportional to the product of the concentrations of two reactants. For a second-order reaction, the exponent x is equal to 2, and $r = k[A]^2$ or $r = k[A][B]$.

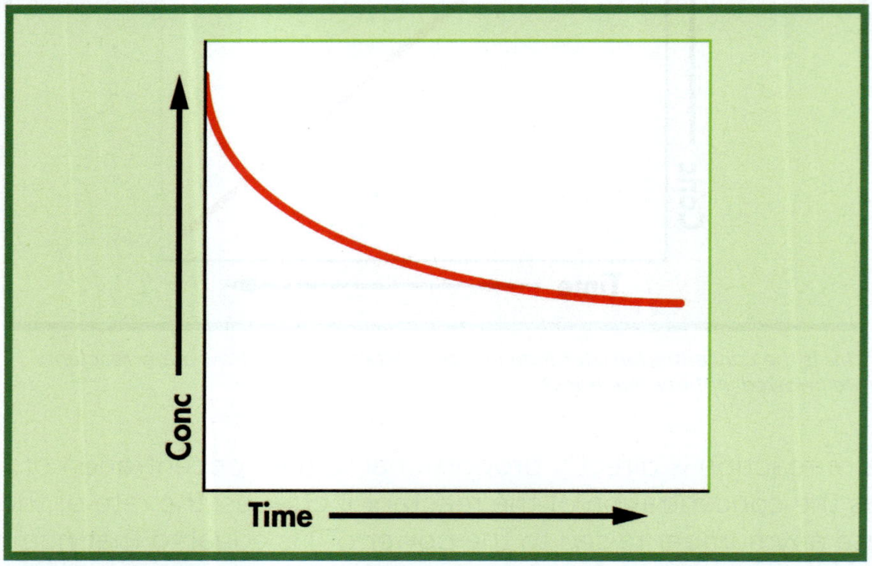

This is the concentration over time for a second-order reaction. How much greater is reaction rate affected by the reactant concentration in these reactions?

The overall order of a chemical reaction is found by adding the orders of each reactant. For example, if the order of the reaction with respect to the first reactant is 0 and the order of the reaction with respect to the second reactions is 2, the overall order of the reaction is 2.

The Order of Reactions: Sample Problem

Write the rate law equation and determine the overall order of the chemical reaction between acetone (C_3H_6O) and iodide (I_2), given the experimental data in the equation and in the following table.

$$C_3H_6O + I_2 \rightarrow C_3H_5I + HI$$

Trial	$[C_3H_6O]$	$[I_2]$	Initial Rate of Reaction (mol/L·s)
1	0.1 M	0.1 M	2.52×10^{-2}
2	0.1 M	0.2 M	2.50×10^{-2}
3	0.2 M	0.1 M	5.14×10^{-2}

Review the sample data table. What can you observe about trials 1, 2, and 3?

Solution

Substitute the chemical formulas of the reactants into the rate law equation:

$$r = k[\mathrm{A}]^{x}[\mathrm{B}]^{y}$$
$$r = k[\mathrm{C_3H_6O}]^{x}[\mathrm{I_2}]^{y}$$

Determine *x* and *y* from the experimental data in the table. Comparing the data for trials 1 and 2, note that as the concentration of the acetone is unchanged and the concentration of the iodine is doubled, the rate did not change significantly. This means the reaction is zero-order with respect to iodine. Therefore:

$$r = k[\mathrm{C_3H_6O}]^{x}[\mathrm{I_2}]^{0}$$

Because the reaction rate more than doubled when the concentration of acetone was doubled in trial 3, the reaction is first order with respect to acetone. Therefore:

$$r = k[\mathrm{C_3H_6O}]^{1}[\mathrm{I_2}]^{0}$$

Any number to the zero power equals 1. Therefore

$$r = k[\mathrm{C_3H_6O}]^{1} = k[\mathrm{C_3H_6O}]$$

The orders with respect to $[C_3H_6O]$ and $[I_2]$, 1 and 0 respectively, have a total value of 1. Therefore, the overall order of this reaction is 1.

Consider the Explain Question:

What factors influence a reaction rate, and what is their impact on reaction rate?

Go online to complete the scientific explanation.

Check for Understanding:

What are the factors that influence reaction rate?

STEM in Action

STEM and Reaction Rate

An industrial chemist is a chemist who works to develop and manufacture products and processes for the better good of society and the environment. One important development by industrial chemists was the use of catalysts. World history would have been different without catalysts because wars would have been less destructive, transportation would be much slower, and Earth's population would be much smaller. These surprising claims are true because most industrially produced chemicals depend on catalytic action, thanks to the work of industrial chemists.

Some of the more important industrial chemicals produced by catalyzed reactions are explosives, fertilizers, gasoline, and polymers. Ammonia (NH_3) is manufactured by the Haber process, which would not be possible without an iron catalyst. Nitric acid (HNO_3) is produced by a different reaction, which is catalyzed by platinum (Pt). Together, these chemicals are used to manufacture the powerful explosive ammonium nitrate (NH_4NO_3). This one compound is said to have changed the course of World War I. The historical importance of ammonium nitrate as a fertilizer is even greater. Without this and other fertilizers, the world's population would be much smaller due to agricultural limitations. Sulfates, another important class of fertilizers, are produced from sulfuric acid (H_2SO_4), which results from a reaction catalyzed by vanadium (V) oxide (V_2O_5).

NASA engineers have also been working hard to develop low-temperature oxidation catalysts throughout the past two decades. They were initially intended to support space-based CO_2 lasers but are now being used in several different facets of everyday life. These catalysts are used to sense and remove two deadly gases, carbon monoxide and formaldehyde, from houses and other buildings. They can also be used in automotive gas and diesel catalytic converters to improve their efficiency. Engineers continue their work in these low-temperature catalysts for use in commercial and industrial applications.

The Haber process is the reaction in which nitrogen gas reacts with hydrogen gas, producing ammonia. This reaction normally uses an iron catalyst to speed it up.

$$N_2(g)+H_2(g) \rightarrow NH_3(g)$$

Now, complete the activity about the Haber process that follows.

Use the graph to answer the following questions.

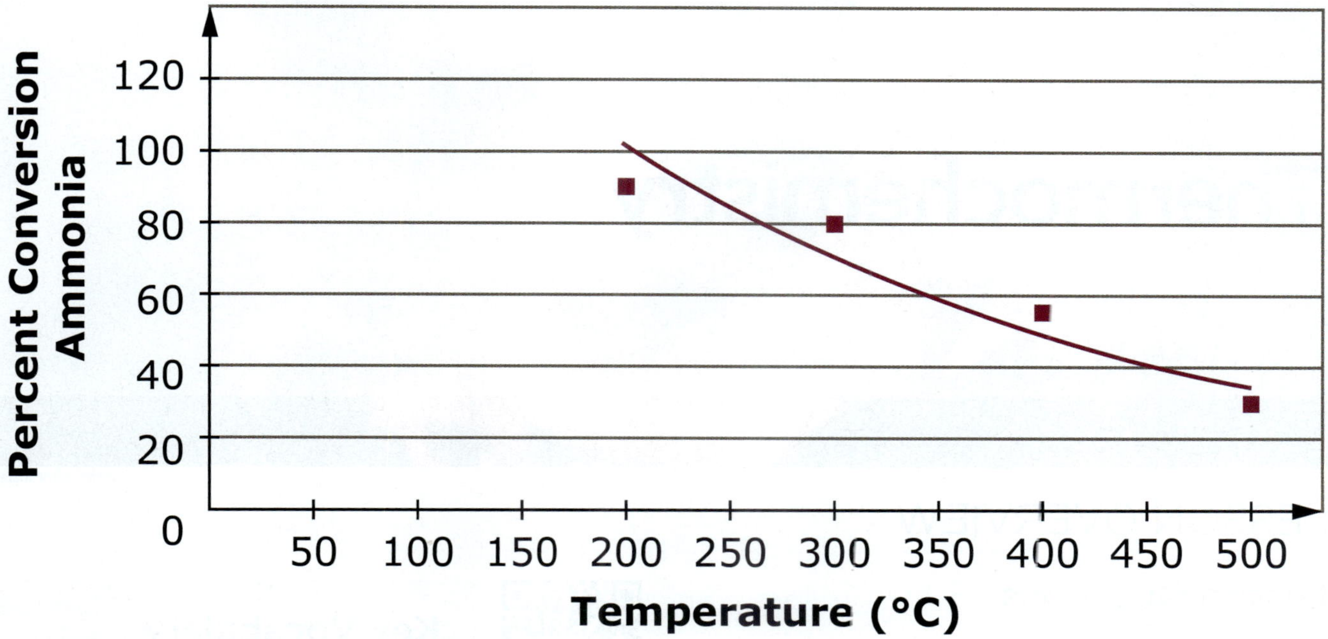

1. What does the graph show about the effect of temperature on the Haber process?
2. This process is normally carried out at a temperature of 400°C. Provide an explanation as to why this temperature is chosen when a lower temperature gives a higher percent conversion to ammonia.

Thermochemistry

photo: Getty Images

LESSON OVERVIEW

Lesson Questions

- How do you distinguish between exothermic and endothermic reactions and processes?
- What is the relationship between a system and its surroundings?
- What is the relationship between physical changes and changes in energy?
- What is the relationship between chemical changes and changes in energy?

Lesson Objectives

By the end of the lesson, you should be able to:

- Distinguish between exothermic and endothermic reactions and processes.
- Explain the relationship between a system and its surroundings.
- Explain the relationship between physical changes and changes in energy.
- Explain the relationship between chemical changes and changes in energy.
- Determine the change in enthalpy of a chemical reaction.

Key Vocabulary

Which terms do you already know?

- ☐ **activation energy**
- ☐ **average kinetic energy**
- ☐ **calorimetry**
- ☐ **chemical potential energy**
- ☐ **combustion reaction**
- ☐ **conduction**
- ☐ **convection**
- ☐ **critical point**
- ☐ **deposition (phase change)**
- ☐ **electromagnetic radiation**
- ☐ **endothermic**
- ☐ **energy**
- ☐ **enthalpy**
- ☐ **exothermic**
- ☐ **first law of thermodynamics**
- ☐ **heat energy**
- ☐ **Hess's law**
- ☐ **Joule**
- ☐ **kinetic energy**
- ☐ **Le Chatelier's principle**
- ☐ **metric**
- ☐ **non-spontaneous**
- ☐ **phase diagram**
- ☐ **second law of thermodynamics**
- ☐ **specific heat**
- ☐ **spontaneous**
- ☐ **standard heat of formation**
- ☐ **sublimation**
- ☐ **surroundings**
- ☐ **system**
- ☐ **temperature**
- ☐ **triple point**

ENGAGE

After a long, strenuous workout, you probably feel exhausted and sore. Can a heating pad or hot pack work to relieve some soreness? Or should ice be used to help you feel better? Visit the Engage page online to learn more.

Explain Question

How are endothermic and exothermic processes identified, and how does the energy of the system change with each?

EXPLORE

How Do You Distinguish between Exothermic and Endothermic Reactions and Processes?

Releasing Energy

As wood burns, energy is released. Combustion of the wood releases energy in the form of light and thermal energy. Likewise, the synthesis reaction of sodium metal with chlorine gas also releases energy. Chemical reactions that release energy are called exothermic reactions. The combustion reaction, such as that of methane with oxygen or propane with oxygen, is another exothermic reaction. During each reaction, energy is released as light and thermal energy.

photo: Getty Images

What is happening to the energy of reactants and products of the chemical reactions that occur during a fireworks show?

Where does this energy come from? The energy released during a reaction results from a decrease in the chemical potential energy as reactants change into products. Chemical bonds between atoms store chemical potential energy. In an exothermic reaction, the products of the reaction have less chemical potential energy than the reactants. The difference in potential energy is released as light or thermal energy.

Absorbing Energy

Some chemical reactions and physical processes release energy. Others absorb energy. Energy is absorbed during an endothermic reaction or process. Toasting marshmallows and popping corn are endothermic processes. The substances absorb energy as they undergo changes. Boiling water and melting steel are also endothermic. The water or steel must absorb energy to undergo the change. Instant cold packs work on the basis of an endothermic reaction when a solid is dissolved in water. Energy is absorbed by reactants during the reaction, so the pack becomes cold to the touch.

Another example of an endothermic process is photosynthesis. Energy is added to the system from sunlight and helps the plant convert carbon dioxide and water into glucose and oxygen. The energy absorbed remains in the glucose molecules. The decomposition of sodium chloride into sodium and chlorine is the reverse reaction of the exothermic processes described previously. Energy must be absorbed to break the chemical bond between sodium and chlorine. The potential energy stored in the bond is greater than the chemical potential energy of the products, sodium and chlorine. The reactant must absorb the difference in energy to undergo the reaction.

What Is the Relationship between a System and Its Surroundings?

Conservation of Energy

When an exothermic reaction releases energy, people might describe the reaction as "producing" energy. Likewise, an endothermic reaction does not destroy the energy it absorbs. Energy can be transformed from one form into another, but it cannot be created or destroyed during chemical and physical changes.

For exothermic reactions, the chemical energy stored in the bonds of the reactants is transformed into thermal energy that makes the surrounding matter warmer. Sometimes, chemical energy changes into light energy, which happens when you make a light stick glow. The opposite happens during endothermic reactions. Thermal energy, or even light, is transformed into chemical energy that is stored in the bonds of the products. Energy is often transformed to a useful form of energy, but usually some energy also becomes a less useful form.

Energy is transformed from one useful form into another useful form in a car engine when the energy released by burning gasoline is transformed into kinetic energy. However, not all of the thermal energy released by the reaction makes the car move. Some of the thermal energy also heats the parts of the engine, which is why the hood of a car that has just been driven feels warm. In this case, the transformation to thermal energy that heats the car parts is not a useful transformation. People may say that some of the released energy is lost during combustion. This does not mean that it was destroyed. The "lost" energy changed to a form of energy that was not used to make the car run.

Energy in Chemical Reactions

Exothermic and Endothermic Processes

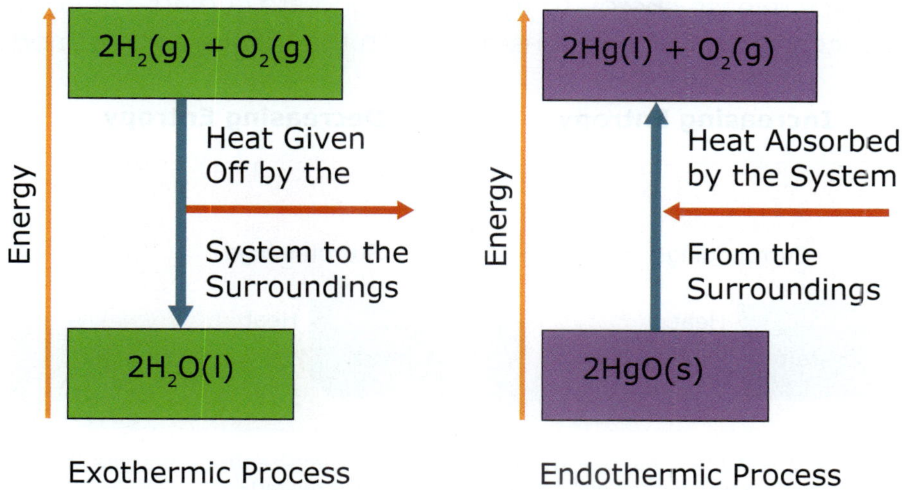

Not all of the energy in a chemical reaction is useful when it transforms. Consider this diagram: Are the energy transformations in a car endothermic or exothermic?

In addition to being transformed from one form into another, energy can be transported from one place to another and transferred between systems. This movement of energy is what makes your hand warm when you touch the hood of a car when exothermic combustion reaction is happening in the engine.

The statement that energy can be transferred or transformed but is not created or destroyed during chemical and physical changes is the first law of thermodynamics. This law is also called the law of conservation of energy.

System vs. Surroundings

To get a more complete view of the energy changes associated with a chemical or physical change, consider two regions. The system is the part studied. Everything outside of the system is the surroundings. A barrier separates the system's interconnected parts from the surroundings. When considering systems, one must consider if the system represented is an open, closed or isolated system.

- In an open system, both mass and energy are exchanged with its surrounding. Matter can be added easily like adding spices to a cooking pot. Energy can be exchanged through heat and work, like turning up the heat on the cooking pot.
- In a closed system, energy is exchanged with its surrounding, but there is no transfer of mass. If one places a lid on the cooking pot, matter cannot be added, but energy, in the form of heat can still be added to the system.
- In an isolated system, there is no transfer of mass or energy across the boundaries of the systems. Real-world isolated systems are rare. One example of an isolated system is an insulated closed container. If hot liquid is poured into the container and the lid is placed on the container, there is no additional energy or mass applied to the system.

Cooling an injury with an ice pack can help reduce swelling. Consider the ice pack as the system—everything outside of the ice pack is the surroundings. The ice warms up and melts within the system because it absorbs energy from the surroundings. The injured leg and the air around the ice pack release energy to the system and are cooled. The total energy of the universe remains unchanged, which is consistent with the first law of thermodynamics.

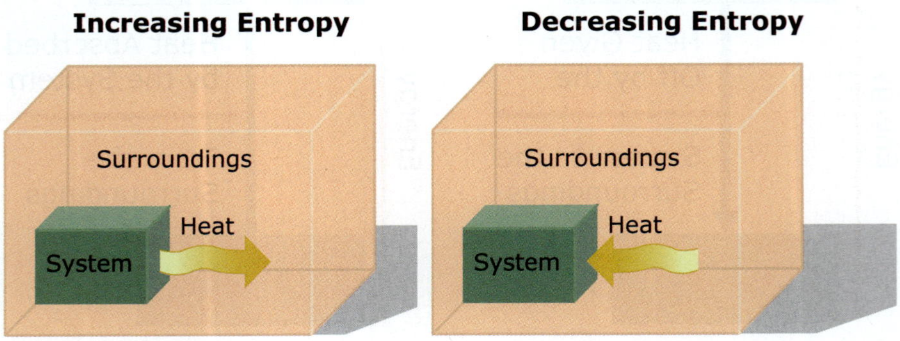

The flow of heat into or out of a system changes the system's entropy. How are the entropy of a system and the entropy of its surroundings related?

From Hot to Cold

Heat transfer occurs from the surroundings into a system during an endothermic change. Heat transfer occurs from the system to the surroundings during an exothermic change. In each case, thermal energy naturally moves from a warmer area to a cooler area during the change. This is just as the second law of thermodynamics would indicate. The second law of thermodynamics states that in any natural process, the total entropy of a system and its surrounding environment must increase.

Energy transfer always favors a more stable state, meaning the energy distribution becomes more uniform. In other words, the warmer parts become cooler and the cooler parts become warmer. For example, when hot water is poured into a cup, the water transfers heat to the cup until the cup and the water are the same temperature. The cup becomes warmer, and the water becomes slightly cooler. In a closed system, the cup and the water would reach a stable state and no more transfer would occur between them. However, a cup sitting on the table is not in a closed system; it is in an open system. An open system allows the exchange of matter and energy with its surroundings. The system including the cup and water will continue to transfer heat to the surrounding air until the cup and the water inside it reach the same temperature as the air.

Temperature is a measure of the average kinetic energy of the particles in matter. The higher the temperature, the faster the particles move. It makes sense to think that energy moves from warm areas to cool ones. In other words, energy moves from an area where particles are moving quickly to an area where they are slower. But how does this heat transfer happen?

Thermal energy will move from a high-temperature object to a low-temperature one. It will end when the objects are at the same temperature (second law). The total energy released by the warmer object will be the same as the total energy absorbed by the cooler one (first law).

Heat transfer happens in three ways: through convection, radiation, or conduction:

- In convection, currents in air, water, or other fluids carry thermal energy from one place to another. These currents form as warmer fluids rise and cooler fluids sink. For example, a warm ocean current allows heat transfer between one part of the ocean and another.
- In radiation, energy travels as electromagnetic radiation. The radiation is in the form of visible or invisible light. For example, a red-hot stove burner transfers heat by radiating infrared and red light energy.
- In conduction, heat transfer occurs between a higher-temperature object and a lower-temperature object that are in contact. The thermal energy is transferred directly through the collision of the moving particles. For example, heat transfer takes place when you put your cold feet on a hot-water bottle.

What Is the Relationship between Physical Changes and Changes in Energy?

Energy Released in Freezing and Condensing

Freezing temperatures can destroy an orange crop. How can a farmer protect the unharvested fruit from freezing? A farmer can protect the fruit by spraying water on and around the trees in the orchard. As water cools and then freezes, it releases energy. Freezing is an exothermic process. Thermal energy released by the water is transferred to the air and the fruit. The fruit is warmed as the water freezes. The ice on the fruit also insulates the fruit against air temperatures below 0°C, the freezing point of water.

The change of a gas to a liquid, or condensing, is also an exothermic physical change. A gas must release energy to change into a liquid. So, when a mirror fogs up, it absorbs the energy released by the condensing water vapor.

Another exothermic process is deposition. Deposition is a physical change that results in a gas changing into a solid, without first changing into a liquid. Deposition can be observed on windows during cold weather when water vapor changes into ice crystals.

Energy Absorbed in Melting and Vaporizing

A change that is exothermic in one direction is endothermic in the reverse direction. Freezing is exothermic, so melting must be endothermic. An ice cube melts when it is taken out of a freezer. Thermal energy flows into the ice cube from the warmer air around it, so it melts. Condensing is exothermic, so vaporizing is endothermic. Before it vaporizes, water usually boils. For example, when water is boiled when cooking pasta, energy must be added from a heat source, such as the burner underneath the pot.

Calorimetry Calculations

To understand how substances change temperature with gain or loss of energy, chemists measure the amount of energy that is absorbed or released by a substance. They use a calorimetry to determine specific heat and other quantities related to this process. The chemist conducts the energy transfer in a calorimeter. For example, the chemist puts a sample of hot metal in cool water inside a calorimeter. The metal and water reach the same temperature. Then, the chemist finds the change in energy of the water from the temperature change. The decrease in energy of the metal must equal the increase in energy of the water for energy to be conserved. The chemist can use these data to calculate the specific heat of the metal. The equation that relates the heat (q) that a substance absorbs or releases to its specific heat (c), mass (m), and temperature change (ΔT) is:

$$q = cm\Delta T$$

Using this equation, chemists can determine the change in energy as measured in Joules (J).

A calorimetry reaction that is exposed to its surroundings occurs at constant pressure. The thermal energy absorbed or released by the surroundings is called the enthalpy. Its value is the same as q, but its sign is opposite:

$$\Delta H = -cm\Delta T$$

If a substance releases heat, then q is positive, ΔH is negative, and the reaction is exothermic. If a substance absorbs heat, then q is negative, ΔH is positive, and the reaction is endothermic.

When a hot metal gives off heat when placed in cool water, then ΔH is negative, because heat energy is lost by the metal. It is similar to purchasing an item from the store. If a person pays for an item in cash, then a specific amount of money is lost from the wallet. That same amount of money is added to the amount of money in the cash register. Cash flow for the shopper is negative, but it is positive for the store. Heat, q, is the same way. Heat released by a system is $-q$, something has to gain that heat and will be the same number but a positive value. In the example shown, the heat lost by the metal, $-q$, must be equal in value to the heat gained by the water, but positive.

Calorimetry Calculations: Sample Problem

A calorimeter holds 145 grams water at 22.0°C. A sample of hot aluminum is added to the water, and the final temperature of the water and aluminum is 28.0°C. How much heat was released by the aluminum to make this change? The specific heat of water is 4.18 J/g°C.

Solution

The given information includes:

$$m = 145\ g$$
$$\Delta T = 6.0°C$$
$$c = 4.19\ J / g°C$$

Substitute into the equation and solve for *q*:

$$q = -Cm\Delta T$$
$$q = -(4.19 J / g°C)(145\ g)(6.0°C)$$
$$q = -3{,}637$$

Rounding to the correct number of significant digits, the heat released is -3.6×10^3 Joules.

Changing Temperature, Phase Change, and Specific Heat

Temperature is related only to the kinetic energy of particles, but particles also have potential energy. Thermal energy is the sum of the kinetic energy and potential energy of the particles within a substance. When a substance is heated, the thermal energy of that substance is increased. Usually, this thermal energy changes to kinetic energy of the particles. The increase in the particles' kinetic energy increases the temperature of the substance.

The addition or removal of thermal energy in a substance can result in a phase change, like when a solid turns into a liquid. During a phase change, the thermal energy changes to potential energy of the particles. The increase in the particles' potential energy changes the phase of the substance.

When a hot object is in contact with a cold object, thermal energy flows from the hot object to the cold object. If an object absorbs thermal energy, its temperature increases. So, the cold object becomes warmer. If an object releases thermal energy, its temperature decreases. So, the hot object becomes colder. Eventually, the two objects will reach the same temperature, and the flow of thermal energy will stop.

Think about baking a cheese pizza. The hot air in the oven transfers energy to the pizza and bakes it, raising its temperature. When the pizza is taken out, all of its ingredients are at the same temperature. After a few minutes, the crust is cool enough to touch, but the cheese is still hot. Why is the cheese hotter than the crust when they both came out of the oven at the same time?

Specific heat is the amount of energy needed to change the temperature of 1 gram of a material by 1°C. When a substance with a high specific heat changes temperature, it absorbs or releases a large amount of energy per gram. Likewise, a substance with a low specific heat absorbs or releases less energy per gram for the same change in temperature of a substance with higher specific heat. So, the pizza effect is explained because the cheese has a higher specific heat than the crust. The cheese remains hotter than the crust because more energy per gram of cheese must be released compared with the same change in temperature for an equal mass of crust.

Another example of specific heat is observed at the beach. During the hot summer months, the sand along the water is hotter than the adjacent water. This is because the specific heat of the sand is less than the specific heat of the water. Less energy from the sun is required to increase the temperature of the sand than is required to increase the temperature of the water.

Phase Diagrams

Matter can exist in a variety of phases depending upon the conditions of temperature and pressure. For example, water changes from ice to liquid and then to steam when its temperature changes. Substances can also be made to change state by changing pressure. For example, when compressed enough, the gases in air will eventually turn into a liquid. The actual phase a substance exists in can therefore be plotted using two axes—one for temperature (*x*) and one for pressure (*y*). The diagrams produced using these axes are called phase diagrams.

The diagram shows a typical simple phase diagram for a pure substance.

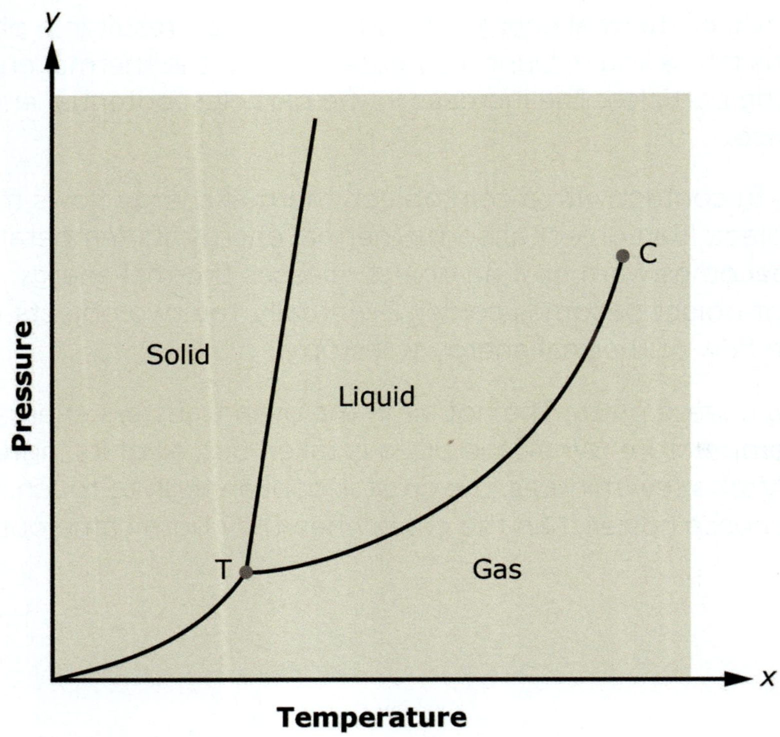

In a typical phase diagram, temperature is plotted on the x-axis and pressure is plotted on the y-axis. How do the lines represent where phases exist in equilibrium?

The diagram is split into three areas that represent the three phases of solid, liquid, and gas. The lines plotted on the graph, therefore, represent the conditions of temperature and pressure where more than one phase can exist. They are lines of equilibrium between two phase changes. There are many different sets of conditions that lead to a phase change of a substance. Whenever the equilibrium lines are crossed, at different temperatures and pressures, the substance undergoes a phase change. A substance can turn from a solid to a liquid (melting), from a liquid to vapor (evaporation), and from a solid to a gas (sublimation). At these same temperatures and pressures, this substance can also turn from a liquid to a solid (freezing), a vapor to a liquid (condensation), and a gas to a solid (deposition). At just one set of conditions, the substance can exist in all three forms. For this reason, the particular temperature and pressure is called a triple point and is marked T.

In the previous phase diagram, the critical point is marked with the letter C. The temperature and pressure coordinates for the critical point are known as the critical temperature and, pressure respectively. Below the critical point, it is possible to condense a gas into a liquid by increasing the pressure. Above the critical point, it is impossible to condense the gas by increasing the pressure. The more pressure applied, the more compressed the gas gets, but it does not condense. This is because the gas particles are too energetic for intermolecular attraction to enable a liquid to be formed.

Sample Problem: Interpreting Phase Diagrams

Let's look at the phase diagram for carbon dioxide and use it to answer some questions.

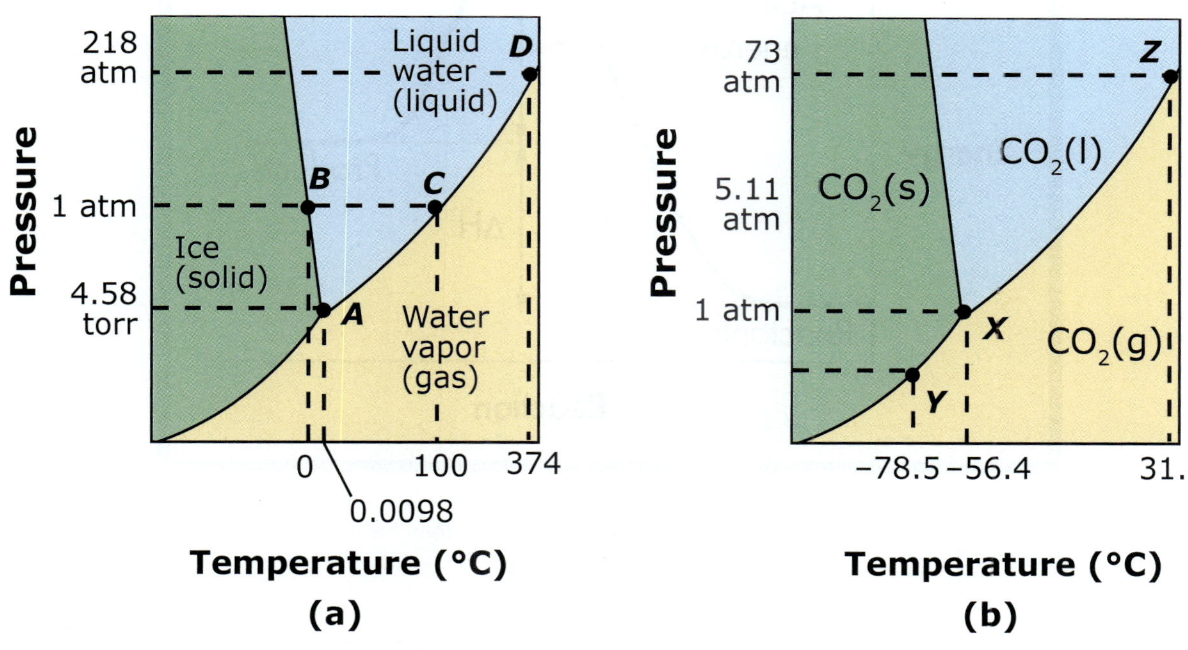

- At about what temperature and pressure do the three phases of carbon dioxide exist?

  At about 57°C, 5 atmospheres (the triple point).

- What is the approximate temperature and pressure of the critical point?

  About 31°C, 73 atmospheres.

- What is/are the phase(s) of carbon dioxide at 0°C and a pressure of 10 atmospheres (atm)?

  Carbon dioxide is a gas.

- At about what temperature would carbon dioxide at 100 atm pressure exist as both a solid and a liquid?

  About –58°C.

What Is the Relationship between Chemical Changes and Changes in Energy?

Activation Energy

Burning a strip of magnesium metal is an exothermic reaction. As the reactants change into products, there is a decrease in enthalpy as energy is released. This change in enthalpy is the heat of reaction. Enthalpy (H) is the amount of energy exchanged between a system and its surroundings. This energy is usually thermal energy. The energy of the product, magnesium oxide, is lower than the combined energy of the reactants.

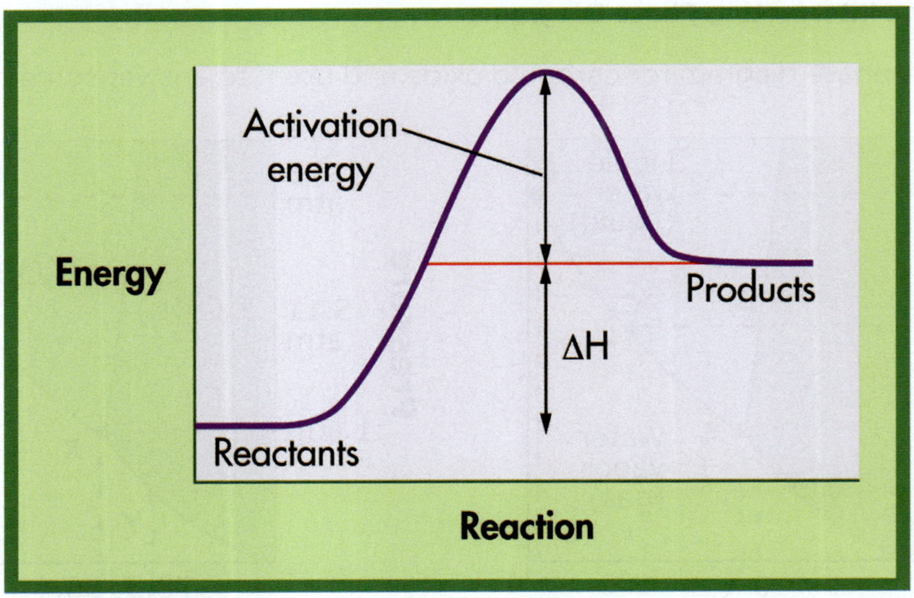

To start the exothermic chemical reactions of a sparkler, a flame must first supply the activation energy. Why is the energy of the products lower than the energy of the reactants?

Although the reaction is exothermic overall, some energy must be supplied to the magnesium to begin the reaction. Activation energy is the energy needed to begin a reaction. In the case of magnesium, an open flame will supply the energy.

What about the reverse reaction? In the reverse direction, the activation energy is much higher than in the forward direction because the products are at a lower energy than the reactants. In the reverse direction, the reaction's change in enthalpy has the same value but the opposite sign as that of the forward reaction. If the forward reaction is exothermic, the reverse reaction is endothermic, but the amount of the enthalpy change is the same. Only the sign is reversed.

Hess's Law

In calorimetry, we could do experiments to calculate the heat associated with chemical and physical changes, but not all reactions happen that quickly. There are two other methods to calculate heat changes with chemical reactions. One is Hess's law which will be examined here. The method that will be looked at in the next section is Heats of Reactions. Imagine two hikers climbing to the top of a mountain. One hiker goes straight up the slope of the mountain. The other reaches the top by a path that meanders up and down the sides of the mountain. In the end, the net change in position of each hiker is the same.

Chemical reactions present a similar situation. Hess's law states that the enthalpy change of a reaction is the same no matter how many intermediate steps occur. Like the hikers, the reaction will have the same net enthalpy change whether it occurs as a single reaction or as the result of several reactions.

When combining equations, keep in mind the following points:

- If you reverse an equation, change the sign of the enthalpy change from negative to positive, or positive to negative.
- If you multiply the coefficients in an equation, multiply the enthalpy change.

Example: Determine the enthalpy for the reaction of sulfur with oxygen to form sulfur trioxide:

$$2S(s)+3O_2(g)\rightarrow 2SO_3(g)\Delta H=?$$

The following reactions are known and can be used.

1. $S(s)+O_2(g)\rightarrow SO_2(g)\Delta H=-297\ kJ$
2. $2SO_3(g)\rightarrow 2SO_2(g)+O_2(g)\Delta H=198\ kJ$

Multiply the first reaction (a) by 2:

$$2S(s)+2O_2(g)\rightarrow 2SO_2(g)\Delta H=2(-297\ kJ)$$

Reverse the second (b) reaction:

$$2SO_2(g)+O_2(g)\rightarrow 2SO_3(g)\Delta H=-198\ kJ$$

Finally, combine the two reactions, making sure when adding reactions a and b, add up to make the wanted reaction and then add the enthalpies:

$$2S(s)+3O_2(g)\rightarrow 2SO_3(g)\Delta H=-792\ kJ$$

In this example, notice that when the second equation was reversed, the enthalpy changed from 198 kJ to –198 kJ.

Heat of Reactions

It's practically impossible to gather data on the enthalpy change of every reaction. There are too many reactions, and the change in enthalpy of a reaction is subject to a number of factors, including temperature. To help standardize enthalpy change values, chemists have assembled data for many reactions. In these standardized reactions, the reactants and products are at specified thermodynamic conditions of 25°C and 1 atm. (These conditions differ from the standard temperature and pressure used in the study of gases.) The symbol for the enthalpy change of a reaction at standard conditions is ΔH°rxn. The symbol ° indicates standard conditions.

To make it easier, chemists use tables of the heats of reactions. The heat of reaction is the change in the enthalpy of a chemical reaction that occurs at a constant pressure in which 1 mole of a substance is formed from its elements. These values are called standard heats of formation, ΔH°f. The standard heat of formation of an element is defined as 0 kilojoules per mole.

Using standard heats of formation, the heat of reaction is calculated by subtracting the total heat of formation of the reactants from the total heat of formation of the products. The formula for calculating the heat of a reaction can be written:

$$\Delta H^\circ_{rxn} = \Delta H^\circ_{products} - \Delta H^\circ_{reactants}.$$

Because the standard heats of formation are expressed in kilojoules per mole, the heat of formation for a specific compound must be multiplied by the coefficient of the substance in the balanced equation.

The use of standard heats of formation combines the equations for the formation of products in the reaction. Each of these equations is multiplied by the coefficient in the desired equation, and the equations for the reactants are reversed. Hence, their total enthalpy change is subtracted. The effect is the same as breaking down the reactants to elements and then combining the elements to form the products.

Standard Heat Table	
Substance	**$\Delta H^\circ f$ (kJ/mol)**
$C_2H_5OH(l)$	-277.0
$O_2(g)$	0
$CO_2(g)$	-393.5
$H_2O(g)$	-241.8

Table comparing different standard heats of formation. What standard conditions are assumed?

Standard Heats of Formation: Sample Problem

Calculate the standard heat of reaction for the complete combustion of etnanol, $C_2H_5OH(l)$.

$$C_2H_5OH(l)+3O_2(g)\rightarrow 2CO_2(g)+3H_2O(g)$$

Solution

The standard heat of reaction is calculated using Hess's law:

$$\Delta H^\circ_{rxn} = \Delta H^\circ_{products} - \Delta H^\circ_{reactants}$$

For this reaction, the equation becomes:

$$\Delta H^\circ_{rxn} = [2\ mol \Delta H^\circ_f(CO_2(g)) + 3\ mol \Delta H^\circ_f(H_2O(g))] - [1\ mol \Delta H^\circ_f(C_2H_5OH(l)) + 3\ mol \Delta H^\circ_f(O_2(g))]$$

Substitute the values from the table into the equation, and then solve:

$$\Delta H^\circ_{rxn} = [2\ mol(-393.5\ kJ/mol) + 3\ mol(-241.8\ kJ/mol)] - [1\ mol(-277.0\ kJ/mol) + 3\ mol(0\ kJ/mol)]$$
$$\Delta H^\circ_{rxn} = -1235.4\ kJ$$

Consider the Explain Question:

How are endothermic and exothermic processes identified, and how does the energy of the system change with each?

Go online to complete the scientific explanation.

Check for Understanding:

Can you describe the energy changes of a system and surroundings associated with exothermic changes and with endothermic changes?

STEM in Action

STEM and Thermochemistry

Efforts to develop clean, renewable energy resources have often focused on the sun. Photovoltaic cells, such as those that power solar-powered calculators, convert light energy directly into electrical energy. Typical solar panels only generate electricity during daylight hours. During the night, electricity comes from the utility grid. Solar engineers have been hard at work creating systems to store heat absorbed during the day to provide a steady stream of electricity day or night.

Using a different approach to solve the problem, scientists have used nanotechnology to develop super-thin solar films. These films have embedded nanoantennas, which absorb infrared energy released by the earth for several hours after the sun sets. Because they absorb energy from both sunlight and the earth itself, these tiny cells are more efficient than conventional solar cells.

Power plants have been built around the world, which store energy and generate electrical energy even when the sun is not shining. One method designed by engineers uses large arrays of mirrors to shine sunlight on a tower. In the tower, a liquid of molten salts absorbs the energy and is heated to over 1,000°C. The hot, molten liquid is stored in well-insulated tanks. When energy is needed, thermal energy from the salt is transferred to water, which becomes superheated steam. The steam drives turbines that generate electrical energy. The cooled liquid is stored in another tank for return to the tower where it is reheated. Molten salt thermal energy storage is the most economical method to store solar energy.

Recently, companies have completed experimental grid-style projects, which use batteries to store energy from solar panels. Using this new battery technology initially developed for electric cars, the batteries allow power systems to dispense electricity in the evening after it was stored by solar panels during the day.

photo: Getty Images

This concentrating solar power tower in Spain uses molten salts to capture the sun's energy. How could this process advance the solar power industry?

Ammonia (NH_3) is a common substance used in refrigeration. Use the phase diagram to answer the questions.

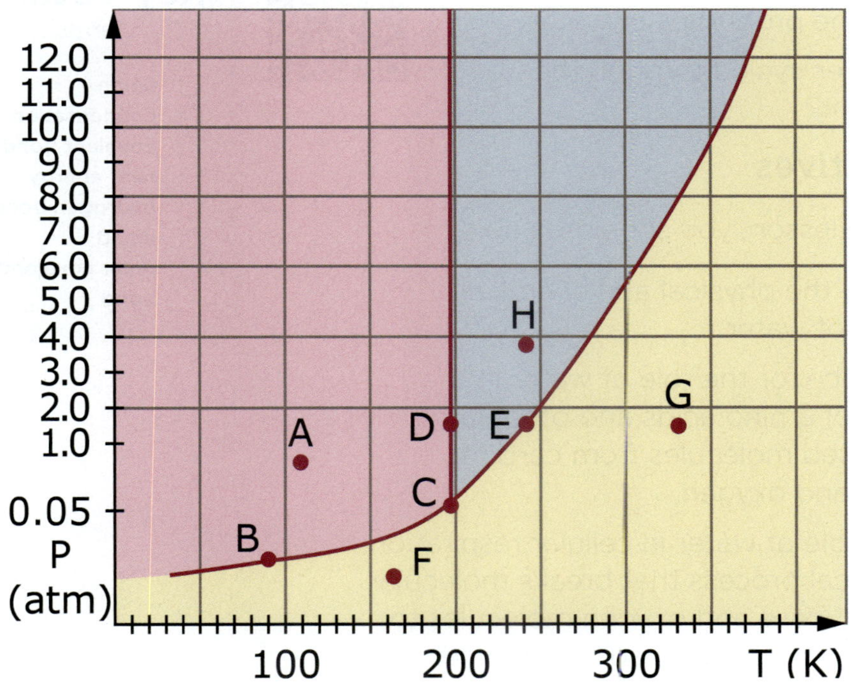

1. What letter represents the solid phase? Liquid? Gas?
2. What is the triple point of ammonia? Explain how you determined this.
3. What would happen at 1.0 atm if ammonia was cooled from 300 K to 210 K?
4. Explain why you think ammonia makes a good refrigerant.

Water

photo: Getty Images

LESSON OVERVIEW

Lesson Questions

- What are the properties of water?
- How is water cycled through the environment?

Lesson Objectives

By the end of the lesson, you should be able to:

- Summarize the physical and chemical properties of water.
- Cite evidence for the role of water in the formation of amino acids and other large carbon-based molecules from carbon, hydrogen, and oxygen.
- Show the role of water in cellular respiration as a chemical process that breaks molecular bonds in glucose and oxygen molecules to form new compounds, resulting in a net transfer of energy.
- Design and evaluate solutions to real-world problems using the engineering approach, prioritizing criteria and trade-offs to account for relevant constraints.

Key Vocabulary

Which terms do you already know?

- ☐ **anion**
- ☐ **condensation**
- ☐ **covalent bond**
- ☐ **heat energy**
- ☐ **hydrogen bond**
- ☐ **liquid**
- ☐ **polar covalent bond**
- ☐ **solvent**
- ☐ **specific heat**
- ☐ **surface tension**
- ☐ **universal solvent**
- ☐ **water**

ENGAGE

What are some ways that water sustains life today? What kinds of species live in or depend on water? Visit the Engage page online to learn more.

Explain Question

In what ways do the properties of water affect life on Earth?

EXPLORE

What Are the Properties of Water?

The Water Molecule

A water molecule consists of one oxygen atom and two hydrogen atoms. In this molecule, H_2O, the hydrogen-oxygen bond is a polar covalent bond. In a polar covalent bond, the electrons shared between the atoms tend to spend more time closer to the nucleus of the more electronegative atom. Therefore, in water, the electrons tend to stay near the oxygen atom. This polarity means that, in the water molecule, the region of the oxygen atom is slightly negative and the regions of the hydrogen atoms are slightly positive.

Due to the polarity of water molecules, they bond with one another through hydrogen bonds. A hydrogen bond between water molecules is an attractive force between the hydrogen atom of one water molecule and the oxygen atom of another water molecule. Hydrogen bonds between water molecules are weaker bonds than are the polar covalent bonds within water molecules. Hydrogen bonding between water molecules is a dynamic process. Hydrogen bonds that connect water molecules are constantly forming, breaking, and reforming.

Two water molecules can dissociate to form one hydronium ion and one hydroxide ion. A hydronium ion (H_3O^+) contains three hydrogen atoms and one oxygen atom and is positively charged. A hydroxide ion (OH^-) contains one hydrogen atom and one oxygen atom and holds a negative charge. The chemical reaction of water dissociation to form these ions is as follows:

$$2H_2O \leftrightarrow H_3O^+ + OH^-$$

This reaction proceeds back and forth in both directions. In its liquid state, water molecules are constantly switching back and forth between these forms.

Properties of Water: Stickiness

Water is different from other substances in many ways. Notably, it exists in all three states of matter within a narrow range of temperature—0 to 100 degrees Celsius—on Earth's surface. Most water exists as a liquid or ice on Earth's surface. A small amount exists as water vapor in the air.

Its polar covalent bonding, together with its hydrogen bonding, gives water unique properties, including cohesion, the ability to moderate temperature, the ability to dissolve many other materials, and the ability to expand upon freezing.

Cohesion refers to the attraction of two like materials. In the case of water, cohesion refers to the binding of water molecules via hydrogen bonds. In other words, water molecules cohere to one another. Cohesion is very important to life on Earth. One useful outcome of cohesion is that it helps water travel from the roots upward in plants. Starting at the roots of the plant, water molecules cohere and travel upward in columns in the plant's vascular tissue. When one molecule of water evaporates from the leaf, it pulls the water molecule lying below it to the surface of the leaf, and in this way water travels upward from roots to leaves, driven by evaporation.

Water molecules can also stick to other materials and surfaces; the attraction of two different materials is called adhesion. For example, adhesion is also critical to plants because it allows water to stick to cell walls. This helps prevent water from falling downward with gravity. It is also important deep inside of your lungs, where a wet surface is essential for gas exchange.

Another important property of water is its high surface tension. Due to hydrogen bonding, it is relatively difficult to "break" the surface of liquid water compared with other liquids. Water's high surface tension can be demonstrated by filling a container with water. Water can actually be filled slightly above the height of the container without spilling because of the hydrogen bonds between water molecules that hold the molecules together. High surface tension allows some insects to move on the surface of water. The insects' legs do not produce enough pressure to break water's surface tension.

Properties of Water: The "Universal Solvent"

Another important property of water is its ability to dissolve more substances than any other liquid, making it known as the universal solvent. A solvent is any fluid in which one or more substances can be dissolved. When water dissolves a substance, water molecules surround a molecule of a substance and form a shell of hydration. This shell effectively dissolves the substance.

Water's ability to act as a solvent for many substances is crucial to life for many reasons. For example, the water inside cells must be able to dissolve the proteins, carbohydrates, ions, and other molecules necessary for cellular function. Water readily dissolves ions and polar molecules. It does not dissolve certain other molecules such as lipids. This is why oil and water do not mix well. Also, because water dissolves many substances, water provides an important medium for molecules to combine and form new molecules.

photo: Getty Images

How was water able to shape natural rocks over the years?

Properties of Water: Expansion Upon Freezing

Water has the unique property of expanding upon freezing. It is one of the few substances on Earth that is denser in its liquid form than its solid form. Again, this expansion property is due to the polarity and hydrogen bonding of water molecules. Normally, liquid water molecules constantly form and re-form their hydrogen bonds with one another. The molecules constantly move closer and farther from one another. When temperatures reach freezing, however, all water molecules form a hydrogen-bonded crystal lattice. The configuration of this lattice holds water molecules stretched out from one another, creating a less dense structure than liquid water. Thus, ice floats in water because frozen water is less dense than liquid water.

The ability of ice to float in water may not seem very important to life on Earth. However, imagine what would happen to large bodies of water in cold climates if this property of water did not exist. Instead of floating, ice would sink. There would be no suitable liquid water underneath the ice to house aquatic life. Eventually, these bodies of water would most likely freeze over completely, making them an uninhabitable environment for most organisms.

Properties of Water: Specific Heat and the Ability to Moderate Temperature

Water has the extraordinary ability to moderate temperatures on Earth. It does this by absorbing thermal energy from warmer air and releasing thermal energy into cooler air. Large bodies of water temper the climate around them. They hold local temperatures within a relatively small range by heating or cooling the air through energy absorption and release.

The ability of water to moderate temperatures is made possible by its very high heat capacity. Heat capacity of a material is a property of the material that is often referred to as the material's "specific heat." Specific heat is a measure of how much energy is required to change the temperature of a given mass of a substance by a set amount. The specific heat of water is 1 calorie/gram °C. This means that 1 calorie of energy is required to change the temperature of 1 gram of water by 1 degree Celsius. (When specific heat is described in SI units, specific heat is described as a change in Joules [J] per Kelvin [K].) The specific heat of water is very high relative to other substances. Water requires a great deal of energy to change its temperature.

The outcome of a high specific heat is that water maintains its temperature within a relatively narrow range, even when absorbing large amounts of energy. As a result, large bodies of water can absorb and release a lot of energy from the surrounding air, moderating air temperatures with relatively small changes to their own temperatures. Similarly, water inside the cells of an organism can absorb and release heat while maintaining fairly stable temperatures within the organism.

How Is Water Cycled through the Environment?

Water is essential for the survival of all living things on the planet. While it is true that water covers a large percentage of Earth's surface, much of this water is not available for plants and animals that live on land because it is either saltwater or frozen in glaciers.

The Water Cycle

The water cycle is made up of various processes that move water through the atmosphere and biosphere. These processes are evaporation, condensation, precipitation, and transpiration. All of these processes involve changes in the state of water.

Evaporation occurs when liquid water changes into its gaseous form: water vapor. As water on Earth's surface—in lakes, rivers, oceans, and soil—warms, it evaporates. Water also evaporates from the leaves of plants in the process of transpiration. The water vapor rises into the atmosphere. As it rises, it cools. When it cools enough to change from a gas to a liquid, in a process called condensation, it collects in droplets. Clouds and fog are made up of many of these water droplets.

As the water droplets become larger, they may fall to the ground as rain. If the air is very cold, water in the air may freeze and fall as snow, hail, or sleet. Rain, hail, snow, and sleet are all forms of precipitation. Precipitation occurs when liquid or solid water falls from the atmosphere to Earth's surface. Some of the precipitation falls into bodies of water. Some falls on land, where it either flows into a body of water or soaks into the ground and is used by plants and stored as groundwater.

Driving the Water Cycle

Because water's state changes are crucial to the water cycle, and state changes occur when temperature of the water changes, it is important to recognize that this cycle needs an energy source. The main energy source for the water cycle is the sun; it is the sun's energy that causes liquid water to evaporate into water vapor, and solid water (ice) to melt into liquid water.

Gravity is another driver of the water cycle. Earth's gravitational force pulls precipitation to the ground. Once on Earth's surface, gravity's effects depend on the surface on which the precipitation falls. Some precipitation falls directly into bodies of water, and some falls on land. If rain falls on soil or sand, the water will soak into the ground, where it can be absorbed by the roots of plants. If rain falls on rocky areas, gravity pulls much of the water along the ground, so that the water flows downhill, creating systems of rivers and streams that provide water to animals and plants that live along the shore. Eventually, most of the water in rivers and streams will flow all the way to the ocean.

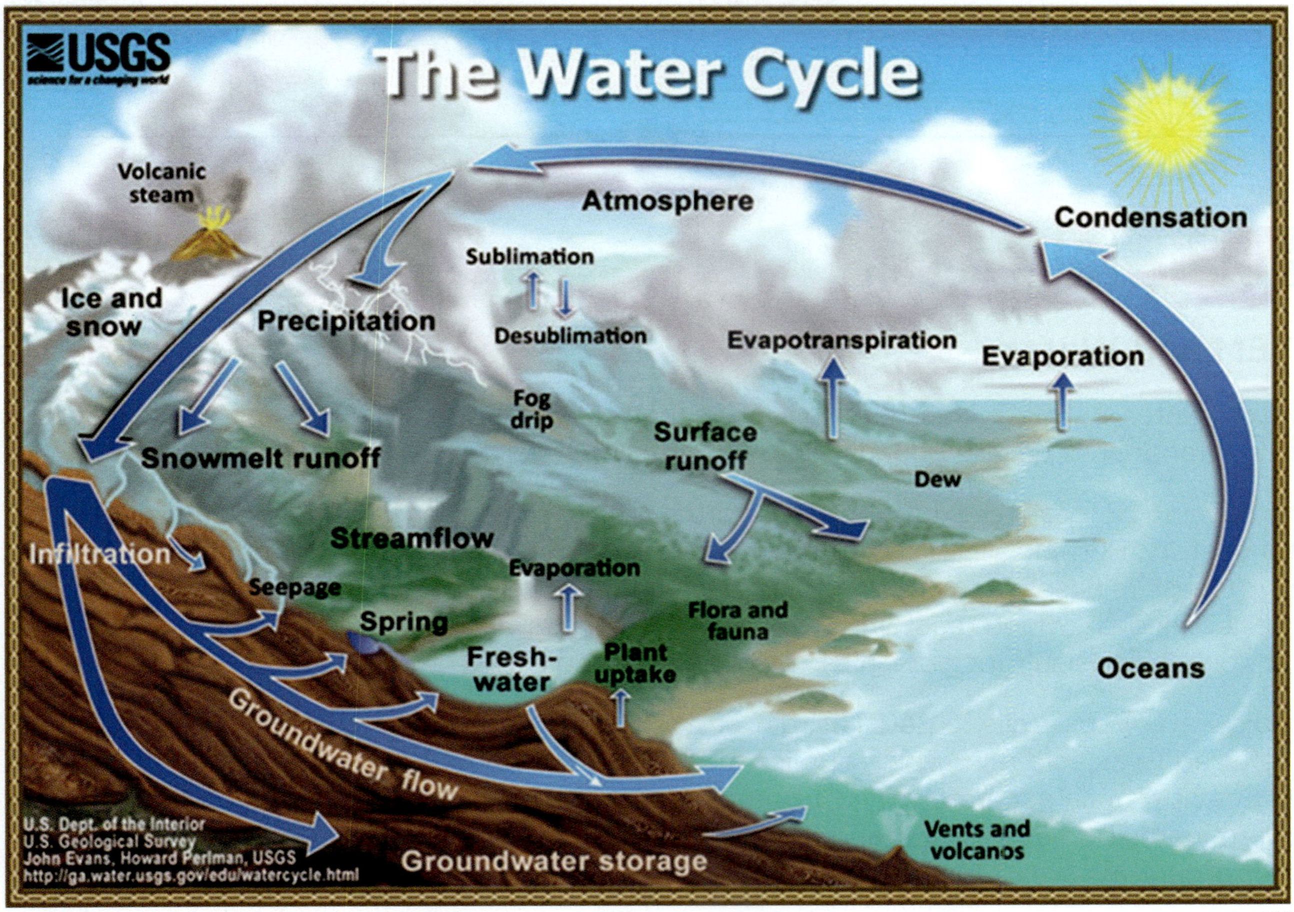

photo: USGS

The water cycle is the process by which water moves through Earth's systems. What happens to water as it moves from the ocean to the clouds to the ground?

The sun's energy is also the cause of weather systems that bring precipitation to almost all of the parts of Earth—even those far from the ocean. Sunlight is absorbed by water, land, and the air around Earth. But not all of Earth receives the same amount of sunlight, causing Earth to heat unevenly. Because heat flows from areas of greater warmth to cooler areas, moving air—wind—moves Earth's atmosphere in currents. Air masses, driven by convection, are a powerful force that moves water across the planet. As an air mass warms, it becomes less dense. The warmer air mass will be pushed upward by the cooler air below it. As the rising air cools, it begins to sink. The rising and falling air creates convection currents that continuously circulate air as wind.

Consider the Explain Question:

In what ways do the properties of water affect life on Earth?

Go online to complete the scientific explanation.

Check for Understanding:

How could you design an experiment to investigate the properties of water?

STEM in Action

STEM and Water

Clearly, water is vital to life on Earth. Unfortunately, waste and pollution from human activity has contaminated Earth's oceans, lakes, and rivers.

The burning of fossil fuels has increased dramatically since the industrial revolution of the 19th century. Burning oil, coal, and natural gas powers most of our energy needs. However, this process releases many harmful gases—including carbon dioxide, sulfur dioxide, and nitrogen oxides—which react with water vapor in the atmosphere. The products of such reactions include sulfuric acid, nitric acid, and carbonic acid.

Acidic water vapor falls to Earth as rain, snow, fog, or other precipitation. This phenomenon is referred to as acid rain, which applies to rain below a normal pH of approximately 5.6. Scientists can accurately measure the acidity of rain using a pH meter. Acid rain contaminates bodies of water and is detrimental to wildlife. Land animals, insects, fish, and other organisms are all harmed and killed by acidic lakes and rivers.

A great deal of scientific research has been conducted to determine the causes and impacts of acid rain. During this time, the government has passed numerous laws in an effort to reduce the burning of fossil fuels. Limits have been imposed on the amount of carbon and sulfur emissions that factories can produce.

Marine biologists are particularly concerned about how the increased concentration of carbon dioxide in the atmosphere is acidifying the oceans. Many ocean-dwelling organisms are very sensitive to changes in acidity. Organisms that make shells or skeletons out of calcium carbonate are seriously impacted by ocean acidification. For example, corals cannot make the skeletons that are the basis of coral reefs.

Today, many scientists and bioengineers are researching alternative energy sources that will reduce our dependence on fossil fuels, thus reducing the amount of acid rain that falls to Earth. Alternative energy sources include wind, solar, and geothermal energy.

photo: Paul Fuqua

How acidic do you think rain can be?

Analyzing Effects of Acid Rain

Read the article "Surviving Acid Rain" on the STEM in Action Page online, and then respond to the following two prompts:

1. What is acid rain, and how is it contributing to the deterioration of natural and manmade environments?
2. Describe a device that would allow you to measure the acidity of rain. How would the design work, and what would it look like?

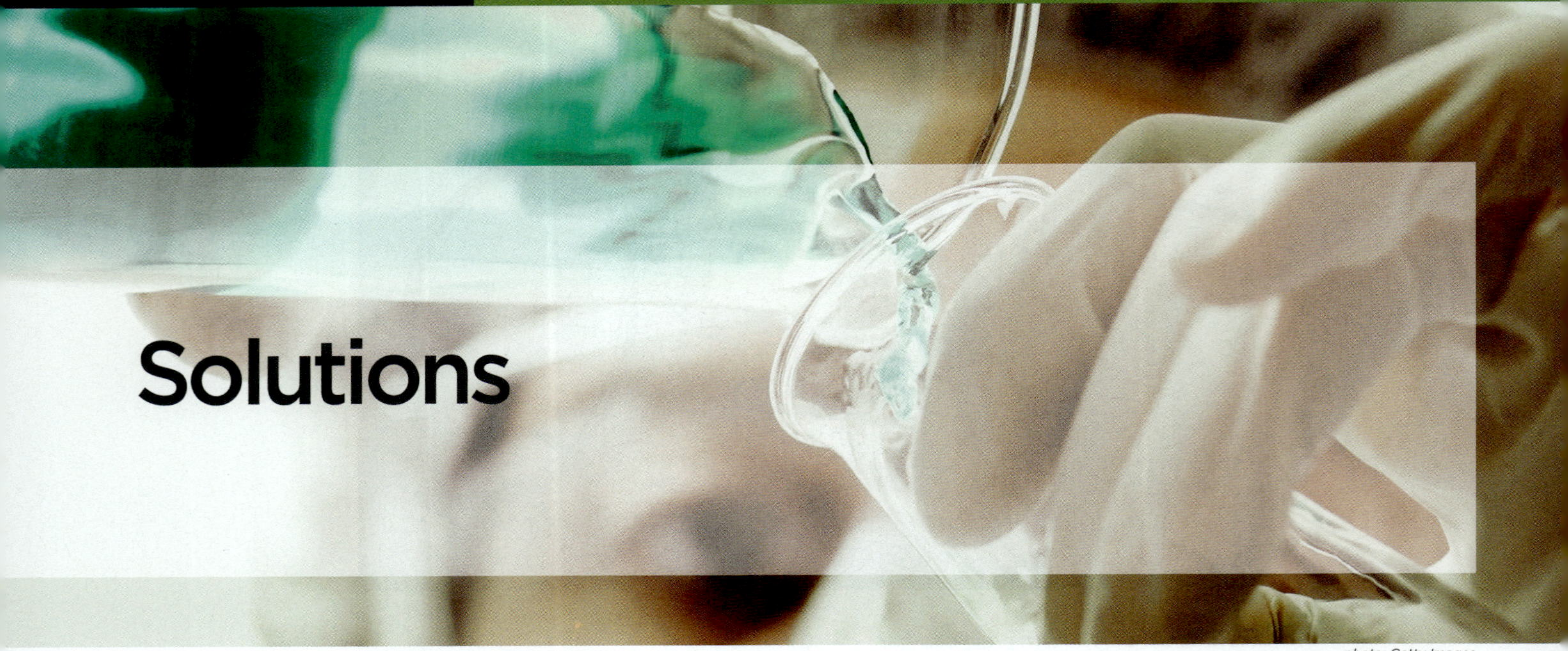

Solutions

photo: Getty Images

LESSON OVERVIEW

Lesson Questions

- What are the various parts of a solution?
- How do various factors affect the solubility of a solute?
- How are concentrations of solutions using molarity, mass percent, volume percent, and parts per million (ppm) calculated?
- What are the various colligative properties of solutions?

Lesson Objectives

By the end of the lesson, you should be able to:

- Distinguish between the parts of a solution.
- Summarize factors that affect the solubility of a solute.
- Calculate concentrations of solutions using molarity, mass percent, volume percent, and parts per million (ppm).
- Describe the various colligative properties of solutions.

Key Vocabulary

Which terms do you already know?

- [ ] **boiling point elevation**
- [ ] **colligative property**
- [ ] **concentration**
- [ ] **dilute solution**
- [ ] **dissolve**
- [ ] **electrolyte**
- [ ] **freezing**
- [ ] **freezing point depression**
- [ ] **heterogeneous**
- [ ] **homogeneous**
- [ ] **ideal solution**
- [ ] **mass percent**
- [ ] **molality**
- [ ] **molarity**
- [ ] **mole**
- [ ] **mole fraction**
- [ ] **nonpolar covalent bond**
- [ ] **osmotic pressure**
- [ ] **parts-per-million (ppm)**
- [ ] **polarity**
- [ ] **precipitate**
- [ ] **solubility**
- [ ] **solubility product constant (Ksp)**
- [ ] **solute**
- [ ] **solution**
- [ ] **solvent**
- [ ] **surface area**
- [ ] **universal solvent**
- [ ] **volume percent**

ENGAGE

What happens to your body when you are very sick or have just completed a long, rigorous workout? How can you help your body replenish important nutrients that have been lost? Visit the Engage page online to learn more.

Explain Question

What factors influence the rate at which a solute dissolves?

EXPLORE

What Are the Various Parts of a Solution?

Solutions

Solutions are homogeneous mixtures in which the components are distributed evenly and will not separate. Colloids and suspensions are heterogeneous mixtures, not solutions. They are composed of small particles distributed throughout a medium, but the particles are large enough that they can often be filtered out. They are not truly dissolved.

Solutions can be solid, gaseous, or liquid. For example, steel is a solid solution in which carbon is mixed evenly with iron. (Solid solutions that contain metals are called alloys.) Air is a solution that contains primarily gases, but might also contain tiny particles in solid or liquid form. In chemistry, many of the solutions discussed are in liquid form.

There are two necessary components to every solution. A solute is the substance that is dissolved in a solution. The solvent is the substance that dissolves the solute. Usually, there is more solvent than solute in a solution. For example, if you add one spoonful of sugar to a jug of water, the water is the solvent and the sugar is the solute. When two liquids are mixed together, the solute is the substance of the lesser quantity. For example, rubbing alcohol is a liquid mixture of 70 percent isopropyl alcohol and 30 percent water. In this solution, water is considered the solute.

photo: Getty Images

What can you do to cause the sugar to dissolve faster in this coffee?

Aqueous Solutions

Many aqueous solutions are formed in nature, where water commonly dissolves other substances in the environment. In fact, water is often referred to as the universal solvent because so many other substances will dissolve in it. Water is unique because it naturally exists in all three states of matter—solid, liquid, and gas. Water's unique properties are primarily caused by the polarity of the water molecule. The positive and negative areas of charge are attracted to many other types of particles, and can dissolve them easily.

An electrolyte is any substance that dissolves to form free ions, which make the solution electrically conductive. Batteries contain electrolytes because they allow negative charge (electrons) to flow from one location to another in an electric current. The human body needs electrolytes to support cellular processes, such as metabolism and muscle function. Electrolytes also help keep the body's fluid levels in balance.

A nonelectrolyte is a solute that dissolves in water to form a nonconductive solution. Sugar is a nonelectrolyte. When mixed with water, sugar creates a nonconductive solution.

How Do Various Factors Affect the Solubility of a Solute?

Solubility Rate

The solubility of a substance indicates how much of that substance will dissolve in a specified amount of solvent under certain conditions. Solubility is usually measured in terms of concentration at a given temperature and pressure.

There are many factors that will increase or decrease the rate of solubility. As the surface area of a solute is increased, the rate of dissolving increases. Surface area is the total area a substance has on its exterior. For example, a teaspoon of sugar contains about the same mass of sugar found in a sugar cube. However, the teaspoon of loose sugar will dissolve faster because it has more surface area available to interact with the solvent.

As the temperature of a solvent is increased, the rate of dissolving usually increases. This occurs because temperature is related to the kinetic energy of the particles involved. The more kinetic energy solvent particles have, the faster they move. The faster they move, the more they can interact with a solute. Kinetic energy can also be added to the system by stirring or shaking it.

Chemical Makeup and Solubility

If the solute is a gas, increasing the temperature has the opposite effect on the solubility. The increased kinetic energy of the solution allows the gas particles to move more freely, and they are more likely to escape from the solution. The solubility of the gas will decrease.

Carbonated beverages exhibit another solubility rule known as Henry's law. This law states that if temperature does not change, the amount of gas that dissolves in a liquid is proportional to the partial pressure of that gas in equilibrium with that liquid. For example, inside an unopened can of soda, the pressure is much higher than that of the surrounding area. This pressure keeps carbon dioxide gas dissolved in the liquid soda solution. When the can is opened and pressure is decreased, the carbon dioxide escapes the solution.

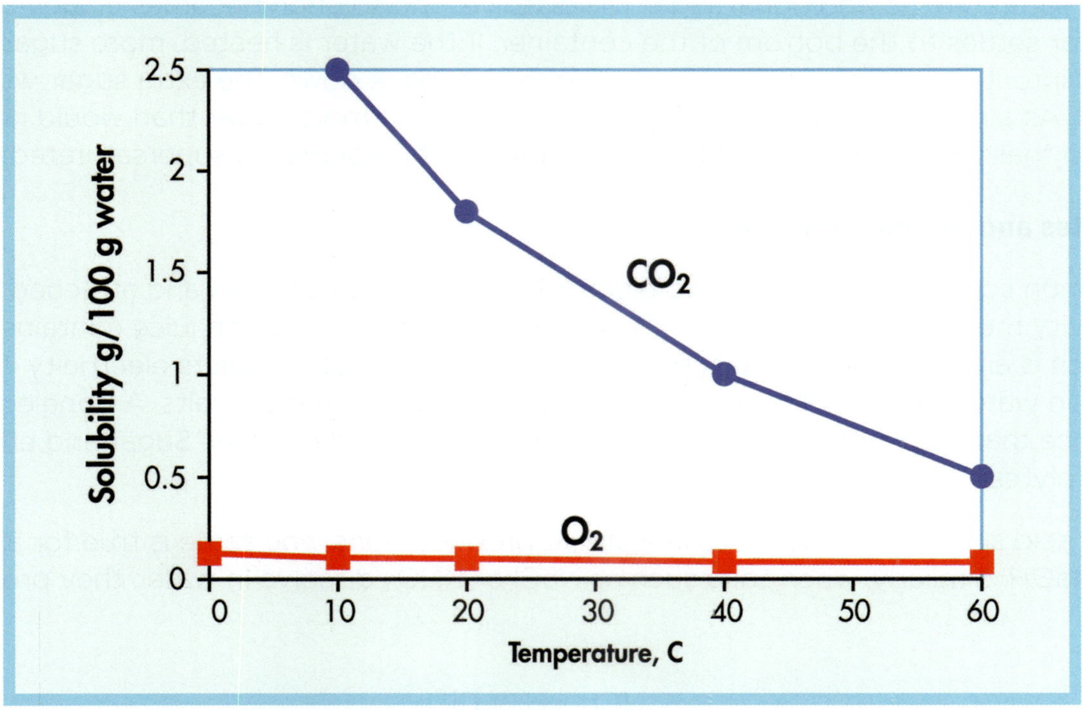

The solubility of gas in water decreases as the temperature of the water increases. Why do CO_2 and O_2 have such different solubility?

The chemical makeup of the solvent and solute are also related to solubility. Remember that some molecules have a negative or positive charge because the electrons are not distributed evenly around them. These are referred to as polar molecules. Solutes of a specific polarity (negative or positive) will only dissolve in solvents with a similar polarity. The rule for this is "like dissolves like." For example, the oil and vinegar in salad dressing do not mix because vinegar is polar and oil is nonpolar. However, water and vinegar will form a solution because they are both polar.

Saturated, Unsaturated, and Supersaturated Solutions

When a large amount of salt is mixed into cold water, some of it does not dissolve. In fact, the salt remains solid and settles on the bottom of the container.

Because of the way in which particles react, a solvent can dissolve only a certain amount of a solute. A saturated solution is a solution in which no more solute can be dissolved. In fact, if additional solute were added, it could cause some dissolved solute to become solid and settle out of the solution. The solid that forms and separates from a solution is known as a precipitate.

An unsaturated solution is a solution that contains less solute than can be dissolved under existing conditions. If more solute is added, it will simply dissolve into the solvent.

Under special conditions, a saturated solution can be made to dissolve more solute. Most often, this requires the addition of heat or pressure to the system. For example, water at room temperature will dissolve a certain amount of sugar. However, when the solution becomes saturated, no additional sugar will dissolve. The solution becomes cloudy until the extra sugar settles to the bottom of the container. If the water is heated, more sugar can be dissolved into it. When the solution is allowed to cool back down, the extra sugar will remain dissolved. As a result, the final, cooler solution will contain more sugar than would normally dissolve in that volume of water at that temperature. This solution is supersaturated.

Electrolytes and Nonelectrolytes

In a common science experiment, two electrodes are stuck in a lemon and attached to a conductivity meter. Amazingly, the lemon conducts electricity! Lemon juice contains citric acid, which is an electrolyte. An electrolyte is a substance that conducts electricity if it dissolves in water. Acids and bases are electrolytes, and so are most salts. A nonelectrolyte is a substance that does not conduct electricity when dissolved in water. Sugar and ethanol are nonelectrolytes.

When an acid such as HCl dissolves in water, it produces ions. The same is true for a base such as NaOH. Similarly, when salts such as NaCl or KNO_3 dissolve in water, they produce ions.

$$\mathrm{HCl} \rightarrow \mathrm{H}^+ + \mathrm{Cl}^-$$

$$\mathrm{NaOH} \rightarrow \mathrm{Na}^+ + \mathrm{OH}^-$$

$$\mathrm{NaCl} \rightarrow \mathrm{Na}^+ + \mathrm{Cl}^-$$

$$\mathrm{KNO_3} \rightarrow \mathrm{K}^+ + \mathrm{NO_3^-}$$

These ions are the reason why solutions of electrolytes can conduct electricity. Negative and positive ions move in the solution. They can carry electrical charge from one electrode to the other. Strong acids and bases are strong electrolytes because they dissociate completely into ions. Weak acids and bases are weak electrolytes because they dissociate only partly.

The conductivity of a solution is a measure of how well it conducts electricity. Conductivity depends on how many ions are present in the solution. For one electrolyte, a concentrated solution usually has a higher conductivity than a dilute solution.

Conductivity can be used to measure the concentration of a solution. Different electrolytes have different conductivities. The relationship between concentration and conductivity can be found for each electrolyte. (This data can be measured in the laboratory or obtained from reference books.) Then, this relationship can be used to identify an unknown concentration by measuring its conductivity.

How Are Concentrations of Solutions Using Molarity, Mass Percent, Volume Percent, and Parts per Million (ppm) Calculated?

Concentrations of Solutions

The concentration of a solution is a measure of how much solute is dissolved in a given amount of solvent. There are three ways the concentration of a solution can be described.

The first method for describing concentration is known as molarity. Molarity (*M*) gives the number of moles of solute divided by the total number of liters of the solution:

$$M = \frac{\text{moles of solute}}{\text{liters of solution}}$$

Another way to describe solution concentration is the mole fraction. The mole fraction (*X*) of a solution is the ratio of moles of the solute to the total number of moles in a solution. Because mole fraction describes a percentage, it can never be greater than 1:

$$X = \frac{\text{moles of solute}}{\text{moles of solute and solvent combined}}$$

The final way to describe concentration is molality. Molality (*m*) is the number of moles of solute dissolved in 1 kilogram of the solvent:

$$m = \frac{\text{moles of solute}}{\text{1 kg of solvent}}$$

These methods of calculating concentration are most useful when solute concentrations are quite high.

Concentrations of Solutions: Sample Problem

A solution is formed by combining 0.250 moles of sodium chloride (NaCl) in 3.61 liters of water. Assume that the addition of NaCl does not significantly change the volume of the solution. The volume of the solution, then, is 3.61 liters. Compare the molarity and the molality of NaCl in the solution. What is the mole fraction of NaCl in the solution? Assume that 1 liter of water has a mass of 1 kilogram.

Solution

First, to calculate the molarity of the solution, substitute the given information into the molarity ratio:

$$\begin{aligned} M &= \frac{\text{moles of solute}}{\text{liters of solution}} \\ &= \frac{0.250\ \text{mol}}{3.61\ \text{L}} \\ M &= 0.0693\ \text{mol/L} \end{aligned}$$

Notice that two extra digits are included because this is an intermediate calculation. Carrying extra digits can avoid round-off errors in the final answer. The molarity of the NaCl in the solution is 0.0693 moles per liter. Notice that a volume of 3.61 liters of solution is used in the calculation. The actual volume would be slightly higher due to the addition of NaCl. Notice also that the answer is expressed with the correct number of significant digits, which in this case is three.

To find molality, substitute the given information into the molality ratio:

$$\begin{aligned} m &= \frac{\text{moles of solute}}{\text{kg solvent}} \\ &= \frac{0.250\ \text{mol}}{3.61\ \text{kg}} \\ m &= 0.0693\ \text{mol/kg} \end{aligned}$$

$$X = \frac{\text{moles of solute}}{\text{moles of solute and solvent combined}}$$

The value for the moles of solute NaCl is given as 0.250 moles. The total number of moles of solute and solvent combined must next be calculated.

The solution contains 3.61 kilograms of water, but the number of moles of water must be determined. Because the molar masses of hydrogen and oxygen are given in grams, it is convenient to convert the mass of water in the solution to grams, as well.

$$3.16\ \cancel{\text{kg}} \times \frac{1{,}000\ \text{g}}{1\ \cancel{\text{kg}}} = 3160\ \text{g}$$

The mass of the water in grams is 3,610 grams. This value can be used to determine the moles of water in the solution.

From the periodic table, we can find the following atomic masses:

$$\begin{aligned} \text{hydrogen (H)} &= 1.008 \text{ g} \\ \text{oxygen (O)} &= 16.00 \text{ g} \end{aligned}$$

These masses of 2 moles of hydrogen and 1 mole of oxygen are added together to find the molar mass of water.

$$(2 \times 1.008 \text{ g/mol}) + 16.00 \text{ g/mol} = 18.016 \text{ g/mol}$$

Next, we determine the number of moles in 3,160 grams of water.

$$3160 \cancel{g} \times \frac{1 \text{ mol}}{18.016 \cancel{g}} = 175.39964 \text{ mol}$$

The total moles of the solute and solvent equals:

$$0.250 \text{ mol NaCl} + 200.377 \text{ mol H2O} = 200.627 \text{ mol total}$$

The mole fraction of NaCl in the solution is:

$$\begin{aligned} X &= \frac{\text{moles of solute}}{\text{moles of solute and solution}} \\ &= \frac{0.250 \text{ mol}}{175.64964 \text{ mol}} \\ X &= .0014253 \end{aligned}$$

Note that extra digits were used during this last set of calculations so that the moles of solute and total moles in the solution could be properly compared. Notice that two extra digits are carried. Carrying extra digits can avoid round-off errors in the final answer. The final answer, however, should again have four significant digits. The mole fraction of NaCl in the solution is about 0.14 percent.

Using Molarity to Calculate the Dilutions of Solutions

A dilution is making a solution that has smaller amounts of solute than the original solution. In other words, it has a lower concentration than that of the original. Dilution is accomplished by adding more solvent to the solution to create a smaller ratio of solute to solvent.

To find a dilution of an existing solution of known concentration, the following formula can be used:

$$\text{Molarity1} \times \text{Volume1} = \text{Molarity2} \times \text{Volume2}, \ (M_1V_1 = M_2V_2)$$

Molarity1 is the known concentration of the initial solution and Volume1 is the known volume of the initial solution. Either the desired molar concentration must also be known or the new volume required. These would be Molarity2 and Volume2 values, respectively. The fourth and final value can then be calculated.

An example of how this is done is as follows:

You have five liters of a 2 M hydrochloric acid solution. What would the final concentration be if you added enough water to get a final volume of 20 liters?

$$M_1 = 2\text{ M},\ V_1 = 5\text{ liters},\ M_2 = \text{unknown, and } V_2 = 20\text{ liters}$$

Rearranging the formula to solve for the variable in question would be

$$M_2 = \frac{M_1V_1}{V_2}.$$

Substituting in the known values would give:

$$M_2 = \frac{2\text{M} \times 5\text{L}}{20\text{ liters}}.$$

Solving the equation would give a value of 0.5 M.

Additional Methods for Calculating Concentrations

In the following sample problem, concentration was calculated using molarity, molality, and mole fraction. Other methods of calculating concentrations are related to the mass and volume of the solution's components instead of molar values.

Mass percent is the mass of a given solute divided by the mass of the solution. The mass percent of each component in a solution should add up to equal 100 percent:

$$\text{mass \%} = \frac{\text{mass of solute}}{\text{mass of solution}} \times 100\%$$

Another method, volume percent, is often used to describe the concentrations of liquid solutions. For example, you may have seen 70 percent isopropyl alcohol in a first aid kit. It is made using a ratio of 70 parts isopropyl alcohol and 30 parts water. Because volumes do not always increase when solutions are formed, the volume percent concentrations of each solute in a solution may not add up to be exactly 100 percent.

If a solution is formed from liquid or gas components, the volume of the solute is used to calculate volume percent concentration:

$$\text{volume \%} = \frac{\text{volume of solute}}{\text{volume of solution}} \times 100\%$$

Molarity is generally used when the concentration of a solute is very high. Using another method, referred to as parts per million (ppm), is better suited for solutions with low concentrations of a solute. This concentration is calculated using ratio of the mass of solute to the mass of solution, then applying the ratio to a quantity of one million particles.

In other words, the calculation shows how many parts of solute would be found in one million parts of solution:

$$\text{ppm} = \frac{\text{mass of solute}}{\text{mass of solution}} \times 10^6$$

Calculating Concentrations: Sample Problem

A solution is formed when 5.000 grams of sodium bicarbonate, $NaHCO_3$, is added to 2.000 liters of water. Calculate the mass percent of sodium bicarbonate in solution. What is the concentration of sodium bicarbonate, in parts per million? Assume 1 liter of water has a mass of 1 kilogram.

Solution

To calculate the mass percent of sodium bicarbonate in solution, the mass of the solution must first be determined. The mass of the solute is given as 5.000 grams. The mass of the solvent must be determined in kilograms, and then converted to grams.

$$\text{mass of solvent} = 2.000\ \cancel{\text{L}} \times \frac{1.0\ \text{kg}}{1\ \cancel{\text{L}}} = 2.000\ \text{kg}$$

$$\text{mass of solvent} = 2.000\ \cancel{\text{kg}} \times \frac{1{,}000\ \text{g}}{1\ \cancel{\text{kg}}} = 2{,}000\ \text{g}$$

The total mass of the solution, then, combines the two values.

$$\text{mass of solute} + \text{mass solution} = \text{total mass of solution}$$

$$5.000\ \text{g} + 2.000 \times 10^3\ \text{g} = 2{,}005\ \text{g}$$

Next, we substitute the correct values into the equation for mass percent:

$$\begin{aligned} \text{mass percent} &= \frac{\text{mass of solute}}{\text{mass of solution}} \times 100\% \\ &= \frac{5.000 \text{ g}}{2{,}005 \text{ g}} \times 100\% \\ &= 0.2494\% \end{aligned}$$

The mass percent of the sodium bicarbonate in the solution is 0.2494 percent.

To determine the concentration of the $NaHCO_3$ solution in parts per million, the mass values used above are substituted into this equation:

$$\begin{aligned} \text{parts per million} &= \frac{\text{mass of solute}}{\text{mass of solution}} \times 10^6 \\ &= \frac{5.000 \text{ g}}{2{,}005 \text{ g}} \times 10^6 \\ &= 2{,}494 \text{ ppm} \end{aligned}$$

For every million parts of solution, there are 2,494 parts of solute.

What Are the Various Colligative Properties of Solutions?

Colligative Properties

Colligative properties depend on the number of particles of solute present in a solution. They are not determined by the properties of the solute or the solvent, and always follow some specific patterns.

The primary property, which is affected by colligative factors, is vapor pressure. In a liquid, vapor pressure is caused by particles escaping and changing to a gas state. If the solution is in a closed container, they cannot escape the entire system, so they occupy the space above the solution and below the top of the container. As temperature increases, more particles evaporate and vapor pressure increases.

When a solute is added to a solvent, it becomes more difficult for the particles of the solvent to evaporate because there is less surface area available. In other words, some of the surface is blocked by the solute particles. The more solute particles there are, the more vapor pressure decreases.

The change in vapor pressure affects two specific colligative properties. First, when a solid solute is added to a liquid to create a solution, the boiling point of that solvent will increase. This colligative property is described as boiling point elevation. Because particles of solute on the surface of the solution make it more difficult for evaporation to occur, boiling requires extra energy.

Second, adding solid solute to a solvent will decrease the freezing point of the solvent used in the solution. This colligative property is known as freezing point depression. Freezing point depression is also proportional to the number of solute particles in the solution and is independent of the type of particle. The solute particles interfere with the solvent's formation of crystals. The freezing process does not occur until the temperature is lowered further.

To calculate changes in boiling or freezing point, the concentration of the solution must be known. Molality is proportional to the change in boiling and freezing points, but the proportional relationship depends on the specific pure solvent or solution that is present. Constants for boiling point elevation (K_b) and freezing point depression (K_f) have been determined through research.

Based on this proportional relationship, the change in temperature required for boiling can be expressed by the equation:

$$\Delta T_b = iK_b m$$

Similarly, the change in temperature required for freezing is calculated using the equation:

$$\Delta T_f = iK_f m$$

The Greek symbol delta, Δ, represents "change." The m in each equation is the molality. The van 't Hoff factor, i, is determined by the solute's behavior in the particular solvent. For example, solutes that do not dissociate (or break apart into two components) in water, $i = 1$. For solutes that completely dissociate into two ions, $i = 2$. Sodium chloride, NaCl, completely dissociates into two ions, Na+ and Cl–. Some solutes completely dissociate into three ions, and for those $i = 3$. For example, $CaCl_2$ completely dissociates into three ions, one Ca^+, and two Cl^-. The values of i continue to increase as the number of ions created during dissociation increases.

Calculating Freezing Point: Sample Problem

A solution is made with 12.55 grams of sodium chloride, NaCl, and 180.4 milliliters of water. The water is 35.0°C. What is the freezing point of this solution? Assume the NaCl dissolves completely in the water. The density of water at 35.0°C is 0.9941 grams per milliliter, and the K_f of H_2O is 1.858°C kilograms per mole.

Solution

To find the freezing point of the solution, use the formula:

$$\Delta T_f = iK_f m$$

First, it is known that i = 2 because two ions (Na+ and Cl–) form when NaCl dissociates.

The value K_f of water is given as 1.858°C kilograms per mole.

The value of m must be calculated. This is the molal concentration of the solution, or:

$$m = \frac{\text{moles of solute}}{\text{1 kg of solvent}}$$

To find the moles of NaCl, first use atomic masses from the periodic table to find the molar mass of NaCl.

$$\text{Sodium (Na)} = 22.99\text{ g}$$
$$\text{Chlorine (Cl)} = 35.45\text{ g}$$

$$\text{molar mass} = 22.99\text{ g / mol} + 35.45\text{ g / mol}$$
$$= 58.44\text{ g / mol}$$

One mole of NaCl has a mass of 58.44 grams.

The solution contains 12.55 grams of NaCl. This is divided by the molar mass to find the number of moles present:

$$\text{number of moles} = 12.55\ \cancel{\text{g}} \times \frac{1\text{ mol}}{58.44\ \cancel{\text{g}}} = 0.214750\text{ mol}$$

There are 0.214750 moles of solute present. Now, the total moles of solute and solvent must be found. The volume of the water in the solution is 180.4 milliliters, and the density of the water is given as 0.9941 grams per milliliter:

$$\text{mass of solvent} = 0.9941\frac{\text{g}}{\cancel{\text{mL}}} \times 180.4\ \cancel{\text{mL}} = 179.336\text{ g}$$

Again, two extra digits are carried. Because the molality ratio uses the unit of kilograms, the mass of the water must be converted from grams:

$$\text{mass of solvent} = 179.3\cancel{\text{g}} \times \frac{1\text{ kg}}{1{,}000\ \cancel{\text{g}}} = 0.179336\text{ kg}$$

The values can then be substituted into the molality ratio to find *m*.

$$m = \text{moles of solute kg solvent} = \frac{0.214750 \text{ mol NaCl}}{0.179336 \text{ kg } H_2O} = 1.19747 \text{ mol/kg}$$

Now, all of the values can be substituted into the formula to find the freezing point depression.

$$\Delta T_f = iK_f m$$

$$\Delta T_f = 2 \times 1.858°C \ \frac{\cancel{kg}}{\cancel{mol}} \times 1.19747 \frac{\cancel{mol}}{\cancel{kg}}$$

$$= 4.44980°C$$

Rounding to four significant digits, ΔT_f = 4.450°C. The freezing point of pure water is 0°C. Apply the freezing point depression as calculated above to determine the new freezing point:

$$0°C - 4.450°C = -4.450°C$$

Adding 12.55 grams of sodium chloride to 180.4 milliliters of water will decrease the freezing point to more than 4.4 degrees lower than normal.

Osmotic Pressure

The movement of solutions through structures is part of many natural processes. Plant and animal cells, for example, rely on the movement of solutions through their bodies. This allows transport of nutrients, waste, and other biochemicals. These cells are surrounded by membranes that control which substances enter and exit cells. Only solute particles of a certain size can pass through the membrane. Typically, this means that solutes are filtered out by the membrane. Pure solvent passes through it. This movement of a solvent across a membrane is called osmosis.

Think about a solution of sugar and water that is passed through a semipermeable membrane. The sugar molecules will only be found on one side of the membrane (the solution side). Pure water will be on the other side (the solvent side).

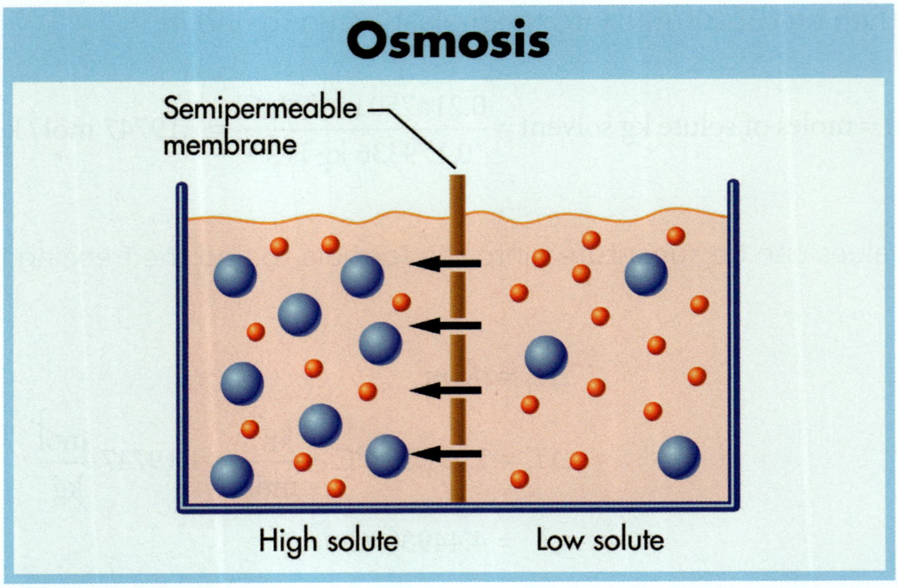

A semipermeable membrane will allow only particles of certain sizes to pass through. What causes the particles to move through the membrane?

The water molecules move back and forth through the membrane. More of them tend to move from the solvent side to the solution side. This creates osmotic pressure on the solution side of the membrane. This could eventually increase so that osmosis stops completely. Factors such as temperature also affect osmotic pressure. This is because the energy of the particles will affect how they move through the semipermeable membrane.

While osmosis is a complex process, the key is to remember that only water can pass through the "wall." This is because the sugar molecules are too large. The more sugar molecules in the solution, the more osmotic pressure increases.

Consider the Explain Question:

What factors influence the rate at which a solute dissolves?

Go online to complete the scientific explanation.

Check for Understanding:

What are three factors that impact the rate at which a solution can be made?

STEM in Action

STEM and Solutions

Certain fuels, such as jet fuel, diesel fuel, and gasoline, are a common source of groundwater contamination. When a spill occurs, some of the organic components leak into the groundwater. The most common contaminants are benzene, toluene, ethylbenzene, and xylene. Even though these compounds are slightly soluble in water, this process can be hazardous. Over time, these compounds will degrade into other compounds, one of which is hydrogen gas. The concentration of hydrogen gas in a groundwater sample can be easily calculated using Henry's law and will further help a chemist determine what treatments are appropriate and the fate of contaminants at that site.

Environmental chemists use a method known as the bubble stripping method to determine the dissolved hydrogen concentrations in well water. The method is based on the principle that gases will undergo a partitioning between a vapor phase and a liquid phase that are in contact with each other. At equilibrium, the partitioning of a gas between a vapor and liquid phase can be quantified by applying Henry's law.

photo: Getty Images

How can chemists determine the extent of contaminated water?

Examine the graph and use it to answer the questions that follow.

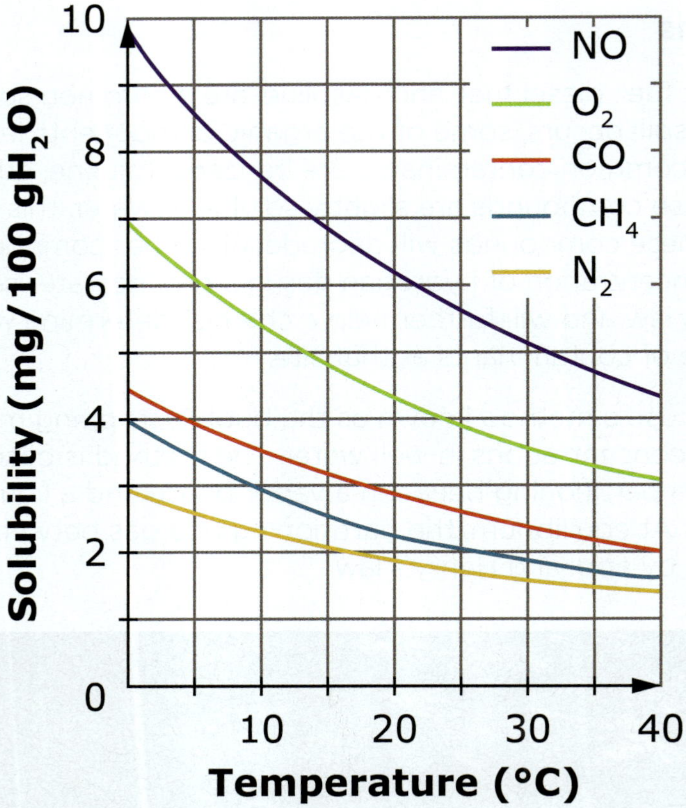

1. What is the solubility of nitrogen gas (in mol/L) at 25°C?
2. The Henry's law constant can be found using the equation: $P = K_H C$ where C is the concentration of the liquid in M, and P is the partial pressure of the gas above the liquid. Calculate the Henry's law constant for nitrogen gas at a constant temperature 25°C at a partial pressure of 1 atm.
3. When people go scuba diving, they must be careful to descend (and especially ascend) very carefully. Use the Internet to research the bends. Then, use Henry's law to explain this illness.

Acids, Bases, and Salts

photo: Getty Images

LESSON OVERVIEW

Lesson Questions

- What are similarities and differences among the Arrhenius, Brønsted-Lowry, and Lewis acid-base theories?
- What determines if a precipitate will form when solutions are combined, and how are net ionic equations written?
- What are the similarities and differences between acid and base strength and concentration?
- How are hydronium ions formed?
- What products are formed in acid-base reactions?
- What occurs during the neutralization process?
- What are pH and pOH?
- What is the correlation between logarithmic pH scale changes and corresponding changes in concentration?
- How is K_w (water dissociation constant) related to pH, pOH, concentration of hydrogen ions, and concentration of hydroxide ions?

Key Vocabulary

Which terms do you already know?

- ☐ acid dissociation constant (Ka)
- ☐ amphoterism
- ☐ aqueous solution
- ☐ Arrhenius acids and bases
- ☐ Brønsted-Lowry acids and bases
- ☐ buffer
- ☐ buffer solution
- ☐ concentration
- ☐ conjugate acid
- ☐ conjugate base
- ☐ dissociation
- ☐ electron pair acceptor
- ☐ electron pair donor
- ☐ equilibrium
- ☐ hydrogen ion
- ☐ hydronium ion
- ☐ hydroxide ion
- ☐ indicator
- ☐ ion product constant of water
- ☐ ionization
- ☐ K_w
- ☐ Lewis acids and bases
- ☐ logarithmic
- ☐ molarity
- ☐ neutralization
- ☐ pH
- ☐ pOH
- ☐ protic acid
- ☐ salt
- ☐ strong acid
- ☐ strong base
- ☐ titration
- ☐ weak acid
- ☐ weak base

Lesson Objectives

By the end of the lesson, you should be able to:

- Compare and contrast the Arrhenius, Brønsted-Lowry, and Lewis acid-base theories.
- Predict whether a precipitate will form when solutions are combined and write a net ionic equation for a precipitation reaction.
- Compare and contrast acid or base strength and concentration.
- Explain how hydronium ions are formed.
- Identify products formed in acid-base reactions.
- Describe the neutralization process.
- Explain pH and pOH.
- Correlate between logarithmic pH scale changes and corresponding changes in concentration.
- Relate K_w (water dissociation constant) to pH, pOH, concentration of hydrogen ions, and concentration of hydroxide ions.

ENGAGE

Acids and bases may sound familiar, but what exactly are they? Visit the Engage page online to learn more.

Explain Question

Explain why acetic acid, $HC_2H_3O_2$, is defined as an acid, and describe both conceptually and using a chemical equation what happens when $HC_2H_3O_2$, a weak acid, reacts with sodium hydroxide, NaOH, a strong base.

EXPLORE

What Are Similarities and Differences among the Arrhenius, Brønsted-Lowry, and Lewis Acid-Base Theories?

Acid and Base Theories

There are many theories describing the properties of acids and bases. Each theory has a particular application. The original theory of acids and bases was proposed by Svante August Arrhenius around 1884. His definition was specific to hydrogen and hydroxide ion concentrations in solution.

An Arrhenius acid is any substance that increases the number of hydrogen ions (H^+) in an aqueous solution. In the following example, hydrochloric acid is an Arrhenius acid because hydrogen ions are formed when it is added to water.

$$HCl\ (aq) \rightarrow H^+\ (aq) + Cl^-\ (aq)$$

Common Acids	
Anion	**Corresponding Acid**
fluoride (F^-)	hydrofluoric acid (HF)
chloride (Cl^-)	hydrochloric acid (HCl)
bromide(Br^-)	hydrobromic acid (HBr)
iodide(I^-)	hydroiodic acid (HI)
cyanide (CN^-)	hydrocyanic acid (HCN)
sulfide (s^2)	hydrosulfuric acid (H_2S)

Acids are formed from common anions. Why are acids formed from anions and not cations?

An Arrhenius base is any substance that increases the number of hydroxide ions (OH^-) in an aqueous solution. For example, sodium hydroxide is an Arrhenius base in an aqueous solution. Sodium hydroxide produces hydroxide ions when added to water.

$$NaOH\,(aq) \rightarrow Na^+\,(aq) + OH^-\,(aq)$$

Cation	Corresponding Base
lithium (Li^+)	$LiOH$
potassium (K^+)	KOH
sodium (Na^+)	$NaOH$
rubidium (Rb^+)	$RbOH$
NH_2^+	NH_3
calcium (Ca^{2+})	$Ca(OH)_2$

Bases are formed from common cations. Why are bases formed from cations and not anions?

Not all acids and bases produce hydrogen or hydroxide ions in solutions. The Brønsted-Lowry definition of acids and bases was created to solve this issue. It specifically describes the gain or loss of hydrogen ions (protons). Johannes Nicolaus Brønsted and Thomas Martin Lowry proposed their version of acid-base theory around 1923.

A Brønsted-Lowry acid is any substance that can donate a hydrogen ion (proton). For example, hydrochloric acid was described as an Arrhenius acid. Hydrochloric acid is also a Brønsted-Lowry acid in an aqueous solution because it donates a hydrogen ion.

$$HCl(aq) \rightarrow H^+(aq) + Cl^-(aq)$$

A Brønsted-Lowry base is any substance that can accept a hydrogen ion (proton). For example, ammonia is a Brønsted-Lowry base in an aqueous solution. Ammonia can accept a hydrogen ion.

$$NH_3(aq) + H^+(aq) \rightarrow NH_4^+(aq)$$

Gilbert Lewis proposed a different mechanism for acid-base theory, also around 1923. The Lewis theory is based on the transfer of electrons.

A Lewis acid is an electron pair acceptor. A Lewis base is an electron pair donor. Water is both a Lewis acid and a Lewis base. For example, boron trifluoride can accept electron pairs, whereas fluorine can donate electron pairs.

$$BF_3(aq) + F^-(aq) \rightarrow BF_4^-(aq)$$

What Determines If a Precipitate Will Form When Solutions Are Combined, and How Are Net Ionic Equations Written?

Precipitation Reactions

Acid-base reactions are not the only reactions that occur in aqueous solutions. A precipitation reaction is a reaction in an aqueous solution that results in an insoluble product. In other words, the product of the reaction cannot dissolve in water. Instead, it precipitates out of the solution in solid form.

A precipitation reaction can sometimes occur when two salts are mixed together in an aqueous solution. For example, the two salts, barium chloride and sodium sulfate, undergo a precipitation reaction when mixed together in water:

$$BaCl_2(aq) + Na_2SO_4(aq) \rightarrow BaSO_4(s) + 2NaCl(aq)$$

The term (aq) shows that most of the compounds are dissolved in water. The term (s) shows that $BaSO_4$ is not dissolved but is in a solid state. As the reaction proceeds, barium sulfate precipitates out of the solution as a white powder.

Writing the equation in terms of the ions involved gives a clearer picture of what is happening in this precipitation reaction. This is called the ionic equation:

$$Ba^{2+}(aq) + 2Cl^-(aq) + 2Na^+(aq) + SO_4^{2-}(aq) \rightarrow BaSO_4(s) + 2Na^+(aq) + 2Cl^-(aq)$$

Because the sodium and chloride ions are on both sides of the equation, they can be removed from the equation. Ions that are present on both sides of an equation are called spectator ions. Once the spectator ions have been removed from the equation, what remains is called a net ionic equation:

$$Ba^{2+}(aq) + SO_4^{2-}(aq) \rightarrow BaSO_4(s)$$

How can you tell if a reaction involving ions is a precipitation reaction or not? Solubility rules are guidelines that help to identify insoluble compounds. A list of solubility rules is given in the following table.

Soluble compounds have these ions:	Insoluble compounds have these ions:
• Cl^-, Br^-, I^-, NO_3^-, HCO_3^-, ClO3-, SO_4^{2-} • Li^+, Na^+, K^+, Rb^+, Cs^+, NH_4^+	• CO_3^{2-}, PO_4^{3-}, CrO_4^{2-}, S_2^-, OH^-
Except for these combinations of ions:	**Except for these combinations of ions:**
Compounds containing Ag^+, Hg_2^{2+}, or Pb^{2+} **AND** Cl^-, Br^-, or I^- are **insoluble** (AgCl is insoluble) Compounds containing Ag^+, Ca^{2+}, Sr^{2+}, Ba^{2+}, or Pb^{2+} **AND** SO_4^{2-} are **insoluble** (CaSO4 is insoluble)	Compounds containing Li^+, Na^+, K^+, Rb^+, NH_4^+ **AND** CO_3^{2-}, PO_4^{3-}, CrO_4^{2-}, S^{2-} are **soluble** (Na_2CO_3 is soluble) Compounds containing Na^+, K^+, Rb^+, Ba2+ **AND** OH^- are **soluble** (NaOH is soluble)

Using the Solubility Rules: Sample Problem

When a solution of $Pb(NO_3)_2$ is added to a solution of NaI, a reaction occurs to produce PbI_2 and $NaNO_3$.

(a) Write the chemical equation and identify any insoluble products. Is this a precipitation reaction or not?

(b) Write the net ionic equation.

Solution

(a) Write the chemical equation and balance it.

$$Pb(NO_3)_2(aq) + 2NaI(aq) \rightarrow PbI_2\ (?) + 2NaNO_3\ (?)$$

The solubility rules state that compounds with Cl^- are usually soluble except when combined with Pb^{2+} ions, so PbI_2 is insoluble.

The solubility rules state that products containing NO_3^- ions and Na^+ ions are soluble, so $NaNO_3$ is soluble.

This is a precipitation reaction because the two reactants are dissolved in water, but one of the products is insoluble.

Chemical equation with states:

$$Pb(NO_3)_2(aq) + 2NaI(aq) \rightarrow PbI_2(s) + 2NaNO_3(aq)$$

(b) ionic equation:

$$Pb^{2+}(aq) + 2NO_3^-(aq) + 2Na^+(aq) + 2I^-(aq) \rightarrow PbI_2(s) + 2Na^+ + 2NO_3^-(aq)$$

Removing the spectator ions that are present on both sides leaves the net ionic equation:

$$Pb^{2+}(aq) + 2I^-(aq) \rightarrow PbI_2(s)$$

What Are the Similarities and Differences between Acid and Base Strength and Concentration?

Acid or Base Strength

According to the Arrhenius definition of acids and bases, the strength of an acid or base is defined as the degree of ionization or dissociation of the acid or base in aqueous solution. Ionization is the process by which an atom becomes an ion through the addition or loss of an electron. Dissociation is the process by which a substance breaks up into smaller particles. All acids dissociate, but only some will ionize.

The strengths of acids and bases can be measured using the equilibrium constant expressions K_a (for acid strength) and K_b (for base strength). Strong acids and bases dissociate almost completely in water. This means that almost all of the particles in the acid or base break up into smaller particles. Weak acids and bases dissociate slightly in water. Most of the particles that make up a weak acid or weak base do not split into smaller particles.

During the process of dissociation, if the acid dissociates completely, it is considered a strong acid. If it does not completely dissociate, then the strength can be determined by finding the acid dissociation constant (K_a). To begin, it is helpful to study the acid-base reaction in the form of an equation:

$$HA = H^+ + A^-$$

In this equation, HA represents the acid, H^+ represents protons that are produced, and A^- represents the base.

Using the concentrations of protons $[H^+]$, the conjugate base $[A^-]$, and the acid [HA], the acid dissociation constant can be calculated:

$$K_a = \frac{[H^+][A^-]}{[HA]}$$

The strength of an acid is measured using the acid dissociation constant (K_a). The higher the K_a value, the stronger the acid.

Similarly, if the base dissociates completely, it is considered a strong base. If it does not completely dissociate, then the strength can be determined by finding the base dissociation constant (K_b).

In this equation, B represents the base, OH^- represents the hydroxide ions that are produced, and BH^- represents the conjugate acid.

$$B = OH^- + BH^+$$

Using the concentrations of the base [B], hydroxide ions [OH^-], and conjugate acid [B], the dissociation constant for the Arrhenius base can be calculated:

$$K_b = \frac{[HB^+][OH^-]}{[B]}$$

The strength of a base is measured using the base dissociation constant (K_b). The higher the K_b value, the stronger the base.

According to the Brønsted-Lowry definition of acids and bases, an acid that loses a proton during dissociation is called a conjugate base. A base that gains a proton during dissociation is called a conjugate acid. In the Brønsted-Lowry model, strength of the acid is determined by how readily the acid gives up a proton. Conversely, the strength of a base is determined by how readily it accepts the proton. In the Brønsted-Lowry model, an obvious inverse relationship can be seen between the acid and base in the acid-base pair. The stronger an acid is, the weaker its conjugate base is. While the stronger a base is, the weaker its conjugate acid.

It should be noted that some substances may act as either an acid or a base; it depends upon what other substances are around them. These substances are said to be amphoteric. Amphoterism occurs when a substance can act as either an acid or a base. Whether an amphoteric substance acts as an acid or base depends on the solution it is in. Water is the most common amphoteric molecule.

In the ionization of hydrochloric acid, water acts as a base:

$$\underset{\text{Acid}}{HCl(aq)} + \underset{\text{Base}}{H_2O(l)} \rightarrow \underset{\text{Acid}}{H_3O^+(aq)} + \underset{\text{Base}}{HSO_4^-(aq)}$$

However, water acts as an acid in the following reaction:

$$\underset{\text{Base}}{NH_3(aq)} + \underset{\text{Acid}}{H_2O(l)} \rightarrow \underset{\text{Acid}}{NH_4^+(aq)} + \underset{\text{Base}}{OH^-(aq)}$$

If water reacts with a compound that is a stronger acid than itself, then water acts as a base. If water reacts with a compound that is a stronger base than itself, then water acts as an acid.

How Are Hydronium Ions Formed?

The Formation of Hydronium Ions

A protic chemical is one that acts as a proton donor. When a protic acid is added to water and releases a hydrogen ion, the hydrogen ion immediately bonds with a nearby water atom. This forms H_3O^+, an ion known as hydronium.

Hydronium is a water molecule with one extra hydrogen ion attached. Hydronium is sometimes referred to as a proton because it can act as a hydrogen or proton donor in solution. Hydronium carries a positive charge and is usually an intermediate product of reactions.

Water, naturally and continuously, dissociates into hydroxide and hydronium ions. The ions then reassociate and the reactions are repeated continually.

$$2H_2O(aq) \rightleftarrows OH^-(aq) + H_2O^+(aq)$$

What Products Are Formed in Acid-Base Reactions?

Products of Acid-Base Reactions

In a strong acid-strong base reaction, a salt and water will be formed as the products. Because both reactants are strong, they will dissociate completely. For example, hydrochloric acid and sodium hydroxide will form sodium chloride (a salt):

$$HCl(aq) + NaOH(aq) \rightarrow NaCl(aq) + H_2O(aq)$$

In a weak acid-strong base reaction, the resulting solution will be basic, and there will likely be a soluble salt. The strong base will dissociate completely. The weak acid will only dissociate a little. For example, acetic acid and potassium hydroxide will produce potassium acetate (a salt):

$$HC_2H_3O_2(aq) + KOH(aq) \rightarrow H_2O(aq) + KC_2H_3O(aq)$$

photo: Getty Images

This student is neutralizing strong acid with a strong base. What pH would you expect the final solution to have?

In a strong acid-weak base reaction, the resulting solution will be acidic, and there will likely be a soluble salt. The strong acid will dissociate completely. The weak base will only dissociate a little. For example, hydrochloric acid and ammonia form ammonium chloride (a salt):

$$HCl(aq) + NH_3(aq) \rightleftarrows NH_4Cl(aq)$$

In a weak acid-weak base reaction, the resulting solution will be neutral. The reactants are weak, so they will only dissociate a little. For example, ammonia is a weak base, and water is a weak acid.

$$NH_3(aq) + H_2O(aq) \rightleftarrows NH_4 + (aq) + OH^-(aq)$$

Acid-Base Reactions: Sample Problem

Ammonium is a weak base that dissolves in water. What are the products formed in this reaction?

$$NH_4^+ + OH^-$$

Solution

Ammonium is a cation and has an extra hydrogen atom. Water will take that hydrogen and form hydronium. This reaction will also form ammonia.

$$NH_4^+ + OH^- \rightleftarrows NH_3 + H_2O$$

What Occurs during the Neutralization Process?

Buffers and Neutralization

One way that chemists can maintain a constant concentration of the pH of a solution is through the use of buffer solutions. Buffer solutions are solutions that can resist a change in pH when small amounts of acid or base are added to it. Acidic buffer solutions have a pH of less than seven, while basic buffer solutions have a pH of over seven.

Acidic buffer solutions are made from a weak acid and one of its salts. This will allow the buffer solution to resist against a higher concentration of hydroxide ions (OH^-) being produced in the solution.

If a small amount of base is added to an acetic acid-sodium acetate solution, the OH^- ions of the base react with and remove hydronium ions to form nonionized water molecules. The acetic acid molecules then ionize and restore the equilibrium concentration of hydronium ions.

$$CH_3COOH(aq) + H_2O(l) \rightleftarrows CH_3COO^-(aq) + H_3O^+(aq)$$

If a small amount of acid is added to an acetic acid-sodium acetate solution, the acetate ion will react with most of the added hydronium ions to form the nonionized acetic acid molecules.

$$CH_3COO^-(aq) + H_3O^+(aq) \rightleftarrows CH_3COOH(aq) + H_2O(aq)$$

Conversely, basic or alkali buffer solutions are made from a weak acid and one of its salts. This enables the buffer to resist the higher concentration of hydrogen ions (H^+) being produced in the solution.

Neutralization is a chemical reaction. During neutralization, an acid and a base react to form water and a salt. The resulting solution is neutral, having the properties of neither an acid nor a base.

The concentration can be expressed in normality (N). Normality has the unit of moles per liter. Remember that molarity describes the moles of solute per liter of solution. Normality is very similar, but instead it describes the moles of each ion in the solution.

For example, consider a solution that has a concentration (molarity) of 1 mol/L, or 1 M NaOH. Each NaOH dissociates to form one Na^+ ion and one OH^- ion. The normality of Na^+ ions in the solution would be 1 mol/L, or 1 N. The normality of the OH^- ions would also be 1 N.

In contrast, a solution with a concentration of 1 M H_2SO_4 would produce H^+ ions and SO_4^{2-} ions. The normality of the H^+ would be 2 mol/L, or 2 N, and the normality of the SO_4^{2-} would be 1 mol/L, or 1 N. Notice that the normality is always a whole number because ionic compounds dissociate into a whole number of ions.

An acid and a base neutralize when they are combined because all of the ions that are present reorganize themselves to form neutral components. The number of hydrogen ions must be made to equal the number of hydroxide ions in the solution so that they form water molecules. The remaining ions bond to form salts.

Neutralization calculations are done using this formula:

$$N_a V_a = N_b V_b$$

Where

- N_a is the normality of the acid
- V_a is the volume of the acid
- N_b is the normality of the base
- V_b is the volume of the base

Given the volume and concentration of an acid (or a base), you can calculate how much base (or acid) is needed to neutralize the solution.

Neutralization: Sample Problem

What volume of 2.0 M H_2SO_4 (strong acid) is required to neutralize 250 mL of 0.50 NaOH (strong base)?

Solution

The equation to determine the amount needed for neutralization can be rearranged to solve for the value needed:

$$N_a V_a = N_b V_b$$

$$V_a = \frac{N_b V_b}{N_a}$$

In the acid, there are two hydrogen ions produced for every molecule of H_2SO_4. There are 2 moles of acid in 1 liter of the solution, which means there are 4 moles of hydrogen ions in 1 liter of solution:

$$N_a = 2.0\ \text{mol/L} \times 2 = 4\ \text{mol/L}\ H^+$$

In the base, there is one hydroxide ion produced for every molecule of NaOH. There are 0.5 moles of base in 1 liter of solution, which means there are 0.5 moles of hydroxide ions in the solution:

$$N_b = 0.50\ \text{mol/L} \times 1 = 0.50\ \text{mol/L}\ OH^-$$

The volume of the base is given as 250 milliliters but must be converted to liters to be consistent with the normality values:

$$V_b = 250\,\text{m}\cancel{\text{L}} \times \frac{1\text{L}}{1{,}000\,\text{m}\cancel{\text{L}}}$$

$$V_b = 0.250\,\text{L NaOH}$$

Next, the values can be substituted into the equation:

$$V_a = \frac{N_b V_b}{N_a}$$

$$V_a = \frac{0.50\,\text{N} \times 0.250\,\text{L}}{4.0\,\text{N}}$$

$$V_a = 0.03125\,\text{L}$$

Finally, the volume should be converted back to milliliters:

$$0.03125\,\text{L} \times \frac{1{,}000\,\text{mL}}{1\,\text{L}} = 31\,\text{mL}$$

Therefore, 31 mL of 2.0 M H_2SO_4 is required to neutralize 250 mL of 0.50 NaOH.

What Are pH and pOH?

pH and pOH

It is often helpful to know about the strength of an acid or base before you use a chemical or combine it with another substance. Also, knowing whether an unknown substance is an acid or a base can be helpful in identifying it. The most common method for describing the strength of an acid or base is known as the pH scale.

The term *pH* is a mathematical way of comparing the strength of acids and bases. The difference of one on the pH scale indicates a tenfold difference in the strength of the compounds. An acid with the pH of one is 10 times as strong as an acid with the pH of two.

To calculate pH, the strength or concentration, often known as molarity (M) of the acid must be known. By taking the negative logarithm of the molarity of the acid, the pH can be found:

$$pH = -\log[\text{molarity}]$$

For example, the pH of an acid solution with the concentration of 0.01 M would be two:

$$pH = -\log[\text{molarity}]$$
$$pH = -\log[0.01]$$
$$pH = 2$$

The pH scale ranges in values from one to 14. One is the most acidic, seven is neutral, and 14 is the most basic. These values represent the quantity of hydrogen ions in a solution.

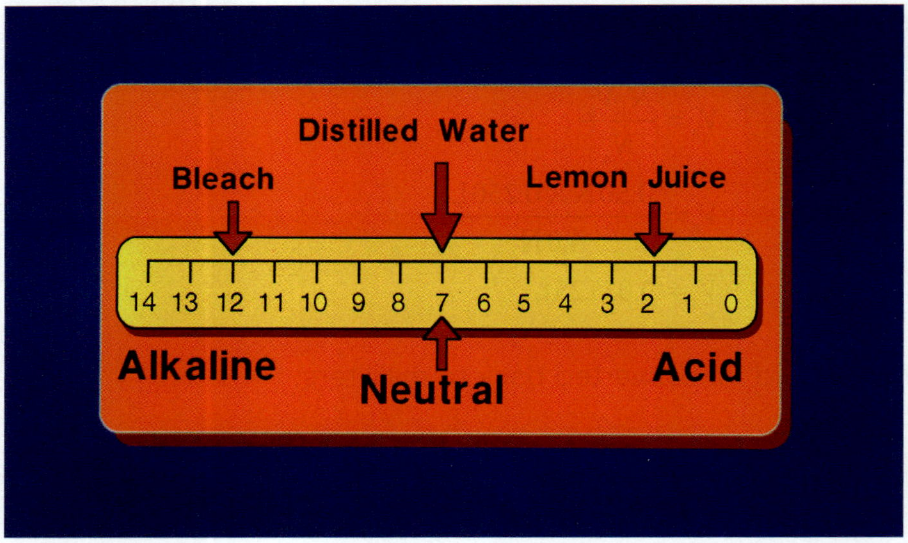

photo: Paul Fuqua

The pH scale covers both acids and bases. What does a higher number mean?

Another method for describing the strength of acids and bases is the pOH scale, which describes the quantity of hydroxide ions in the solutions. It runs in the opposite direction as the pH scale. On the pOH scale, one is the most basic, and 14 is the most acidic.

Because the pH and pOH of a solution always total 14, one can be calculated from the other:

$$pH + pOH = 14$$

Indicators are useful in determining the pH of a solution. They will change color when exposed to an acid or a base of a certain strength. Indicators are often used during acid-base titrations. Some common indicators are phenolphthalein, thymol blue, and methyl orange.

Red Cabbage Indicator

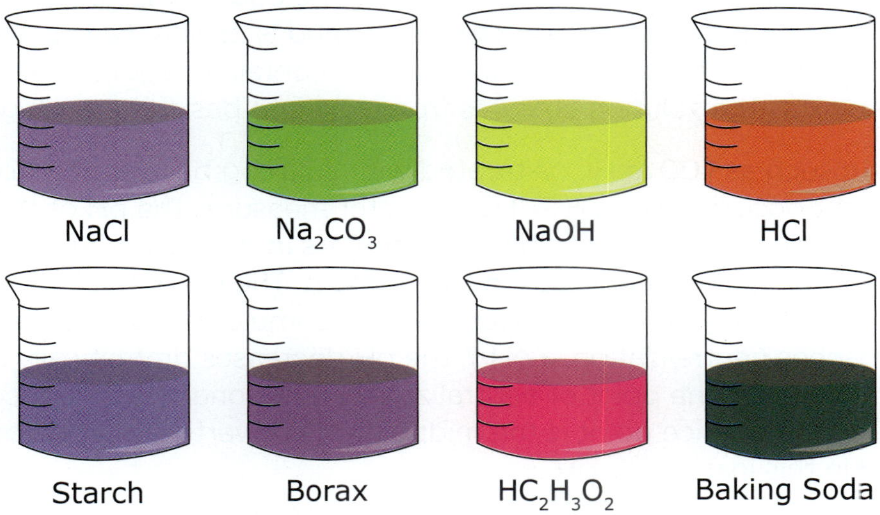

Red cabbage juice contains a natural pH indicator that changes color with the acidity of the solution. Which colors indicate acids? Which colors indicate bases?

Titration

Acids and bases react with each other to form a salt and water by a process called neutralization. Chemists use acid-base titrations to study neutralization reactions.

Titration is a method to calculate the concentration of a sample by analyzing neutralization during the reaction. A solution of known concentration (the titrant, or standard solution) is added in incremental measured amounts to a second solution of known volume but unknown concentration (the titer). The titrant is added until the reactants are neutralized. This is called the equivalence point. Indicators allow an observer to detect changes in the solution's pH during titration by changing color. Eventually, a color change occurs with the addition of a single drop, indicating the end point of the reaction.

The volume of the unknown solution necessary to complete the reaction allows its concentration to be calculated. The end point in a titration indicates when the equivalence point has been reached and can be detected by some form of indicator which changes color or a pH meter.

An appropriate acid-base indicator is used to show when neutralization has occurred. In the laboratory, bromothymol blue is often used for strong acid-strong base neutralizations with endpoints around pH 7.

Indicators that change color at pH lower than seven are useful in determining the equivalence point in strong acid-weak base titrations. The equivalence point of a strong acid-weak base titration is acidic because the salt formed is itself a weak acid. Thus, the salt solution has a pH lower than seven. Methyl-orange can be used as an indicator for these reactions.

Indicators that change color at pH higher than seven are useful in determining the equivalence point in weak acid-strong base titrations. The equivalence point of a weak acid-strong base titration is basic because the salt formed is itself a weak base. Thus, the salt solution has a pH higher than seven. Solutions that contain phenolphthalein turn from colorless to deep pink as the solution changes from acidic to basic at a pH higher than seven.

When a strong acid such as 1.00 M HCl is titrated with a strong base such as 1.00 M NaOH, the equivalence point occurs about pH 7. If a pH meter measures the pH of the solution during the titration, a titration curve can be produced. As in the figure "Strong Acid-Strong Base," the pH of the solution changes during the titration. The pH of the initial base solution is high. As the acid is added, the pH increases because some of the base is neutralized. As the titration approaches neutralization, a pH 7, the pH decreases dramatically as the OH^- ions are used up. Once past the point of neutralization, additional acid produces a further decrease in pH. The equivalence point is the midpoint of the vertical portion of the pH titration curve and in this instance is pH 7.

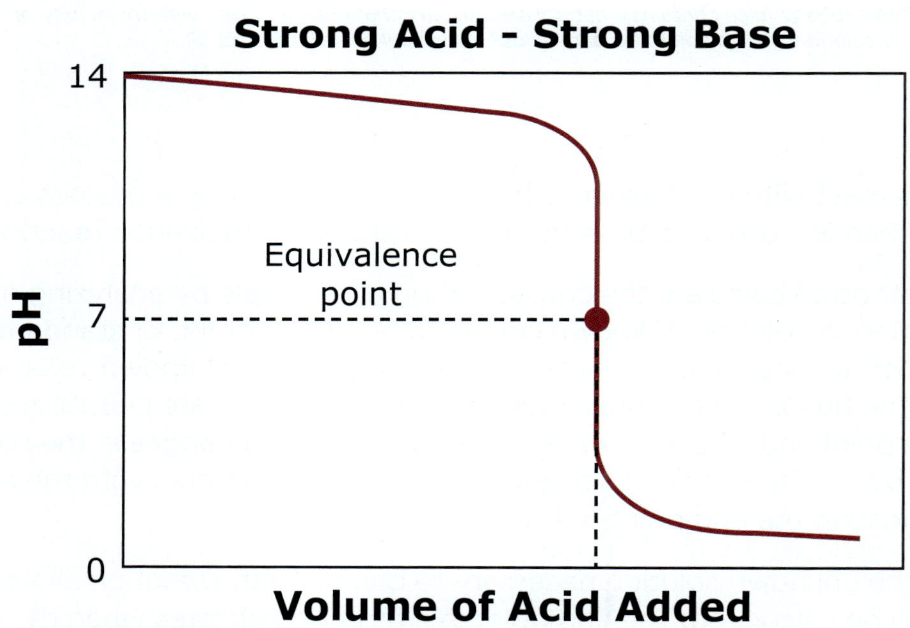

This graph shows the equivalence point. Is this point acidic or basic?

When a strong acid is titrated with a weak base, the resulting solution will be acidic. The strong acid dissociates completely while the weak base will only dissociate a little.

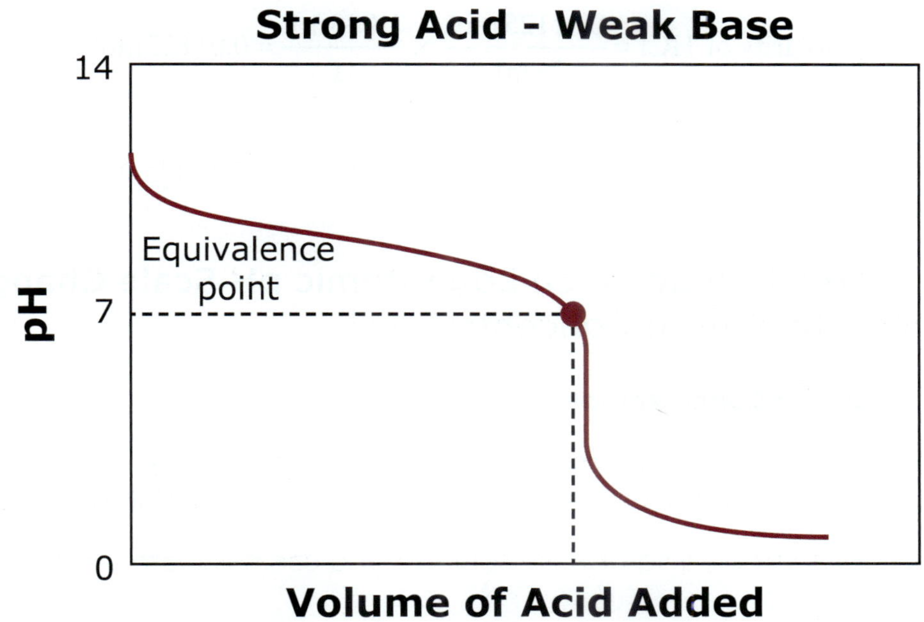

Find the equivalence line on this graph. Does this scenario result in a basic or acidic solution?

Buffer solutions contain a weak acid and its conjugate base (or weak base and its conjugate acid). A buffer can accept the addition of a strong acid or base without a significant change in pH. Buffers are common in many products where pH must remain stable over time. Blood also contains buffers that resist change in pH to keep the body working properly.

pH and pOH: Sample Problem

Using titration, a 75 mL solution of 0.15 M NaOH is found to neutralize a 25 mL sample of HCl. What is the concentration of the HCl?

Solution

Every mole of NaOH will produce 1 M of OH^-. Because there is 0.15 M NaOH, there will be $[OH^-]$ = 0.15 M:

$$\text{molarity} = \text{moles} / \text{volume}$$

$$\text{moles} = \text{molarity} \times \text{volume}$$

$$\text{moles}\left[\, OH^- \right] = 0.15\,\text{M} \times 75\,\text{mL} \times \frac{1\,\text{L}}{1{,}000\,\text{mL}} = 0.01125\,\text{mol}$$

When a base neutralizes an acid, the number of moles of OH^- is equal to the number of moles of H^+. Therefore, the number of moles of H^+ = 0.01125 M.

Every mole of HCl will produce one mole of H^+, so the number of moles of HCl equals the number of moles of H^+:

$$\text{Molarity of HCl} = \frac{0.01125\,\text{mol}}{25\,\text{mL}} \times \frac{1{,}000\,\text{mL}}{1\,\text{L}} = 0.01125\,\text{mol}$$

For a 25 mL sample of HCl to neutralize a 75 mL solution of 0.15 M NaOH, a molarity of 0.45 M is required.

What Is the Correlation between Logarithmic pH Scale Changes and Corresponding Changes in Concentration?

Logarithmic pH Scale and Concentration

The pH scale is logarithmic. A logarithmic scale is based on the power of 10. Therefore, a solution with a pH of one is 10 times more concentrated than a solution with a pH of two.

If you know the pH of a strong acid or base solution, you can calculate the concentration of the acid or the base:

$$M = \frac{1}{10^{\text{pH}}}$$

where *M* is the molarity.

There is a correlation between logarithmic pH scale changes and corresponding changes in concentration. Concentration refers to the amount of hydronium ions in a solution. A low pH indicates a solution has a high concentration of hydronium ions. A high pH indicates that a solution has a low concentration of hydronium ions.

The pH can be calculated when the hydronium ion concentration is known:

$$\text{pH} = -\log\left[\text{H}^+\right]$$

This is also true for pOH, where $[OH^-]$ is the hydroxide concentration:

$$\text{pOH} = -\log\left[\text{OH}^-\right]$$

For example, a glass of orange juice has a hydronium ion concentration of [H+] = 0.0003:

$$\text{pH} = -\log\left[\text{H}^+\right] = -\log[0.0003] = 3.522$$

This pH is below seven; therefore, the orange juice is acidic.

How Is K_w (Water Dissociation Constant) Related to pH, pOH, Concentration of Hydrogen Ions, and Concentration of Hydroxide Ions?

The Water Dissociation Constant

The water dissociation constant (K_w) describes the degree to which water dissociates into hydrogen and hydroxide ions at various temperatures. K_w is related to pH, pOH, concentration of hydrogen ions, and concentration of hydroxide ions.

At standard conditions, K_w is always calculated as follows:

$$K_w = [H^+][OH^-] = 1.0 \times 10^{-14}$$

The acid and base dissociation constants can be used to determine K_w as well:

$$K_w = K_a \times K_b$$

Salts also dissociate into ions when placed in water. The K_a or K_b value of the salt when dissolved in water determines if the salt is acidic or basic.

The Water Dissociation Constant: Sample Problem

Complete the following table:

pH	[H⁺]	[OH⁻]	pOH	Kw
2.5				

Solution

$$K_w = [H^+][OH^-] = 1.0 \times 10^{-14}$$

$$pH + pOH = 14$$

$$pOH = -\log[OH^-]$$

$$pH = -\log[H^+]$$

$$pH + pOH = 14$$

$$pOH = 14 - pH = 14 - 2.5 = 11.5$$

From pH = –log[H⁺], we know that $\frac{1}{10^{pH}}$=M of hydronium ions.

From pOH = –log[OH+], we know that $\frac{1}{10^{pH}}$=M of hydronium ions.

$$\frac{1}{10^{11.5}} = 3.16227766 \times 10^{-12}\,\mathrm{M\,OH^-}$$

$$K_w = \left[H^+\right]\left[OH^-\right] = 3.16 \times 10^{-3} \times 3.16 \times 10^{-12} = 1.00 \times 10^{-14}$$

K_w checks out, so we know our calculations are correct.

pH	$[H^+]$	$[OH^-]$	pOH	K_w
2.5	3.16×10^{-3}	3.16×10^{-12}	11.5	1.00×10^{-14}

Consider the Explain Question:

Explain why acetic acid, $HC_2H_3O_2$, is defined as an acid, and describe both conceptually and using a chemical equation what happens when $HC_2H_3O_2$, a weak acid, reacts with sodium hydroxide, NaOH, a strong base.

Go online to complete the scientific explanation.

Check for Understanding:

What are the quantitative values that describe acids and bases?

STEM in Action

STEM and Acids, Bases, and Salts

Imagine a farmer sprinkling salt on farmland to enhance the growth of crops. Do you think it would work? It may sound strange, but that's exactly what many farmers do. The salt they use isn't the table salt sodium chloride, of course. The sodium ions released by table salt would be harmful to crops. Instead, farmers use other types of salts that provide the nutrients nitrogen, phosphorus, and potassium that crops need to grow. One example of a salt in commercially available fertilizers for crops is ammonium nitrate. Ammonium nitrate encourages plant growth by slightly acidifying the soil to a pH of around 6 to 7.

Although salt-based fertilizers allow plants to grow faster and larger, they also have disadvantages. Because these salts are fairly soluble in water, runoff containing high levels of fertilizer can flow into nearby waterways. In high enough concentrations, the runoff causes algae to grow wild. This large amount of algae reduces the concentration of oxygen in the water available to plants and fish. It also feeds some plants and weeds so well that they grow and overtake native species. This results in a population decline of native species.

Environmental engineers have developed a variety of technologies and procedures to combat nutrient pollution of waterways by salt-based fertilizers. One method is to build a vegetative swale alongside farmland. These U-shaped channels, filled with vegetation, are designed to catch and filter the runoff of the fertilizers. Another method is to build a detention basin to hold and filter runoff. Agricultural researchers have also developed methods that farmers can implement to reduce the detrimental effects of salt-based fertilizers. For instance, farmers are encouraged to plant cover crops in winter to reduce runoff from fields, and vegetation planted alongside streams near farmland can hinder the flow of salt-based fertilizer runoff into the water. Careful soil testing is required to ensure correct application rates for fertilizers.

photo: Paul Fuqua

Salt-based fertilizers sprayed onto a field encourage plant growth. What are some methods that this farmer can take to avoid polluting nearby waterways with runoff from the fertilizers?

If soil testing reveals high acidity, lime (a base) can be added to neutralize the soil. The following graph shows the results that a soil analyst obtained from comparison titrations of a soil sample and an acid. The base was the same in both titrations. Answer the questions to interpret the results of the analysis.

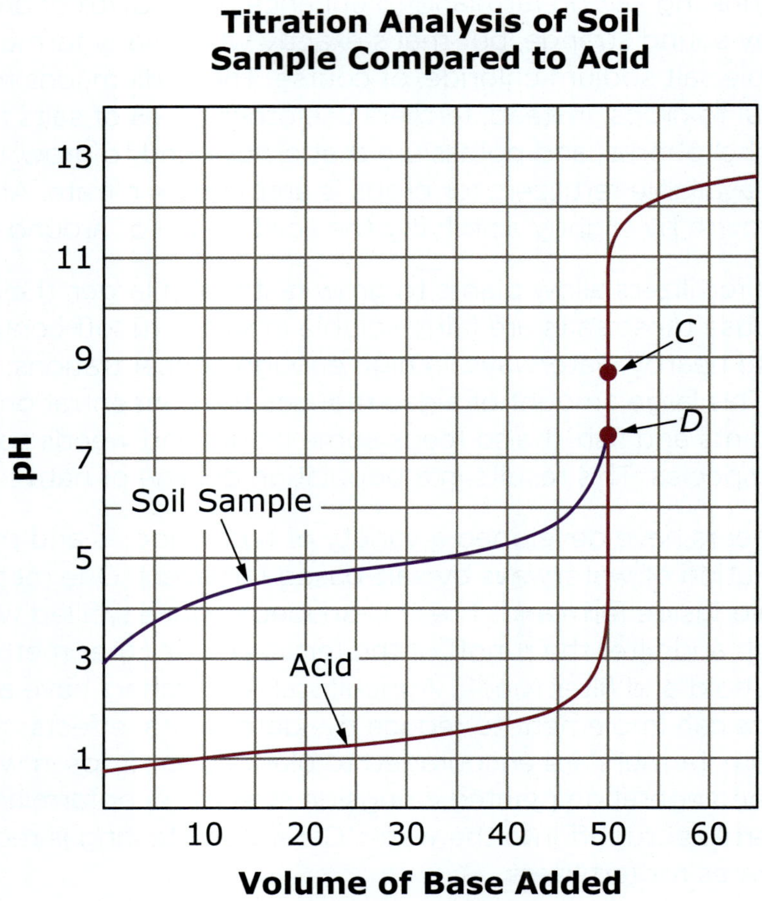

1. According to the graph, what is different about the acidity of the soil sample compared with the acid that the analyst tested? Explain how you know.
2. What accounts for the difference in the shapes of the graphs of the soil sample and the acid? In your answer, explain what is happening to the particles in each acid-base solution.
3. What is the significance of the points labeled C and D? What is different about the two points?

